Primer to the Immune Response
Academic Cell Update

Primer to the Immune Response
Academic Cell Update

Tak W. Mak and Mary E. Saunders

Contributors

Maya R. Chaddah

Wendy L. Tamminem

ELSEVIER

AMSTERDAM • BOSTON • HEIDELBERG • LONDON
NEW YORK • OXFORD • PARIS • SAN DIEGO
SAN FRANCISCO • SINGAPORE • SYDNEY • TOKYO

Academic Press is an imprint of Elsevier

Academic Press is an imprint of Elsevier
30 Corporate Drive, Suite 400, Burlington, MA 01803, USA
525 B Street, Suite 1800, San Diego, California 92101-4495, USA
84 Theobald's Road, London WC1X 8RR, UK

Notices
Knowledge and best practice in this field are constantly changing. As new research and experience
broaden our understanding, changes in research methods, professional practices, or medical
treatment may become necessary.

Practitioners and researchers must always rely on their own experience and knowledge in
evaluating and using any information, methods, compounds, or experiments described herein. In
using such information or methods they should be mindful of their own safety and the safety of
others, including parties for whom they have a professional responsibility.

To the fullest extent of the law, neither the Publisher nor the authors, contributors, or editors,
assume any liability for any injury and/or damage to persons or property as a matter of products
liability, negligence or otherwise, or from any use or operation of any methods, products,
instructions, or ideas contained in the material herein.

Library of Congress Cataloging-in-Publication Data
Application submitted.

British Library Cataloguing-in-Publication Data
A catalogue record for this book is available from the British Library.

ISBN: 978-0-12-384743-0

For information on all Academic Press publications
visit our Web site at www.elsevierdirect.com

Printed in Canada
10 11 12 13   8 7 6 5 4 3 2 1

Preface

Academic Cell: How We Got Here

In speaking with professors across the biological sciences and going to conferences, we, the editors at Academic Press and Cell Press, saw how often journal content was being incorporated in the classroom. We understood the benefits students were receiving by being exposed to journal articles early: to add perspective, improve analytical skills, and bring the most current content into the classroom. We also learned how much additional preparation time was required on the part of instructors finding the articles, then obtaining the images for presentations and providing additional assessment.

So we have collaborated to offer instructions and students a solution, and *Academic Cell* was born. We offer the benefits of a traditional textbook (to serve as a reference to students and framework to instructors) but we also offer much more. With the purchase of every copy of *Academic Cell* book, students can access an online study guide containing relevant, recent Cell Press articles and providing bridge material to help ease them into the articles. In addition, the images from the articles are available in PowerPoint and we have optional test bank questions.

We plan to expand this initiative, as future editions will be further integrated with unique pedagogical features incorporating current research from the pages of Cell Press journals into the textbook itself.

Primer to the Immune Response
Academic Cell Update Edition

Immunology is a complex subject. Over my many years as a research scientist, I have enjoyed the challenge of sharing the mysteries of immunology with students of all ages and educational levels. For younger students, I have tried to explain difficult concepts using "real world" analogies, such as "The Adventures of Tommy the T Cell", who stands fast against marauders of disease.

For the reader of an immunology textbook – that is, for a true student of immunology – explanations necessarily have to become more sophisticated. In writing *Primer to the Immune Response*, Dr. Mary Saunders and I were extremely happy to have created a textbook that presents complex immunological concepts to the novice student in a clear, interesting and effective way. Reading of this book will firmly ground a student in the basics of the subject, but there remains a desire to connect this information with immunology in the "real world". This means going beyond the facts and figures to give the student a taste of how this body of knowledge was created in the first place, and how it continues to expand on a daily basis in research labs around the world. It also means giving the student a look at how today's new ideas and observations can translate into medical treatments that improve the health and lives of people everywhere. With this desire to establish a "real-world connection" in mind, I am pleased to introduce the new Cell Press *Study Guide for Primer to the Immune Response*, written by Dr. Bradley Jett.

In each chapter of the Study Guide, Dr. Jett has selected a journal article from the current scientific literature that complements the corresponding chapter of *Primer to the Immune Response*. In each case, after offering a concise, easy-to-read summary of the chapter contents, he has provided thought-provoking questions and scenarios that relate to the journal article and challenge the student to think like a scientist. In this way, the Study Guide provides a bridge that encourages the reader to cross over from the world of the "immunology student" into the world of the "immunologist".

By using the Study Guide as part of their immunology course work, students will grasp immunological concepts more fully. They will also face head-on the sometimes frustrating, sometimes exhilarating fact that the field of immunology is never static. The science constantly moves forward, driven by the novel, exciting and sometimes controversial work of front-line researchers. It is into this fascinating world that the Study Guide immerses the budding immunologist.

It is my hope that students and instructors will take advantage of this excellent Study Guide and use it as an enriching companion to *Primer to the Immune Response*. They will benefit, and I will feel I have taken another step in my evolution as an educator. And to think it all started with 'Tommy the T cell'.

For access to the Study Guide visit http://booksite.academicpress.com/Mak/primerAC/

Tak W. Mak
Toronto, Canada
July 2010

Preface to the *Immunomovie*

Immunology is like a very large and complex city. There are a few major well-known streets surrounded by many smaller winding roads. The latter are seldom traveled, and individuals who attempt to do so frequently become lost. Some neighborhoods are new, some are old, and others are constantly being torn down and reconstructed. Many buildings have multiple names and although the directions to find these buildings are perfectly understandable to some, they can be hard to follow by others. In trying to describe the "metropolis of immunology" to students, each teacher will take a different approach and bring a unique and personal bias to the task. In every case, the challenge is to connect the dots so that students can make their own maps and navigate the city themselves.

Another challenge facing anyone teaching immunology in the twenty-first century is that students are more media-savvy than ever before. Unlike their teachers, today's students have grown up learning from the Internet and watching endless numbers of videos. These tools define how modern students navigate the world and process information. About 7 years ago, I realized that "things had changed" in the world of learning and I decided to create a new teaching tool. With the help of many around me who gave generously of their time, their ideas, their voices, their interpretations, and their talents at drawing story boards and recording voices, I gradually created an animated video about immunology. The resulting *Immunomovie* teaches students in an entertaining and humorous way about how the immune system works. The *Immunomovie* is not meant to replace textbooks, journal articles, reviews, lectures, seminars or labs: it is simply a means of gathering the information gained from these various sources and placing it in the Big Picture in a way that is both entertaining and easily remembered. Since the *Immunomovie* presents the Big Picture, it is best watched after the basics have been absorbed from either an inspiring text or a scintillating series of lectures. Feel free to dip into a single episode, or to watch the entire extravaganza from start to finish (about 60 minutes).

Many topics in the *Immunomovie* are necessarily oversimplified, and inaccuracies may have crept in either because prevailing hypotheses have changed or because one explanation had to be selected over another to tell the story. Your favorite cytokine, CD antigen or signaling pathway might be missing and for that I apologize. Please also keep in mind that we are not professional animators and that we were often limited in how we could best depict a particular immunological process. Rest assured it is my intention to continually revise and update the *Immunomovie* and improve it technically as I become better at this new way of teaching. For the moment, I ask students to "cut us some slack" and to view the current version of the *Immunomovie* with a sense of humor and with the objective of getting the grand view of immunological events. For me personally, creating the *Immunomovie* was great fun and a wonderful learning experience. It is my hope that students will both enjoy our creation and learn from it. If something said or portrayed in the *Immunomovie* disagrees with a course's prescribed text or lecturer's comments, I do apologize, but I also advise students to realize that the field of immunology is a work in progress with many holes and disagreements and infusions of new information every day.

With that, I invite you to enjoy your voyage into one of the most fascinating disciplines in biomedical science.

Dr. Ellen Vitetta
Professor and Director, Cancer Immunobiology Center
University of Texas Southwestern Medical Center,
Dallas, Texas, USA

Access to the *Immunomovie* is included with the purchase of *Primer to The Immune Response*.

http://booksite.academicpress.com/Mak/primerAC/, see inside front cover for password.

Biographies

Authors

Tak Wah Mak

Tak W. Mak is the Director of the Campbell Family Institute for Breast Cancer Research in the Princess Margaret Hospital, Toronto, and a University Professor of the Departments of Medical Biophysics and Immunology, University of Toronto. He was trained at the University of Wisconsin in Madison, the University of Alberta, and the Ontario Cancer Institute. His research interests center on immune recognition and regulation as well as cell survival and cell death in normal and malignant cells. He gained worldwide prominence in 1984 as the leader of the group that first cloned the genes of the human T cell antigen receptor. His more recent work includes the creation of a series of genetically altered mice that have proved critical to understanding intracellular programs governing the development and function of the immune system, and to dissecting signal transduction cascades in various cell survival and apoptotic pathways as well as tumorigenesis. Dr. Mak holds honorary doctoral degrees from universities in North America, Asia and Europe, is an Officer of the Order of Canada, and has been elected a Foreign Associate of the National Academy of Sciences (U.S.) as well as a Fellow of the Royal Society of London (U.K.). Dr. Mak has won international recognition as the recipient of the Emil von Behring Prize, the King Faisal Prize for Medicine, the Gairdner Foundation International Award, the Sloan Prize of the General Motors Cancer Foundation, the Novartis Prize in Immunology, the Paul Ehrlich Prize, and the Ludwig Darmstaedter Prize.

Mary Evelyn Saunders

Mary E. Saunders holds the position of Scientific Editor for the Campbell Family Institute for Breast Cancer Research, Toronto. She completed her B.Sc. degree in Genetics at the University of Guelph, Ontario, and received her Ph.D. in Medical Biophysics at the University of Toronto. Dr. Saunders works with Dr. Mak and members of his laboratory on the writing and editing of scientific papers for peer-reviewed journals as well as on various book projects. She takes pride and pleasure in producing concise, clear, highly readable text and making complex scientific processes readily understandable.

Contributors

Wendy Lynn Tamminen

Wendy L. Tamminen completed her B.Sc. degree in Chemistry and Biochemistry at McMaster University, Hamilton, Ontario, and received her Ph.D. in Immunology from the University of Toronto. She taught immunology at the undergraduate level for several years to students in both the biomedical sciences and medicine at the University of Toronto, where her teaching skills were recognized with an Arts and Science Undergraduate Teaching Award. As a writer, editor, and lecturer, Dr. Tamminen's main interest is the communication of scientific concepts to both science specialists and non-specialists. www.wendytamminen.com

Maya Rani Chaddah

Maya R. Chaddah graduated with a double major in Human Biology and Spanish, followed by an M.Sc. in Immunology at the University of Toronto. In 1996, Ms. Chaddah started a business focused on the writing and editing of scientific and medical publications. Her expertise has grown to include scientific and medical illustration, and she continues to produce a variety of communications for diverse audiences in the public and private sectors. www.mayachaddah.com

Study Guide Author

Bradley D. Jett

Dr. Bradley Jett is the James Hurley Professor of Biology at Oklahoma Baptist University in Shawnee, Oklahoma. He completed his B.S. degree in Biology from OBU, followed by his Ph.D. in Microbiology and Immunology from the University of Oklahoma College of Medicine. After a postdoctoral fellowship at Washington University in St. Louis, Missouri, he joined the faculty at the University of Oklahoma College of Medicine. His research interests are primarily focused on host-parasite relationships. Much of his published work relates to the virulence factors of Gram-positive bacteria such as *Enterococcus, Staphylococcus,* and *Bacillus,* as well as the host immunological responses to these infections. In his current full-time, undergraduate teaching position at his alma mater he has been awarded Oklahoma Baptist University's Promising Teacher Award and the Distinguished Teaching Award.

Immunovie Creator

Ellen S. Vitetta

Dr. Ellen Vitetta is a Professor of Microbiology, the Director of the Cancer Immunobiology Center, the holder of the Sheryle Simmons Patigian Distinguished Chair in Cancer Immunobiology, and a Distinguished Teaching Professor at the University of Texas Southwestern Medical Center in Dallas, Texas. She and her colleagues were the first to describe IgD on the surface of murine B cells and she was the co-discoverer of IL-4. Over the past two decades, she has been involved in translational ("bench to bedside") research developing antibody-based immunotoxins that can destroy cancer cells and HIV-infected cells. These novel therapeutics have now been evaluated in over 300 patients. Dr. Vitetta is a member of the National Academy of Sciences, the Institute of Medicine, and the American Association of Arts and Sciences. In 1994, she served as the President of the American Association of Immunologists. She has taught immunology to medical and graduate students at UT Southwestern for 17 years and has won numerous Faculty Teaching Awards, including the Distinguished Educator Award. In 2006, Dr. Vitetta was inducted into the Texas Women's Hall of Fame. In 2007, the American Association of Immunologists awarded her with its highest honor, the Lifetime Achievement Award.

Acknowledgments

The authors are indebted to the following individuals for the reviewing of one or more draft chapters of this book.

Sonia Bakkour
Mount Holyoke College
South Hadley, Massachusetts, USA

Daniel Bergey
Black Hills State University
Spearfish, South Dakota, USA

Wiebke Bernhardt
European Patent Office, Biotechnology
Munich, Germany

Bruce Blazar
University of Minnesota
Minneapolis, Minnesota, USA

James R. Carlyle
Sunnybrook Research Institute
Toronto, Ontario, Canada

Radha Chaddah
Neurobiology Research Group, University of Toronto
Toronto, Ontario, Canada

Vijay K. Chaddah
Grey Bruce Regional Health Centre
Owen Sound, Ontario, Canada

Dominique Charron
Institut Universitaire d'Hématologie (IUH), Hôpital Saint Louis
Paris, France

Irvin Y. Chen
David Geffen School of Medicine at UCLA
Los Angeles, California, USA

Mark S. Davis
University of Evansville
Evansville, Indiana, USA

David Glick
Kings College
Wilkes-Barre, Pennsylvania, USA

Dale Godfrey
University of Melbourne
Parkville, Australia

Pazanta Goolsby
Winston-Salem State University
Winston-Salem, North Carolina, USA

Douglas R. Green
St. Jude Children's Research Hospital
Memphis, Tennessee, USA

Andrew Greene
Ashland University
Ashland, Ohio, USA

Carla Guthridge
Cameron University
Lawton, Oklahoma, USA

Zhenyu Hao
The Campbell Family Institute for Breast Cancer
Toronto, Ontario, Canada

William Heath
The Walter and Eliza Hall Institute
Victoria, Australia

Jules Hoffmann
Institute de Biologie Moleculaire et Cellulaire, CNRS
Strasbourg, France

Kristin Ann Hogquist
Center for Immunology, University of Minnesota
Minneapolis, Minnesota, USA

Jacob M. Hornby
Lewis-Clark State College
Lewiston, Idaho, USA

Robert D. Inman
Toronto Western Hospital
Toronto, Ontario, Canada

Brad Jett
Oklahoma Baptist University
Shawnee, Oklahoma, USA

Edward C. Kisailus
Canisius College
Buffalo, New York, USA

Joseph Larkin
University of Florida
Gainesville, Florida, USA

Jennifer K. Leavy
Georgia Institute of Technology
Atlanta, Georgia, USA

Robert Lechler
King's College London
London, England, UK

Eddy Liew
University of Glasgow
Glasgow, Scotland, UK

Bernard Malissen
Centre d'Immunologie de Marseille-Luminy
Marseille, France

John Martinko
Southern Illinois University Carbondale
Carbondale, Illinois, USA

Ruslan Medzhitov
Howard Hughes Medical Institute, Yale University School of Medicine
New Haven, Connecticut, USA

Mark Minden
Princess Margaret Hospital
Toronto, Ontario, Canada

Thierry Molina
Universite Paris-Descartes, Hôtel Dieu
Paris, France

David Nemazee
Scripps Research Institute
La Jolla, California, USA

Judith Ochrietor
University of North Florida
Jacksonville, Florida, USA

Pamela Ohashi
The Campbell Family Institute for Breast Cancer Research
Toronto, Ontario, Canada

Marc Pellegrini
The Campbell Family Institute for Breast Cancer Research
Toronto, Ontario, Canada

John Palisano
Sewanee, The University of the South
Sewanee, Tennessee, USA

Margaret A. Reinhart
University of the Sciences in Philadelphia
Philadelphia, Pennsylvania, USA

Michael R. Roner
The University of Texas at Arlington
Arlington, Texas, USA

Noel R. Rose
Johns Hopkins Center for Autoimmune Disease Research
Baltimore, Maryland, USA

Lawrence E. Samelson
Center for Cancer Research, National Cancer Institute
Bethesda, Maryland, USA

Daniel N. Sauder
Robert Wood Johnson School of Medicine
New Brunswick, New Jersey, USA

Kathy Schaefer
Randolph College
Lynchburg, Virginia, USA

Warren Strober
NIAID, National Institutes of Health
Bethesda, Maryland, USA

Susan Sullivan
Louisiana State University at Alexandria
Alexandria, Louisiana, USA

Clara Toth
Saint Thomas Aquinas College
Sparkill, New York, USA

John Trowsdale
Cambridge Institute for Medical Research
Cambridge, England, UK

Ellen Vitetta
Cancer Immunology Center, University of Texas Southwestern Medical Center
Dallas, Texas, USA

Helen Walter
Mills College for Women
Oakland, California, USA

Peter A. Ward
University of Michigan Health Systems
Ann Arbor, Michigan, USA

Tania Watts
University of Toronto
Toronto, Ontario, Canada

Ben Whitlock
University of Saint Francis
Joliet, Illinois, USA

Hans Wigzell
Microbiology and Tumor Biology Centre, Karolinska Institute
Stockholm, Sweden

David Williams
University of Toronto
Toronto, Ontario, Canada

Gillian E. Wu
York University
Toronto, Ontario, Canada

Juan-Carlos Zúñiga-Pflücker
Sunnybrook Research Institute
Toronto, Ontario, Canada

Contents

PART I
BASIC IMMUNOLOGY

1

Introduction to the Immune Response

WHAT'S IN THIS CHAPTER?

Books must follow sciences, and not sciences books.
Francis Bacon

A. Historical Orientation

What is immunology? Simply put, *immunology* is *the study of the immune system*. The *immune system* is a *system of cells, tissues and their soluble products that recognizes, attacks and destroys foreign entities that endanger the health of an individual*. The normal functioning of the immune system gives rise to *immunity*, a word derived from the Latin *immunitas*, meaning "to be exempt from." This concept originated in the 1500s, before the causes of disease were understood. Survivors who resisted death during a first exposure to a devastating disease and who did not get sick in a subsequent exposure were said to have become "exempt from" the disease, or "immune."

In 1796, the English physician Edward Jenner carried out experiments that solidified the birth of immunology as an independent science. At that time, smallpox was a disfiguring and often fatal disorder that decimated whole villages (Plate 1-1). Jenner observed that dairymaids and farmers lacked the pock-marked complexions of their fellow citizens, and wondered whether those who worked with cattle might be resistant to smallpox because of their close contact with livestock. Cows of that era often suffered from cowpox disease, a disorder similar to smallpox but much less severe. In an experiment that would be prohibited on ethical grounds today, Jenner deliberately exposed an 8-year-old boy to fluid from a cowpox lesion. Two months later, he inoculated the same boy with infectious material from a smallpox patient (Plate 1-2). In this first example of successful vaccination, the boy did not develop smallpox. Jenner's approach to smallpox prevention was quickly adopted in countries throughout Europe. As is discussed in more detail in Chapter 14, the modern story of smallpox vaccination and the global eradication of this scourge is one of the most successful public health endeavors in history.

Jenner's work was the first controlled demonstration of the immune response but little was understood at that time about the cellular and molecular mechanisms underlying the observed immunity. In Jenner's day, the cause of infectious disease was still a mystery and theories of that time did not envision the transmission of disease-causing germs. In 1884, Robert Koch proposed the *germ theory of disease* that stated that microbes invisible to the naked eye were responsible for specific illnesses, and indeed the first human disease-causing organisms or *pathogens* were identified in the late 1800s. At about the same time, Louis Pasteur applied Jenner's immunization technique for smallpox to the prevention of various animal diseases. Pasteur demonstrated that inoculation with a pathogen that had been weakened in the laboratory could protect against a subsequent exposure to the naturally occurring pathogen. It was Pasteur who coined the term *vaccination* (from the Latin *vaccinus*, meaning "derived from cows") for this procedure, in honor of Jenner's work. The research of Pasteur and other investigators spurred the evolution of immunology as a science distinct from (but related to) the established fields of microbiology, pathology, biochemistry and histology. Today, the core of modern immunology can be defined as *the study of the cells and tissues that mediate immunity and the investigation of the genes and proteins underlying their function*. Selected landmark discoveries in immunology and immunologists who have won Nobel Prizes for their work are featured in Appendices A and B.

B. The Nature of the Immune Response

The immune system has evolved to counteract assault on the body by non-self entities that may compromise an individual's health. Such entities include pathogens, inert injurious materials (such as splinters) and threats generated within an individual's own body (such as cancers). However, since infectious agents, including bacteria, viruses, parasites and fungi, are literally everywhere on Earth, the immune system is primarily

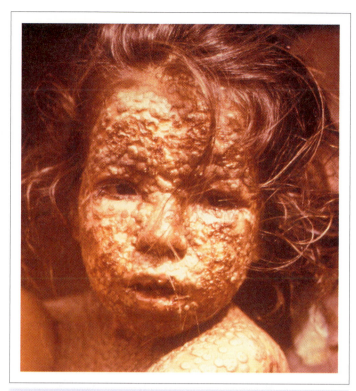

Plate 1-1 Smallpox
[Reproduced by permission of the Public Health Image Library, CDC]

Humans are surrounded by other organisms—in the air, in the soil, in the water, on the skin, and on the *mucosae*, the protective layers of epithelial cells that line the gastrointestinal, urogenital and respiratory tracts. While most of these organisms are harmless and some are even beneficial, some are pathogenic. Like all species, pathogens live to reproduce. In order to reproduce, however, many must penetrate a host's body or one of its component cells. *Infection* is defined as *the attachment and entry of a pathogen into the host*. Once inside the body or cell, the pathogen replicates, generating progeny that spread into the body in a localized or systemic fashion. The manner of this replication determines whether the pathogen is considered extracellular or intracellular.

Extracellular pathogens, such as certain bacteria and parasites, do not need to enter cells to reproduce. After accessing the body, these organisms replicate first in the interstitial fluid bathing the tissues and may then disseminate via the blood (Fig. 1-1). *Intracellular pathogens*, such as viruses and other bacteria and parasites, enter a host cell, subvert its metabolic machinery, and cause it to churn out new virus particles, bacteria, or parasites. These pathogens may then also travel systemically by entering the blood. For both intracellular and extracellular pathogens, if the infectious agent overwhelms normal body systems or interferes with cellular functions, the body becomes "sick." This sickness is manifested as a set of

Plate 1-2 Smallpox Immunization by Edward Jenner
[Reproduced by permission of Wellcome Library, London]

occupied with containing attacks from this quarter. Indeed, despite the successful eradication of smallpox and the development of antibiotics, infectious diseases remain among the biggest killers on the planet. Long-standing disorders such as malaria and tuberculosis, as well as "emerging" diseases caused by new pathogens such as West Nile virus and severe acute respiratory syndrome (SARS) virus, present ongoing challenges to the immune system.

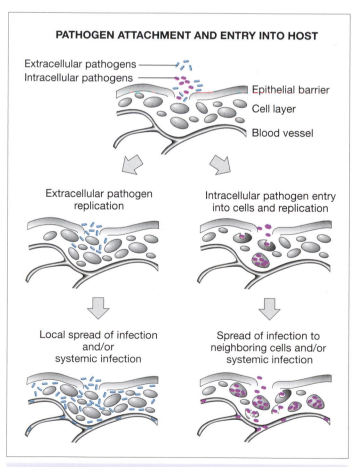

Fig. 1-1 Establishment of Infection by Extracellular and Intracellular Pathogens

characteristic clinical symptoms that we call an "illness" or "disease".

Our bodies are under constant assault by harmful microbes. Yet, most of the time, assaults by these organisms are successfully repelled and disease is prevented. This resistance is due to both basic and sophisticated immune responses that combat pathogens. Simply put, the job of the immune response is to "clean up" infections in the interstitial fluid, tissues and blood, and to destroy infected host cells so that neighboring host cells do not share their fate. Because pathogens are constantly evolving mechanisms to evade or block immune defenses, the immune system must constantly adapt to maintain its effectiveness. It is a continual horse race as to which will be the more successful mechanism: the body's immune surveillance or the pathogen's invasion and infection strategy. We note here that the immune response itself may also cause some collateral damage to tissues as part of the larger battle against pathogens but such *immunopathic* effects are usually short-lived in an otherwise healthy individual.

At the turn of the twentieth century, there were two schools of thought on what mechanisms underlay immune responses. One group of scientists believed that immunity depended primarily on the actions of cells that destroyed or removed unwanted material from the body. This clearance process was referred to as *cell-mediated immunity*. However, another group of researchers was convinced that soluble molecules in the serum of the blood could directly eliminate foreign entities without the need for cellular involvement. In this case, the clearance process was referred to as *humoral immunity*, a term derived from the historical description of body fluids as "humors". Today, we know that both cell-mediated and humoral responses occur simultaneously during an immune response and that both are often required for complete clearance of a threat.

The cells responsible for cell-mediated immunity are collectively called *leukocytes* ("leuko," white; "cyte," small body, i.e., a cell) or *white blood cells*. However, "blood cell" is a bit of a misnomer, because a majority of leukocytes reside in tissues and specialized organs, and move around the body through both the blood circulation and a system of interconnected vessels called the *lymphatic system*. The soluble molecules responsible for humoral immunity are proteins called *antibodies*, and antibodies are secreted by a particular type of leukocyte. The production of these antibodies and the mounting of cell-mediated immune responses depend on an elaborate signaling system by which leukocytes communicate with each other as well as with other cell types in the body. This signaling is mediated by small secreted proteins called *cytokines*, which are mainly produced by leukocytes.

C. Types of Immune Responses: Innate and Adaptive

Humans are capable of two types of immune responses: the *innate* (or *natural*) and *adaptive* (or *acquired*) responses.

Whereas the innate response is always involved in repelling an invader, the adaptive response is mounted only as needed. In both cases, the response works to clear the body of unwanted entities. However, the *innate immune response* involves recognition that is either totally *non-specific* or is *broadly specific* and *non-selective*, while the *adaptive immune response* involves recognition that is *uniquely specific* and *selective*. Leukocytes involved in the innate response recognize a limited number of molecular patterns common to a wide variety of pathogens. These molecular structures are called *pathogen-associated molecular patterns* (PAMPs) (Fig. 1-2). In contrast, leukocytes of the adaptive response recognize unique molecular structures that are present on or derived from a single pathogen and are not shared with any other infectious agent. Historically, these unique structures were called *antigens* because they were first identified as components of pathogens that could be bound by antibodies; that is, these components induced "antibody generation." The term "antigen" is now used to refer to molecular structures that are targeted by either a humoral or cell-mediated adaptive response. From the point of view of the body's leukocytes, a complex pathogen represents a collection of many different PAMPs and antigens.

Innate responses occur all over the body, primarily just below the skin surface and at the mucosae where microbes commonly attempt to gain access. Most of the time, the innate response successfully eliminates the invader via several mechanisms that are introduced here and discussed in detail in Chapter 3. Only those entities that succeed in overwhelming the innate defenses evoke adaptive immune responses that are initiated in specialized tissues called *lymphoid tissues* (see Ch. 2). Elements of an innate immune response can be found in all multicellular organisms, whereas the mechanisms of the more recently evolved adaptive immune response are present only in the higher vertebrates (fish and above). A Table of Comparative Immunology is presented in Appendix C that summarizes the immune system elements present in an evolutionarily diverse range of animals.

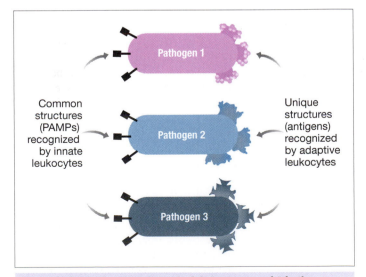

Common structures (PAMPs) recognized by innate leukocytes

Unique structures (antigens) recognized by adaptive leukocytes

Pathogen 1

Pathogen 2

Pathogen 3

Fig. 1-2 Molecular Structures Recognized during Innate and Adaptive Immune Responses

Table 1-1 Comparison of Innate and Adaptive Immune Responses	
Innate	**Adaptive**
Pre-existing and induced mechanisms	All mechanisms induced
Broad range of pathogens recognized by a small number of recognition structures of limited diversity	Broad range of pathogens recognized by an almost infinite number of extremely diverse adaptive receptors
Response most often triggered by binding of pathogen-associated molecular patterns (PAMPs) to pattern recognition receptors (PRRs)	Response triggered by binding of pathogen antigens to B cell receptors (BCRs) and pathogen antigen fragments to T cell receptors (TCRs)
No memory of pathogens	Pathogens remembered by memory cells
Level of defense is similar upon repeated exposures to the same pathogen	Level of defense is more vigorous and finely tuned with repeated exposures to the same pathogen

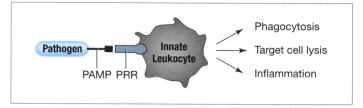

Fig. 1-3 **PAMP Recognition and Effector Functions of Innate Leukocytes**

The general characteristics of innate and adaptive immune responses are described in the next two sections and summarized in Table 1-1.

I. GENERAL FEATURES OF INNATE IMMUNITY

Innate immunity is established by immune response mechanisms that do not make fine distinctions among invading pathogens and do not undergo fundamental permanent change as a result of exposure to a pathogen. A second exposure to an infectious agent provokes a response of a very similar magnitude and character as the first exposure; that is, there is no long-term "memory" of a previous exposure to that agent. The first level of innate defense is mediated by a collection of physical, chemical and molecular barriers that exclude foreign material in a way that is totally non-specific and requires no induction. These barriers include *anatomical barriers* and *physiological barriers*. An example of an anatomical barrier is intact skin, while the low pH of stomach acid and the hydrolytic enzymes in body secretions are examples of physiological barriers.

Should barrier defense prove insufficient, other forms of innate defense are induced. A key player in the induced innate response is *complement*, a complex system of enzymes that circulates in the blood in an inactive state. Once activated by a pathogen attack, the complement system contributes directly and indirectly to innate and adaptive mechanisms of pathogen destruction. In addition to complement, there is an array of innate leukocytes that take action when invaders breach anatomical and physiological barriers. Innate leukocytes possess receptors called *pattern recognition receptors* (PRRs) that recognize PAMPs (Fig. 1-3). The binding of a PAMP to a PRR triggers a clearance response that generally takes the form of

phagocytosis, *target cell lysis*, and/or the induction of *inflammation*.

i) Phagocytosis
Leukocytes frequently use a sophisticated means of engulfing pathogens or their components that is called *phagocytosis* ("cell eating"). Phagocytosis is carried out primarily by three types of PRR-bearing leukocytes: *neutrophils*, *macrophages* and *dendritic cells* (DCs). Neutrophils engulf and destroy bacteria and other pathogens. Macrophages and DCs engulf not only pathogens but also dead host cells, cellular debris and host macromolecules. The phagocytosis of foreign entities by macrophages and DCs allows these cells to "present" unique antigens from this material on their cell surfaces in such a way that the antigens can be recognized by cells of the adaptive immune response. The adaptive leukocytes are then activated and mediate antigen-specific protection.

ii) Target Cell Lysis
Cancer cells and cells infected with intracellular pathogens frequently express molecules on their cell surfaces that mark them as "target cells" for immune destruction. When these molecules are recognized and bound by various receptors expressed by innate leukocytes, complex processes are initiated that result in the lysis of the cancer cell or infected cell. These types of innate responses are carried out primarily by neutrophils, macrophages and another type of leukocyte called a *natural killer cell* (NK cell).

iii) Inflammation
When leukocytes phagocytose and digest foreign material or cell debris, they produce chemical signals that promote wound healing and secrete cytokines that summon other leukocytes to the same location. The resulting influx of first innate and later (if necessary) adaptive leukocytes into the site of injury or infection is called *inflammation* or an *inflammatory response*. The redness and swelling we commonly associate with inflammation are the outward physical signs of this response.

II. GENERAL FEATURES OF ADAPTIVE IMMUNITY

In contrast to the innate immune response, the adaptive immune response involves recognition and effector mechanisms that are highly specific for unique antigens expressed by the particular entity triggering the response. Adaptive responses are mediated by a class of leukocytes called *lymphocytes*. Lymphocytes are categorized as either *B lymphocytes* (B cells) or

T lymphocytes (T cells). The fundamental functional and developmental characteristics of lymphocytes are responsible for *specificity, division of labor, memory, diversity* and *tolerance* of the adaptive response. Each of these important concepts is introduced here as a prelude to more detailed discussions of these topics in later chapters.

i) Specificity

The selective specificity of an adaptive immune response for a particular antigen is determined by the nature of the antigen-specific receptors expressed on the surfaces of T and B lymphocytes. Each lymphocyte expresses thousands of identical copies of a unique antigen receptor protein. Interaction of these antigen receptor molecules with antigen triggers activation of the lymphocyte. In contrast to the broad recognition mediated by innate leukocytes, antigen binding by a lymphocyte antigen receptor usually hinges on the presence of a unique molecular shape unlikely to appear on more than one pathogen.

The antigen receptors on the surface of a B cell are called *B cell receptors* (BCRs), whereas those on the T cell surface are called *T cell receptors* (TCRs). In both cases, the antigen receptor is itself a complex of several proteins. Some of these proteins interact directly and specifically with antigen, while others convey the intracellular signals triggered by antigen binding into the cytoplasm and nucleus of the lymphocyte. These signals are critical for sustaining the activation of the lymphocyte and generating an appropriate effector response that eliminates the antigen.

ii) Division of Labor

"Division of labor" in the adaptive response refers to the different but equally important contributions of B lymphocytes and two types of T lymphocytes. These two major T cell subsets are called *cytotoxic T cells* (Tc) and *helper T cells* (Th). The roles of B, Tc and Th cells in the adaptive response are defined by the distinct mechanisms by which they recognize and respond to antigen (Table 1-2).

B cells recognize antigen in a way that is fundamentally different from that used by Tc and Th lymphocytes. The BCR of a B cell binds directly to an intact antigen, which may be either a soluble molecule or a molecule present on the surface of a pathogen (Fig. 1-4A). Activation of a B cell in this way causes it to proliferate and produce identical daughter effector

cells called *plasma cells* (PCs) that secrete vast quantities of specific antibody. An *antibody* is a protein that is a *modified, soluble form of the original B cell's membrane-bound antigen receptor.* Antibody molecules enter the host's circulation and tissues, bind to the specific antigen, and mark it for clearance from the body, establishing a humoral response. Since antibodies are present in the extracellular milieu, such a response is effective against extracellular pathogens. However, antibodies are unable to penetrate cell membranes and so cannot attack an intracellular pathogen once it has entered a host cell.

Unlike BCRs, TCRs are unable to recognize whole, native antigens. Rather, a TCR recognizes a bipartite structure displayed on the surface of a host cell. This bipartite structure is made up of a host membrane protein called an *MHC molecule* (see Chapter 6) bound to a short *peptide* derived from a protein antigen. This complex is referred to as a *peptide–MHC complex* (pMHC). Tc and Th cells are activated by the binding of their TCRs to pMHCs "presented" on the surface of a DC. DCs can acquire extracellular pathogen antigens by phagocytosis, and intracellular pathogen antigens either by infection or by phagocytosis of the debris of dead infected cells. Peptides from these antigens are bound to one of two types of MHC molecules, either *MHC class I* or *MHC class II*, and then displayed on the DC surface for inspection by T cells.

Tc cells recognize antigenic peptides that the DC has presented on MHC class I molecules (Fig. 1-4B). The activated Tc cell proliferates and differentiates into identical effector cells called CTLs (*cytotoxic T lymphocytes*) that can lyse any host cell displaying the same peptide–MHC class I complex recognized by the original Tc cell. MHC class I molecules are expressed on almost all body cells, making them all potential targets for CTL-mediated lysis. However, unlike DCs, most host cells cannot phagocytose extracellular entities. It is therefore only after infection with an intracellular pathogen that a host cell displays an antigenic peptide–MHC class I complex that can be recognized by the TCR of the CTL. The cell-mediated immunity directed against such infected host cells is critical to the clearance of intracellular pathogens.

Th cells recognize antigenic peptides that a DC has presented on MHC class II molecules (Fig. 1-4C). The activated Th cell proliferates and differentiates into identical Th effector cells that can recognize the same peptide–MHC class II structure that activated the original Th cell. However, MHC class II molecules are expressed only by certain cell types that can

---

Table 1-2 Division of Labor in Adaptive Immune Responses

Lymphocyte	Antigen Recognized	Pathogens Combated	Mechanism Used by Effector Cells
B cell	Whole soluble antigen or whole antigen present on a pathogen surface	Extracellular pathogens	Antibody-mediated clearance of pathogen
Tc cell	Antigen peptide + MHC*	Primarily intracellular pathogens	Lysis of infected host cells to prevent pathogen spread
Th cell	Antigen peptide + MHC	Many extracellular and intracellular pathogens	Secretion of cytokines required for activation of B cells and Tc cells, and stimulation of macrophages

*MHC, major histocompatibility complex; see Ch. 2.

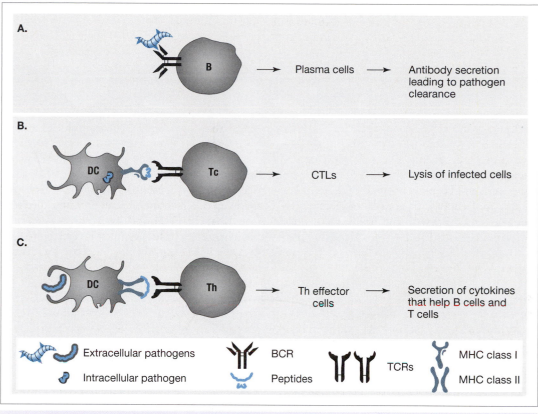

Fig. 1-4 Antigen Recognition and Effector Functions of B, Tc and Th Cells

act as specialized *antigen-presenting cells* (APCs), including DCs, macrophages and activated B cells. In the course of a pathogen attack, APCs can acquire the pathogen and display peptides derived from its antigens on MHC class II. When the TCR of a Th effector cell recognizes this pMHC structure, the cell is stimulated to produce cytokines that assist B cells and Tc cells to become fully activated. In this way, Th cells contribute to both humoral and cell-mediated adaptive responses, and are thus crucial for protection against many extracellular and intracellular pathogens. Cytokines secreted by Th cells also stimulate macrophage activities. The combined efforts of all three types of lymphocytes may be necessary to eliminate a particularly wily invader.

iii) Immunological Memory
The innate immune response deals with a given pathogen in a very similar way each time it enters the body. In contrast, the adaptive immune response can "remember" that it has seen a particular antigen before. This *immunological memory* means that an enhanced adaptive response is mounted upon a second exposure to a given pathogen, so that signs of clinical illness are mitigated or prevented. In other words, immunity is achieved because the body has effectively "adapted" its defenses and "acquired" the ability to exclude this pathogen.

Immunological memory arises in the following way. A constant supply of resting B and T lymphocytes is maintained throughout the body, each expressing its complement of unique antigen receptor proteins. When a pathogen antigen enters the body for the first time, a process of *clonal selection* takes place in which only those lymphocytes bearing receptors specific for that antigen are triggered to respond (Fig. 1-5). In other words, only those lymphocyte clones "selected" by the antigen leave the resting state and proliferate, undergoing clonal expansion to generate daughter cells all expressing the same BCR or TCR as the original parent lymphocyte. The differentiation of these daughter cells gives rise to short-lived *effector cells* equipped to eliminate the pathogen, as well as long-lived *memory cells* that persist in the tissues essentially in a resting state until a subsequent exposure to the same pathogen. The attack on the pathogen by this first round of effector cells is called the *primary immune response*. The second (or subsequent) time that the particular pathogen enters the body, it is met by an expanded army of clonally selected, antigen-specific memory cells that undergo much more rapid differentiation into effector cells than occurred during the first antigen encounter. The result is a stronger and faster *secondary immune response* that eliminates the pathogen efficiently, mitigating or preventing disease. New populations of memory cells are also produced during the secondary response, ensuring that the host maintains long-term or even lifelong immunity to that pathogen. The generation of protective memory cells is the basis of vaccination; a healthy person is immunized with a vaccine containing pathogen antigens in order to provoke the production of memory cells that will prevent disease from developing if the individual is ever naturally infected by that pathogen.

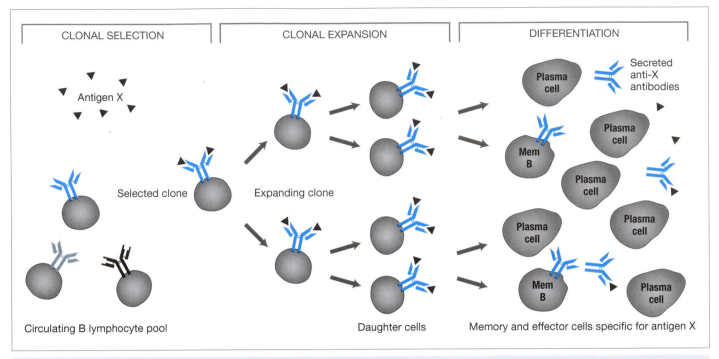

CLONAL SELECTION | CLONAL EXPANSION | DIFFERENTIATION

Antigen X

Selected clone

Expanding clone

Circulating B lymphocyte pool

Daughter cells

Secreted anti-X antibodies

Plasma cell

Mem B

Plasma cell

Plasma cell

Plasma cell

Plasma cell

Mem B

Plasma cell

Memory and effector cells specific for antigen X

Fig. 1-5 **Clonal Selection and Generation of Memory and Effector Cells, as Shown for B Cells**

iv) Diversity

The degree of diversity of antigens recognized also distinguishes the adaptive response from the innate response. Whereas the innate immune system exhibits a genetically fixed and finite capacity for antigen recognition, the recognition capacity of the adaptive immune system is nearly limitless. In fact, our bodies are capable of recognizing totally synthetic antigens that do not occur in nature. This huge diversity arises from the combined effects of several mechanisms that affect the genes encoding the antigen receptors. Some of these mechanisms operate before a lymphocyte encounters antigen, and some after (see Ch. 4 and 8).

The primary source of antigen receptor diversity is a gene rearrangement process that occurs during the development of B cells and T cells prior to encounter with antigen. The genes encoding antigen receptors are not individual continuous entities. Rather, the BCR and TCR genes are assembled from a large collection of pre-existing gene segments by a mechanism called *somatic recombination*. A single, random combination of gene segments is thus created in each developing lymphocyte. As a result, a lymphocyte population is generated in which the antigen receptor proteins are vastly diverse in specificity because they are encoded by hundreds of thousands of different DNA sequences. In the case of B cells, additional mutational mechanisms operate that result in further structural diversification of antibody proteins. The sheer numbers of B and T lymphocyte clones generated in a healthy individual guarantee that there will be at least one clone expressing a unique receptor sequence for every antigen encountered during

the host's life span. These clones, with their array of antigen receptor specificities, are collectively called the individual's lymphocyte *repertoire*.

v) Tolerance

The generation of lymphocyte clones that can theoretically recognize any antigen in the universe raises the question of how the body avoids lymphocyte attacks on molecules present in its tissues. This avoidance is called *tolerance*, the fifth aspect in which the adaptive response exhibits more refinement than the innate response. As stated earlier, the specificity of each antigen receptor is randomly determined by somatic recombination during early lymphocyte development. By chance, some of the genetic sequences produced will encode receptors that recognize self molecules (*self antigens*). These lymphocytes must be identified as recognizing self and then either removed from the body entirely, or at least inactivated, to ensure that an individual has an effective lymphocyte repertoire that does not attack healthy tissue. The multiple mechanisms involved in establishing tolerance are grouped into two broad stages. The first stage, called *central tolerance*, occurs during early lymphocyte development. The mechanisms of central tolerance are designed to eliminate clones that recognize self antigens, thus establishing a lymphocyte repertoire that targets "non-self" (see Ch. 5 and 9). In the second stage, called *peripheral tolerance*, any B and T lymphocytes that recognize self but somehow escaped the screening of central tolerance and completed their development are functionally silenced by another set of inactivating mechanisms (see Ch. 10).

D. Interplay between the Innate and Adaptive Responses

The immune system of a vertebrate host under pathogen attack responds in three phases, two of which involve innate immunity and the last of which requires the adaptive response. First to offer immediate protection are the innate physical barriers. The components making up these barriers are *non-inducible* in that they pre-exist and do not develop or change in response to the presence of antigen (Fig. 1-6). Should the barriers be penetrated, the second, *inducible* phase of innate defense becomes effective 4–96 hours after a pathogen enters the body. Innate leukocytes activated by the binding of PAMPs present on the attacking pathogen work quickly to eliminate the invader, using the mechanisms of phagocytosis, target cell lysis and inflammation. Complement activation also contributes to pathogen clearance at this stage. In many cases, an infection is completely controlled by the two phases of innate defense before adaptive immunity is even triggered. However, the cells of the innate response are ultimately limited both in numbers and in recognition capacity. Should the pathogen thwart innate immunity, the third and final phase of host defense is mediated by the lymphocytes of the adaptive response. T and B lympho-

cytes specifically recognize components of a particular pathogen, proliferate, and differentiate into large numbers of daughter effector cells that eliminate the pathogen via humoral and cell-mediated mechanisms. Because it takes time for clonally selected lymphocytes to be activated and respond to a persistent threat, adaptive responses are not usually observed until at least 96 hours after the initial infection.

The second and third phases of immune defense are more tightly interwoven than it may first appear. Innate and adaptive immunity do not operate in isolation, and each depends on or is enhanced by elements of the other. The lymphocytes of the adaptive response require the involvement of cells and cytokines of the innate system to become activated and undergo differentiation into effector and memory lymphocytes. Conversely, activated lymphocytes secrete products that can stimulate and improve the effectiveness of innate leukocytes.

The elements of a three phase immune response are presented in Figure 1-7. Pathogens that manage to penetrate the body's physical barriers (skin, mucosae, enzymes etc.) and avoid non-specific destruction by complement are met by innate leukocytes. Using broadly specific control mechanisms, these cells attempt to limit the pathogen's spread. If the innate leukocytes are unable to eradicate the threat, the lymphocytes of adaptive immunity generate humoral and cell-mediated responses against unique pathogen antigens. The interplay between innate and adaptive immunity is sustained by cytokine signaling and through direct intercellular contacts between innate and adaptive leukocytes. By cooperating in this way, innate and adaptive immunity mount an optimal defense against pathogens.

0–4 hrs   4–96 hrs   > 96 hrs

Pathogen entry

PHASE 1
NON-INDUCED INNATE RESPONSE
NON-SPECIFIC

- Preformed defenses (skin barrier, mucosal barrier, pH, saliva proteases)

PHASE 2
INDUCED INNATE RESPONSE
BROADLY SPECIFIC

- Complement activation
- Phagocytosis
- Target cell lysis
- Inflammation

PHASE 3
INDUCED ADAPTIVE RESPONSE
UNIQUELY SPECIFIC

- B cells (antibodies)
- Th cells (cytokines)
- Tc cells (cytolysis)

Fig. 1-6 **The Three Phases of Host Immune Defense**

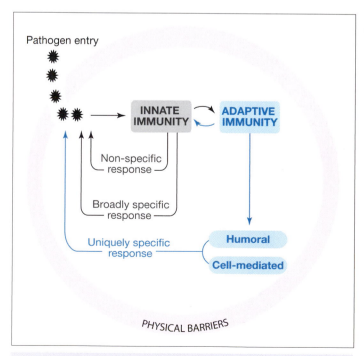

Fig. 1-7 **Interplay between Innate and Adaptive Immunity in Host Defense**

E. Clinical Immunology

When the immune system is functioning normally, harmful entities are recognized and the host is protected from external attack by pathogens and internal attack by cancers. Some localized tissue damage may result from the inflammatory response that develops as innate leukocytes work to eliminate the threat, but these immunopathic effects are limited and controlled. As well, the tolerance of the healthy immune system prevents cells of the adaptive response from attacking normal self tissues.

What happens when the immune system malfunctions, or when its normal functioning has undesirable effects? There are several instances in which inappropriate actions of the immune system can result in pathologic consequences that tip the balance from health to disease (Fig. 1-8). Firstly, the normal attack of a healthy immune system on a foreign tissue transplant meant to preserve life results in *transplant rejection* and deleterious consequences for the transplant patient. Secondly, when the tolerance of an individual's adaptive immune response fails, self tissues are attacked and may result in *autoimmune disease*. Thirdly, an immune response that is unregulated or too strong can result in *hypersensitivity* to an entity. In this situation, the inflammation that accompanies the response causes significant collateral tissue damage. When an individual's adaptive immune system responds inappropriately to an antigen that is generally harmless, the immune response is manifested clinically as a form of hypersensitivity called *allergy*. Each of these outcomes is addressed in later chapters.

Sometimes the immune system malfunctions because it is incomplete, either because of *primary immunodeficiency* or *acquired immunodeficiency*. In primary immunodeficiencies, the failure of the immune system is congenital; that is, an affected individual is born with a genetic defect that impairs his/her ability to mount innate and/or adaptive immune responses. In an acquired immunodeficiency, an external factor such as a nutritional imbalance or a pathogen may cause the loss of an immune system component. For example, in *acquired immunodeficiency syndrome* (AIDS), the *human immunodeficiency virus* (HIV) destroys the T lymphocytes needed to fight infection. In both primary and acquired immunodeficiencies, patients show a heightened susceptibility to recurrent infections and sometimes tumors. The prognosis for most of these diseases is very poor and treatment options are limited. A full discussion of HIV and AIDS appears in Chapter 15, and various primary immunodeficiencies are discussed in focus boxes embedded in the chapter where the normal immune system component is discussed.

The remainder of this book will take the reader on a logical tour of the immune response. Chapter 2 continues our basic immunology section, providing detailed descriptions of the various cells, tissues and cytokines involved in mediating immune responses. Chapters 3–12 cover all aspects of the innate and adaptive immune responses in depth. The chapters in our clinical immunology section, Chapters 13–20, address immunological principles not only of practical value to the more medically oriented reader but also of interest to all who seek to understand the pivotal role of the immune system in health and disease. It is our hope that the reader will come away with a solid understanding of the cellular and molecular mechanisms underlying immunity and how these mechanisms can go awry to cause disease. We also provide a glimpse of how researchers seek to manipulate these mechanisms with the goal of ensuring good health for all.

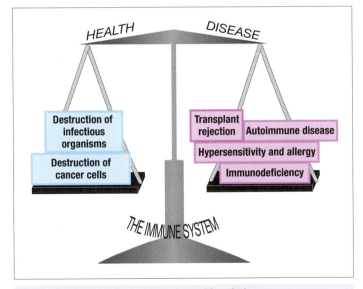

Fig. 1-8 **Role of the Immune System in Health and Disease**

CHAPTER 1 TAKE-HOME MESSAGE

- The immune system is the central player in the maintenance of human health.

- Immune responses depend on the coordinated action of leukocytes that travel throughout the body to recognize and eliminate extracellular and intracellular pathogens, toxins and cancerous cells.

- Cytokines are secreted intercellular messenger proteins that mediate complex interactions among leukocytes.

- Adaptive responses are mounted after innate responses have failed to remove an unwanted entity, and are dependent on cytokines produced during the innate response.

- Complete protection from and clearance of unwanted entities often involves both innate and adaptive immune responses.

- Some innate immune mechanisms require no induction and are completely non-specific, whereas others are inducible and involve broadly specific recognition of molecular patterns.

- Adaptive immunity involves the selective activation of lymphocytes and is highly specific for the particular inciting entity.

- Elements of innate immunity influence adaptive immunity and vice versa.

- Malfunctions of the immune system can cause cancer, autoimmune disorders, allergy, hypersensitivity, transplantation rejection and immunodeficiency.

DID YOU GET IT? A SELF-TEST QUIZ

Section A

1) Can you define these terms? immunology, immunity, pathogen

2) What was Koch's theory and why was it important?

3) Who coined the term "vaccination" and on what basis?

Section B

1) Can you define these terms? mucosae, leukocyte, cytokine

2) What is the difference between infection and disease?

3) What is the major difference between extracellular and intracellular pathogens?

4) Distinguish between cell-mediated and humoral immunity.

Section C

1) Can you define these terms? phagocytosis, inflammation, PAMP, PRR, MHC, lymphocyte, TCR, BCR, antibody, clonal selection, tolerance, self antigen, immunological memory

2) What is a major difference between innate and adaptive immune responses?

3) Can you give two examples of innate barrier defenses?

4) How does target cell lysis help protect the body?

5) Describe five important characteristics of the adaptive response.

6) How do Th, Tc and B cells differ in antigen recognition?

7) How does each lymphocyte subset contribute to immunity against extracellular and/or intracellular pathogens?

8) Why is the secondary immune response stronger and faster than the primary response?

9) What are the two stages of tolerance?

Section D

1) Can you describe the three phases of immune defense and the components that mediate them?

2) How do the innate and adaptive responses influence each other and why is this important?

Section E

1) Can you describe four examples of diseases arising from immune system failure?

2

Components of the Immune System

No country can act wisely simultaneously in every part of the globe at every moment of time.

Henry Kissinger

The immune system is not a discrete organ, like a liver or a kidney. It is an integrated partnership, with contributions by the circulatory system, lymphatic system, various lymphoid organs and tissues, and the specialized cells moving among them. In this chapter, we describe the components of the immune system, how communication is carried out among these components, and how leukocyte movement underlies immune responses.

A. Cells of the Immune System

I. TYPES OF HEMATOPOIETIC CELLS

Mammalian blood is made up of *hematopoietic cells* (red and white blood cells) that are carried along in a fluid phase called *plasma*. As introduced in Chapter 1, the white blood cells are infection-fighting *leukocytes*, while the red blood cells are *erythrocytes* that carry oxygen in the blood to the tissues. All hematopoietic cells are generated in the bone marrow from a common precursor called the *hematopoietic stem cell* (HSC) in a process called *hematopoiesis*. Immunologists generally classify hematopoietic cells not by their red or white color but by their developmental pathway from the HSC during hematopoiesis. One developmental pathway gives rise to hematopoietic cells of the *myeloid* lineage. Myeloid cells include *neutrophils, basophils, eosinophils, monocytes* and *macrophages*. Neutrophils, basophils and eosinophils are all *granulocytes*, which are myeloid cells that harbor intracellular granules containing microbe-killing molecules. Erythrocytes and *megakaryocytes* (the source of platelets for blood clotting) are technically myeloid cells but are often considered a separate lineage. The primary functions of erythrocytes and megakaryocytes are non-immunological and so will not be discussed further in this book. A second developmental pathway from the HSC produces cells of the *lymphoid* lineage. Lymphoid

cells include *T* and *B lymphocytes* and their effector cells, *natural killer* (NK) *cells*, and *natural killer T* (NKT) *cells*. *Dendritic cells* (DCs) can arise from either the myeloid or lymphoid pathways. A third developmental pathway from the HSC gives rise to *mast cells*. The physical characteristics and functions of these cell types are summarized in Figure 2-1.

Different hematopoietic cell types tend to reside and function in different body compartments. Hematopoietic stem cells and early hematopoietic progenitors derived from HSCs are concentrated in the bone marrow, while lymphocytes move constantly between the blood and the lymphoid organs and tissues (Fig. 2-2). Other hematopoietic cell types tend to concentrate in either the blood or the tissues, but may move between these compartments during an inflammatory response to infection or injury. These cells are attracted from the circulation to the site of inflammation by *chemokines*. Chemokines are specialized cytokines that draw leukocytes toward a site of attack in a process called *chemotaxis*. As chemokines diffuse into the tissues surrounding an inflammatory site, a concentration gradient is created with the highest chemokine concentration at the site of injury or attack, and steadily decreasing concentrations at increasing distances from this area. Leukocytes express chemokine receptors that cause these cells to migrate up a gradient toward the source of the chemokine. Chemokines are secreted by injured or infected cells, as well as by other leukocytes responding to the invasion.

II. CELLS OF THE MYELOID LINEAGE

i) Neutrophils

Neutrophils are the most common leukocytes in the body and constitute the majority of cells activated during an inflammatory response. These cells were originally called *polymorphonuclear* (PMN) leukocytes because of the appearance under the light microscope of their irregularly shaped, multilobed nuclei. The average adult human possesses approximately 50 billion

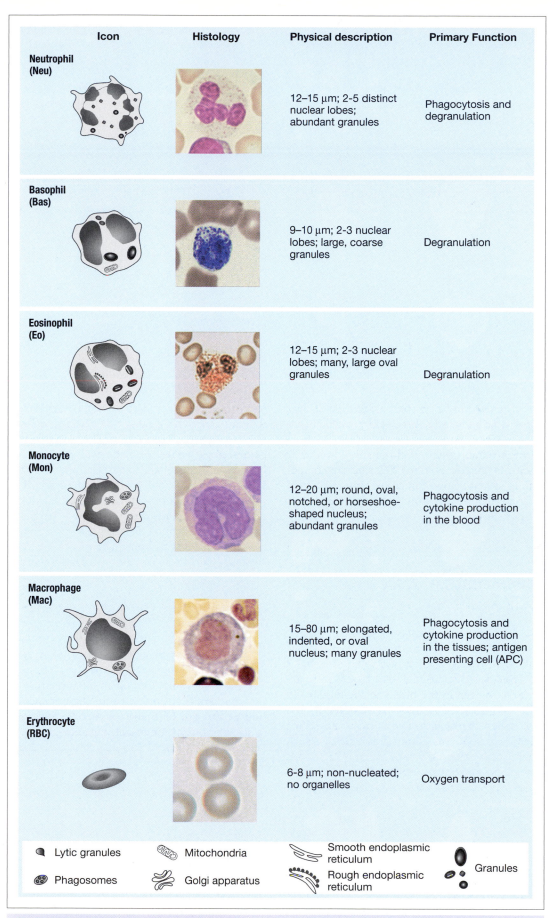

	Icon	Histology	Physical description	Primary Function
Neutrophil (Neu)			12–15 µm; 2-5 distinct nuclear lobes; abundant granules	Phagocytosis and degranulation
Basophil (Bas)			9–10 µm; 2-3 nuclear lobes; large, coarse granules	Degranulation
Eosinophil (Eo)			12–15 µm; 2-3 nuclear lobes; many, large oval granules	Degranulation
Monocyte (Mon)			12–20 µm; round, oval, notched, or horseshoe-shaped nucleus; abundant granules	Phagocytosis and cytokine production in the blood
Macrophage (Mac)			15–80 µm; elongated, indented, or oval nucleus; many granules	Phagocytosis and cytokine production in the tissues; antigen presenting cell (APC)
Erythrocyte (RBC)			6-8 µm; non-nucleated; no organelles	Oxygen transport

Lytic granules   Mitochondria   Smooth endoplasmic reticulum

Phagosomes   Golgi apparatus   Rough endoplasmic reticulum   Granules

Fig. 2-1 (Part 1) **Characteristics of Hematopoietic Cells**
[Micrographs of plates reproduced by permission of D.C. Tkachuk, J.V. Hirschmann and J.R. McArthur, *Atlas of Clinical Hematology* 2002, W.B. Saunders Company.]

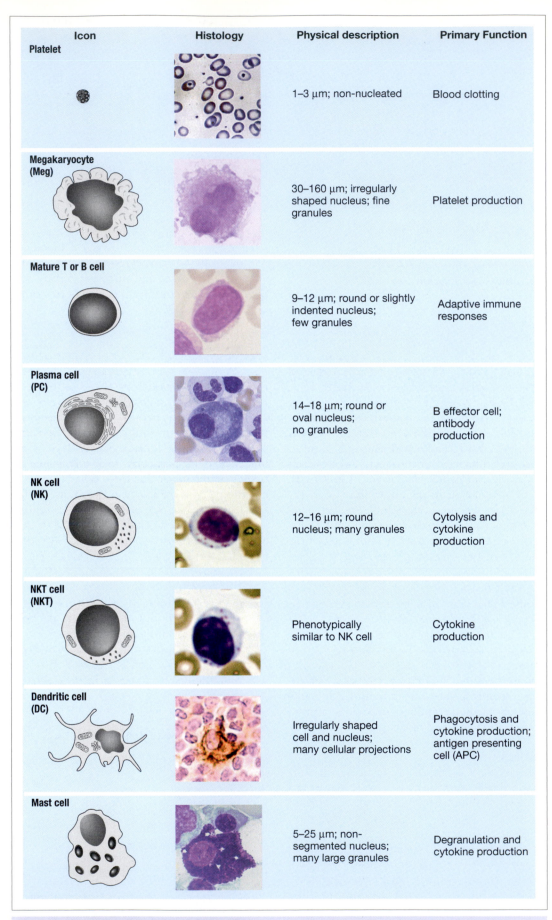

Icon	Histology	Physical description	Primary Function
Platelet		1–3 µm; non-nucleated	Blood clotting
Megakaryocyte (Meg)		30–160 µm; irregularly shaped nucleus; fine granules	Platelet production
Mature T or B cell		9–12 µm; round or slightly indented nucleus; few granules	Adaptive immune responses
Plasma cell (PC)		14–18 µm; round or oval nucleus; no granules	B effector cell; antibody production
NK cell (NK)		12–16 µm; round nucleus; many granules	Cytolysis and cytokine production
NKT cell (NKT)		Phenotypically similar to NK cell	Cytokine production
Dendritic cell (DC)		Irregularly shaped cell and nucleus; many cellular projections	Phagocytosis and cytokine production; antigen presenting cell (APC)
Mast cell		5–25 µm; non-segmented nucleus; many large granules	Degranulation and cytokine production

Fig. 2-1 (Part 2) **Characteristics of Hematopoietic Cells**
[Micrographs of plates reproduced by permission of D.C. Tkachuk, J.V. Hirschmann and J.R. McArthur, *Atlas of Clinical Hematology* 2002, W.B. Saunders Company.]

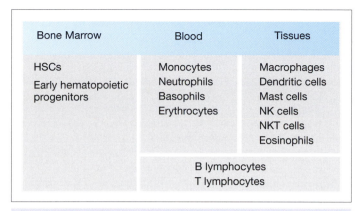

Bone Marrow	Blood	Tissues
HSCs	Monocytes	Macrophages
Early hematopoietic progenitors	Neutrophils	Dendritic cells
	Basophils	Mast cells
	Erythrocytes	NK cells
		NKT cells
		Eosinophils
	B lymphocytes	
	T lymphocytes	

Fig. 2-2 **Compartmentalization of Hematopoietic Cells**

mature neutrophils in the circulation at any given moment; however, each neutrophil has a life span of only 1–2 days.

As early as 30 minutes after an acute injury or the onset of infection, neutrophils enter the site of attack in response to local chemokine gradients. Once in the inflammatory site, the neutrophils are activated by the binding of their pattern recognition receptors (PRRs) to microbial pathogen-associated molecular patterns (PAMPs). The activated neutrophils immediately phagocytose the pathogen or its products and sequester them in intracellular vesicles called *phagosomes* (see Ch. 3). Cytoplasmic granules in the neutrophil then fuse with the phagosomes, and within seconds, the granule contents are released into these vesicles and destroy the engulfed entity. Neutrophils can also release the destructive contents of their granules extracellularly in a process called *degranulation*. If an inflammatory response is particularly long or intense, nearby healthy host cells may be liquefied as a consequence. The accumulation of dead host cells, degraded microbial material, and fluid forms *pus*, a familiar sign of infection.

Neutrophils are primarily a component of the innate response, since these cells recognize pathogens with only broad specificity. However, neutrophils are also indirectly linked to the adaptive response because, in addition to PRRs, neutrophils express surface receptors able to bind to antibody molecules. These receptors facilitate antibody-mediated antigen clearance by a mechanism described in Chapter 5.

ii) Basophils and Eosinophils

Basophils and eosinophils are named for the colors of their cytoplasmic granules when stained blood smears are viewed under the microscope. Basophilic granules react with basic dyes such as hematoxylin and stain a dark blue color. Eosinophilic granules stain reddish with acidic dyes such as eosin. Although eosinophils and basophils can capture microbes by phagocytosis, degranulation is the primary means by which these cells defend the body. Eosinophilic granules are filled with highly basic proteins and enzymes that are effective in the killing of large parasites, whereas basophilic granules contain substances that are important for sustaining the inflammatory response. Basophils are present in the body in very low numbers (1% of all leukocytes) and reside primarily in the blood until

they move into the tissues during inflammation. In contrast, the vast majority of mature eosinophils (4% of all leukocytes) reside in the connective tissues. Eosinophils also play a role in allergy (see Ch. 18).

iii) Monocytes

Among the largest cells resident in the blood circulation are the monocytes. The principal functional features of these phagocytes are their numerous cytoplasmic lysosomes filled with hydrolytic enzymes, and their abundance of the organelles required for the synthesis of secreted and membrane-bound proteins. Monocytes circulate in the blood at low density (3–5% of all blood leukocytes) for approximately 1 day before entering the tissues and maturing further to become macrophages. In cases of infection, monocytes can also give rise to certain types of DCs.

iv) Macrophages

Macrophages are generally several times larger than the monocytes from which they are derived and display further enhancements of the protein synthesis and secretion machinery. These long-lived (2–4 months) powerful phagocytes reside in all organs and tissues, usually in sites where they are most likely to encounter foreign entities. Subtle differences in the morphology and function of macrophages develop as a result of the influence of a particular microenvironment, giving rise to tissue-specific names for these cells (Table 2-1). For example, macrophages in the liver are known as *Kupffer cells*, whereas those in the bone are called *osteoclasts*. Regardless of their location, macrophages constantly explore their tissue of residence and tirelessly engulf and digest not only foreign entities but also spent host cells and cellular debris.

Macrophages play key roles in both the innate and adaptive immune responses. Macrophages express a large battery of PRRs, meaning that these cells are readily activated by direct contact with PAMP-bearing pathogens or their products. Host tissue breakdown products and proteins of the complement or the blood coagulation systems can also activate macrophages during the innate response. Activated macrophages phagocytose and eliminate pathogens as well as secrete chemokines

Table 2-1 **Macrophages Named by Tissue**

Macrophage Location	Name
Liver	Kupffer cells
Kidney	Mesangial phagocytes
Brain	Microglia
Connective tissues	Histiocytes
Bone	Osteoclasts
Lung	Alveolar macrophages
Spleen	Littoral cells
Joints	Synovial A cells

that draw neutrophils and other innate and adaptive leukocytes to a site of inflammation. Macrophages also produce cytokines and growth factors that stimulate innate leukocytes as well as cells involved in wound healing.

With respect to the adaptive response, macrophages produce several cytokines that influence lymphocyte activation, proliferation and effector cell generation. Most importantly, macrophages are one of the few cell types that can function as an antigen-presenting cell (APC). As introduced in Chapter 1, APCs take up a protein antigen, digest it into peptides, and combine the peptides with MHC class II molecules. These pMHC complexes are then displayed on the APC surface for inspection by Th cells. Recognition of a pMHC by the T cell receptor (TCR) of a Th cell triggers the activation of the latter. Unlike DCs, macrophages cannot activate *naïve* T cells (T cells that have never encountered their specific antigen before) but do activate effector and memory T cells. Macrophages also express receptors that can bind to antibodies, allowing these cells to avidly take up and dispose of antigens that have been coated in soluble antibody (see Ch. 5). With respect to cell-mediated immunity, macrophages can respond to cytokines secreted by activated T cells by becoming *hyperactivated*. Hyperactivated macrophages gain enhanced antimicrobial and antiparasitic activities and new cytolytic capacity, particularly against tumor cells. The mutual stimulation of T cells and macrophages is vital, since it amplifies both the innate effector mechanisms of macrophage action and the adaptive effector mechanisms of the cell-mediated and humoral responses.

III. CELLS OF THE LYMPHOID LINEAGE

i) T and B Lymphocytes

Resting, mature B and T lymphocytes that have not interacted with specific antigen are said to be *naïve, virgin* or *unprimed*. These cells have a modest life span (up to a few weeks) and are programmed to die through a process called *apoptosis* (see later) unless they encounter specific antigen. As outlined in Chapter 1, T and B lymphocytes recognize and bind antigens using antigen receptors. A mature T cell bears ~30,000 copies of a particular TCR on its cell surface, and each B lymphocyte carries close to 150,000 identical copies of its BCR. In both cases, the antigen-binding site of the receptor protein is designed to recognize a particular antigen shape, and only those antigens binding with adequate strength to a sufficient number of receptors will trigger activation of the lymphocyte and initiate an adaptive response. Although it is often said that "one lymphocyte recognizes one antigen", a T or B cell can in fact recognize more than one antigen because different molecules closely related in shape to the "ideal antigen" will fit to some degree into the antigen-binding cleft of the receptor. This very small collection of antigens will therefore be recognized by the lymphocyte with a strength proportional to the quality of the fit. The antigen receptor is said to *cross-react* with such antigens.

Binding of specific antigen activates a lymphocyte and stimulates it to progress through cell division. The transcription of numerous genes is triggered, causing the progeny cells to

Table 2-2 Effector and Memory Lymphocytes

Type of Lymphocyte	Effector Cell	Effector Action	Memory Cell
B cell	Plasma cell	Antibody production	Memory B cell
Tc cell	CTL	Target cell cytolysis	Memory Tc cell
Th cell	Effector Th cell	Cytokine production	Memory Th cell

undergo morphological and functional changes that result in the production of *lymphoblasts*. This conversion of the resting lymphocyte into a lymphoblast occurs within 18–24 hours of antigen receptor activation. Lymphoblasts, which are larger and display more cytoplasmic complexity than their resting cell counterparts, undergo rapid cell division and differentiation into short-lived effector cells and long-lived memory cells (Table 2-2). For B cells, the fully differentiated effector cell is an antibody-secreting plasma cell. For Tc cells, the fully differentiated effector cell is a cytotoxic T cell (CTL) capable of target cell cytolysis and cytotoxic cytokine secretion. For Th cells, the fully differentiated T helper effector cell secretes copious amounts of cytokines supporting both B and Tc responses. The integrity of these effectors and the lymphocyte compartment as a whole is vital for the maintenance of effective adaptive immunity. Primary immunodeficiencies arising from genetic defects that are particularly devastating to lymphocytes are described in Box 2-1.

B and T lymphocytes are difficult to distinguish morphologically. However, immunologists have devised a means of using antibodies to detect differences in the expression of certain cell surface proteins called *CD markers* (see Box 2-2). For example, B cells typically express a surface protein called CD19 but not a protein called CD3, while the reverse is true for T cells. T cell subsets are also distinguished by CD marker expression, as Th cells generally express CD4 while Tc cells usually express CD8. Because they bind to invariant regions of the same MHC molecule that presents antigenic peptide to the T cell's TCRs, CD4 and CD8 are known as *coreceptors*.

ii) Natural Killer Cells, γδ T Cells and Natural Killer T Cells

NK cells, γδ T cells and NKT cells are three types of lymphoid cells that make their primary contributions to innate, rather than adaptive, immunity. None of these cell types expresses the highly diverse and specific antigen receptors found on B and T cells; nor do they generate memory cells upon activation. Each cell type is described briefly here, and more on their mechanisms of antigen recognition and clearance can be found in Chapter 11.

NK cells are large cells found in the blood and lymphoid tissues and morphologically resemble effector T cells. Like CTLs, NK cells are capable of cytolytic killing of target cells but this destruction is achieved without the fine specificity of a T lymphocyte. NK cells are particularly important for their ability to recognize and kill many virus-infected and tumor cells. Furthermore, when cultured in the presence of sufficient quantities of the cytokine *interleukin-2* (IL-2), NK cells can be induced to differentiate into *lymphokine-activated killer* (LAK) cells. (Lymphokines are cytokines synthesized by lymphocytes.) At least *in vitro*, LAK cells have even greater powers of target cell recognition and cytolysis than NK cells. Because activated T cells are the principal producers of IL-2, the development of NK cells into LAK cells may be another example of the interplay between the innate and adaptive responses. In addition, NK cells carry receptors that allow them to lyse cells that have been coated with soluble antibody (see Ch. 5).

$\gamma\delta$ T lymphocytes are a type of T lymphocyte that supports the innate (rather than the adaptive) response due to the nature of its TCR. The vast majority of T cells in the body are $\alpha\beta$ *T cells*, so termed because their TCRs are made up of a TCRα and a TCRβ chain (TCR$\alpha\beta$; see Ch. 8). The CD4$^+$ Th and CD8$^+$ Tc cells of the adaptive response are $\alpha\beta$ T cells expressing $\alpha\beta$ TCRs. In contrast, $\gamma\delta$ T cells express $\gamma\delta$ TCRs made up of two slightly different chains, TCRγ and TCRδ (TCR$\gamma\delta$). $\gamma\delta$ T lymphocytes never express $\alpha\beta$ TCRs, and $\alpha\beta$ T cells never express $\gamma\delta$ TCRs. Unlike the enormous diversity of $\alpha\beta$ TCRs, $\gamma\delta$ TCRs bind to only a limited set of broadly expressed antigens that may or may not be presented by molecules other than MHC class I and II. As well, $\gamma\delta$ T cells often reside in the

mucosae where they can respond rapidly to an incipient pathogen attack. As is detailed in Chapter 11, activated $\gamma\delta$ T cells proliferate and differentiate into effectors that are capable of cytolysis or cytokine secretion. The cytokines secreted by $\gamma\delta$ T cells may also influence $\alpha\beta$ T cells, establishing a link to the adaptive response.

NKT cells combine features of both $\alpha\beta$ T lymphocytes and NK cells. Like an $\alpha\beta$ T cell, a given NKT cell expresses many copies of a single $\alpha\beta$ TCR on its cell surface. However, unlike the diversity of the $\alpha\beta$ T cell repertoire, most NKT clones express highly similar TCR$\alpha\beta$ molecules, such that the NKT TCR is described as being "semi-invariant". Like the $\gamma\delta$ TCR, the semi-invariant NKT TCR recognizes only a small collection of antigens. Once activated by engagement of the semi-invariant TCR, NKT cells immediately secrete cytokines that can help support the activation and differentiation of B and T cells; that is, NKT cells do not have to first differentiate into effector cells like $\gamma\delta$ and $\alpha\beta$ T cells do. NKT cells resemble NK cells morphologically and express some of the same CD markers. However, it remains unclear whether an NKT cell can carry out target cell cytolysis *in vivo* like an NK cell can. The rapidity of the NKT response and the limited range of antigens to which these cells respond mark them as innate leukocytes. However, the influence of NKT-secreted cytokines on B and T lymphocyte functions indicates that NKT cells constitute yet another important link between innate and adaptive immunity.

IV. DENDRITIC CELLS

In general, DCs are irregularly shaped, short-lived cells that exhibit long, fingerlike membrane processes resembling the dendrites of nerve cells. DCs are important because they are the only type of APC that can activate naïve Tc and Th cells. There are many different subtypes of DCs with characteristics that depend on where in the body they develop and whether the individual is healthy at the time or experiencing an infection (see Ch. 7). These DC subsets show subtle differences in their cytokine production, responses to cytokines, and influence on T cell responses.

Both myeloid and lymphoid precursors present in the bone marrow can give rise to numerous DC subtypes. Interestingly, it appears that some DCs can also arise from precursors present in a lymphoid organ called the *thymus* (see later), whereas others can arise from more differentiated cell types in the blood or tissues. Some DCs remain in the lymphoid tissues and monitor antigens brought into these structures, while other DCs patrol throughout the body's peripheral tissues. In all cases, the function of the DC is to capture, process and display peptides derived from proteins in the surrounding microenvironment. If the individual is healthy, the DCs maintain their *immature state* and the pMHCs that they present (which contain self peptides) do not activate the T cells that the DCs encounter. However, if a DC engulfs a foreign entity during an infection, the DC undergoes *maturation* and acquires the ability to activate naïve T cells. The pMHCs presented by the DC contain foreign peptides complexed to either MHC class I or II and so are recognized by naïve Tc or Th cells, respectively. This recognition initiates an adaptive response.

An unrelated cell type called the *follicular dendritic cell* (FDC) is important for B cell responses. These cells excel at trapping antigens on their surfaces, where these entities are readily bound by the BCRs of B cells. FDCs are abundant in the secondary lymphoid organs where B cell activation occurs (see later). The lineage of FDCs remains unclear, with some FDCs appearing to be of non-hematopoietic origin.

V. MAST CELLS

Mast cells are important for defense against worms and other parasites. Mast cells resemble basophils in that they contain basic-staining cytoplasmic granules that harbor an array of pro-inflammatory substances. The degranulation of mast cells is rapidly triggered by tissue invasion or injury, resulting in a flood of cytokines and other molecules that initiate the inflammatory response. Mast cells differ from basophils in several ways: (1) Mast cell granules are smaller and more numerous than those of basophils. (2) Unlike basophils, mast cells are rarely found in the blood and preferentially reside in the con-

nective tissues and gastrointestinal mucosa. (3) Mast cells are not of the myeloid or lymphoid lineage and take an independent path in their differentiation from HSCs. Importantly, the degranulation of mast cells and basophils is associated with hypersensitivity and allergy (see Ch. 18).

VI. HEMATOPOIESIS

As introduced earlier, the generation of all red and white blood cells in the body is called *hematopoiesis*. Hematopoiesis is a continual process that ensures an individual has an adequate supply of erythrocytes and leukocytes throughout life. In humans, all hematopoietic cells arise from HSCs that can first be detected in embryonic structures at 3–4 weeks of gestation (Fig. 2-3). The HSCs migrate to the human fetal liver at 5 weeks of gestation, and liver hematopoiesis completely replaces embryonic hematopoiesis by 12 weeks of gestation. At 10–12 weeks of gestation, some HSCs or their descendants commence migration to the spleen and bone marrow. The fetal spleen transiently produces blood cells between the third and seventh months of gestation. The bone marrow of the long bones (tibia, femur) assumes an increasingly important role by about the fourth month and takes over as the major site of hematopoiesis during the second half of gestation. By birth, virtually all hematopoiesis occurs in the marrow of the long bones. After birth, the activity of the long bones steadily declines and is replaced in the adult by the production of hematopoietic cells in the axial skeleton—the sternum, ribs and vertebrae as well as the pelvis and skull. If the bone marrow suffers an injury in an adult, the liver and spleen may resume some level of hematopoiesis.

Hematopoietic stem cells occur at a frequency of just 0.01% of all bone marrow cells. However, HSCs are capable of tremendous proliferation and differentiation in response to an increased demand by the body for hematopoiesis. Hematopoietic stem cells are said to be both *multipotent* and *self-renewing*. Multipotency means that an HSC can differentiate into any one of a variety of hematopoietic cell types, including lymphocytes, macrophages, DCs etc. Self-renewal means that,

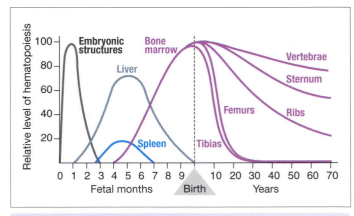

Fig. 2-3 **Sites of Hematopoiesis in Humans**
[Adapted from Klein J. and Hořejší V. (1997). *Immunology*, 2nd ed. Blackwell Science, Osney Mead Oxford.]

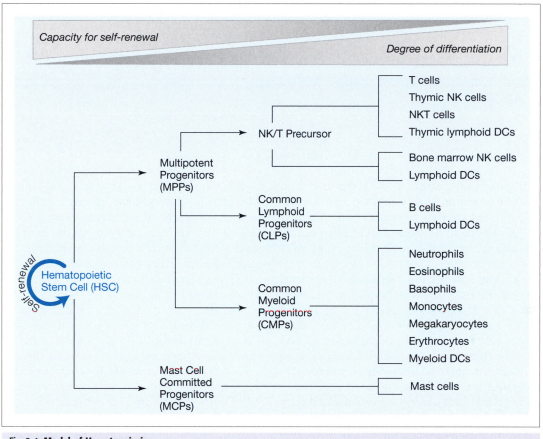

Fig. 2-4 **Model of Hematopoiesis**

instead of differentiating into a cell such as a lymphocyte or a macrophage, the HSC can generate more HSCs. Thus, when stimulated to divide, HSCs in the bone marrow will either self-renew or commit to generating cells of the myeloid lineage *(myelopoiesis)*, cells of the lymphoid lineage *(lymphopoiesis)*, or cells of the mast cell lineage.

Development from HSCs to mature blood cells is a continuum, and the identifiable intermediates progressively lose their capacities to self-renew and differentiate into multiple cell types. Immunologists differ widely on what these cells are called and how to identify them. Figure 2-4 shows one model of how a multipotent HSC can either self-renew, or produce early progenitors sometimes called the *multipotent progenitor* (MPP) and the *mast cell committed progenitor* (MCP). The MPP gives rise to three precursors of more limited potential called the *NK/T precursor*, the *common lymphoid progenitor* (CLP), and the *common myeloid progenitor* (CMP). Although these early progenitors are thought to retain some self-renewal capacity, their primary function is to differentiate into various types of mature hematopoietic cells. For example, NK/T precursors that enter the thymus give rise mainly to T cells and NKT cells, whereas NK/T precursors that remain in the bone marrow primarily generate NK cells. CLPs were originally thought to give rise to both B and T cells (but not myeloid cells); however, recent work has clarified that CLPs are

restricted to the generation of B cells. In parallel, CMPs generate granulocytes and other myeloid cells but not lymphoid cells. DCs can be generated from NK/T precursors, CLPs or CMPs but only mast cells arise from MCPs. Once a mature hematopoietic cell of any lineage is generated, its high level of specialization means that it can be stimulated to divide and produce effector cells but can no longer give rise to precursors or even mature cells like itself. A primary immunodeficiency that appears to result from a defect in early hematopoiesis is described in Box 2-3.

Box 2-3 **Wiscott–Aldrich Syndrome**

Wiscott–Aldrich syndrome (WAS) is an X-linked, clinically heterogeneous primary immunodeficiency caused by various mutations in the multifunctional WAS protein (WASP). WASP is involved in both the signal transduction and actin polymerization required for hematopoietic cell differentiation and activation. Classical WAS patients show a characteristic constellation of immunodeficiency, eczema and platelet deficiency, and are highly susceptible to infection, tumorigenesis, autoimmunity and allergy. Non-classical WAS patients have only some of these disease features. In classical WAS patients, the few T cells present show abnormal proliferation and impaired responses to antigens. Antibody responses, megakaryocyte differentiation and leukocyte chemotaxis are also defective.

VII. APOPTOSIS

The body needs a huge number of hematopoietic cells to maintain its ability to mount effective immune responses. For example, the average healthy adult human maintains an estimated 10^{12} lymphocytes and an estimated 5×10^{10} circulating neutrophils. However, to ensure that the required hematopoietic cells are always fresh and fully functional, leukocytes are programmed to die after only days or months of patrolling the body for pathogens. To meet the resulting demand for the replacement of the spent hematopoietic cells, the bone marrow is a prodigious cell-generating factory and produces HSCs and hematopoietic progenitors at a constant and rapid rate. This balance between hematopoiesis and the "programmed cell death" (*apoptosis*) of leukocytes is essential for the maintenance of *homeostasis*, the body's "steady state". Tight regulatory controls ensure that neither a dangerous accumulation nor a deficiency of a particular hematopoietic cell type occurs.

Apoptotic cell death differs in several important ways from *necrosis*, which is cell death caused by tissue injury or pathogen attack. A cell that has been injured lyses in an uncontrolled way and becomes necrotic. Unlike a normal cell (Plate 2-1A), a necrotic cell (Plate 2-1B) loses the integrity of its nucleus and releases its intracellular contents, which can damage nearby cells and provoke an inflammatory response. In contrast, apoptosis is a controlled form of cell death that follows specific biochemical paths and results in specific morphological changes to only those cells intentionally marked for "suicide". An apoptotic cell (Plate 2-1C) undergoes an orderly demise in which it cleaves its own chromosomal DNA, fragments its nucleus, degrades its proteins and shrinks its cytoplasmic volume. Eventually the cell is reduced to an *apoptotic body*, which is a small membrane-bound structure in which the remaining organelles such as lysosomes and mitochondria are held in intact form. The rapid phagocytosis of the apoptotic bodies by macrophages prevents the leakage of deadly lysosomal contents (including hydrolytic enzymes and oxidative molecules) that would otherwise precipitate neighboring cell breakdown and an inflammatory response. Apoptosis occurs when it is advantageous to the body for a cell to die, such as during embryonic development, or in order to remove expended cells. In the case of the hematopoietic system, apoptosis removes aged and spent cells as quickly as the bone marrow creates new ones, maintaining immune system homeostasis.

B. How Leukocytes Communicate

I. INTRACELLULAR COMMUNICATION: SIGNAL TRANSDUCTION

The binding of a ligand to its receptor on a leukocyte is the initiating signal indicating that a response is required. Such ligands include antigens, PAMPs, cytokines and molecules present in the membranes of other host cells. The cell surface receptors that bind to these ligands usually consist of one or more transmembrane proteins. The *extracellular domains* of these molecules bind the ligand, the *transmembrane domains* anchor the receptor chains in the plasma membrane, and the *cytoplasmic domains* transduce intracellular signals into the interior of the cell (Fig. 2-5). However, a signal is not actually

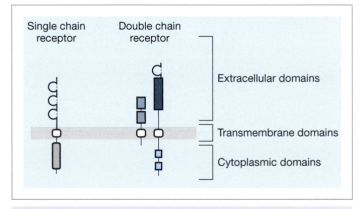

Fig. 2-5 **Examples of Cell Surface Receptor Structure**

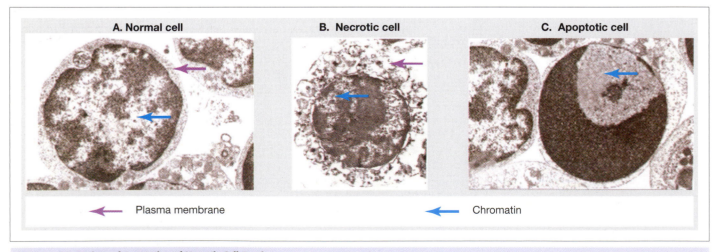

Plate 2-1 **Comparison of Apoptotic and Necrotic Cell Death**
[Reproduced by permission of Andrea Cossarizza, University of Modena and Reggio Emilia, Italy.]

generated until the binding of the ligand induces the aggregation of two or three receptor molecules.

When a ligand binds to the extracellular domain of a complete receptor molecule, a change (usually conformational) is induced in the cytoplasmic domains of the component receptor chains. This change induces the phosphorylation (or dephosphorylation) of other proteins, particularly *protein tyrosine kinases* (PTKs) associated with the cytoplasmic domain of the receptor. PTKs are enzymes that, when activated, phosphorylate the tyrosine residues of substrate proteins. Within seconds of ligand molecules binding to the receptor aggregate, the PTKs associated with the receptor "tail" are activated and initiate a complex phosphorylation/dephosphorylation cascade involving other PTKs and additional substrates (Fig. 2-6). This cascade of specific enzyme activation eventually triggers the release of mediators such as calcium ions from their sequestration in special intracellular storage structures. When the response required by the cell takes the form of degranulation or actin cytoskeleton reorganization, the end point of the signaling pathway is in the cytoplasm and does not involve the nucleus. However, the proteins needed to carry out an effector action are often not routinely present as part of a cell's "housekeeping" metabolism, so that new synthesis of specialized proteins is needed. New protein synthesis requires new transcription, so that the enzymatic cascade and mediator release continue until one or more *nuclear transcription factors* capable of entering the nucleus are stimulated. The activated nuclear transcription factors induce the transcription of the previously silent genes encoding the proteins responsible for the required effector action.

The intracellular signaling pathways in lymphocytes are among the most complex identified. The binding of antigen to TCRs or BCRs involves multiple receptor proteins and the recruitment of numerous PTKs and other molecules (see Ch. 5 and 9). The resulting cascade of PTK activation, calcium ion release and nuclear transcription factor activation leads to new molecules appearing on the lymphocyte surface, including receptors for cytokines that stimulate cellular proliferation and the differentiation of effector cells. In this way, antigen binding is translated into the production of cells equipped to eliminate the entity. The complexity of these signaling pathways has no doubt evolved because of the fine degree of control necessary for regulating an immune response and channeling the power of effector lymphocytes.

II. INTERCELLULAR COMMUNICATION: CYTOKINES

i) The Nature of Cytokines

In both innate and adaptive immunity, there are many situations when leukocytes must communicate with one another in order to achieve a response. This intercellular communication is most often mediated by *cytokines*, which are low molecular weight peptides or glycoproteins of diverse structure and function. Well over 100 cytokines have been identified since the discovery of the first such molecules in the 1950s. Cytokines are secreted primarily, but not exclusively, by leukocytes, and regulate not only immune responses but also hematopoiesis, wound healing and other physiological processes. Cytokines are induced in both the adaptive and innate immune responses, and cytokines secreted in one type of response frequently stimulate the secretion of cytokines influential in the other type. In this context, cytokines act to intensify or dampen the response by stimulating or inhibiting the secretory activity, activation, proliferation or differentiation of various cells. This regulation is achieved by the intracellular signaling that is triggered when a cytokine binds to its receptor on the membrane of the cell to be influenced.

While several structural families can be discerned among cytokines, molecules of similar function can have very different structures, and those of similar structure can be very different in function. As well, although cytokines are genetically unrelated, many appear to be functionally redundant; that is, the same biological effect may result from the action of more than one cytokine. This overlap ensures that a critical function is preserved should a particular cytokine be defective or absent. A given cytokine can be produced by multiple cell types, may be *pleiotropic* (act on many different cell types), or have several different effects on the same cell type. Cytokines can also affect

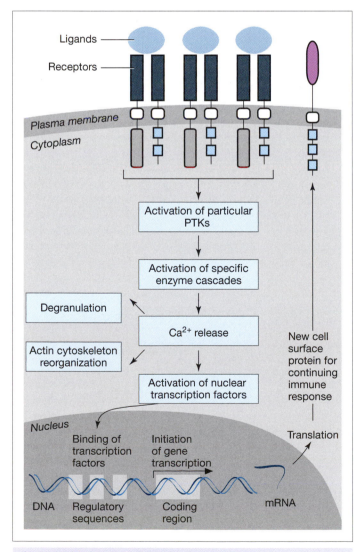

Fig. 2-6 **Model of an Intracellular Signaling Pathway**

Table 2-3 Principal Functions of Important Cytokines

Cytokine (Symbol)	Principal Functions	Cytokine (Symbol)	Principal Functions
Interferon (IFNα, IFNβ)	• Has antiviral activity • Has antiproliferative activity • Promotes inflammation	Interleukin-9 (IL-9)	• Stimulates differentiation of erythroid, myeloid and neuronal precursors • Stimulates proliferation and differentiation of mast cells • Promotes defense against parasitic worms
Interferon (IFNγ)	• Has antiviral activity • Has antiproliferative activity • Promotes inflammation • Influences T cell subset differentiation and macrophage activation	Interleukin-10 (IL-10)	• Damps down inflammation • Has immunosuppressive effects
Interleukin-1 (IL-1)	• Promotes inflammation and induces fever • Regulates early innate response	Interleukin-11 (IL-11)	• Stimulates megakaryocyte production • Stimulates proliferation of hematopoietic cells • Has anti-inflammatory effects
Interleukin-2 (IL-2)	• Stimulates B cell activation and T, B and NK cell proliferation • Stimulates B cell and CTL differentiation • Acts as T cell chemoattractant • Enhances monocyte responses • Promotes tolerance and homeostasis	Interleukin-12 (IL-12)	• Stimulates T cell subset differentiation • Induces production of IFNγ and other cytokines • Promotes cytotoxicity of CTLs and NK cells
Interleukin-3 (IL-3)	• Stimulates mast cell growth • Promotes antiparasitic response of basophils and mast cells • Promotes myelopoiesis	Interleukin-15 (IL-15)	• Stimulates NK and γδ T cell development and proliferation • Stimulates mast cell proliferation • Promotes T cell activation, proliferation, differentiation, memory cell formation, homing and adhesion
Interleukin-4 (IL-4)	• Counters some effects of interferon • Promotes T cell subset differentiation • Promotes humoral response • Stimulates mast cell growth • Regulates allergic response	Interleukin-17 (IL-17)	• Promotes inflammation • Stimulates neutrophil production
Interleukin-5 (IL-5)	• Promotes survival, differentiation and chemotaxis of eosinophils • Enhances growth and differentiation of B cells and CTLs • Promotes histamine release in mast cells	Interleukin-18 (IL-18)	• Promotes cytotoxicity in NK cells • Stimulates T cell proliferation • Enhances IFNγ production
Interleukin-6 (IL-6)	• Promotes hematopoiesis and inflammation • Induces fever • Influences nervous and endocrine systems • Stimulates B cell and T cell subset differentiation	Interleukin-23 (IL-23)	• Stimulates T cell subset proliferation and differentiation • Stimulates DCs to produce IFNγ
Interleukin-7 (IL-7)	• Stimulates B and T cell development • Stimulates generation and maintenance of memory T cells	Tumor necrosis factor (TNF)	• Mediates inflammation • Promotes immunoregulatory, cytotoxic, antiviral, coagulatory and hematopoietic effects • Can trigger either apoptotic or cell survival signaling
Interleukin-8 (IL-8)	• Acts as chemokine for neutrophil chemotaxis • Promotes inflammatory response of neutrophils • Upregulates adhesion molecules on neutrophils • Activates endothelial cells, eosinophils and monocytes • Stimulates T cell activation and mobility	Transforming growth factor β (TGFβ)	• Promotes anti-inflammatory responses • Has immunosuppressive effects • Promotes T cell subset differentiation • Acts as chemoattractant for T cells, monocytes and neutrophils • Promotes angiogenesis

the action of other cytokines, and may behave in a manner that is *synergistic* (two cytokines acting together achieve a result that is greater than additive) or *antagonistic* (one cytokine inhibits the effect of another). Among the most important cytokines are the *interleukins* (IL-1, IL-2 etc.), the *interferons* (IFNα, IFNβ and IFNγ), *tumor necrosis factor* (TNF), and *transforming growth factor beta* (TGFβ). The functions of some of these molecules are described briefly in Table 2-3. A more comprehensive listing of cytokines and their receptors, production and functions appears in Appendix E.

Cytokines share several properties with hormones and growth factors, the other major communication molecules of

Table 2-4 Cytokines versus Growth Factors and Hormones

Property	Cytokine	Growth Factor	Hormone
Solubility	Soluble	Soluble	Soluble
Receptor required?	Yes	Yes	Yes
Site of production	Many cell types with wide tissue distribution	Several cell types with moderate tissue distribution	Specialized cell types in glands
Expression	Induced or upregulated	Constitutive	Induced
Range of effect	Autocrine or paracrine	Paracrine or endocrine*	Endocrine
Cell types influenced	Several	Broad range	Very limited

*In this context, "endocrine" means systemic effects.

the body, but also differ from them in important ways (Table 2-4). Like hormones and growth factors, cytokines are soluble proteins present in very low amounts that exert their effects by binding to specific receptors on the cell to be influenced. However, cytokines differ from hormones and growth factors in their sites of production, modes of operation and range of influence. Hormones are induced by specific stimuli and are synthesized in specialized glands. Hormones tend to operate in an *endocrine* fashion (over substantial distances or systemically) and influence only a very limited spectrum of target cells. Growth factors are usually produced constitutively and by individual cells (both leukocytes and non-leukocytes) rather than by glands. Many but not all growth factors can be detected at significant levels in the circulation, and a broad range of cell types is influenced. Taking the middle road, cytokines are synthesized under tight regulatory controls by a sizable variety of leukocyte and non-leukocyte cell types, and generally exert their effects in a fashion that is either *autocrine* (upon themselves) or *paracrine* (upon nearby cells).

ii) Production and Control of Cytokines

Most cytokines are synthesized primarily by activated Th cells in response to antigen, or by activated macrophages in response to the presence of microbial or viral products. More modest contributions are made by other leukocytes and some non-leukocyte cell types. Several means are used to regulate cytokine production and effects. The half-lives of cytokines and their mRNAs are generally very short, meaning that new transcription and translation are required when an inducing stimulus is received. The effect is one of a transient flurry of cytokine production followed by resumption of a resting state in the absence of fresh stimulus. The influence of cytokines is also curtailed by controls on receptor expression. Cells lacking expression of the appropriate receptor may be sitting in a pool of cytokine but are unable to respond to it. Lastly, since most cytokines act only over a short distance, only those cells that are in the immediate vicinity of a producing cell (and express the required receptor) will be influenced.

iii) Functions of Cytokines

In general, there are three broad categories of cytokine function: regulation of the innate response, regulation of the adap-

tive response, and regulation of the growth and differentiation of hematopoietic cells. Many cytokines mediate aspects of all three, resulting in a constellation of effects on the proliferation, differentiation, migration, adhesion and/or function of a range of cell types. Some events that can be attributed at least in part to cytokine action include pro-inflammatory and anti-inflammatory responses, antiviral responses, growth and differentiation responses, cell-mediated immune responses, humoral immune responses and chemotaxis.

The intracellular signaling triggered by the binding of a cytokine to its specific receptor results in new gene transcription and changes to cellular activities. The affinity of binding between cytokines and their receptors is high ($K = 10^{10}$–10^{12} M^{-1}), so that picomolar concentrations of cytokine are often sufficient to produce a physiological effect. A frequent result of cytokine action is the induction of expression of another cytokine or its receptor, creating a cascade of receptivity and a coordinated response. It is therefore not unusual for one cytokine to induce the expression of a nuclear transcription factor needed to bind to the promoter of another cytokine gene and spark production of this second cytokine. Alternatively, one cytokine may repress the expression of another cytokine or its receptor, creating an antagonistic situation.

Because the immune system cannot operate without the signals delivered by cytokines, their activities often overlap. The sharing of protein chains by different cytokine receptors forms the basis for at least some of this observed functional redundancy (Fig. 2-7). For example, the interleukins IL-2 and IL-15 show an overlap in function in that both molecules stimulate the proliferation of T lineage cells. The interleukin-2 receptor (IL-2R) contains three chains: the unique IL-2Rα chain that binds to the IL-2 molecule, and the IL-2Rβ and γ$_c$ chains that convey intracellular signals to the cell's nucleus. The interleukin-15 receptor (IL-15R) contains a unique IL-15Rα chain that binds to the IL-15 molecule but also the IL-2Rβ and γ$_c$ chains. In fact, the γ$_c$ chain (the "common gamma chain") is shared quite widely, also being present in the receptors for IL-4, IL-7, IL-9 and IL-21. In addition to multiple cytokines having the same effect, more than one effect of an individual cytokine may be felt by a given cell over a period of time, with some aspects being felt immediately or within minutes, and others delayed by hours or even days. *In vivo*, a

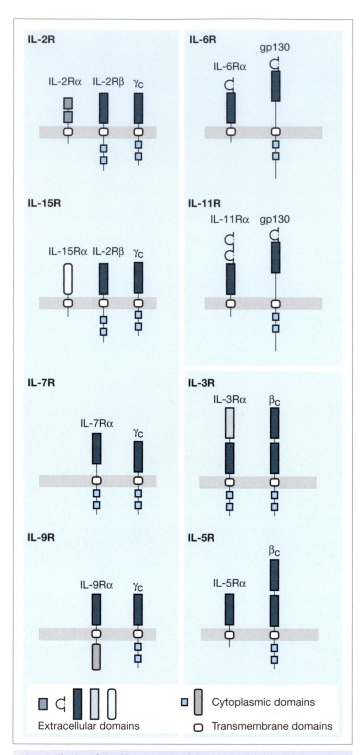

IL-2R

IL-2Rα IL-2Rβ γc

IL-15R

IL-15Rα IL-2Rβ γc

IL-7R

IL-7Rα γc

IL-9R

IL-9Rα γc

IL-6R

IL-6Rα gp130

IL-11R

IL-11Rα gp130

IL-3R

IL-3Rα βc

IL-5R

IL-5Rα βc

Extracellular domains | Cytoplasmic domains | Transmembrane domains

Fig. 2-7 Sharing of Cytokine Receptor Chains

cell may be exposed to a complex mixture of cytokines for an extended period, making the cellular outcome difficult to predict.

iv) Rationale for Cytokine Network Complexity

Why should evolution have created this complex web of cytokine interactions? Cytokines are a remarkably flexible means of controlling the immune response, which can be harmful to the host if too vigorous or too long in duration. Positive and negative feedback loops, agonistic (activating) and antagonistic (inhibiting) relationships and redundant functions exist among the cytokines to provide a fine level of control over the powerful cells and effector mechanisms that can be unleashed to contain a pathogen attack. Without such controls, extreme immunopathic damage to host tissues or even autoimmune disease could result. In most cases, such multilevel structuring also ensures that no matter what strategy the pathogen uses to invade the host and avoid immune surveillance, a cytokine can be induced to mobilize an effective response. If this cytokine fails, there are others with overlapping functions that can fill the gap.

C. Lymphoid Tissues

I. OVERVIEW

A *lymphoid tissue* is simply a tissue in which lymphocytes are found. Lymphoid tissues range in organization from diffuse arrangements of individual cells to encapsulated organs (Fig. 2-8). *Lymphoid follicles* are organized cylindrical clusters of lymphocytes that, when gathered into groups, are called *lymphoid patches*. *Lymphoid organs* are usually groups of follicles that are surrounded, or encapsulated, by specialized supporting tissues and membranes.

Lymphoid tissues are classified as being either *primary* or *secondary* in nature (Table 2-5). The *primary lymphoid tissues* in mammals are the bone marrow and the thymus. It is in these sites that lymphocytes develop and central tolerance is established; that is, most lymphocytes recognizing self antigens are deleted and only cells recognizing foreign antigens are permitted to mature. All lymphocytes arise from HSCs in the bone marrow and B cells largely mature in that site, whereas newly produced T cells migrate from the bone marrow to the thymus and mature in this location. The naïve T and B cells then leave the primary lymphoid tissues and migrate through the blood and lymph (see later) to take up residence in the secondary lymphoid tissues.

Secondary lymphoid tissues are those in which mature, naïve lymphocytes recognize antigen and become activated,

Table 2-5 Primary and Secondary Lymphoid Tissues

Primary Lymphoid Tissues	Secondary Lymphoid Tissues
Bone marrow Thymus	Lymph nodes Spleen MALT (mucosa-associated lymphoid tissue) • NALT (nasopharynx-associated lymphoid tissue) • BALT (bronchi-associated lymphoid tissue) • GALT (gut-associated lymphoid tissue) SALT (skin-associated lymphoid tissue)

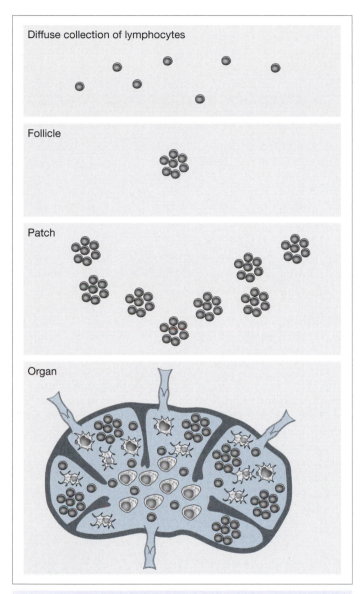

Fig. 2-8 **Lymphoid Tissue Organization**

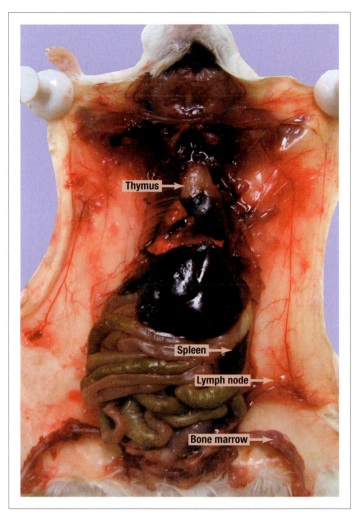

Plate 2-2 **Mouse Primary and Secondary Lymphoid Organs and Tissues**
[Reproduced by permission of Alejandro Ceccarelli, Department of Pathobiology, University of Guelph.]

undergoing clonal selection and proliferation followed by differentiation into effector and memory cells. The secondary lymphoid tissues provide sites for both antigen accumulation and the gathering of leukocytes (primarily B cells, T cells and APCs) that must cooperate to mount an optimal adaptive response. Once generated, the effector cells migrate from the secondary lymphoid tissues to infected organs or tissues to wage war on invading pathogens. Since pathogens can attack at any location, the secondary lymphoid tissues are widely distributed throughout the body's periphery. Secondary lymphoid tissues in mammals include the *lymph nodes*, the *spleen*, the *skin-associated lymphoid tissue* (SALT), and the *mucosa-associated lymphoid tissue* (MALT) present in the bronchi (BALT), nasopharynx (NALT) and gut (GALT). The lymph nodes collect antigen moving through the body in the lymphatic system, the spleen traps antigen circulating in the blood, and the SALT and MALT deal with antigens attempting to

penetrate through the skin or mucosae, respectively. A photograph of a dissected mouse showing the positions of the major lymphoid organs and tissues as they sit in the body is shown in Plate 2-2, and a diagram of human lymphoid tissues appears in Figure 2-9.

II. PRIMARY LYMPHOID TISSUES

i) The Bone Marrow

The bone marrow is the primary site of hematopoiesis in the adult human. The total bone marrow cell population consists of about 60–70% myeloid lineage cells, 20–30% erythroid lineage cells, 10% lymphoid lineage cells, and 10% other cells such as stromal cells, adipocytes (fat cells) and mast cells. Many of these other cells are vital for hematopoiesis because they secrete the cytokines and growth factors that are required for blood cell maturation.

The compact outer matrix of a bone surrounds a *central cavity* (also called the *medullary cavity*) (Fig. 2-10A). Within

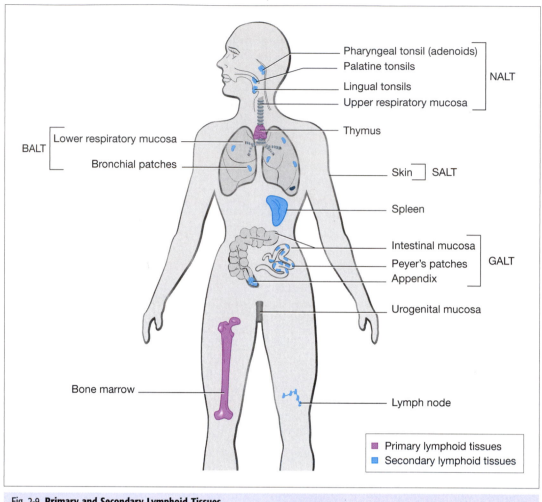

Fig. 2-9 Primary and Secondary Lymphoid Tissues

the central cavity is the marrow, which may appear red or yellow in color. The *yellow marrow* is usually hematopoietically inactive but contains significant numbers of adipocytes that act as an important energy reserve. *Red marrow* is hematopoietically active tissue that gains its color from the vast numbers of erythrocytes produced. The bones of an infant contain virtually only red marrow, but as the child grows, the demand for hematopoiesis slackens slightly such that the number of bones with active red marrow declines. In the adult, only a few bones retain red marrow, including the sternum, ribs, pelvis and skull.

Underneath the outer layer of compact bone, the central cavity is composed of an inner layer of "spongy" bone with a honeycomb structure. The honeycomb is made up of thin strands of bone called *trabeculae*. The red marrow lies within the cavities created by the trabeculae. A hematopoietically active bone is nourished by one or more nutrient arteries that enter the shaft of the bone from the exterior, as well as by arteries that penetrate at the ends of the bone. Branches of nutrient arteries thread through the *Haversian canals* of the compact bone to reach the central cavity. The circulatory loop is completed when the branches of the nutrient arteries make contact with small vascular channels called *venous sinuses*.

The venous sinuses eventually feed into nutrient veins that exit the shaft of the bone.

Within a trabecular cavity, the network of blood vessels surrounds groups of developing hematopoietic cells (Fig. 2-10B). It is here that the HSCs either self-renew or commence differentiation into red and white blood cells of all lineages. Tucked into the spaces between blood vessels are interconnecting stromal cells and fibers that form a framework supporting the developing hematopoietic cells. As the hematopoietic cells reach maturity, they squeeze between the endothelial cells lining the venous sinuses, enter a nutrient vein and finally join the blood circulation. Erythrocytes remain in the circulation, while leukocytes are distributed between the blood and the tissues.

ii) The Thymus

The thymus is a lymphoid organ located above the heart, and it is in the thymus that immature T cells complete their development. Hematopoiesis in the bone marrow generates NK/T precursors that leave the bone marrow and enter the blood circulation. The epithelial cells of the thymus secrete chemokines that specifically attract these precursors from the blood into the thymus, where many of them become *thymocytes*. During their proliferation and maturation in the thymus,

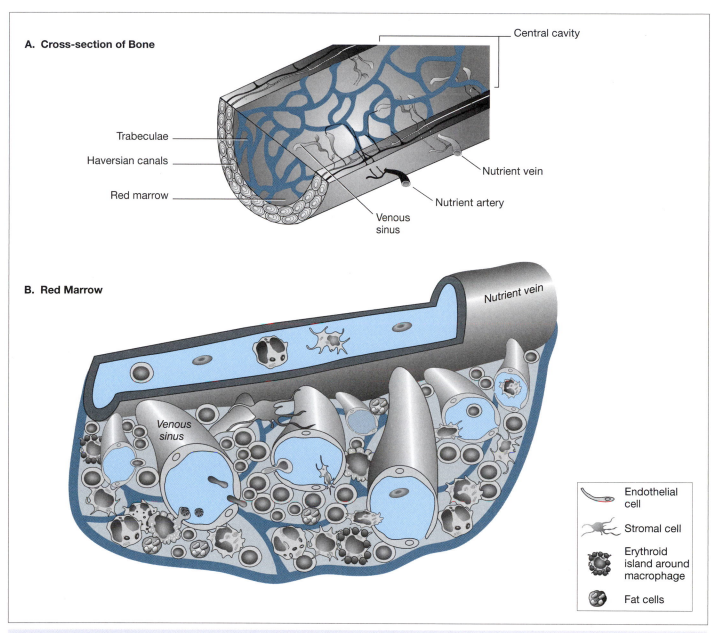

A. Cross-section of Bone

Central cavity

Trabeculae

Haversian canals

Red marrow

Nutrient vein

Nutrient artery

Venous sinus

B. Red Marrow

Nutrient vein

Venous sinus

Endothelial cell

Stromal cell

Erythroid island around macrophage

Fat cells

Fig. 2-10 **The Bone Marrow**

thymocytes undergo a process called *thymic selection* that determines the specificity of the mature T cell repertoire. Thymic selection has two parts: *positive selection* and *negative selection*. Positive selection ensures that only thymocytes expressing TCRs with at least some binding affinity for the host's MHC molecules survive. Negative selection ensures that thymocytes with TCRs that recognize pMHC complexes in which the peptide is derived from a self antigen are eliminated. As a result, self-reactive clones are deleted from the T cell repertoire and only the T cells most likely to recognize foreign entities survive (see Ch. 9). Only about 1% of all thymocytes survive positive and negative thymic selection, making the thymus a site of both tremendous T cell proliferation and T cell apoptosis.

Structurally, the thymus is a bilobed organ in which each lobe is encapsulated and composed of multiple lobules (Fig. 2-11A). These lobules are separated by connective tissue strands again called trabeculae. As thymocytes mature through various stages, they generally move from the densely packed outer *cortex* into the sparsely populated inner *medulla* of a lobule. Along the way, the developing thymocytes interact with and are supported by *cortical thymic epithelial cells* (cTECs), *medullary thymic epithelial cells* (mTECs), thymic DCs and macrophages (Fig. 2-11B). In the outer cortex, specialized epithelial cells called *nurse cells* form large multicellular complexes that envelope up to 50 maturing thymocytes within their long processes. At the junction of the cortex and medulla, thymic DCs begin to take on the nurturing role. Epithelial cells

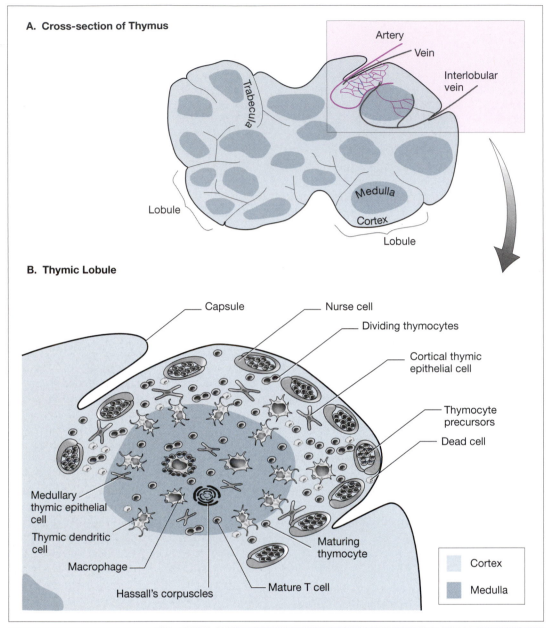

A. Cross-section of Thymus

Artery
Vein
Interlobular vein
Trabecula
Lobule
Medulla
Cortex
Lobule

B. Thymic Lobule

Capsule
Nurse cell
Dividing thymocytes
Cortical thymic epithelial cell
Thymocyte precursors
Dead cell
Medullary thymic epithelial cell
Thymic dendritic cell
Macrophage
Hassall's corpuscles
Mature T cell
Maturing thymocyte

Cortex
Medulla

Fig. 2-11 **The Thymus**
[With information from Klein J. and Hořejší V. (1997). *Immunology*, 2nd ed. Blackwell Science, Osney Mead Oxford.]

that have exhausted their supportive function within the thymus degenerate, forming the *Hassall's corpuscles* found in the medulla.

After puberty, the thymus starts to regress in a process called *thymic involution*. Most of the lymphoid components of the thymus are eventually replaced with fatty connective tissue, greatly decreasing thymic production of mature T cells. However, certain T cell subsets can mature outside the thymus in parts of the intestine. This *extrathymic development* provides for a limited amount of mature T cell generation in the adult. The development of T cells is described in detail in Chapter 9.

III. SECONDARY LYMPHOID TISSUES

i) MALT and SALT

The mucosa-associated lymphoid tissues (MALT) and the skin-associated lymphoid tissues (SALT) are the first elements of the adaptive response encountered by a pathogen that has overwhelmed the body's passive anatomical and physiological barriers. The leukocyte populations that make up the MALT and SALT are situated at the most common points of antigen entry, behind the mucosae of the respiratory, gastrointestinal and urogenital tracts, and just below the skin (refer to Fig. 2-9). Thus, the MALT and SALT protect the body at all surfaces that interface with the external environment. Populations of mature T and B lymphocytes, γδ T cells, DCs, NK and NKT cells and macrophages are positioned in these locations to counter pathogens as they breach a body surface.

The MALT is subdivided by body location and includes the NALT, BALT and GALT. The nasopharynx-associated lymphoid tissue (NALT) is located in the upper respiratory tract. The bronchi-associated lymphoid tissue (BALT) is located in the lower respiratory tract and the lungs. The gut-associated lymphoid tissue (GALT) is located in the digestive tract. Throughout most of the MALT, the leukocytes are dispersed in diffuse masses under a mucosal epithelial layer. In some cases, slightly more organized collections of cells exist, such as the Peyer's patches of the GALT. In other cases, the cells are organized into discrete structures such as the appendix in the GALT and the tonsils of the NALT. The SALT comprises small populations of leukocytes resident in the epidermis (upper layer) and dermis (lower layer) of the skin. Dendritic cells known as *Langerhans cells* are scattered throughout the epidermis; these cells function as APCs and secrete cytokines drawing lymphocytes to the area. The underlying dermis is dominated by T cells, dermal DCs and macrophages. How these elements act to eliminate antigens is discussed in more detail in Chapter 12.

ii) The Lymphatic System

All cells in the body are bathed by nutrient-rich interstitial fluid. This fluid is blood plasma that, under the pressure of the circulation, leaks from the capillaries into spaces between cells (Fig. 2-12). Ninety percent of this fluid returns to the circulation via the venules, but 10% filters slowly through the tissues and eventually enters networks of tiny channels known as the *lymphatic capillaries*, where it becomes known as *lymph*. The overlapping structure of the endothelial cells lining the lymphatic capillaries creates specialized pores that allow microbes, leukocytes and large macromolecules to also pass into the

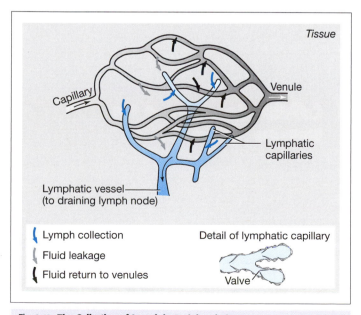

Fig. 2-12 **The Collection of Lymph in Peripheral Tissues**

lymphatic capillaries. Valves in the lymphatic capillaries ensure that the lymph and its contents move only forward as the lymphatic capillaries collect into progressively larger *lymphatic vessels*. These vessels in turn connect with one of two large *lymphatic trunks* called the *right lymphatic duct* and the *thoracic* (or *left lymphatic*) *duct*. The entire network of vessels and ducts that collects and channels the lymph and its contents throughout the body is known as the *lymphatic system* (Fig. 2-13). The right lymphatic duct drains the right upper body, while the entire lower body drains into the *cisterna chyli* at the base of the thoracic duct. Lymph from the left upper body also enters the thoracic duct. The right lymphatic duct empties the lymph into the *right subclavian vein* of the blood circulation, while the thoracic duct connects with the *left subclavian vein*. As a result, the lymphatic system is connected to the blood circulation.

iii) Lymph Nodes

As lymph flows through the lymphatic vessels, it passes through the *lymph nodes*. The lymph node is the major site for the interaction of lymphocytes with antigen during a primary adaptive response. Antigen that finds its way past the body's innate defenses and escapes MALT and SALT is collected in the lymph and brought to the nearest (*draining*) lymph node. Lymph nodes occur along the entire length of the lymphatic system but are clustered in a few key regions. For example, the *cervical* lymph nodes drain the head and neck, while the *popliteal* nodes drain the lower legs (refer to Fig. 2-13).

Lymph nodes are bean-shaped, encapsulated structures 2–10 mm in diameter that contain large concentrations of lymphocytes, FDCs and APCs (Fig. 2-14). Lymph enters a lymph node through several *afferent* lymphatic vessels. It then passes through the *cortex*, *paracortex* and *medulla* of the node, and exits on the opposite side through a single *efferent* lymphatic vessel. The cortex contains large numbers of resting B cells, FDCs and macrophages arranged in lymphoid follicles. The paracortex is home to many T cells and thymic DCs, while the medulla becomes well-stocked with antibody-secreting plasma cells during an adaptive response.

iv) The Spleen

Most antigens escaping the innate immune response, MALT and SALT make their way into the tissues, are collected in the lymphatic system and are channeled into local lymph nodes. However, there are several ways by which an antigen can access the blood circulation: (1) Sometimes an antigen is introduced directly into the blood, as during drug injection or insect and snake bites. (2) Overwhelming local infection at skin and mucosal sites can result in penetration of underlying blood vessels by the pathogen. (3) Systemic infection that cannot be contained by the lymph nodes pours into the efferent lymph, so that antigens are eventually dumped into the blood. Fortunately, we have the *spleen*, an abdominal organ that traps blood-borne antigens. The entire blood volume of an adult human courses through the spleen four times daily.

The structural framework of the spleen is created by a network of reticular fibers (again called trabeculae) that branch out of the *hilus* (base) of the spleen and connect to a thin

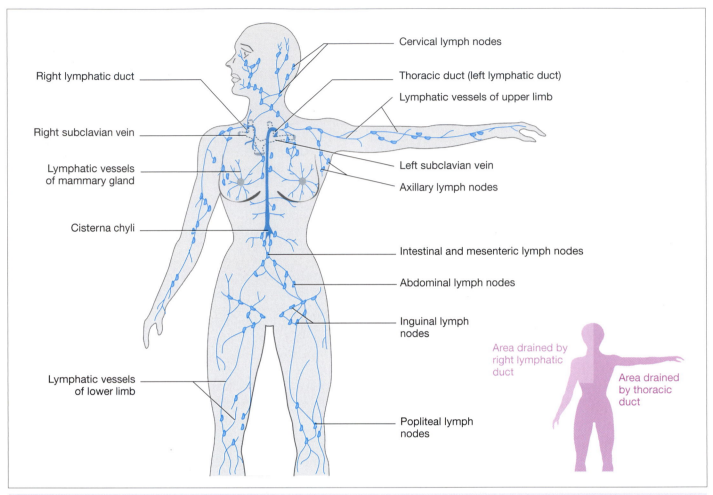

Fig. 2-13 **Major Vessels and Nodes of the Lymphatic System**

exterior capsule (Fig. 2-15A). Antigen carried in the blood circulation enters the spleen by a single artery called the *splenic artery*. The splenic artery branches into a network of smaller arteries that travel through the organ via the hollow spaces created by the trabeculae. These smaller arteries branch into *arterioles* that penetrate the trabeculae. The arterioles in turn branch off into capillaries, some of which connect directly with the venous sinuses (Fig. 2-15B). Other capillaries are open-ended and empty into the *splenic cords*, which are collagen-containing lattices through which blood percolates before entering the venous sinuses. Blood collecting in the venous sinuses empties into small veins that connect with the *splenic vein* exiting the spleen.

Each arteriole in the spleen is encased by a *periarteriolar lymphoid sheath* (PALS), a cylindrical arrangement of tissue populated primarily by mature T cells but also containing low numbers of plasma cells, macrophages and conventional DCs. The tissue surrounding the PALS is filled with lymphoid follicles containing resting B cells and macrophages. Surrounding the follicles is the *marginal zone*, which contains particular B cell subsets. *White pulp* is the name given to that part of the spleen containing the splenic arterioles with their PALS, the follicles and the marginal zone. *Red pulp*, named for its abun-

dance of erythrocytes, consists of the splenic cords and the venous sinuses and surrounds the white pulp. The red pulp functions chiefly in the filtering of particulate material from the blood and in the disposal of senescent or defective erythrocytes and leukocytes.

D. Cellular Movement in the Immune System

I. LEUKOCYTE EXTRAVASATION

A unique feature of the immune system is that its constituent cells are not fixed in a single organ but instead move around the body in the blood circulation to tissues where they are needed. To exit from the blood into a tissue under attack, leukocytes carry out a migration process called *extravasation* (Fig. 2-16). Extravasation has two phases: *margination*, in which leukocytes adhere to the endothelial cells lining the postcapillary venules, and *diapedesis*, in which leukocytes pass between the endothelial cells and through the basement membrane, emigrating into the tissues. The margination phase can

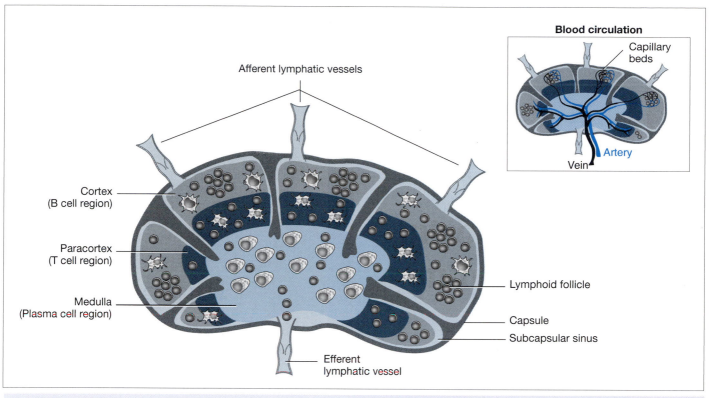

Afferent lymphatic vessels

Blood circulation

Capillary beds

Artery

Vein

Cortex
(B cell region)

Paracortex
(T cell region)

Medulla
(Plasma cell region)

Lymphoid follicle

Capsule

Subcapsular sinus

Efferent
lymphatic vessel

Fig. 2-14 **The Lymph Node**

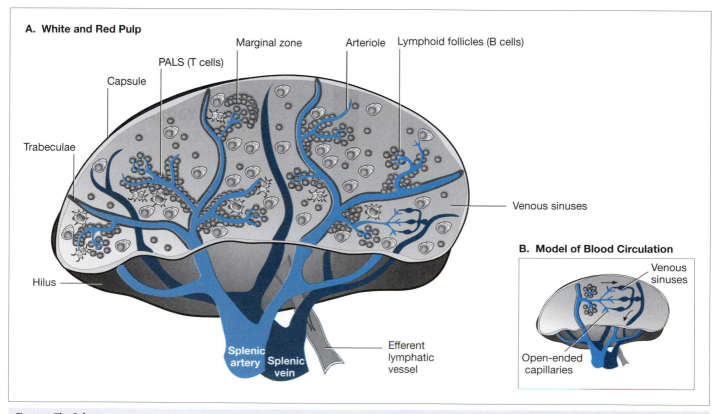

A. White and Red Pulp

Capsule

PALS (T cells)

Marginal zone

Arteriole

Lymphoid follicles (B cells)

Trabeculae

Venous sinuses

Hilus

B. Model of Blood Circulation

Venous sinuses

Open-ended capillaries

Splenic artery

Splenic vein

Efferent lymphatic vessel

Fig. 2-15 **The Spleen**

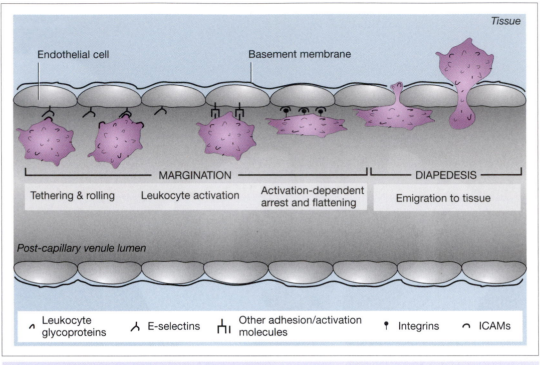

Fig. 2-16 Leukocyte Extravasation

be broken down into three steps: *tethering and rolling, activation,* and *activation-dependent arrest and flattening.*

The margination phase of extravasation starts with leukocytes circulating through a postcapillary venule. In these small vessels, the blood flow is relatively slow, shear forces are decreased, and the electrical charge on the surfaces of the endothelial cells lining the venule is low. Endothelial cells express adhesion molecules called *E-selectins* that bind to complementary glycoproteins present on the surface of a passing leukocyte, "tethering" it to the venule wall. However, these bonds are relatively weak and can be reversed in seconds, so that the leukocyte "rolls" across several endothelial cells, increasingly slowed by the sequential binding. During tethering and rolling, intracellular signaling is delivered that activates the leukocyte and causes it to display new adhesion/ activation molecules on its surface. Among these molecules are the *integrins*, which bind to *intercellular adhesion molecules* (ICAMs) expressed by the endothelial cells. Integrin-mediated binding is strong enough to flatten the leukocyte onto the endothelial surface, halting its movement and resulting in a state of so-called *activation-dependent arrest.* The leukocyte is then ready for the second phase of migration, diapedesis. The flattening of the leukocyte permits it to insert a pseudopod between two endothelial cells of the venule lining and squeeze between them. The leukocyte then secretes enzymes that digest the basement membrane supporting the endothelium, allowing the leukocyte to complete its migration through the venule wall into the tissue. Once the leukocyte has entered the tissue, the binding of additional integrin molecules to components of the extracellular matrix allow it to migrate through the tissue.

How do leukocytes know where to extravasate out of the blood into a tissue? A tissue suffering from trauma or infection induces the local release of inflammatory molecules that activate local endothelial cells and upregulate their expression of selectins and ICAMs. Passing leukocytes in the blood are then readily tethered exactly where they are needed, and migrate out of the blood through the activated endothelium. The cells then follow chemotactic gradients to the inflammatory site.

II. LYMPHOCYTE RECIRCULATION

Although all leukocytes can extravasate into inflammatory sites, only lymphocytes can recirculate in the absence of inflammation. *Lymphocyte recirculation* is a process by which most resting T cells and some B cells migrate continually between the blood and secondary lymphoid tissues (Fig. 2-17). This strategy helps to solve the problem of getting the right lymphocyte to the right place at the right time to meet the right antigen. The total cellular mass of lymphocytes in the human body (approximately 10^{12} cells) is the same as that of the liver or the brain. However, because the number of lymphocyte specificities is so large, only a very small number of individual cells (one naïve T cell in 10^5) exists to deal with any particular foreign substance. Nevertheless, because naïve cells recirculate from blood to the secondary lymphoid organs to lymph and back to the blood on average once or twice each day, the chance of any one of these cells meeting its specific antigen is greatly increased.

During recirculation, resting mature lymphocytes extravasate into the secondary lymphoid tissues where they can easily sample antigens that have been collected and concentrated at

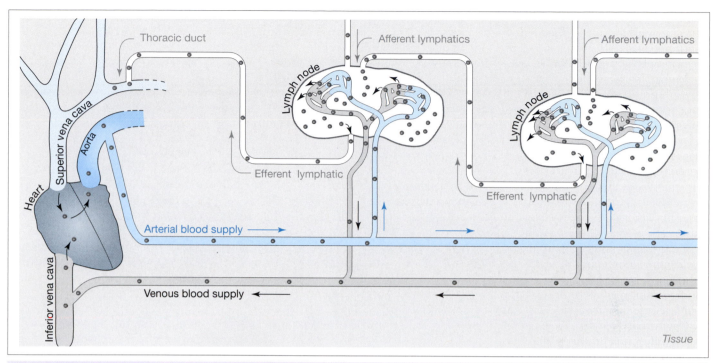

Fig. 2-17 Schematic Representation of Lymphocyte Recirculation

these sites. To facilitate this extravasation, all secondary lymphoid tissues (except the spleen) contain specialized postcapillary venules which, due to their plump, cuboidal appearance, are called *high endothelial venules* (HEVs). The HEVs have specialized features that are induced and maintained by cells and factors present in the afferent lymph coming into lymphoid tissues. In particular, the endothelial cells lining the HEVs constitutively express high levels of adhesion molecules that promote lymphocyte extravasation. Indeed, it has been estimated that, throughout the body, a total of 5×10^6 lymphocytes exit the blood through the HEVs each second. Such a high migration rate means that, on average, a given lymphocyte can travel through every lymph node in the body at least once a day. Thus, naïve lymphocytes can efficiently and thoroughly survey sites of antigen collection in the host on a continuous basis.

Most lymphocytes that enter a given lymph node fail to encounter their specific antigens and exit directly via the efferent lymphatic without undergoing activation. These cells eventually rejoin the blood circulation via the connections between the lymphatic system and the left and right subclavian veins. However, a few lymphocytes may find their specific antigens deposited in the cortex or paracortex of the lymph node. A B cell may recognize antigen held by FDCs in the cortex, whereas a T cell may recognize a pMHC complex on the surface of a DC in the paracortex. Engagement of the antigen receptors then activates the B and T cells. In the case of B lymphocytes, the effector plasma cells and memory B cells generated when antigen is encountered in the lymph node tend to remain in the node and do not recirculate. Instead, the antibodies secreted by the plasma cells diffuse through the medulla of the node and exit it via the efferent lymphatic, eventually entering the

blood to provide systemic protection. In contrast, if a naïve T cell is activated in the lymph node, the effector and memory T lymphocytes exit the lymph node via the efferent lymphatic and join the blood circulation. However, as is described later, these cells will then migrate to peripheral tissues under attack and not back to the lymph node.

In the spleen, the situation is slightly different. Unlike a lymph node, the spleen has no afferent lymphatic or HEVs, so that lymphocytes must enter the spleen via the splenic artery. The cells exit the blood via capillary beds in the marginal zone, and then migrate to the follicles and PALS. Antigen in the blood percolating through this area is taken up by resident DCs and used to activate T cells in the PALS. Alternatively, macrophages patrolling the red pulp can collect filtered antigen and convey it to lymphocytes in the PALS. Lymphocytes that are not activated exit the spleen via the efferent lymphatic.

III. LYMPHOCYTE HOMING

Different T cell subsets show different recirculation patterns that correlate with the expression of varying adhesion molecules both on the surfaces of the lymphocytes and on the HEVs of various tissue sites. Adhesion molecules expressed only on HEVs from particular sites in the body are called *vascular addressins*, because these molecules direct lymphocytes with the complementary adhesion molecules to those specific locations ("addresses"). In these cases, the complementary adhesion molecules on the lymphocytes are called *homing receptors*. Some homing receptors are widely expressed on multiple lymphocyte subsets, while others are found predominantly on a specific lymphocyte subset. Particular lymphoid organs and tissues display distinct addressins, attracting only those lymphocytes with the appropriate homing receptor.

As well, naïve T cells express different homing receptors than do effector and memory T cells. For example, L-selectin, a member of the selectin family of adhesion molecules, is highly expressed on naïve T cells and readily binds to the vascular addressin GlyCAM-1 (glycosylation-dependent cellular adhesion molecule 1), which is preferentially expressed by HEVs in the lymph nodes. However, effector and memory T cells express very little L-selectin, and so no longer recirculate through the lymph nodes. Instead, these cells express homing receptors such as VLA-4 (very late antigen 4) that mediate their migration to peripheral tissues where inflammation has been initiated. The endothelial cells of the postcapillary venules in these sites express new vascular addressins like VCAM-1 (vascular cellular adhesion molecule 1) that bind to VLA-4 and permit the extravasation from the blood of effector and memory lymphocytes as well as neutrophils and monocytes. Because this leukocyte influx includes effector and memory lymphocytes, such sites of inflammation are sometimes called *tertiary lymphoid tissues*. If the inflammation does not resolve and becomes chronic, the postcapillary venules in the affected site take on the characteristics of HEVs, inducing lymphocytes to traffic through the area on a constant basis.

Although effector T cells generally die in the peripheral tissue in which they have been sent to combat an invader, memory T cells generated in a primary response express homing receptors that allow them to return repeatedly (after the inflammation is resolved) to monitor the peripheral tissue in which they first encountered antigen. For example, memory T lymphocytes expressing LPAM-1 (lymphocyte Peyer's patch adhesion molecule 1) are drawn to MAdCAM-1 (mucosal addressin cellular adhesion molecule 1)-expressing endothelial cells in the intestinal Peyer's patches. Likewise, memory T lymphocytes of the MALT and SALT recirculate in the mucosae and skin, respectively, under the influence of mucosal and cutaneous addressins. This tactic makes evolutionary sense because the triggering antigen is likely to enter the body in the same fashion in a subsequent infection. The restricted homing of memory cells thus increases their chances of being in the right place at the right time to encounter the specific antigen, contributing to the shorter lag time of the secondary immune response.

Now that the reader is familiar with the cells and tissues of the immune system, we turn in the next chapter to innate immunity, to describe in detail how most foreign entities attacking our bodies are defeated.

CHAPTER 2 TAKE-HOME MESSAGE

- The immune system is a partnership of the blood circulation, the lymphatic system, various lymphoid organs and tissues and the hematopoietic cells moving among them.

- Hematopoietic cells in the bone marrow give rise to myeloid, lymphoid and mast cell progenitors that differentiate into the mature leukocytes that populate the body.

- The primary lymphoid tissues are the bone marrow and the thymus. The secondary lymphoid tissues are the lymph nodes, the spleen, the MALT and the SALT. Lymphocytes develop in the primary lymphoid tissues and are activated by antigen in the secondary lymphoid tissues.

- The binding of antigen, PAMPs, cytokines or other ligands to the appropriate cell surface receptors triggers intracellular signaling that results in particular cytoplasmic or nuclear outcomes with the potential to alter cell behavior.

- The primary function of cytokines is to mediate intercellular signaling between leukocytes. The induced innate and adaptive responses are largely dependent on this signaling.

- Extravasation of leukocytes is mediated by the binding of complementary adhesion molecules that are upregulated in response to inflammation.

- Lymphocyte recirculation allows lymphocytes to continuously patrol the body's sites of antigen entry and collection, even in the absence of inflammation.

Section A.I

1) What are hematopoietic cells? What are granulocytes?

2) Name four types of myeloid cells and three types of lymphoid cells.

3) Can you describe how chemotaxis works?

Section A.II

1) Can you define these terms? phagosome, degranulation, APC

2) Describe the functions of four types of myeloid cells.

3) How are neutrophils linked to the adaptive response?

4) How are macrophages linked to the adaptive response?

5) How do macrophages become hyperactivated and why is this important?

Section A.III

1) Can you define these terms? cross-reacting antigen, lymphoblast

2) What are CD markers and why are they useful?

3) Describe the functions of four types of lymphoid cells.

4) Give two examples of primary immunodeficiencies caused by defects in DNA repair.

5) Why are γδ T cells and NKT cells considered innate leukocytes?

Section A.IV

1) Why are DCs particularly important for primary adaptive responses?

2) In what areas of the body are DCs generated?

3) What is the function of a DC?

4) Distinguish between immature and mature DCs.

Section A.V

1) Can you define these terms? hematopoiesis, myelopoiesis, lymphopoiesis, pluripotent, self-renewing

2) How does the location of hematopoiesis shift as a human fetus develops into an adult?

3) How does an HSC differ from an early progenitor?

Section A.VI

1) Can you define these terms? homeostasis, apoptotic body

2) How do apoptosis and necrosis differ?

Section B.I

1) What are the the three principal domains constituting many cell surface receptors?

2) Describe three elements of an intracellular signaling cascade.

3) Intracellular signaling can lead to a cytoplasmic or a nuclear outcome. Give an example of each.

Section B.II

1) In the context of cytokines, can you define these terms? endocrine, paracrine, autocrine, pleiotropic, synergistic, antagonistic

2) What are two differences between cytokines, growth factors and hormones?

3) How is the power of cytokines controlled?

4) Name the three principal categories of cytokine function.

5) Why do cytokines overlap in function, and how is this overlap often achieved?

Section C.I

1) Can you distinguish between lymphoid follicles, lymphoid patches and lymphoid organs?

2) What are the primary and secondary lymphoid tissues?

3) What are two major differences between the primary and secondary lymphoid tissues?

Section C.II

1) Can you define these terms? red marrow, trabeculae, thymocyte, thymic involution, PALS, afferent, efferent, marginal zone

2) What are positive and negative thymic selection and why are they important?

Section C.III

1) Can you describe MALT, SALT, BALT, NALT and GALT?

2) How does the lymphatic system connect with the blood circulation?

3) Describe the key functions of lymph nodes and the spleen.

Section D.I

1) Can you define these terms? margination, diapedesis, ICAMs

2) How does inflammation promote leukocyte extravasation?

Section D.II

1) Where are the HEVs and how do they promote lymphocyte recirculation?

2) How do leukocyte extravasation and lymphocyte recirculation differ?

3) Describe antigen collection in the lymph nodes and the spleen.

Section D.III

1) Can you define these terms? homing receptor, vascular addressin

2) What is the effect of inflammation on vascular addressin expression?

3) How do memory lymphocyte homing patterns contribute to the efficiency of secondary adaptive responses?

3

Innate Immunity

WHAT'S IN THIS CHAPTER?

A year that's good is evident by its spring.
Persian

Innate immunity is conferred by both non-inducible and inducible mechanisms. With respect to the former, the body has natural anatomical and physiological barriers that act non-specifically to prevent infection. If these barriers are penetrated, certain soluble molecules and cell surface receptors on innate leukocytes are able to recognize invading entities in a broadly specific way and trigger various inducible innate mechanisms. Innate defenses do not generally undergo permanent change or establish long-term memory following exposure to a pathogen, and a second encounter provokes a response of a magnitude and character very similar to that triggered by the first encounter. The host uses innate immunity to rapidly generate an initial response to almost any pathogen, either eliminating it or at least containing it until the slower but more focused adaptive immune response can develop. It is the cytokines produced by innate leukocytes

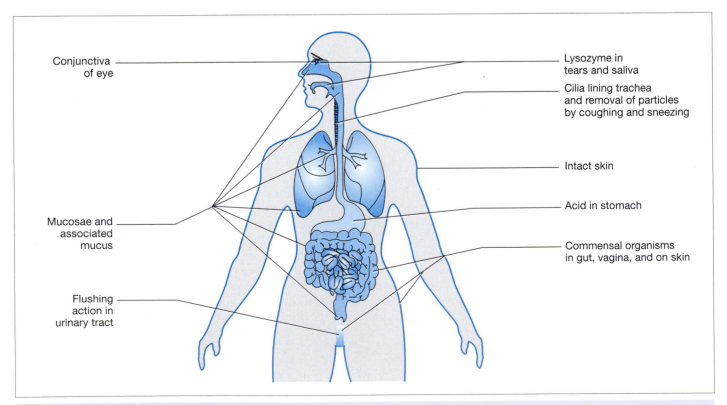

Fig. 3-1 **Anatomical and Physiological Barriers**

locked in battle with the pathogen that play key roles in recruiting and activating the highly specific lymphocytes of adaptive immunity.

A. Non-Induced Innate Mechanisms

Several anatomical and physiological barriers substantially decrease the likelihood of infection and reduce its intensity should it occur. Anatomical barriers include structural elements such as the skin and mucosae that physically prevent access through the body surfaces and orifices. Physiological barriers include the actions of body structures (such as sneezing) or substances produced by tissues (such as tears and mucus) that reinforce the anatomical barriers. These elements are illustrated in Figure 3-1.

Intact skin provides a tough surface that is relatively difficult for microorganisms to penetrate. The skin is composed of three layers: the outer *epidermis*, the underlying *dermis* and the fatty *hypodermis* (Fig. 3-2A). Cells at the exterior surface of the epidermis are dead and filled with *keratin*, a protein that

confers water resistance. Other features of the epidermis that discourage infection are its lack of blood vessels and its rapid turnover: complete renewal of human outer skin occurs every 15–30 days. Below the epidermis lies the dermis, which contains all of the blood vessels and other tissues necessary to support the epidermis. Among these are the sebaceous glands. These glands produce *sebum*, an oily secretion with a pH of 3–5 that inhibits the multiplication of most microbes. (We note here that the dermis also contains populations of leukocytes that can both defend the dermis and migrate into the epidermis; the activation of these cells is considered part of induced innate immunity.) Below the dermis lies the hypodermis, which provides a barrier of fat that impedes pathogens. Failure of the skin as a defense occurs when it is breached, as in the case of wounds or insect bites (Fig. 3-2B).

Other areas of the body that come into contact with the outside world have developed different anatomical barriers. Instead of dry skin, which would not permit the passage of air or food, the surfaces of the gastrointestinal, respiratory and urogenital tracts are covered with the thin layer of mucosal epithelial cells known as a mucosa. Similarly, the eye possesses a delicate epithelial layer called the *conjunctiva* that lines the eyelids and protects the exposed surface of the eye while allowing the passage of light. The conjunctiva of the eye and the mucosae of internal body tracts are defended from microbial penetration by the flushing action of secretions such as mucus, saliva and tears. Mechanical defense in the lower respiratory and gastrointestinal tracts is derived from the sweeping action of tiny oscillating hairs called *cilia* on the surface of the mucosal epithelial cells. Vomiting, sneezing, coughing and bowel activity also serve to expel mucus-coated microbes. Failure of these innate measures allows pathogens to penetrate the epithelial layer (Fig. 3-3). Infection can also occur when a microorganism possesses a cell surface molecule that permits it to attach to surface proteins of conjunctival or mucosal epithelial cells. This attachment circumvents the barriers offered by the tears or mucus and facilitates the entry of the pathogen directly into the epithelial cells.

A microorganism that manages to evade the external and internal barriers of the body may succumb to a hostile physi-

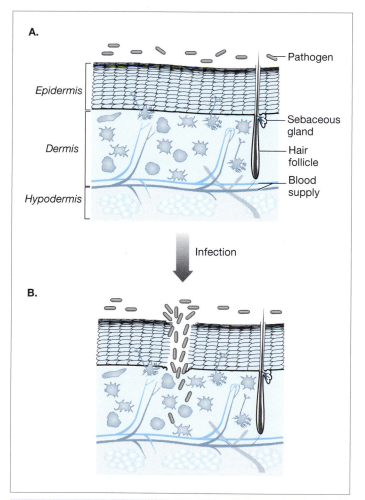

Fig. 3-2 **The Skin as an Innate Barrier**

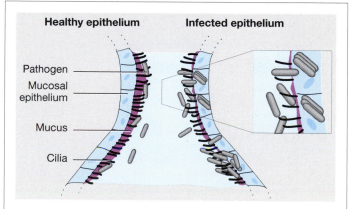

Fig. 3-3 **The Respiratory Mucosal Epithelium as an Innate Barrier**

ological environment, such as a body temperature too high or a stomach pH too low to permit replication. Body fluids often contain antibacterial or antiviral substances. For example, *lysozyme*, an enzyme that hydrolyzes bacterial cell walls, is found in tears and mucus. Another physiological barrier is provided by the billions of benign bacteria and fungi that inhabit the mouth, the digestive and respiratory systems and the skin surface. These beneficial microbes are collectively known as *commensal organisms*. Commensal organisms compete with pathogens for available resources, and may activate leukocytes or secrete antimicrobial toxins. More on the immune defense of the muscosae and the skin appears in Chapter 12.

B. Induced Innate Mechanisms

If a pathogen or foreign macromolecule penetrates the body's anatomical barriers and escapes destruction by physiological defenses, the induced innate response is next in line to try to eliminate the invader. "Induction" implies that a recognition event has occurred and has sparked a response. In the case of the induced innate response, the recognition event is broadly specific and mediated through receptors that carry out *pattern recognition*. Pattern recognition triggers one or more of several mechanisms designed to contain the assault, including complement activation, the inflammatory response and cellular internalization mechanisms such as phagocytosis. These mechanisms are discussed in detail here. In addition, as is described in later chapters, most activated innate leukocytes secrete copious cytokines and some trigger the cytolysis of infected cells.

I. PATTERN RECOGNITION MOLECULES

As introduced in Chapter 1, PAMPs are highly conserved and repetitive structures that are common to a wide variety of microbes or their products but are not usually present in host cells. Common PAMPs include pathogen cell wall components, such as lipopolysaccharide (LPS), peptidoglycan and bacterial lipoproteins, as well as viral DNA and RNA genomes. Proteins that recognize these PAMPs are called *pattern recognition molecules* (PRMs). A PRM may be fixed in a leukocyte's plasma membrane as a pattern recognition receptor (PRR) or may be a soluble molecule present in the interstitial spaces or circulation (Table 3-1). In contrast to the genes encoding the TCRs and BCRs of lymphocytes, the genes encoding PRRs are

Table 3-1 Pattern Recognition Molecules

	Examples of Ligands	Location	Primary Function
PRRs			
Toll-like receptors		Monocytes, macrophages, neutrophils, immature DCs, NK cells, some T and B cell subsets, some non-leukocytes	Activation of phagocytes, induction of pro-inflammatory cytokines, APC preparation
TLR2	Bacterial lipoteichoic acids		
TLR3	Viral dsRNA		
TLR4	Bacterial LPS		
TLR5	Bacterial flagellin		
TLR6	Mycoplasma lipopeptides		
TLR7, 8	Viral ssRNA		
TLR9	Bacterial DNA with unmethylated CpG		
TLR11	Protozoan profilin protein		
Scavenger receptors	Bacterial and fungal cell wall components	Macrophages, monocytes, DCs, hepatic endothelial cells	Phagocyte activation
NK receptor	Antigens from infected, stressed or cancerous cells	NK cells	Target cell lysis, pro-inflammatory cytokine secretion
NKT receptor	Glycolipid antigens	NKT cells	Cytokine secretion
γδ TCR	Antigens from infected, stressed or damaged host cells	γδ T cells	Target cell lysis, pro-inflammatory cytokine secretion
Soluble PRMs			
Collectins (MBL)	Microbial polysaccharides	Blood plasma (secreted by hepatocytes)	Complement activation, opsonized phagocytosis
Acute phase proteins (CRP)	Microbial polysaccharides	Blood plasma (secreted by hepatocytes)	Complement activation, opsonized phagocytosis
NOD proteins	Bacterial peptidoglycan derivative	Cytoplasm	Cytokine production

fixed in the germline DNA configuration and do not undergo somatic recombination. Thus, compared to the almost infinite diversity of lymphocyte receptors, innate leukocytes exhibit a fairly limited repertoire of receptor specificities.

i) PRRs

ia) Toll-like receptors. The *Toll* gene was originally discovered in *Drosophila*, where the transmembrane Toll protein plays a dual role in embryonic development and antifungal immunity. In mammals, the Toll-like receptors (TLRs) are key PRRs that are structurally similar to *Drosophila* Toll and mediate antimicrobial innate defense. The TLRs are most highly expressed by monocytes, macrophages, DCs and neutrophils. Some TLRs are also expressed by NK cells, T and B cell subsets and certain non-hematopoietic cells. Different TLRs recognize different microbial structures. For example, TLR2 recognizes bacterial lipoteichoic acids, while TLR4 binds to bacterial LPS. TLR3 binds to double-stranded RNA (dsRNA), while TLR7 binds to viral single-stranded RNA (ssRNA). Binding of the correct PAMP to a TLR initiates intracellular signaling that leads to phagocytosis, cellular activation, and/or the production of the pro-inflammatory cytokines IL-1, IL-6, IL-8, IL-12, TNF and IFNα/β. TLR engagement on macrophages also induces the synthesis of an enzyme called *inducible nitric oxide synthase* (iNOS) that produces the powerful antimicrobial agent *nitric oxide* (NO) (see later). Finally, TLRs are linked to the adaptive response because TLR engagement on macrophages and DCs prepares these cells to function as APCs, and promotes their migration to the lymph nodes where mature T and B cells congregate.

As well as responding to PAMPs, TLR4 and TLR2 may recognize stress molecules produced by a host when it experiences trauma that does not involve a pathogen. Such stress molecules may include the *heat shock proteins* (HSPs), which are molecules upregulated by a distressed or dying host cell in response to many assaults (not just increased temperature). The binding of HSP60 or HSP70 to TLR2 or TLR4 appears to trigger intracellular signaling that leads to cytokine secretion promoting inflammation. The resulting influx of leukocytes into the site of injury may assist in wound healing.

ib) Scavenger receptors. Scavenger receptors (SRs) are expressed primarily on phagocytes and bind to a wide variety of lipid-related ligands derived either from pathogens or from host cells that are damaged, apoptotic or senescent (old). Damaged or dying host cells undergo deleterious changes to their membranes such that internal phospholipids that are normally hidden become exposed. The SRs then recognize molecular patterns that are not present on healthy host cells. Binding of a ligand to an SR triggers phagocytosis of the microbe or dying host cell.

ic) Receptors of NK, NKT and $\gamma\delta$ T cells. As mentioned in Chapter 2, defense mechanisms mediated by NK, NKT and $\gamma\delta$ T cells are considered to be part of the induced innate response because these cells respond much faster than T and B lymphocytes to a threat and carry out ligand recognition

that lacks the fine specificity of TCRs and BCRs. For this reason, the receptors by which NK, NKT and $\gamma\delta$ T cells carry out their innate functions can be considered PRRs. NK, NKT and $\gamma\delta$ T cells are discussed in further detail later and in Chapter 11.

ii) Soluble PRMs

iia) Collectins. Collectins float freely in the blood and other body fluids. These proteins are made up of a collagen domain fused to a *lectin* (carbohydrate-binding) domain. Different collectins recognize different patterns of carbohydrate structures on microbial surfaces. These patterns are clearly distinct from those present on eukaryotic cells, so that the host's cells are ignored. Once bound by a ligand, collectins generally mediate pathogen clearance via complement activation, *opsonization* (see later), or by aggregating microbial cells together (*agglutination*). Perhaps the best-known collectin is the *mannose-binding lectin* (MBL) that triggers the lectin pathway of complement activation (see later).

iib) Acute phase proteins. Early in a local inflammatory response, macrophages activated by engagement of their TLRs secrete cytokines that trigger hepatocytes to produce *acute phase proteins*. At least some of these soluble proteins, particularly *C-reactive protein* (CRP), bind to a wide variety of bacteria and fungi through recognition of common cell wall components that are not present on host cells. Once deposited on the surface of a microbe, these soluble molecules can activate complement and stimulate phagocytosis.

iic) NOD proteins. The *nucleotide-binding oligomerization domain* (NOD) proteins are cytoplasmic molecules that detect PAMPs of intracellular pathogens once these invaders have accessed a host cell's interior. These soluble sensor proteins are structurally related to the TLRs but bind to bacterial peptidoglycan-derived structures not recognized by the TLRs. Ligand binding to a NOD protein triggers intracellular signaling that causes the infected cell to produce cytokines (such as TNF, IL-1, IL-18 and IFNα/β) that promote the inflammatory response.

II. THE COMPLEMENT SYSTEM

The complement system is a vital physiological element that must be induced to carry out its protective functions. Complement was named for its ability to assist, or be "complementary" to, antibodies involved in the lysing of bacteria.

i) Nature and Functions

Complement is not a single substance but rather a collection of about 30 serum proteins that make up a complex system of functionally related enzymes. These enzymes are sequentially activated in a tightly regulated cascade to carry out complement's functions. Complement activation has four principal outcomes: (1) lysis of pathogens; (2) opsonization (coating) of foreign entities to enhance phagocytosis; (3) clearance of *immune complexes* (soluble lattices of antigen bound to antibody); and (4) the generation of peptide by-products that are involved in the inflammatory response. All these functions are discussed here and in later chapters.

The complement system can be activated via three biochemical pathways: the *classical* pathway, the *lectin* pathway and the *alternative* pathway (Fig. 3-4). All three pathways result in the production of the key complement component *C3b* and the assembly on the pathogen surface of a structure called the _membrane attack complex_ (MAC). The classical pathway is triggered when antigen present on a pathogen surface binds to antibody, and this antibody in turn is bound by complement component *C1*. In the lectin pathway, complement activation is initiated by the direct binding of MBL to certain carbohydrates on the surface of a pathogen; no antibody is involved. The alternative pathway is activated when complement component C3 spontaneously hydrolyzes and then interacts with certain enzymatic factors to produce C3b. This C3b can bind to almost any carbohydrate or protein on the surface of a pathogen; again, no antibody is involved. The attachment of C3b to a pathogen surface leads to the formation of the MAC. The MAC then bores a hole in the outer membrane of the pathogen, triggering an osmotic imbalance that lyses the invader.

The classical pathway of complement activation was discovered first but probably evolved last. Due to its involvement of antibody, the classical pathway has the advantage of greater specificity than the alternative and lectin pathways but the disadvantage of tardiness due to its dependence on an adaptive response. The lectin and alternative pathways are less specific but very rapid and can activate complement during the lag phase required for B cell activation and antibody synthesis. In addition, because it involves antibody, the classical pathway links the innate and adaptive responses. The alternative and lectin pathways are purely elements of innate immunity.

Many pathogens with membranes can become targets for MAC assembly and thus subject to destruction via MAC-mediated lysis. Most bacteria are susceptible to MAC-mediated killing, although some of these organisms protect their membranes by covering them with thick polysaccharide capsules. Some enveloped virus particles and virus-infected host cells that express viral antigens in their membranes (and thus become coated by antiviral antibodies) are also subject to MAC-mediated lysis. Even foreign erythrocytes introduced into a host by a blood transfusion may be destroyed by MAC formation if the erythrocytes bear surface antigens recognized by host antibodies. However, most uninfected nucleated cells resist MAC-mediated lysis because of the action of host regulatory proteins (see later).

Amplification is a key feature of complement activation. The cascade is made up of a large number of sequential activation steps, some of which generate multiple products. Since many of these products are themselves enzymes that trigger subsequent steps, an exponential increase in activated molecules is generated. A huge amplification of an initial signal can be achieved, making it possible to generate a massive response from a single triggering event within a short period of time. However, the complement system is also highly sensitive, since a deficiency of any one enzyme can halt the progression of the cascade completely. With respect to sources of complement components, most are synthesized principally by hepatocytes in the liver but are also secreted by macrophages, monocytes and epithelial cells of the urogenital and gastrointestinal tracts. Cytokines produced during the inflammatory response (such as TNF and IL-1) can induce the synthesis and secretion of some complement proteins by other types of host cells.

ii) Nomenclature

Biochemists have adopted certain conventions in naming the complement components. Because they were identified first, the enzymes of the classical pathway are designated by the letter C and the numbers 1–9 (which reflects their order of discovery, not their order of action). Once a protein is cleaved, the resulting peptide fragments are designated as "a" and "b," where "a" indicates the smaller product and "b" indicates the larger one. Generally speaking, the "b" fragments contribute to the next enzymatic activity in the cascade (e.g., C3b), while the "a" fragments (e.g., C5a) are involved in the inflammatory response (see later). The exception is C2, where C2a (not C2b) is the larger fragment and contains the enzymatic domain. Often several complement components must associate to form an active complex, such as C4b, C2a and C3b coming together to form C4b2a3b. The components of the alternative pathway were discovered later historically and are called "factors". The factors are assigned a single letter (Factor B, Factor D etc.) and are often abbreviated simply as B, F, D, H and I. The lectin pathway shares most of the components of the classical pathway but differs in the enzymes that initiate the cascade (see later).

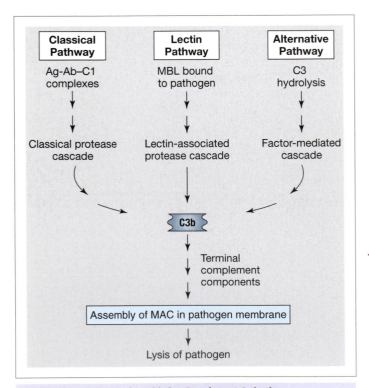

Fig. 3-4 **Three Pathways for Initiating Complement Activation**

iii) Complement Activation Pathways

All three pathways of complement activation achieve the same goal of generating the key molecule C3b. It is C3b that allows the *terminal complement components* (which are the same in all three pathways) to come together and assemble the MAC. The steps of the classical, alternative and lectin complement activation pathways are summarized in Figure 3-5 and Table 3-2. Primary immunodeficiencies that arise from defects in the complement system are outlined in Box 3-1.

iiia) Classical pathway. Complement component C1 circulates in the plasma as a huge, inactive protein complex contain-

only classical pathway required specific Ab. (IgG, IgM)

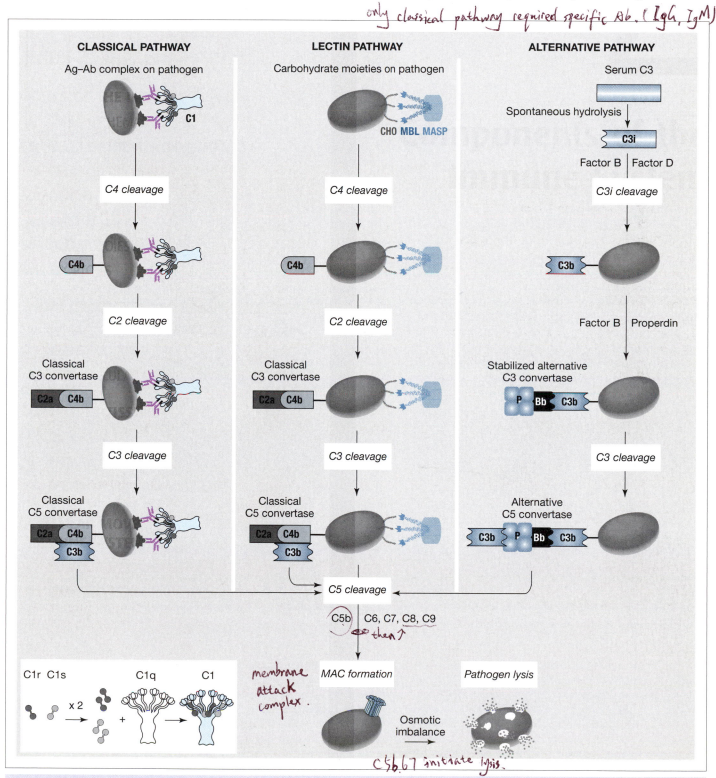

Fig. 3-5 **Formation of Convertases and MAC during Complement Activation**

Table 3-2 Components of Complement Activation Pathways

	Classical Pathway	Lectin Pathway	Alternative Pathway
Initiating components	C1 complex	MBL and MASP* complex	C3i
Start point	Antigen–antibody complex interacts with C1	Microbial carbohydrates interact with MBL	Serum C3 hydrolysis
Early components	C4 C2	C4 C2	Factor B Factor D
Component that attaches to pathogen surface	C4b	C4b	C3b
C3 convertase	C2aC4b	C2aC4b	C3bBbP†
C5 convertase	C2aC4bC3b	C2aC4bC3b	C3bBbPC3b
Terminal components	C5–C9	C5–C9	C5–C9

*MASP, MBL-associated serine protease.
†P, properdin.

self transformal, se

ing one C1q subunit, two C1r subunits and two C1s subunits. The C1q subunit can bind to the *Fc regions* of two antigen-specific antibodies that have bound in close proximity to antigen fixed on the surface of a pathogen. (The Fc region of an antibody is that part of the protein not involved in antigen-binding; see Chapter 4.) This binding activates the C1r subunits such that they activate the C1s subunits. The activated C1s subunits can then cleave serum C4 into C4a and C4b. C4a diffuses away while C4b binds to proteins on the surface of the pathogen and then binds to serum C2. This binding causes C2 to become susceptible to cleavage by C1s, generating C2a and C2b. C2a remains bound to C4b, while C2b diffuses away. The C4bC2a structure is known as the *classical C3 convertase*. The C3 convertase cleaves serum C3 into C3a and the key molecule C3b. C3b binds to C4bC2a to form another enzyme complex called the *classical C5 convertase*.

iiib) Lectin pathway. MBL binds to carbohydrate moieties (CHO) in pathogen proteins. A protease complex called *MBL-associated serine protease* (MASP) associates with MBL and cleaves serum C4 to generate C4a and C4b. C4b then binds to motifs in proteins on the pathogen surface as described for the classical pathway. C2 binds to C4b and is cleaved by the MASP complex such that a classical C3 convertase is formed.

The rest of the pathway is the same as in classical complement activation, such that a classical C5 convertase is formed.

iiic) Alternative pathway. Serum C3 spontaneously hydrolyzes an internal thioester bond to form a product called C3i. Factors B and D can act on C3i to generate C3b. This C3b binds to surface proteins on pathogens and recruits another molecule of Factor B. The C3b–Factor B complex is cleaved by Factor D to generate C3bBb, *the alternative C3 convertase*. This enzyme can generate additional C3b, amplifying the pathway. A protein called *properdin* (P) then binds to the C3 convertase complex, yielding the *stabilized alternative C3 convertase* (C3bBbP). The addition of another C3b molecule yields C3bBbPC3b, the *alternative C5 convertase*.

iv) Terminal Steps

Once a C5 convertase (regardless of derivation) is fixed on a pathogen surface, it cleaves serum C5 to yield C5a (which diffuses away) and C5b (which remains bound to the convertase). The binding of C6 then stabilizes the complex, while the binding of C7 exposes hydrophobic regions that facilitate penetration of the complex into the pathogen membrane. The addition of C8 stabilizes the complex in the membrane and the formation of a pore is initiated. With the addition of

---

Box 3-1 Primary Immunodeficiencies Due to Complement System Defects

Deficiencies for elements of the complement system can lead to primary immunodeficiencies that occur at very low frequency in the general population. Mutations of the mannose-binding lectin (MBL) gene are more common than mutations of other complement system components. Patients with **MBL** deficiency cannot produce the MBL molecule that is critical for triggering the lectin pathway of complement activation. These patients suffer from a variety of recurrent infections in early infancy, before the patient's adaptive immune system has completely matured. Deficiency for **C1 or C4** results in recurrent pyrogenic infections with encapsulated bacteria. (A *pyrogenic* infection induces high fever; a non-pyrogenic infection does not.) In contrast, **C2** deficiency favors susceptibility to non-pyrogenic infections. For unknown reasons, patients lacking C1, C2 or C4 also often suffer from the autoimmune disease *systemic lupus erythematosus* (SLE) (see Ch. 19). Patients lacking **C3, Factor H, Factor I** or **properdin** also have severe recurrent pyrogenic infections but do not show signs of SLE. Increased infection with *Neisseria* bacteria (only) occurs in the absence of **C5, C6, C7** or **C8**, while patients deficient for **C9** are generally asymptomatic.

at least four molecules of C9, the MAC is completed. The pore becomes a tunnel, allowing ions and water molecules to pour through the pathogen membrane into the interior. The pathogen lyses due to osmotic imbalance.

v) Other Roles of C3b

C3b: activated signaling cascade [handwritten annotation]

C3b has several other important functions in addition to initiating MAC assembly. (1) A pathogen coated in C3b binds to *complement receptor 1* (CR1) expressed on the surface of phagocytes (including APCs). These cells then easily engulf and destroy the invader. The C3b is said to be acting as an *opsonin* in this case (see later). By encouraging pathogen uptake by APCs in this way, C3b indirectly enhances antigen presentation to T cells and thus the adaptive response. (2) Soluble antigen–antibody complexes can bind to C1 and trigger the classical complement pathway such that C3b is deposited on the complexes themselves. The C3b then blocks the networking between multiple antigen and antibody molecules that results in the formation of large insoluble immune complex lattices (Fig. 3-6A). The C3b is said to be *solubilizing* the immune complexes, making them easier to clear from the circulation and preventing the damage they might inflict if they accumulated in the small vessels and channels of the body. Erythrocytes expressing CR1 can bind to C3b-coated antigen–antibody complexes and transport them through the circulation to the liver and spleen. Phagocytic cells in these locations then engulf and destroy the antigen–antibody complexes during the routine disposal of red blood cells. (3) C3b also contributes directly to defense against viruses. When these pathogens become coated in C3b, the C3b blocks the binding of the virus to its receptors on a host cell and prevents cell entry (Fig. 3-6B). The C3b is said to have *neutralized* the virus.

vi) Anaphylatoxins

The small fragments (e.g. C3a, C4a, C5a) that are cleaved from various complement components during the activation cascade are called *anaphylatoxins*. Anaphylatoxins are not just inert by-products; they play important roles in inflammation. The anaphylatoxins are so named because, at the high systemic concentrations generated in response to a serious bacterial infection, anaphylatoxins can induce dramatic cardiovascular and bronchial effects that resemble *anaphylaxis* (a severe systemic allergic reaction; see Ch. 18). Anaphylatoxins can trigger the degranulation of mast cells and basophils, resulting in the release of powerful inflammatory mediators (see later). In addition, C5a is a powerful chemoattractant for neutrophils and stimulates the *respiratory burst* (see later) and degranulation of these cells. C5a and C3a also upregulate adhesion molecule expression on neutrophils and endothelial cells, promoting extravasation. Finally, C5a stimulates macrophages and monocytes to secrete increased amounts of the pro-inflammatory cytokines IL-1 and IL-6. Because these cytokines can stimulate the proliferation of activated T cells, C5a plays an indirect role in adaptive immunity. All these effects are mediated by the binding of the anaphylatoxins to specific receptors expressed on various types of leukocytes.

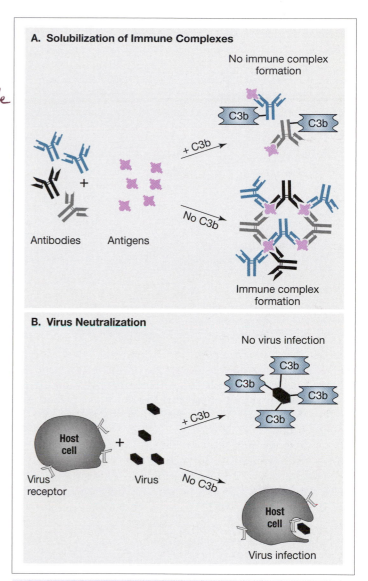

Fig. 3-6 **Some Other Functions of C3b**

vii) Control of Complement Activation

Because of the tremendous number of players involved and their non-specific destructive capacity, the complement cascade is rigidly organized and tightly regulated to minimize damage to host tissues. Firstly, complement enzymes are present in the plasma as *zymogens*; that is, the molecule is inactive until a specific fragment is cleaved off by the enzyme preceding it in the cascade. Secondly, the activated enzymes function for only a short time before being inactivated again. Thirdly, numerous regulatory and inhibitory molecules called *regulator of complement activation* (RCA) proteins ensure that the complement pathways are activated only where and when they should be (Table 3-3). Some of these RCA proteins are soluble inhibitors that circulate in the blood, while others are fixed in the plasma membranes of healthy host cells. In general, both types of RCA proteins act to prevent the assembly of C3 convertases and/or the MAC on healthy host cell surfaces. Occasionally, a partially assembled MAC containing the terminal

Table 3-3 Regulator of Complement Activation (RCA) Proteins

Abbr.	Name of RCA protein	Form	Function
CR1	Complement receptor 1	Membrane-bound	Binds to C4b and C3b to prevent formation of classical and alternative C3 convertases Promotes release of C2a from classical C3 convertases
DAF	Decay accelerating factor	Membrane-bound	Binds to C4b and C3b to prevent formation of classical and alternative C3 convertases Promotes release of C2a from classical C3 convertases
C4bp	C4 binding protein	Soluble	Binds to C4b to prevent C2 binding and formation of classical C3 convertase Promotes degradation of surface-bound C4b by Factor I
I	Factor I	Soluble	Degrades surface-bound C4b
H	Factor H	Soluble	Blocks binding of Factor B to surface-bound C3b Promotes dissociation of alternative C3 convertase Stimulates degradation of C3b by Factor I
MCP	Membrane cofactor protein	Membrane-bound	Blocks binding of Factor B to surface-bound C3b Promotes dissociation of alternative C3 convertase Stimulates degradation of C3b by Factor I
	Vitronectin	Soluble	Binds and inactivates free terminal complexes
	Clusterin	Soluble	Binds and inactivates free terminal complexes
HRF	Homologous restriction factor	Membrane-bound	Binds to C8 and prevents C9 addition
MIRL	Membrane inhibitor of reactive lysis	Membrane-bound	Binds to C8 and C9 and prevents C9 polymerization and pore formation

components C5b67 is released as a *free terminal complex* from the membrane of a pathogen. If not restrained by host regulatory proteins, a free terminal complex could penetrate into the membrane of a nearby healthy host cell and attempt to complete MAC formation.

III. THE CONCEPT OF "DANGER"

As described earlier in this chapter, the engagement of PRMs leads to induced innate responses when these molecules are bound by pathogen structures or host stress molecules released during tissue injury or in response to environmental stress. Some immunologists therefore refer to the ligands that bind to PRMs as "danger signals" (Table 3-4). Danger signals are molecules whose presence alerts the host's innate immune system that something is wrong in the body and remedial action is needed. The appropriate response may range from simple wound healing and debris clearance to innate and then adaptive immune responses to pathogenic threats. Bacterial or viral products (including lipopolysaccharide, peptidoglycan, viral RNA and bacterial CpG dinucleotides) that accumulate in sites of infection and engage the TLRs of innate leukocytes are danger signals, as are products of complement activation. A cell undergoing necrotic death, in which the plasma membrane is breached and harmful internal cell contents spill out into the surrounding tissues, is also a source of danger signals. Lastly, stress molecules or *reactive oxygen intermediates* (ROIs) generated by physically or metabolically stressed cells

Table 3-4 Danger Signals

Danger Signals	Examples
Bacterial products	LPS, peptidoglycan, CpG, mannose-bearing carbohydrates
Viral products	Double-stranded and single-stranded RNA
Complement products	C3b, C4b, iC3b
Reactive oxygen intermediates	H_2O_2, $OH\cdot$, O_2^-
Stress molecules	HSPs, chaperone proteins

also bind to PRMs and are considered by some immunologists to serve as danger signals. Once the PRRs of innate response cells are engaged by danger signals, the cells are activated and carry out the appropriate actions. For example, neutrophils degranulate, macrophages secrete pro-inflammatory cytokines summoning other leukocytes to the scene, $\gamma\delta$ T cells generate effector cells and NK cells enhance their production of IFNγ. As is explained in Chapter 7, the concept of "danger" also figures prominently in the initiation of adaptive responses, as the DCs that present pMHC to naïve T cells are first activated by an encounter with a molecule representing a danger signal.

IV. THE INFLAMMATORY RESPONSE

The *inflammatory response* or *inflammation* is a complex series of non-specific soluble and cellular events that promotes the elimination of a foreign entity. Inflammatory mediators produced during the response both draw innate and adaptive leukocytes to a site of attack and ensure that these cells are fully activated. Inflammation in a given tissue can be induced by pathogen attack, inert tissue injury, products of complement activation, or by cytokines released by innate or adaptive leukocytes that were activated elsewhere.

i) Clinical Signs

The clinical signs of inflammation include localized heat, redness, swelling and/or pain at a site of infection or injury. The heat and redness result from *vasodilation* (an increase in the diameter of local blood vessels), which allows increased blood flow into the affected area. An increase in leukocyte adhesion to local blood vessel walls and increased permeability of the capillaries in this area encourage an influx of leukocytes into the tissue, causing swelling. The pain results not only from the swelling but also from the stimulation of pain receptors in the skin by peptide mediators. The increased permeability of blood vessels also allows large molecules such as antibodies of the adaptive response and enzymes of the blood clotting system to leak into the affected tissue and initiate infection control and tissue repair. Blood clotting enzymes trigger the deposition of *fibrin* to form a blood clot and commence wound healing.

ii) Initiators and Mediators

The actual triggering of inflammation is not well understood. Factors released by injured cells, or pro-inflammatory cytokines produced by activated resident macrophages, are thought to play important roles (Table 3-5). Some factors act directly and locally to induce initial changes in blood vessel diameter and permeability, signaling that an inflammatory response is under way. Among these factors are the *kinins*, a family of small peptides that circulate in the blood in inactive form. Once activated by the blood clotting system or by enzymes released by damaged cells, kinins give rise to potent peptide mediators (including *bradykinin*) that cause vasodilation, increased vascular permeability, smooth muscle contraction and pain.

Other inflammatory mediators work in a more indirect way and, in some cases, their contributions are interconnected. For example, cells infected with viruses are triggered to produce IFNα/β that induce nearby uninfected cells to adopt an "antiviral state" and resist infection (see Ch. 13). IFNγ produced by NK cells and NKT cells activates macrophages. TNF, IL-1 and IL-6 secreted by stimulated NK cells, damaged tissue cells and activated macrophages and endothelial cells perpetuate the inflammatory response. Hepatocytes responding to TNF and IL-6 produce acute phase proteins such as CRP. CRP can initiate the complement cascade via C1 activation in a way that does not involve antibody. The C3b generated by complement activation coats microbes, making them vulnerable to opsonized phagocytosis (see later). C5a and C3a produced during complement activation can provoke the degranulation of mast cells and the consequent immediate release of preformed *heparin* and *vasoactive amines*. Heparin is a blood clotting inhibitor that helps maintain the required influx of soluble factors and leukocytes into the affected area. Vasoactive amines, such as *histamine*, induce vasodilation and increase vascular permeability. Finally, mediators called *leukotrienes* and *prostaglandins* increase vasodilation and promote neutro-

Table 3-5 Major Inflammatory Mediators

Mediator Class	Example	Principal Source	Principal Effect
Kinins	Bradykinin	Blood	Vasodilation and increased vascular permeability, smooth muscle contraction, pain
Pro-inflammatory cytokines	IFNs	Virus-infected cells	Adoption of antiviral state by uninfected cells, activation of macrophages and NK cells
	TNF, IL-1, IL-6	Activated macrophages and endothelial cells, stimulated NK cells, damaged tissue cells	Induction of acute phase protein synthesis by hepatocytes
Acute phase proteins	CRP	Activated hepatocytes	Activation of complement
Complement products	C5a C3a	Complement activation	Leukocyte chemotaxis, mast cell degranulation, smooth muscle contraction
Granule products	Heparin	Degranulation of mast cells, basophils	Prevention of blood clotting
	Histamine		Increased vascular permeability, smooth muscle contraction, chemotaxis
Leukotrienes Prostaglandins	Leukotriene B4 Prostaglandin 1	Phospholipid products of leukocyte membranes	Increased vascular permeability, neutrophil chemotaxis
Chemokines	IL-8, MIP-1α	Activated macrophages, monocytes, lymphocytes, endothelial cells	Leukocyte chemotaxis

phil chemotaxis during the later stages of inflammation. These molecules are not preformed and must be derived from the breakdown of phospholipids in the membranes of activated macrophages, monocytes, neutrophils and mast cells.

iii) Leukocyte Extravasation and Infiltration

The chemical messengers emanating from injured tissues prompt the extravasation of leukocytes from the blood. To aid in margination (refer to Fig. 2-16), endothelial cells respond to tissue damage or pro-inflammatory cytokines by becoming activated and upregulating selectins and other adhesion molecules that can bind to oligosaccharides or adhesion proteins on leukocyte surfaces. Tissue damage also increases the production of *platelet-activating factor* (PAF) on the endothelial cell surface. PAF stimulates the integrin-mediated adhesion of neutrophils to the postcapillary venule wall closest to the site of injury.

Once innate leukocytes have extravasated through the blood vessel wall, they employ chemotaxis to follow a gradient of chemotactic factors and migrate to the site of injury or infection (Fig. 3-7). Chemotactic factors in an inflammatory site can include the complement activation products C5a and C3a, various blood clotting proteins, PAF, leukotriene B4 (LTB4),

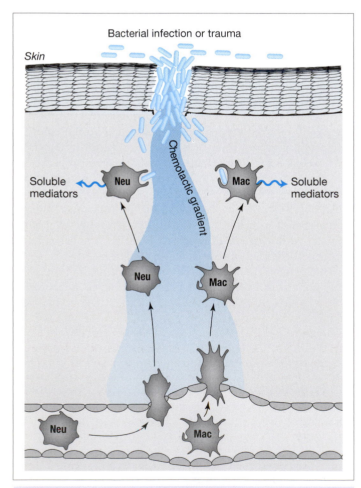

Bacterial infection or trauma

Skin

Chemotactic gradient

Soluble mediators — Neu

Mac — Soluble mediators

Neu   Mac

Neu   Mac

Fig. 3-7 Leukocyte Extravasation and Infiltration

microbial components and chemokines (see Box 3-2). Activated macrophages and neutrophils then vigorously attempt to engulf the foreign entity that has provoked the inflammatory response, and may release antimicrobial peptides called *defensins* that coat and protect body surfaces such as the skin or the intestinal lining. As well, innate leukocytes in an inflammatory site secrete additional chemokines that attract lymphocytes of the adaptive response.

V. CELLULAR INTERNALIZATION MECHANISMS

Engulfment by a cell followed by degradation within that cell disposes of the vast majority of unwanted entities encountered in the body. Primary immunodeficiencies that result from defects in these processes are outlined in Box 3-3.

i) Engulfment

There are three distinct processes for importing extracellular entities into a cell. *Macropinocytosis* and *clathrin-mediated endocytosis* involve the internalization of relatively small soluble macromolecules and are ongoing activities that can be carried out by any cell. *Phagocytosis*, which handles large extracellular particles (including bacteria and viruses), is carried out only by specialized phagocytic leukocytes, predominantly neutrophils, macrophages, monocytes and immature DCs.

ia) Macropinocytosis. In macropinocytosis (Fig. 3-8A), a cell ruffles its plasma membrane to form a small vesicle called a *macropinosome* around a droplet of extracellular fluid. Macropinosomes are highly variable in volume and contain fluid phase solutes, rather than insoluble particles. Macromolecules enter the vesicle with the extracellular fluid on a gradient of passive diffusion. The number of macromolecules caught in the vesicle depends entirely on their concentration. This process provides an efficient but non-specific means of sampling the extracellular environment.

ib) Clathrin-mediated endocytosis. In clathrin-mediated endocytosis (Fig. 3-8B), the uptake of a macromolecule depends on whether it binds to an appropriate cell surface receptor. Such binding triggers the polymerization of *clathrin*, a protein component of the microtubule network located on the cytoplasmic side of the plasma membrane. Invagination of clathrin-coated "pits" internalizes the receptor and its bound ligand into a small clathrin-coated vesicle.

ic) Phagocytosis. Phagocytosis is initiated when multiple cell surface receptors on a phagocyte bind sequentially in a "zippering" manner to ligands present on a large particle, whole microbe or apoptotic host cell (Fig. 3-9). The interactions between these receptors and ligands induce the polymerization of the cytoskeletal protein actin at the site of internalization. The plasma membrane then invaginates and forms a large vesicle called a *phagosome* around the entity to be removed.

Microbes and particles that do not interact with the phagocyte's surface receptors cannot be phagocytosed directly;

Box 3-2 Chemokines

The chemokines are a group of over 40 small (about 8–10 kDa) structurally homologous cytokines that have chemotactic activity. These monomeric proteins are categorized into subgroups based on the presence or absence of one or more intervening amino acids in a particular cysteine-containing motif (see Table). The Cys-Cys chemokines (*CC chemokines*) have no intervening amino acid between the two cysteine residues of the motif. CC chemokines are synthesized principally by activated T cells and tend to act on leukocytes other than neutrophils. Members of the Cys-X-Cys chemokine subgroup (*CXC chemokines*) have one intervening amino acid between the two cysteines. CXC chemokines are produced mainly by activated monocytes and macrophages and primarily stimulate neutrophil migration. Despite its original classification as an interleukin, IL-8 is a CXC chemokine that attracts neutrophils and basophils to inflammatory sites. All chemokines can bind to heparan sulfate in the host cell extracellular matrix and in the endothelial cell layer. This binding has the effect of displaying these chemoattractant molecules where leukocytes can recognize them. Several different chemokine receptors have been identified on various leukocyte subsets, some of which bind multiple chemokines. More on chemokines and their receptors appears in Appendix E.

Chemokine	Source	Chemotactic for
CC Subgroup		
MCP-1/CCL2* (monocyte chemotactic protein)	Monocytes, macrophages, T cells	Monocytes
MIP-1α/CCL3 (macrophage inflammatory protein 1-α)	Monocytes, macrophages, neutrophils, endothelial cells	Monocytes, macrophages, T cells, B cells, basophils, eosinophils
MIP-1β/CCL4 (macrophage inflammatory protein 1-β)	Monocytes, macrophages, neutrophils, endothelial cells	Monocytes, macrophages, naïve Tc cells, B cells
RANTES/CCL5 (regulated on activation, normal T cell expressed and secreted)	T cells, platelets	Monocytes, memory T cells, eosinophils, basophils
CXC Subgroup		
IL-8/CXCL8 (interleukin-8)	Monocytes, macrophages, endothelial cells, fibroblasts, neutrophils	Neutrophils, basophils, T cells
NAP-2/CXCL7 (neutrophil-activating protein)	Platelets	Neutrophils, basophils
GCP-2/CXCL6 (granulocyte chemotatic protein)	Macrophages, fibroblasts, chrondrocytes, endothelial cells	Neutrophils, NK cells
SDF-1/CXCL12 (stromal cell-derived factor)	Bone marrow stromal cells	T cells

*CCL (C-C ligand) and CXCL (C-X-C ligand) are the more recent systematic names for chemokines; see Appendix E.

Box 3-3 Primary Immunodeficiencies Due to Defects in Phagocyte Function or Development

Several primary immunodeficiencies are due to mutations of genes that encode proteins crucial for phagocyte function. **Leukocyte adhesion deficiency type I (LAD-I)** results from mutation of CD18, a protein that is the common subunit of a subset of adhesion molecules called the β2-integrins. Without functional β2-integrins, neutrophils cannot leave the circulation and extravasate into sites of infection or injury. **Leukocyte adhesion deficiency type II (LAD-II)** is caused by an extremely rare mutation of a transporter protein involved in neutrophil extravasation. **Chronic granulomatous disease (CGD)** is a clinically heterogeneous disorder caused by a failure in phagosomal killing. CGD is caused by mutations of the phagosomal NADPH oxidase required for an effective respiratory burst (refer to Fig. 3-12). **Chédiak–Higashi syndrome (CHS)** results from mutation of the *CHS1* gene that encodes a cytoplasmic protein involved in intracellular vacuole and granule fusion. The phagocytes of CHS patients contain abnormal vesicles that cannot fuse properly with phagosomes, so that pathogen killing is impaired. CHS also has a neurological component but its connection with the *CHS1* mutation is unclear. **Severe congenital neutropenia** and **cyclical neutropenia** are innate primary immunodeficiencies characterized by extremely low levels of circulating neutrophils. Both disorders are due to mutations in the neutrophil elastase gene. Elastase is an enzyme crucial for neutrophil differentiation from myeloid precursors.

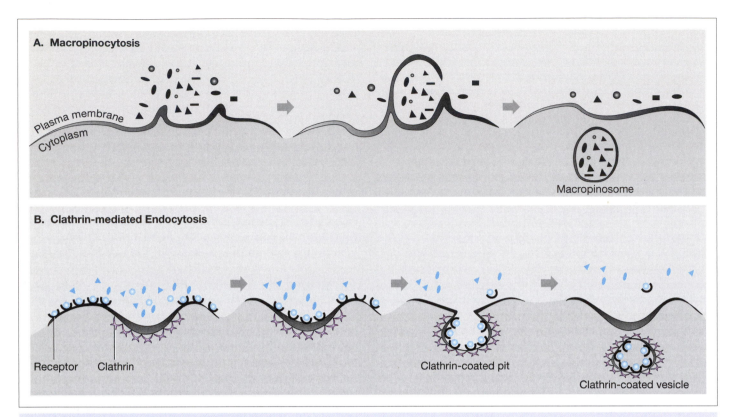

A. Macropinocytosis

Plasma membrane

Cytoplasm

Macropinosome

B. Clathrin-mediated Endocytosis

Receptor   Clathrin

Clathrin-coated pit

Clathrin-coated vesicle

Fig. 3-8 **Macropinocytosis and Clathrin-Mediated Endocytosis**

however, the range of targets is greatly expanded when microbes are coated with *opsonins*. An opsonin is a host-derived protein that binds to the exterior of a microbe and facilitates its engulfment by phagocytes expressing receptors for the opsonin (Fig. 3-10). The most common opsonins are the complement component C3b, which binds to particular motifs in pathogen surface proteins, and antibodies, which recognize pathogen surface antigens. When a microbe coated in C3b binds to CR1 on the surface of a phagocyte, this inter-action triggers phagocytosis of the invader. Similarly, a microbe coated in specific antibody can be bound by a phagocyte's *Fc receptors* (FcRs; receptors that recognize the Fc region of an antibody protein), initiating phagocytosis of the microbe–

antibody complex. When a pathogen is coated with more than one type of opsonin, phagocytosis is even more efficient.

ii) Endocytic Processing

A cell that engulfs a foreign molecule by macropinocytosis or clathrin-mediated endocytosis protects nearby cells from harm but must then dispose of its burden. Macropinosomes or clath-rin-coated vesicles enter directly into the *endocytic processing pathway*, an internal trafficking system for membrane-bound vesicles. The function of the endocytic processing pathway is to degrade the extracellular contents that have been imported into a cell while recycling any receptors and membrane com-ponents that were used for internalization. Entities captured

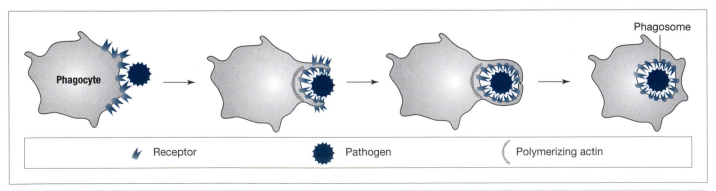

Phagosome

Phagocyte

| Receptor | Pathogen | Polymerizing actin |

Fig. 3-9 **Phagocytosis**

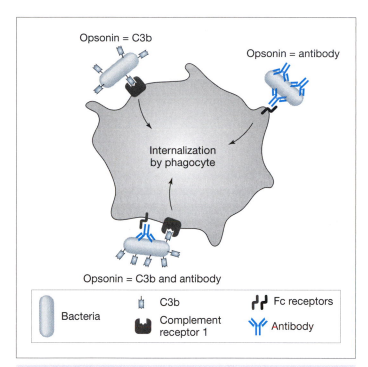

Fig. 3-10 **Opsonization**

by phagocytosis and retained in phagosomes are killed before the phagosome enters the endocytic pathway (see later).

The endocytic pathway has three major membrane-bound vesicular compartments: the *early endosomes, late endosomes* and *lysosomes* (Fig. 3-11). A macropinosome or a clathrin-coated vesicle containing a foreign macromolecule first fuses to an early endosome, which has a mildly acidic environment (about pH 6.5). In the case of clathrin-mediated endocytosis, this increase in pH promotes the dissociation of the receptor proteins from the captured macromolecule. The receptors are then collected in a tubular extension of the early endosome that buds off and fuses with the plasma membrane, returning the receptors to the cell surface. Early endosomes containing captured macromolecules then fuse with Golgi-derived vesicles to create late endosomes, resulting in a further decrease in internal pH. Fusion of late endosomes with additional trans-Golgi vesicles initiates the digestion of the macromolecule. Complete degradation occurs when the late endosome fuses with a lysosome, forming an *endolysosome*. The lysosome has an internal pH of 5 and a cargo of hydrolytic enzymes (including lipases, nucleases and proteases) that degrade the macromolecule to its fundamental components (fatty acids, nucleotides, amino acids, sugars etc.). The products of diges-tion are collected in an *exocytic* vesicle which buds off of the endolysosome. The exocytic vesicle fuses with the plasma membrane and the degraded material is expelled from the cell into the extracellular fluid in a process called *exocytosis*.

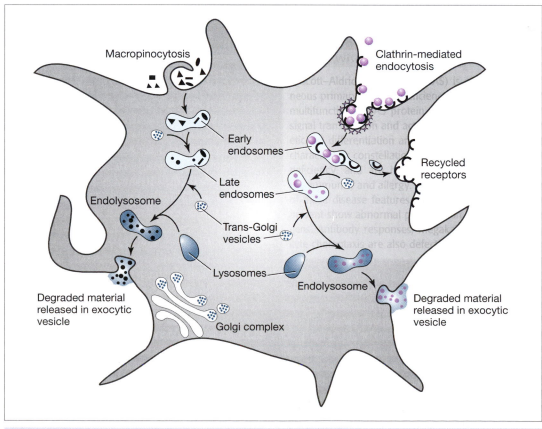

Fig. 3-11 **Model of Endocytic Processing**

iii) Phagosomal Killing and Phagolysosomal Maturation

A phagocyte that has engulfed an entire live microbe and sequestered it in a phagosome must kill the invader before digesting it into macromolecular components. This killing is frequently carried out via oxidation of the pathogen's component molecules. Chemically unstable free radicals present within the phagosome bind to pathogen proteins, oxidizing them such that their properties are altered and their normal functions are impaired. How does this oxidation occur? When a pathogen is captured in a phagosome (Fig. 3-12, #1), an enzyme in the phagosomal membrane called *protein kinase C* (PKC) activates another enzyme in the phagosomal membrane called *NADPH oxidase* (#2). Activation of NADPH oxidase produces an intense, rapid and short-lived increase in O_2 consumption that is called the *respiratory burst* (#3). The respiratory burst generates superoxide anion (O_2^-) that is further reduced to reactive oxygen intermediates (ROIs) such as hydrogen peroxide (H_2O_2) and hydroxyl radicals (OH·) (#4). A phagocyte that has engulfed a foreign entity and become activated also upregulates iNOS and produces NO. When NO reacts with the superoxide anion generated by the respiratory burst, toxic *reactive nitrogen intermediates* (RNIs) are produced (#5). ROIs and RNIs are potent oxidizing agents that interfere with pathogen metabolism and replication, eventually causing pathogen destruction (#6). Should the damaging ROIs or RNIs escape from the phagosome into the cytoplasm, the phagocyte protects itself by activating several enzymes capable of neutralizing these reactive molecules. ROIs and RNIs may also be released extracellularly by phagocytes to combat pathogens in the local environment.

Once a microbe contained within a phagosome is dead, the phagosome enters the endocytic processing pathway and undergoes a stepwise "maturation" to eventually form a *phagolysosome* (Fig. 3-13). Like an endolysosome, the phagolysosome is capable of degrading microbe structures. Maturation involves a series of fusion and fission events between the phagosome and various Golgi-derived vesicles, intracellular vacuoles and granules, early and late endosomes and lysosomes. In many cases, the receptors that initiated the internalization are recycled to the cell surface. The mixing of the contents of successively recruited vesicles and granules drastically decreases the internal pH and alters the protein composition of the maturing phagolysosome, enhancing its ability to degrade the foreign entity. Most of the degradation products are then harmlessly exocytosed out of the phagocyte into the extracellular fluid.

VI. NK, γδ T AND NKT CELL ACTIVITIES

NK, γδ T and NKT cells are leukocytes that bridge innate and adaptive immunity. Although all three of these cell types are closely related to αβ T cells, their responses to injury or

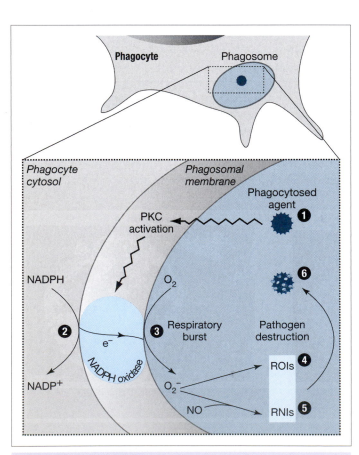

Fig. 3-12 Phagosomal Killing

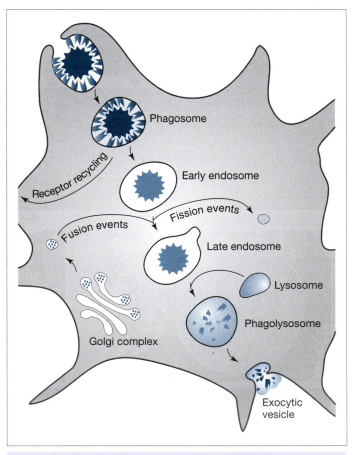

Fig. 3-13 Model of Phagolysosome Maturation in a Phagocyte

infection are rapid and involve broadly specific ligand recognition, marking them as players in innate immunity. The detailed characterization of NK, γδ T and NKT cells has been relatively recent and occurred mainly after the biology of B cells and αβ T cells was well established. Early analyses of NK, γδ T and NKT cell receptors showed that the diversity of these molecules was much less than that of TCRs and BCRs, leading some scientists to conclude that NK, γδ T and NKT cells were lymphocytes with minor roles in the adaptive response. However, more recent work has shown that the relatively limited diversity in ligand binding exhibited by NK, γδ T and NKT cells is essential to their primary physiological function as sentinels on the front lines of innate defense. When the receptors of NK, γδ T and NKT cells are engaged by the appropriate PAMP or stress molecule, these cells produce cytokines that stimulate both the innate and adaptive responses. NK cells and γδ T cells also mediate the cytolysis of infected cells and thus contribute directly to innate defense against certain pathogens. Since the biology of NK, γδ T and NKT cells and their receptors has been elucidated and defined relative to that of αβ T cells (described in Chapter 9), we have reserved a complete discussion of NK, γδ T and NKT cells until Chapter 11.

The innate response deals with foreign entities in a broadly specific manner and is sufficient to counter most of the threats our bodies encounter every day. However, when the innate response is overwhelmed, the more focused mechanisms of adaptive immunity are required. Many of the cells activated in the course of an innate response then act as bridges to the adaptive response, either directly interacting with T cells or producing cytokines needed to support T and B cell activation. The next two chapters, Chapters 4 and 5, discuss the biology of B cells and their genes and proteins, and the mechanisms of the humoral adaptive response.

CHAPTER 3 TAKE-HOME MESSAGE

- **Innate immunity encompasses non-induced and induced mechanisms, including anatomical and physiological barriers, complement activation, inflammatory responses, cytokine secretion, target cell lysis and cellular internalization mechanisms such as phagocytosis.**

- **Collectively, innate mechanisms inhibit pathogen entry, prevent the establishment of infection and clear both host cell and microbial debris from the body.**

- **Some innate mechanisms are completely antigen non-specific, whereas others involve broadly specific pattern recognition by soluble or membrane-fixed molecules.**

- **Innate immunity either succeeds in eliminating the pathogen, or holds infection in check until the slower adaptive immune response can develop.**

- **Innate immune responses also result in the activation of cells that support the adaptive response as well as the production of soluble factors (particularly cytokines) that are critical for the recruitment, activation and differentiation of lymphocytes.**

- **The major types of leukocytes mediating induced innate immunity are neutrophils, DCs, macrophages, mast cells, NK cells, NKT cells and γδ T cells.**

Section A

1) Can you define these terms? hypodermis, lysozyme, commensal organisms

2) Give three examples of non-induced innate defense and explain how they help prevent infection.

Section B.I

1) Can you define these terms? PAMP, PRR, PRM, TLR, stress molecule, HSP, scavenger receptor, CRP, NOD protein

2) How is the induced innate response initiated and how does this differ from the adaptive response?

3) Why is the repertoire diversity of the induced innate response more limited than that of the adaptive response?

4) Name two functions of TLRs in mammals.

5) Name three types of soluble PRMs and describe how they contribute to induced innate defense.

Section B.II–III

1) Can you define these terms? C3 convertase, C5 convertase, Factor B, properdin, MBL, MAC, terminal components, RCA protein.

2) Name four outcomes of complement activation.

3) Describe how the three pathways of complement activation differ in their initiation.

4) Which pathway of complement activation is considered part of the adaptive response and why?

5) Name three functions of C3b.

6) What is an anaphylatoxin and what does it do?

7) Name four RCA proteins and describe how they control the complement cascade.

8) Deficiency for which complement component results in susceptibility to pyrogenic infections?

9) What is a danger signal? Give three examples.

Section B.IV

1) Can you define these terms? inflammatory response, vasodilation, chemotaxis, defensin.

2) Describe three clinical signs of inflammation and how they arise.

3) Can you give four examples of major inflammatory mediators?

4) How does extravasation underpin the inflammatory response?

5) What are two major differences between the CC and CXC chemokines?

Section B.V

1) Can you define these terms? macropinosome, clathrin, endosome, phagosome, phagolysosome, exocytosis, ROI, RNI, iNOS.

2) Can you describe the three mechanisms of cellular internalization and how they differ?

3) What is opsonization and how is it helpful to immune defense? Give two examples of opsonins.

4) Can you describe the endocytic processing pathway?

5) What is the respiratory burst and why is it important?

6) Can you describe three primary immunodeficiencies arising from defects in phagocytosis?

Section B.VI

1) Why are NK, $\gamma\delta$ T and NKT cells considered to bridge the innate and adaptive responses?

4

The B Cell Receptor: Proteins and Genes

WHAT'S IN THIS CHAPTER?

Great acts are made up of small deeds.
Lao Tzu

Adaptive immunity depends in large part on the production and effector functions of antibodies. As introduced in Chapter 1, the binding of specific antigen to the BCR of a B lymphocyte induces the activation, proliferation and differentiation of that cell into both memory B cells and plasma cells that synthesize antibody able to bind to the antigen. The antibody protein itself is a modified form of the membrane-bound BCR expressed by the original B lymphocyte. Both the BCR and the antibody have the same antigenic specificity, and both are forms of *immunoglobulin* (Ig) proteins. This chapter focuses on the structure and function of Ig proteins as well as the Ig genes encoding them.

A. Immunoglobulin Proteins

I. THE NATURE OF IMMUNOGLOBULIN PROTEINS

How did antibodies and immunoglobulins get their names? In the late 1800s, scientists discovered that the serum component of blood could transfer immunity to toxins from an immunized animal to an unimmunized one. The serum preparation from the immunized animal was called the "antiserum", and the active transferable agent in the antiserum that conferred the immunity on the second animal was called an "antitoxin". The label "antitoxin" was changed to "antibody" in the 1930s to account for the fact that transferable agents inducing immunity were also produced in response to substances other than toxins. During the next few decades, scientists made routine use of electrophoresis as a technique for characterizing proteins: a mixture of proteins was introduced into a semisolid medium (such as a starch, agar or polyacrylamide gel) and separated on the basis of size or charge by application of an electric current. When normal serum was electrophoresed, its proteins separated into a large albumin fraction and three smaller α, β and γ globulin protein fractions (Fig. 4-1A). Upon immuniza-

tion, a dramatic increase in the γ globulin peak was observed and attributed to the production of antibodies (Fig. 4-1B), leading to the designation of these proteins first as "gamma globulins" and then as "immunoglobulins" to acknowledge their role in immunity.

II. STRUCTURE OF IMMUNOGLOBULIN PROTEINS

i) Basic Structure

In both humans and mice, all immunoglobulin molecules are large glycoproteins with the same basic core structure: two heavy (H) chains and two light (L) chains that come together to form a Y-shaped molecule (Fig. 4-2). This structure is often abbreviated as "H_2L_2". Genetically, an individual B cell can produce only one kind of H chain and only one kind of L chain, so that the two H chains in a given Ig molecule are identical and the two L chains are identical. In general, the two H chains are joined to each other by two or more covalent disulfide bonds, and each H chain is joined to an L chain by an additional disulfide bond. These disulfide bonds vary in number and position between different types of Ig molecules and among species. The N-termini of the H and L chains come together in H–L pairs to form two identical antigen-binding sites. The much less variable C-terminal end of the Ig molecule contains that part of the antibody that mediates the antibody's effector functions. Brief digestion of an Ig protein with the protease papain results in two identical *Fab* fragments, so called because they retain the antibody's <u>a</u>ntigen-<u>b</u>inding ability (Fig. 4-3). The remaining fragment of the Ig protein is called the *Fc* region because it <u>c</u>rystallizes at low temperature. It is the Fc region that interacts with the body's antigen clearance mechanisms.

ii) Constant and Variable Domains

If the basic structure of all Igs is the same, how do the receptors of different B cells recognize different antigenic structures?

A. Normal Serum

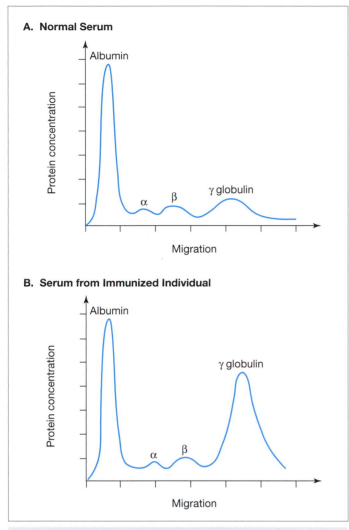

B. Serum from Immunized Individual

Fig. 4-1 Electrophoretic Fractionation of Serum Proteins
[Adapted from Tisselius A. and Kabat E. A. (1939). An electrophoretic study of immune sera and purified antibody preparations. *Journal of Experimental Medicine* **69**, 119–131.]

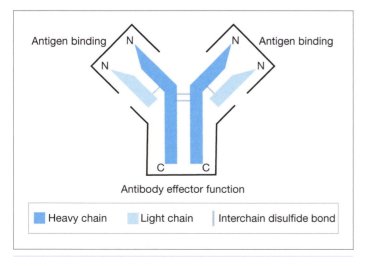

Fig. 4-2 Basic H₂L₂ Structure of the Immunoglobulin Molecule

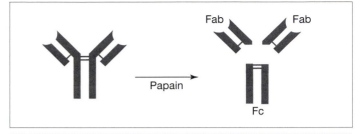

Fig. 4-3 Antibody Fragments Produced by Limited Proteolytic Digestion

Comparison of the complete amino acid sequences of individual Ig molecules reveals a vast diversity in the *variable* (V) *domains* making up the N-termini of the H and L chains (Fig. 4-4A). In contrast, the *constant* (C) *domains* that make up the rest of the H and L chains are relatively conserved. Each L chain contains one V and one C domain, denoted V_L and C_L, respectively. Similarly, each H chain contains one V_H and three or four C_H domains. Within an Ig molecule, the H and L polypeptides are aligned such that the V domains and C domains in the L chain (V_L and C_L) are positioned directly opposite their counterparts in the H chain (V_H and C_H1). Thus, two *V regions* are formed by the pairing of the V_L domain of an L chain with the V_H domain of the H chain with which it is associated. The *C region* of the Ig comprises all the C_L and C_H domains of the H and L chains.

All V and C domains in H and L chains are based on a common structural unit known as an *Ig domain* (Fig. 4-4B). An Ig domain is about 70–110 amino acids in length with cysteine residues at either end that form an intrachain disulfide bond. The amino acid sequences of Ig domains are not identical but are very similar. Within a given Ig domain, the linear amino acid sequence folds back on itself to form an identifiable and characteristic cylindrical structure known as the *immunoglobulin barrel* or *immunoglobulin fold*. Many other proteins involved in intermolecular and intercellular interactions both within the immune system and in other biological processes in the body also contain Ig-like domains with a similar barrel structure. These proteins are all said to be members of the *Ig superfamily*. The presence of the Ig fold in an Ig superfamily member confers enhanced structural stability in hostile environments. Thus, antibodies are remarkably stable under conditions of extreme pH (such as occur in the gut) and are resistant to proteolysis under natural conditions.

As well as stability, Ig molecules are capable of remarkable physical flexibility (Fig. 4-4C). In most Ig molecules, the portion of the H chain between the first and second C_H domains (where the Fab region joins the Fc region at the center of the Y) is somewhat extended and is known as the *hinge region*. The hinge region contains many proline residues, which impart rigidity to the central part of the molecule. However, glycine residues in the hinge region create a flexible secondary structure such that the Fab regions and the Fc region of the Ig molecule may "wag", bend or rotate independently of each other around the proline-stabilized anchor point.

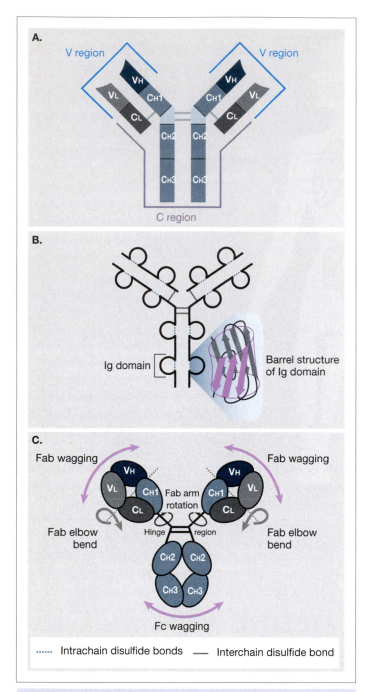

Fig. 4-4 Structural Features of the Immunoglobulin Molecule
[Part C: With information from Brekke O.H. (1996). The structural requirements for complement activation by IgG: does it hinge on the hinge? *Immunology Today* 16, 85–90.]

III. STRUCTURAL VARIATION IN THE V REGION

If one compares the amino acid sequences of the V domains of the L and H chains making up a collection of many different Ig molecules, one finds three short stretches in each chain that exhibit extreme amino acid sequence variability among Ig proteins (Fig. 4-5A). These three short sequences of at least 5–7 amino acids each are called the *hypervariable regions*. The hypervariable regions are responsible for the diversity that allows the total repertoire of Igs to recognize almost any molecule in the universe of antigens. The hypervariable regions are separated by four *framework regions* (FRs) of much more restricted variability. Although the hypervariable regions are not contiguous in the amino acid sequence of the Ig, they are brought together when the Ig protein folds into its native conformation. Each of the two antigen-binding sites of the Y-shaped Ig molecule is then formed by three-dimensional juxtaposition of the three hypervariable regions in the V_L domain with those in the associated V_H domain. Because these sequences result in a structure that is basically complementary to the shape of the specific antigen bound by the Ig, the hypervariable regions are also called *complementarity-determining regions* (CDRs). The CDRs project from the relatively flat surfaces of the framework regions as loops of varying sizes and shapes, and it is these loops that interact with specific antigen (Fig. 4-5B). The framework regions are thought to position and stabilize the CDRs in the correct conformation for antigen binding.

IV. STRUCTURAL VARIATION IN THE C REGION

In contrast to a V domain, the amino acid sequence of a C domain shows very little variation among antibodies because it is the C domain that interacts with the invariant molecules responsible for the body's antigen clearance mechanisms. In this way, antibodies are capable of triggering the same effector action in response to a wide variety of antigens. Nevertheless, an Ig of a given antigenic specificity can display two types of variation in its constant region: it can differ in its *isotype* and in its *structural isoform*.

i) Isotypes

Relatively minor differences in the constant domains of H and L chains give rise to Ig *isotypes*, or constant region classes. Differences in isotype can affect the size, charge, solubility and structural features of a particular Ig molecule. These factors in turn influence where that Ig goes in the body and how it interacts with surface receptors and antigen clearance molecules.

ia) Light chain isotypes. Humans and mice have two types of L chains, the kappa (κ) light chain and the lambda (λ) light chain (Fig. 4-6). All antibodies containing the κ light chain are of the κ isotype, and all antibodies containing a λ light chain are of the λ isotype. In both humans and mice, there is a single major type of κ chain. However, there may be three different λ chains expressed in a mouse, and four to six different λ chains in a human. Although any κ or λ chain can be combined with essentially any type of H chain in an Ig molecule, any one B cell expresses antibodies containing only the κ chain or one type of λ chain. In an individual human's repertoire of antibody-producing B cells, 60% produce antibodies with κ light chains and 40% produce antibodies with λ light chains. Because there are no known functional differences between κ- and λ-containing antibodies, L chains are considered to contribute to the diversity of antibodies but not to their effector functions.

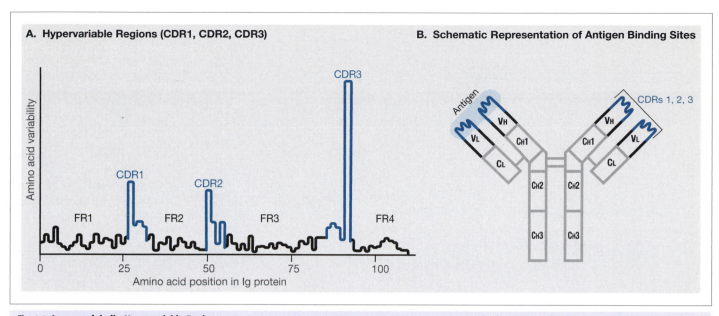

A. Hypervariable Regions (CDR1, CDR2, CDR3)

CDR3

CDR1   CDR2

Amino acid variability

FR1   FR2   FR3   FR4

0   25   50   75   100

Amino acid position in Ig protein

B. Schematic Representation of Antigen Binding Sites

Antigen   CDRs 1, 2, 3

V_H   V_H
V_L   C_H1   C_H1   V_L
C_L   C_L
C_H2   C_H2
C_H3   C_H3

Fig. 4-5 Immunoglobulin Hypervariable Regions
[Part A adapted from Wu T.T. and Kabat E.A. (1970). An analysis of the sequences of the variable regions of Bence Jones proteins and myeloma light chains and their implications for antibody complementarity. *Journal of Experimental Medicine* **132**, 211–250.]

Light chain isotype		% Occurrence in adult	
	κ	Human	60
		Mouse	5
	λ	Human	40
		Mouse	95

Fig. 4-6 Light Chain Isotypes
[With information from Gorman J.R. and Alt F.W. (1998). Regulation of Ig light chain isotype expression. *Advances in Immunology* **69**, 157.]

ib) Heavy chain isotypes. Although the C_H regions of different Ig molecules do not show the sequence variation seen among V_H regions, there are five different types of C_H regions that are distinguished by subtle amino acid differences and allow Igs to engage different antigen clearance mechanisms. The five H chain isotypes are defined by polypeptides called μ, δ, γ, ϵ and α that determine whether an Ig is an IgM, IgD, IgG, IgE or IgA molecule, respectively. The structural differences between H chains are such that the μ and ϵ chains of human IgM and IgE antibodies contain four C_H domains, whereas IgD, IgG and IgA antibodies contain only three C_H domains in their shorter δ, γ and α H chains (Fig. 4-7). The amino acid sequences of the C_H1, C_H3 and C_H4 domains in IgM and IgE correspond to the amino acid sequences of the C_H1, C_H2 and C_H3 domains in IgD, IgG and IgA. Variations among the sequences of the α and γ H chains in humans, and among γ sequences in mice, have given rise to Ig subclasses: IgA1, IgA2, IgG1, IgG2, IgG3 and IgG4 in humans; and IgG1, IgG2a, IgG2b and IgG3 in mice. Thus, a total of nine Ig H chain isotypes can be found in humans, and eight in mice. (It should be noted that isotypes are not strictly correlated across species; for example, human IgG3 is equivalent to mouse IgG2b, not mouse IgG3.) The Ig H chain isotypes also vary in their carbohydrate content. These oligosaccharide side chains, which are generally attached to amino acids in the C_H domains but not in the V_H, V_L or C_L domains, are thought to contribute to the stability of the antibody.

ic) Isotype switching. When a naïve B cell first encounters antigen, its initial set of progeny plasma cells produce only IgM antibodies. However, plasma cells of this clone that are generated later in the response can produce Igs of isotypes other than IgM. These C region changes are due to *isotype switching*, in which the DNA encoding the V region of the Ig H chain protein is reshuffled to combine with the DNA encoding C_H region sequences other than that of IgM (see Ch. 5). Thus, toward the end of the primary response, the B cell clone can switch to making IgG, IgE or IgA antibodies of the same antigenic specificity (because the V region has not changed) (Fig. 4-8). (IgD, which is rarely secreted, is not generated by isotype switching and is discussed in Chapter 5.) Both memory and plasma cells capable of making the new isotypes are generated. In a subsequent response to the antigen, a memory B cell of this clone can be triggered to proliferate and differentiate into plasma cells that secrete antibodies of the new isotype. Indeed, progeny cells of the memory B cell may switch isotypes again as the cycle of proliferation and differentiation repeats.

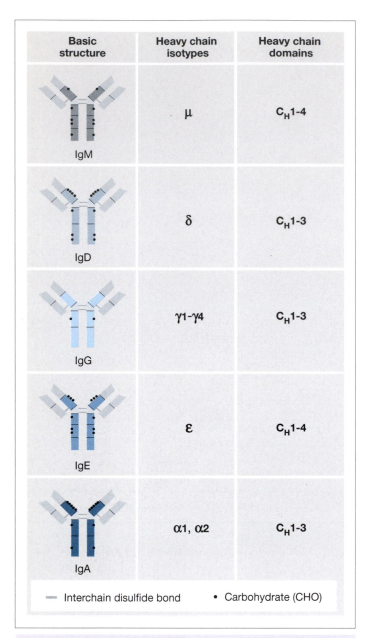

Basic structure	Heavy chain isotypes	Heavy chain domains
IgM	μ	C_H1-4
IgD	δ	C_H1-3
IgG	γ1–γ4	C_H1-3
IgE	ε	C_H1-4
IgA	α1, α2	C_H1-3

— Interchain disulfide bond   • Carbohydrate (CHO)

Fig. 4-7 **Human Heavy Chain Isotypes**

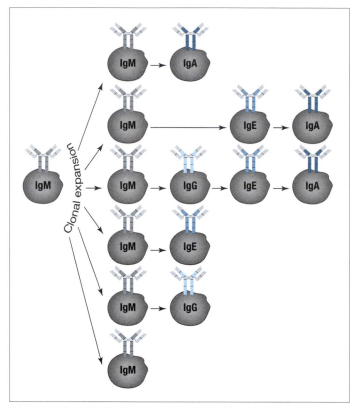

Fig. 4-8 **Examples of Isotype Switching**

This capacity of the host to produce antibodies of different isotypes ensures that all effector mechanisms can be brought to bear to eliminate an antigen.

ii) Structural Isoforms of Immunoglobulins

As noted previously, the basic core structure of any Ig is H_2L_2. Depending on the stage of activation of a given B cell and the cytokine cues it receives from its immediate microenvironment, three different isoforms of this basic core structure can be produced (Fig. 4-9). These structural isoforms, which vary in amino acid sequence at their C-termini (but not at their N-termini), perform distinct functions in the body. *Membrane-bound* Igs serve as part of the BCR complex. *Secreted* (or serum) Igs serve as circulating antibody in the blood. *Secretory*

Igs are secreted antibodies that undergo modifications that enable them to enter and function in the external secretions of the body, such as in tears and mucus. Although their antigenic specificities are identical, a membrane-bound Ig protein does not "turn into" a secreted one, or vice versa. The production of a membrane-bound or secreted Ig is regulated by the B cell at the level of transcription. RNA transcripts of the H chain gene that are slightly different at their 3′ (C-terminal) ends are produced, resulting in different structural forms of the Ig protein (see later). This control step occurs prior to protein synthesis.

iia) Membrane-bound immunoglobulins. The membrane-bound Ig protein (often denoted "mIg") of a B cell is very similar in form to the antibody its progeny cells will later secrete. However, compared to secreted antibody, mIg has an extended C-terminal tail and lacks a short tailpiece (see later). Membrane-bound Ig is ultimately situated in the plasma membrane such that its extracellular domain containing the Fab regions is displayed to the exterior of the cell, and a *transmembrane* (TM) domain extending beyond the last C_H Ig domain spans the plasma membrane and dangles the *cytoplasmic* (CYT) domain of the Ig protein into the cytoplasm (refer to Fig. 4-9A). How is the mIg fixed in the membrane? The Ig H chain gene can be transcribed such that the resulting mRNA includes the sequences that encode the TM domain. The TM domain characteristically contains amino acids with hydrophobic side chains that interact with the lipid bilayer. Thus,

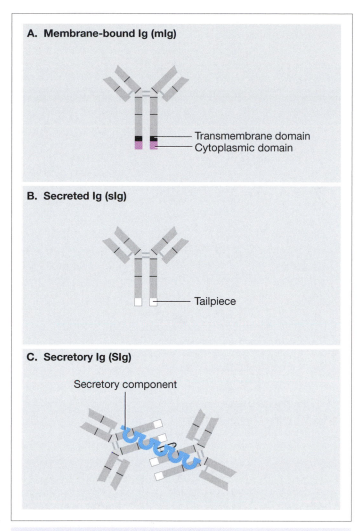

Fig. 4-9 **Structural Isoforms of Immunoglobulins**

A. Membrane-bound Ig (mIg)

Transmembrane domain
Cytoplasmic domain

B. Secreted Ig (sIg)

Tailpiece

C. Secretory Ig (SIg)

Secretory component

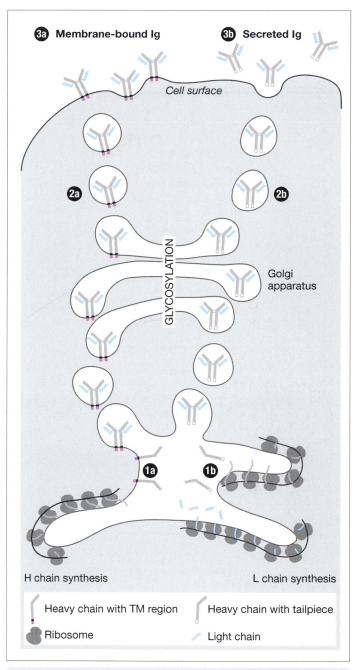

Fig. 4-10 **Synthesis of Membrane-Bound and Secreted Immunoglobulins**

3a Membrane-bound Ig   **3b** Secreted Ig

Cell surface

2a   **2b**

GLYCOSYLATION

Golgi apparatus

1a   **1b**

H chain synthesis   L chain synthesis

Heavy chain with TM region   Heavy chain with tailpiece

Ribosome   Light chain

when the Ig H chain mRNA is translated and the corresponding protein is synthesized on the *rough endoplasmic reticulum* (rER), the hydrophobic residues of the TM domain cause the H chain to become anchored in the ER membrane (Fig. 4-10, #1a). After the L chains associate with the H chains, the anchored mIg passes through the Golgi apparatus and into a vesicle that transports the mIg to the cell surface (Fig. 4-10, #2a). The membrane of the transport vesicle then fuses with the plasma membrane of the B cell, positioning the mIg protein in the membrane (Fig. 4-10, #3a).

iib) Secreted immunoglobulins. The mature plasma cell progeny of an activated B cell secrete the shortest and simplest form of an Ig as antibody, often denoted "sIg" (for "secreted"). An sIg antibody has the same antigenic specificity (same N-terminal sequences) as the mIg in the BCR of the original B cell that was activated by specific antigen, but the C-terminal sequences of the Ig end shortly after the last C_H domain. The mRNA encoding an sIg protein lacks the sequences for the TM and CYT domains and instead specifies a short amino acid sequence called the *tailpiece,* which lies C-terminal to the last

C_H domain and facilitates secretion (refer to Fig. 4-9B). During Ig synthesis in the rER, Ig chains that are synthesized with only the tailpiece (and not the TM domain) are not fixed in the ER membrane (Fig. 4-10, #1b). Rather, these Igs are sequestered as free molecules inside secretory vesicles that bud off from the Golgi (Fig. 4-10, #2b). These secretory vesicles fuse with the plasma membrane of the cell, releasing the free sIg molecules as antibodies into the extracellular environment (Fig. 4-10, #3b).

iic) Polymeric immunoglobulins. The basic H_2L_2 unit of an Ig is sometimes referred to as the Ig "monomer". In most cases, mIg is present in the BCR complex in monomeric form. However, secreted IgM and IgA frequently form soluble polymeric structures because their tailpieces allow interactions between the H chains of several IgM or IgA monomers. Most polymeric forms of IgM and IgA contain *joining* (J) *chains*, which are small acidic polypeptides that bond to the tailpieces of µ or α H chains by disulfide linkages. The J chains are not structurally related to Igs and are encoded by a separate genetic locus. Although a single J chain appears to be able to stabilize the component monomers of polymeric Igs, it is not crucial for the joining event. Typically, five IgM monomers congregate to form a pentamer, whereas two to three IgA monomers form dimers or trimers, respectively (Fig. 4-11).

iid) Secretory immunoglobulins. Most IgA antibodies are found in the external secretions (such as tears, mucus, breast milk and saliva) rather than in the blood. How do they get there? Newly synthesized polymeric sIgA antibodies containing tailpieces and J chains are released into the tissues underlying the mucosae by IgA-secreting plasma cells that have homed to this location (see Ch. 12). The sIgA molecules then bind to a receptor called the *poly-Ig receptor* (pIgR) expressed by mucosal epithelial cells that line body tracts or glandular ducts. The pIgR is present on that surface of the mucosal epithelial cells facing the tissues, rather than the surface facing the lumen of the gland or body tract (Fig. 4-12). The pIgR recognizes the J chain present in polymeric sIgA molecules and binds to the C-terminal domain of sIgA, triggering receptor-mediated endocytosis of the pIgR-bound Ig multimer into a transport vesicle. This vesicle conveys the sIgA through the mucosal epithelial cell to its opposite (lumen-facing) side in a process called *transcytosis*. The membrane of the transport vesicle fuses with the plasma membrane of the epithelial cell at its luminal surface and releases the vesicle contents into the secretions either

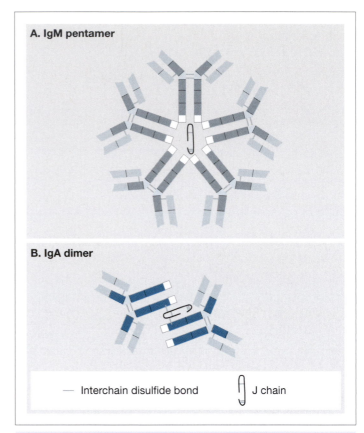

A. IgM pentamer

B. IgA dimer

— Interchain disulfide bond   J chain

Fig. 4-11 **Polymeric Immunoglobulins**

coating the body tract or produced by a gland. However, as the IgA molecule is expelled from the transport vesicle, a part of the pIgR molecule is enzymatically released from the vesicle membrane and remains attached to the polymeric IgA mole-

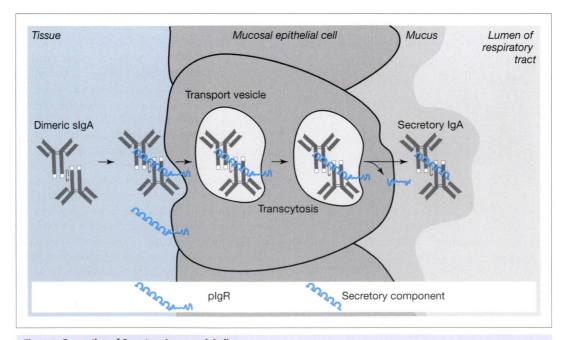

Fig. 4-12 **Generation of Secretory Immunoglobulins**

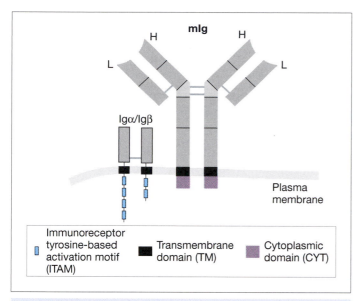

Fig. 4-13 **The BCR Complex**

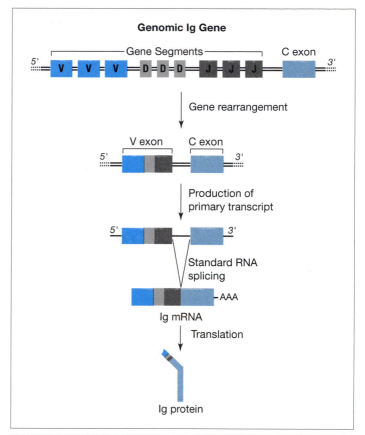

Fig. 4-14 **Overview: From Ig Gene to Ig Protein**

cule as it enters into the secretions. This heavily glycosylated piece is called the *secretory component* and the antibody then becomes known as secretory IgA (SIgA) (refer to Fig. 4-9C).

iii) The BCR Complex

Although an mIg provides the specificity for antigen recognition, the C-terminal cytoplasmic sequences of the Ig protein are so short that other molecules are needed to help convey the intracellular signals indicating that antigen has bound. The tyrosine kinases and phosphatases that carry out the actual enzymatic reactions of intracellular signaling simply cannot bind to the short Ig tails. In addition, the Ig tails contain none of the molecular motifs generally associated with intracellular signaling. Instead, the mIg molecule associates with a disulfide-bonded heterodimer composed of two glycoprotein chains called *Igα* and *Igβ* (Fig. 4-13). These chains, which are expressed only in B cells, are cosynthesized with the mIg molecule and coinserted with it into the rER membrane. An Ig-like domain in the extracellular regions of the Igα and Igβ proteins allows the heterodimer to associate with the C$_H$ domains of any mIg isotype. The long cytoplasmic tails of Igα and Igβ contain tyrosine-rich amino acid motifs called *immunoreceptor tyrosine-based activation motifs* (ITAMs). ITAMs allow the tail of a protein to recruit kinases and other molecules needed to transduce intracellular signaling and activate nuclear transcription factors. These nuclear transcription factors initiate the new gene transcription needed for the activation and differentiation of the B cell.

B. Immunoglobulin Genes

I. OVERVIEW

The antibody repertoire produced over an individual's lifetime includes specificities for essentially any antigen in the universe.

How does the plasma cell population produce this plethora of different antibody proteins, each capable of binding to essentially a single specific antigen? The mIgs in the BCRs already exist on the surfaces of B cells prior to antigen exposure, meaning that it is not antigen that drives the establishment of the antibody repertoire. (Of course, clonal selection by antigen later determines which B cells are activated to produce and secrete antibody.) As well, the antibody repertoire cannot be directly encoded in the DNA of B cells, because the nucleus simply does not have sufficient volume to contain the vast numbers of Ig genes that would be required to individually encode all the Ig proteins of the antibody repertoire. (It turns out that our genomes do not contain any complete Ig genes at all.) Instead, the Ig loci have evolved such that a relatively small amount of DNA can be organized to encode an enormous number of Ig proteins.

In a given developing B cell, the V region of its Ig protein is encoded by *variable* (V) *exon* DNA that is physically separated from the *constant* (C) *exon* DNA encoding the C region of the protein. Moreover, the V exon is assembled from a vast collection of smaller DNA fragments called *gene segments,* two or three of which are physically joined in the genome of each B cell precursor as it matures (Fig. 4-14). There are three types of gene segments: *variable* (V), *diversity* (D) and *joining* (J). These gene segments are randomly brought together by the gene rearrangement process that was introduced in Chapter 1 as somatic recombination and is also known as *V(D)J recombination*. As a result, a huge variety of unique V exons is generated within the B cell precursor population. Within a given developing B cell, the unique V exon and the C exon are transcribed together to give a primary mRNA transcript that is subsequently spliced to generate an mRNA encoding a complete and unique Ig protein.

II. STRUCTURE OF THE Ig LOCI

Ig proteins are encoded by three large genetic loci, the *Igh* (H chain), *Igk* (kappa L chain) and *Igl* (lambda L chain) loci, which are located on different chromosomes (Table 4-1). The complex structures of these loci in the genomes of humans and mice are shown in Figures 4-15 and 4-16, respectively. The *Igk* and *Igl* loci contain V and J segments, while the *Igh* locus contains not only V and J segments but also D gene segments. The estimated numbers of each type of gene segment in humans and mice are given in Table 4-2. In a mature B cell, the V exon encoding the variable region of an H chain is composed of a V, a D and a J gene segment, while the V exon encoding the

variable region of an L chain contains a V and J gene segment. Each type of gene segment contributes coding information for a particular set of amino acids in the V region of an Ig protein.

In contrast to the V exon, the C exon of an Ig polypeptide is not formed by the rearrangement of smaller gene segments. Although the *Igk* locus contains a single Cκ exon, the *Igl* locus contains multiple Cλ loci: seven in human and four in mouse. In a given individual, a variable number of these Cλ exons may be pseudogenes. The *Igh* locus contains multiple C_H exons:

Table 4-1 Chromosomal Location of Ig Genes

Genetic Locus	Chromosome	
	Human	Mouse
Igh	14	12
Igk	2	6
Igl	22	16

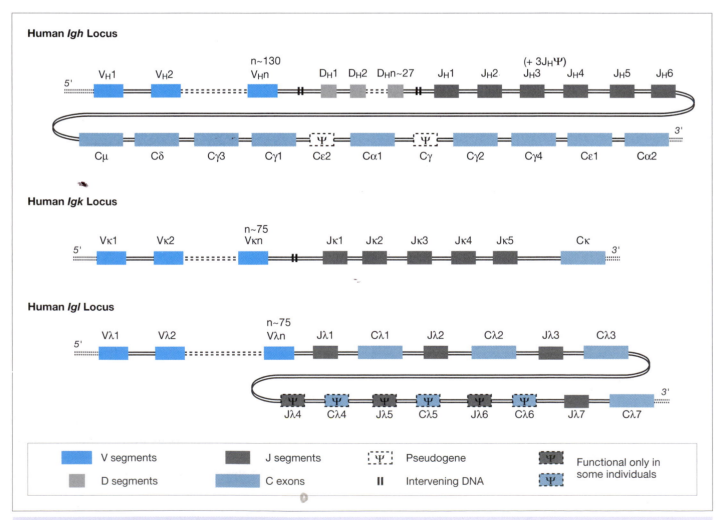

Fig. 4-15 Genetic Organization of the Human Immunoglobulin Loci

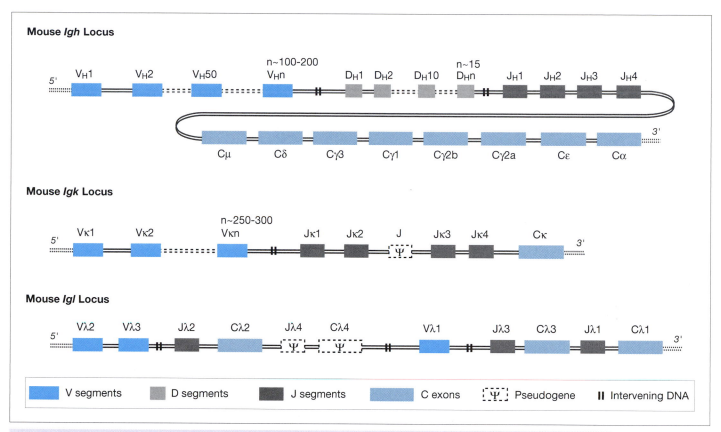

Fig. 4-16 Genetic Organization of the Murine Immunoglobulin Loci

Table 4-2 Estimated Numbers of Gene Segments in Mouse and Human Ig Loci

Gene Segment Type	Number of Gene Segments in Germline*		
	Heavy Chain	Light Chain	
		Kappa	Lambda
Mouse			
V	100–200	250–300	3
D	15	0	0
J	4	4 (1)	3 (1)
Human			
V	130	75	75
D	27	0	0
J	6 (3)	5	7 (3)

*Numbers in parentheses are the numbers of these gene segments that are pseudogenes.

nine in human and eight in mouse. As illustrated in Figures 4-15 and 4-16, the functional exons in the *Igh* loci are designated Cμ, Cδ, Cγ3, Cγ1, Cα1, Cγ2, Cγ4, Cε1 and Cα2 in humans, and Cμ, Cδ, Cγ3, Cγ1, Cγ2b, Cγ2a, Cε and Cα in mice. Each C_H exon encodes the amino acid sequence defining a particular Ig isotype.

III. FROM EXONS TO Ig PROTEINS

As mentioned earlier, somatic or V(D)J recombination is the process that randomly selects and joins together the V and J (and D, if applicable) gene segments of a given locus. V(D)J recombination mediates a physical and permanent juxtaposition at the DNA level that results in a complete V exon. It is this rearrangement in the genomes of individual developing B cells that collectively gives rise to the large, diverse repertoire of Ig proteins. The C exon is joined to the V exon at the RNA level by conventional splicing to produce a translatable Ig mRNA. The Ig mRNA is then translated like any other to generate an H or L protein.

An example of H chain synthesis following V(D)J recombination in the mouse *Igh* locus is shown in Figure 4-17. In this hypothetical situation, the D_H5 gene segment has randomly joined to the J_H3 gene segment, and the V_H15 gene segment has randomly joined to D_H5J_H3 to form the V exon. The V exon is then joined at the RNA level to the C exon. During the synthesis of any one H chain, only the most 5′ of the C exons is normally included in the primary transcript of the rearranged Ig gene. An exception occurs in developing B cells, where both the Cμ and Cδ exons are included in the primary transcript. Through differential processing and RNA splicing of many copies of this primary transcript (see Ch. 5), two populations of mRNAs are produced: one encoding μ H chains and the other specifying δ H chains. Because they share the same V exon, these H chains give rise to IgM and IgD molecules of identical antigenic specificity.

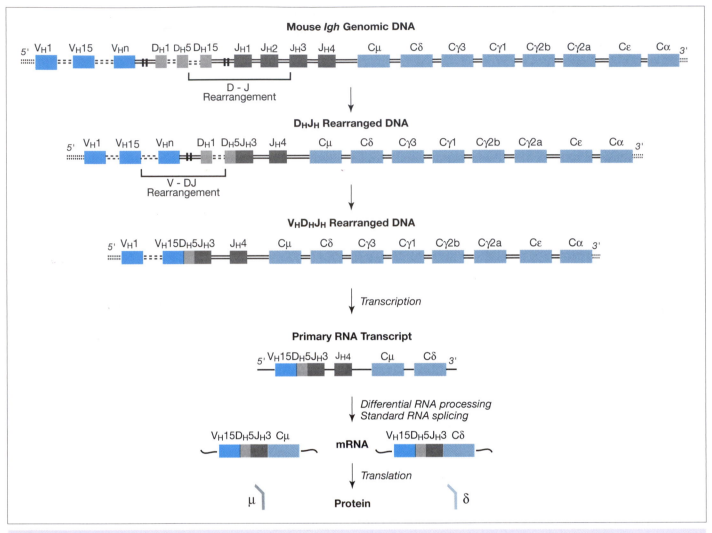

Fig. 4-17 V(D)J Recombination and H Chain Synthesis

IV. V(D)J RECOMBINATION

i) Mechanism

Prior to V(D)J recombination, V, D and J gene segments are distributed over a comparatively large expanse of DNA so that the cell's transcriptional machinery is physically unable to transcribe all the sequences comprising a complete Ig gene. Two specialized recombinase enzymes known as *RAG-1* and *RAG-2* (RAG, recombination activating gene) randomly combine selected V, D and J gene segments in an enzymatic cutting and covalent rejoining of the DNA strand that excludes non-selected V, D and J gene segments and other intervening sequences (refer to Fig. 4-17). This rearrangement creates a VJ or VDJ exon that, because of its proximity to the C exon, allows the transcription of the complete Ig gene followed by translation of the complete H or L chain. The RAG enzymes that carry out this rearrangement are expressed only in developing B and T lymphocytes. Non-lymphocytes never have the capacity for V(D)J recombination, and mature lymphocytes lose it. Furthermore, through developmental restrictions that are not yet understood, B cells do not rearrange their TCR genes, and T cells do not rearrange their Ig genes. Thus, only precursor B cells in the bone marrow are capable of carrying out V(D)J recombination of germline Ig DNA.

ii) Recombination Signal Sequences

V(D)J recombination depends on the cooperative recognition by RAG-1 and RAG-2 of specific *recombination signal sequences* (RSSs) that flank germline V_H segments on their 3' sides, J_H segments on their 5' sides and D_H segments on both sides. Similarly, the V and J segments of the *Igk* and *Igl* loci are flanked by RSSs on their 3' sides and 5' sides, respectively. There are two types of RSSs: the "12-RSS" and the "23-RSS". Both RSSs consist of conserved heptamer and nonamer sequences separated by a non-conserved spacer sequence of either 12 or 23 base pairs of DNA. It is the spacer that identifies the RSS as either a 12-RSS or a 23-RSS. The conserved regions of the 12-RSS and 23-RSS ensure mutually complementary binding. Only when one gene segment is flanked on one side by a 12-RSS, and the other gene segment is flanked by a 23-RSS, can the pair interact and be recognized by the

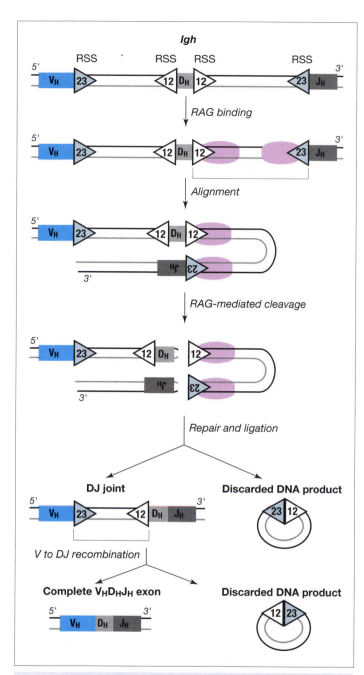

Fig. 4-18 Role of Recombination Signal Sequences in V(D)J Recombination

together to form a circular DNA product that is subsequently lost. The RAG complex then instigates another round of joining by using the 23-RSS on the 3′ side of a V_H segment to join to the 12-RSS on the 5′ side of the D_HJ_H entity to form the complete $V_HD_HJ_H$ exon.

V. ORDER OF Ig LOCUS REARRANGEMENT

The rearrangement of the Ig loci occurs in a particular order. The *Igh* locus is rearranged first, followed by *Igk* and *Igl*. After V(D)J recombination is completed in the *Igh* locus, transcription of the Ig H chain gene can proceed and primary RNA transcripts are produced. The appearance of a newly translated μ chain marks an important checkpoint in Ig synthesis. Further gene segment rearrangements in the *Igh* locus are blocked and somatic recombination of V and J gene segments in the light chain loci is stimulated. If, however, this cycle of *Igh* gene segment joining has led to a μ chain sequence that cannot be translated or is non-functional, the cell continues to randomly join gene segments on the other chromosome. If a functional μ polypeptide still cannot be produced, the cell dies by apoptosis.

When rearrangement at the light chain loci is stimulated by production of a functional μ heavy chain, the *Igk* locus usually goes first. The production of a functional κ light chain appears to shut down any further rearrangement of both the *Igk* and *Igl* loci. However, if a functional κ light chain is not produced from the *Igk* gene on either chromosome, rearrangement is stimulated at the *Igl* locus. If a functional L chain is then produced, the viable H and L chains synthesized by the B cell are assembled into complete Ig molecules that are expressed on the B cell surface in the BCRs. If no functional L chain is produced, no further Ig rearrangements can be initiated and the B cell dies by apoptosis.

VI. ANTIBODY DIVERSITY GENERATED BY SOMATIC RECOMBINATION

There are three sources of variability that contribute to antibody diversity before a given B cell ever encounters its antigen: the existence of multiple V, D and J gene segments in the B cell germline and their combinatorial joining, junctional diversity at the points where the segments are joined and the randomness of heavy–light chain pairing. These elements, which create a unique variable region sequence for the BCR of each B cell prior to its release into the periphery, are discussed in this chapter. Another process called *somatic hypermutation* contributes to diversity in antigen recognition after the naïve B cell is released to the periphery and encounters antigen. This process, which involves point mutations of the rearranged Ig gene, is discussed in Chapter 5.

i) Multiplicity and Combinatorial Joining of Germline Gene Segments

The primary source of diversity in antigen recognition by antibodies is the existence of multiple V, D and J gene segments in the *Igh, Igk* and *Igl* loci (refer to Table 4-2). Because single

RAG-1/RAG-2 complex and participate in V(D)J joining. This requirement has been called the *12/23 rule*. As shown in Figure 4-18, the RAG complex (purple ovals) first brings together the 3′ 12-RSS of a germline D_H segment with the 5′ 23-RSS of a J_H segment so that the complementary sequences of the RSSs are aligned. The RAG complex then cleaves the DNA between these segments such that the D_H and J_H segments are juxtaposed without any RSS between them. The DNA repair machinery of the cell then ligates the DNA strands back together to form a DJ entity with the 12-RSS on its 5′ side (which did not take part in the recombination) left unchanged. The RSS sequences used for the original alignment are joined

V, D and J segments from these collections are randomly chosen and joined together, there are thousands of potential combinations that can be used to create the V_H and V_L exons in a given B cell. Consider the generation of the V exon of an Ig κ chain in a mouse: Each B cell precursor has at its disposal any one of at least 250 $V_κ$ gene segments × 4 (functional) Jκ gene segments = 1000 variable exon sequences. The B cell precursor chooses one combination of Vκ and Jκ segments to fuse covalently and irreversibly to form its Vκ exon. This Vκ exon remains fixed thereafter in the cell's genomic DNA and determines its antigenic specificity and that of its progeny cells for the life of the clone. Similar combinatorial joining can occur in the *Igh* locus, but with an added multiplier effect attributable to the D_H gene segments. Although the D_H gene segments encode only a small number of amino acids, their presence in the *Igh* locus greatly increases the diversity of heavy chains that can be achieved. Consider the following for a mouse Ig heavy chain: At least 100 V_H segments × 15 D_H segments × 4 J_H segments = at least 6000 different V_H exons, as compared to the 400 possible without the D_H segment.

ii) Junctional Diversity

Junctional diversity arises when the fusion of V, D and J segments is not exact. Such "imprecise joining" of gene segments can result from the deletion of nucleotides in the joint region and/or the addition of so-called P and/or N nucleotides during V(D)J recombination (Fig. 4-19). About 80% of antibodies in adult humans show some type of junctional diversity.

iia) Deletion. In the last stages of V(D)J recombination, the DNA strands of the two gene segments to be joined are trimmed by an exonuclease enzyme prior to final linking by DNA ligases. Very often, the exonuclease trims into the coding sequence itself and eliminates a few of the original genomic nucleotides. After DNA ligation is complete, the sequence of the rearranged Ig gene in this region differs slightly from that predicted from the germline sequence. Although each individual change may be subtle, deletion is a common event and so has a significant diversifying effect.

iib) P nucleotide addition. Sometimes the RAG recombinases carrying out V(D)J recombination nick the DNA in the intervening region between gene segments, rather than at the precise ends of the coding sequences of the gene segments. As a result, one strand on each gene segment will have a recessed end while the other strand will have an overhang. The gaps on both strands are filled in with new nucleotides by DNA repair enzymes and the gene segments are ligated together. On the coding strand (only), these inserted nucleotides that modify the joint sequence are called P nucleotides. Although the insertion of P nucleotides can markedly change the amino acid sequence

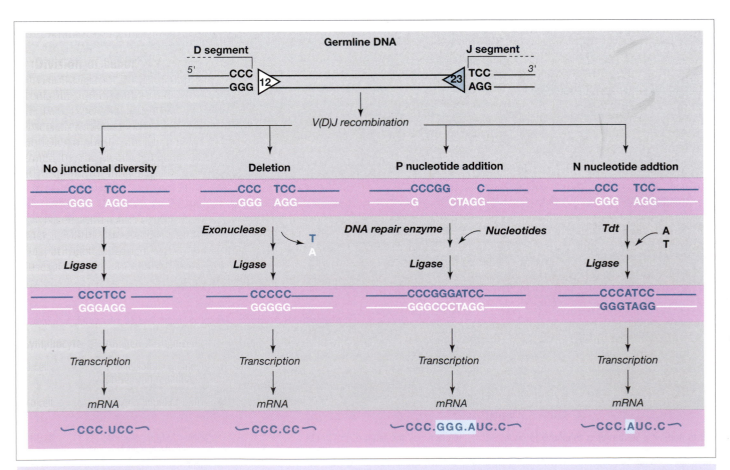

Fig. 4-19 Junctional Diversity

of the Ig protein, it is a relatively rare event and so has a limited impact on diversity.

iic) N nucleotide addition. N nucleotide addition occurs almost exclusively in the *Igh* locus. Usually 3–4 but up to 15 extra nucleotides may be found in VD and DJ joints that do not appear in the germline sequence and are not accounted for by P nucleotides. Sometimes DNA strands are nicked such that they have blunt ends, so that an enzyme called *terminal dideoxy transferase* (TdT) can randomly add "non-templated" N nucleotides onto the ends of these strands before their final ligation. Light chains do not usually undergo N nucleotide addition because the TdT gene is essentially "turned off" at the stage of B cell development when light chain gene rearrangement occurs.

iii) Heavy–Light Immunoglobulin Chain Pairing

The random pairing of H and L chains also contributes to diversity in antigen recognition prior to antigenic stimulation. A mature Ig protein molecule's antigen-binding or antigen-combining site is composed of the V domains of both the L and H chains. Our previous calculations of VJ and VDJ combinatorial joining in the mouse resulted in a total of at least 1000 V_L exons possible for an L chain gene, and at least 6000 V_H exons possible for an H chain gene (not including diversity contributed by deletion and N and P nucleotide addition). Any one B cell synthesizes only one sequence of H chain, and one sequence of L chain, but if one assumes that any of the 6000 H chain genes can occur in the same B cell as any of the 1000 L chain genes, the number of possible L/H chain gene combinations is at least $1000 \times 6000 = $ over 6 million.

iv) Estimates of Total Diversity

The figure most often quoted for the total theoretical diversity of an individual's antibody repertoire is 10^{11} different antigenic specificities. However, it is impossible to accurately quantitate actual repertoire diversity because of the variations introduced by junctional diversity and (later on) somatic hypermutation. It is safe to say that the number of different antibody molecules that an individual can produce is considerably greater than the combinatorial diversity contributed by the selection of different gene segments. However, at any one moment, the actual repertoire available to an individual to counter antigens will be more limited than the theoretical, since a certain proportion of B cells will die before ever encountering their specific antigen. Other B cells will be eliminated before joining the mature B cell pool because their randomly generated BCRs have the potential to recognize self antigens. The removal of these cells from the B cell repertoire is called *B cell negative selection* and is described in Chapter 5.

C. Antigen–Antibody Interaction

I. STRUCTURAL REQUIREMENTS

Whether an Ig protein is membrane-fixed as part of the BCR complex or free as soluble antibody, it is often folded so that

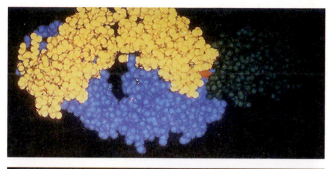
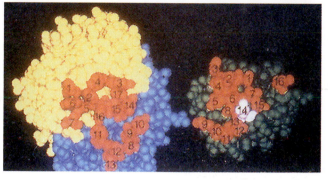

Plate 4-1 X-Ray Crystallography of Fab–Lysozyme Complex
[Reproduced by permission of Amit A.G. *et al.* (1986). Three-dimensional structure of an antigen–antibody complex at 2.8 Å resolution. *Science* **233**, 747–753.]

the CDRs of both V_L and V_H are grouped at the tip of each arm of the Ig "Y". The CDR grouping forms loops that project outward to form an antigen-binding "pocket". While small antigens (such as peptides) can fit securely within this pocket, large macromolecular antigens (such as globular proteins on the surfaces of pathogens) not only project into the V_L–V_H pocket but also make specific surface–surface contacts with the framework regions outside the CDR loops. An example of antigen–antibody binding is given in Plate 4-1, which shows a model derived from X-ray crystallographic studies of the interaction of lysozyme with anti-lysozyme Fab fragment. In the top panel, anti-lysozyme Fab (light chain in yellow, heavy chain in blue) binds to lysozyme (green). A glutamine residue of lysozyme critical for binding to antibody is shown in red. In the bottom panel, the Fab and lysozyme have been separated and rotated to show the contact residues in red (16 for lysozyme, 16 for the Fab spread over all six CDRs). The critical lysozyme glutamine residue appears in white.

The small region of a large antigen that binds to the antigen-binding site of the antibody is called the *antigenic epitope* or *antigenic determinant*. Studies of many antigen–antibody interfaces have revealed that the contact residues of an antigenic epitope are often discontinuous in sequence but contiguous in space and involve a surface area of about 600–900 Å^2. However, since every antibody has a different sequence in its CDRs, the pattern of contact between one antibody and its antigenic epitope is slightly different from that of the next antibody with its epitope. Furthermore, while both the V_L and V_H domains of an Fab are involved in forming the antigen-binding site, not every Fab will use all of its six CDRs to interact with its antigen. In general, antigens form the greatest number of contacts with amino acids in CDR3 of the Ig H chain. CDR3 displays the greatest degree of sequence variability as it is generally encoded by nucleotides at the interfaces between the V, D and J gene segments and is thus the site of junctional diversity.

Although antibody and antigen contact surfaces show a high degree of complementarity prior to binding, there is a degree of flexibility in the antigen-combining site. Upon binding to the antigenic epitope, some antibodies undergo slight conformational modifications to their CDR3 loops and alter the orientations of certain side chains to form a better bond. Immunologists call this phenomenon *induced fit*, since the structure of the antibody is "induced" to fit by the binding of the antigen.

II. INTERMOLECULAR FORCES

The antigen–antibody bond is the result of four types of non-covalent intermolecular forces: hydrogen bonds, Van der Waals forces, hydrophobic bonds and ionic bonds. None of these forces is itself very strong, but because they are all working simultaneously, they combine to forge a very tight bond. The contribution of each type of force to the overall binding depends on the identity and location of the amino acids or other chemical groups in both the antibody and antigen molecules. The more closely the relevant chemical groups can approach one another, the more efficient the binding will be. Similarly, the more complementary the shapes of the antigenic epitope and the antigen-binding site, the more contact sites will simultaneously be brought into close proximity. The number of non-covalent bonds of all types will be increased, resulting in a stronger overall binding. Mapping of binding sites on antibodies has shown that cavities, grooves or planes are often present that correspond to complementary structural features on the antigen. If an antigen approaches whose shape is less complementary to the conformation of the binding site on the antibody, fewer bonds are formed and steric hindrance and repulsion by competing electron clouds are more likely to "push" the prospective antigen away. Only those antigens shaped such that they make a sufficient number of contacts of the required strength will succeed in binding to the antigen-combining site of the antibody molecule.

The strength and specificity of the antigen–antibody bond is often used in the laboratory to isolate antigens from mixtures of proteins (see Appendix F). To release the antigen from the antibody, the non-covalent forces that hold them together are disrupted by the application of high salt concentration, detergent or non-physiological pH.

III. THE STRENGTH OF ANTIGEN–ANTIBODY BINDING

i) Affinity

Immunologists define the *affinity* of an antibody for its antigen as the strength of the non-covalent association between <u>one</u> antigen-binding site (thus, one Fab arm of an antibody molecule) and <u>one</u> antigenic epitope. Affinity is measured in terms of the equilibrium constant (K) of this binding, with a K of 10^9 M^{-1} representing strong affinity and a K of 10^6 M^{-1} representing weak affinity. In a natural immune response to invasion by a complex antigen, many different B cell clones will be activated, leading to a *polyclonal* antibody response. The antibodies generated will be heterogeneous with respect to both specificity and affinity. Not only will antibodies to different epitopes on the antigen be produced, but a single epitope will induce the activation of various B cells whose receptors display a variable accuracy of fit with the epitope. In addition, affinity is not an absolute quality for an antibody: A given antibody may display different affinities for a spectrum of related but slightly different antigens. Importantly, the affinity of antibodies raised during a secondary immune response to a given epitope is increased compared to antibodies induced in the first encounter. This phenomenon, which is called *affinity maturation*, is discussed in Chapter 5.

ii) Avidity

Whereas affinity describes the strength of the bond between one Fab and one antigenic epitope, *avidity* refers to the overall strength of the bond between a multivalent antibody and a multivalent antigen. Microorganisms and eukaryotic cells represent large multivalent antigens when they feature multiple identical protein molecules on their cell surfaces. Other examples of multivalent antigens are macromolecules composed of multiple identical subunits, and macromolecules containing a structural epitope that occurs repeatedly along the length of the molecule. All antibodies are naturally multivalent because they have two or more identical antigen-binding sites. Secreted IgG and IgE molecules have only two Fabs, but polymeric sIgA molecules can exhibit 4 or 6 identical antigen-combining sites, and sIgM molecules are most often found as pentamers with 10 binding sites. Binding at one Fab on a polymeric antibody holds the antigen in place so that other binding sites on the antibody are more likely to bind as well. With each additional binding site engaged by a polymeric antibody, the probability of the simultaneous release of all bonds drops exponentially, resulting in a lower chance of antigen–antibody dissociation and therefore a stronger overall bond. For example, in the case of pentameric IgM, each site in itself may not have a very strong affinity for antigen, but because all 10 sites may be bound simultaneously, IgM antibodies have considerable *avidity* for their antigens. At any given moment, dissociation at any one antibody–antigen combining site may occur, but the complete release of antigen from antibody would require that all 10 binding sites be dissociated from the antigen simultaneously. The practical result of high avidity binding by an antibody is that the association between antigen and antibody is formed more rapidly and stably, and the antigen is thus eliminated from the body more efficiently.

IV. CROSS-REACTIVITY

The specificity of an antiserum as a whole is defined by its component antibodies. Any one epitope of an antigen can induce the production of a collection of antibodies with a range of affinities and avidities for that epitope. As well, an

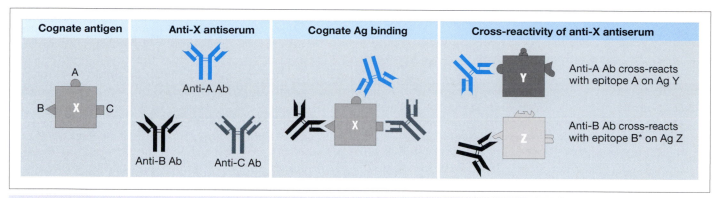

Fig. 4-20 **Antibody Cross-reactivity**

antigen usually exhibits more than one epitope, so that antibodies exhibiting a spectrum of epitope specificities will also be included in the antiserum. The antigen used to immunize an animal and induce it to produce an antiserum is known as the *cognate antigen*. A *cross-reaction* is said to have occurred when an antiserum reacts to an antigen other than the cognate antigen. Cross-reactivity results either when one epitope is shared by two antigens, or when two epitopes on separate antigens are similar in structure.

Figure 4-20 illustrates cross-reactivity. Let us suppose a mouse is immunized with cognate antigen X containing epitopes A, B and C. The polyclonal anti-X antiserum recovered from this mouse contains a mixed population of antibodies, including anti-A, anti-B and anti-C. Let us also suppose that there is a second antigen Y that contains epitope A, and a third antigen Z that contains epitope B* (which is conformationally similar to B). Anti-X antiserum will react with antigen Y, since a proportion of the antibodies (anti-A) will bind strongly to the shared epitope A. Anti-X antiserum will also react with antigen Z, because anti-B antibodies will bind to epitope B*, although probably with a different affinity. Anti-C antibodies will make no contribution to cross-reactivity in either case because they fail to recognize any epitope on antigens Y or Z.

The differing degrees to which cross-reacting antibodies bind to antigen can be understood in the light of the intermolecular binding forces described earlier. An antigen of a shape slightly different from that of the cognate antigen may not be able to form one or more of the multiple types of bonds required for tight binding to antibody; the antibody is thus more easily dissociated from the non-cognate antigen. To continue the previous example, the more closely B* resembles the shape of B, the greater the possible molecular contacts between B* and the binding site of anti-B, and the more likely cross-reactive binding will occur.

This chapter has described the structures of the Ig genes and proteins and the mechanisms of gene segment rearrangement and combinatorial joining that contribute to antibody diversity before an encounter with antigen. The next chapter discusses B cell development, the activation of these cells by antigen, the generation of antibody diversity following activation, and the biological expression of Igs.

- Immunoglobulins (Igs) are antigen-binding proteins produced by B cells. All Igs made by a given B cell have the same antigenic specificity.

- Ig molecules have an H_2L_2 structure in which two identical heavy (H) chains and two identical light (L) chains are held together by disulfide bonds.

- Each Ig chain has an N-terminal variable (V) domain that differs widely in amino acid sequence among B cells. The C-terminal end of each L chain contains a single constant (C) domain of relatively invariant amino acid sequence. Each H chain contains three or four C domains.

- There are two L chain isotypes, κ and λ. There are five H chain isotypes, μ, δ, γ, ε and α, that specify IgM, IgD, IgG, IgE and IgA antibodies.

- There are three isoforms of Ig proteins. Membrane-bound (mIg) molecules have a transmembrane domain and associate with the Igα/Igβ heterodimer to form the BCR complex on the B cell surface. Plasma cells secrete copious amounts of an antibody that is a soluble form (sIg) of the Ig protein. Secretory antibodies (SIg) occur in body fluids.

- The Ig genes are located in the *Igh*, *Igl* and *Igk* loci and contain variable (V) and constant (C) exons.

- V exons are assembled at the DNA level from random combinations of V (variable), D (diversity) and J (joining) gene segments that are flanked by recombination signal sequences (RSSs).

- RAG recombinases carry out V(D)J recombination to join a D segment to a J segment, and then DJ to a V segment to form the H chain V exon. V and J segments are joined in the *Igk* and *Igl* loci to form L chain V exons. V exon RNAs are brought together with C exon RNAs by conventional RNA splicing.

- Successful rearrangement of the *Igh* locus on one chromosome such that a functional μ H chain is produced shuts down V(D)J recombination of the other *Igh* allele. Rearrangement is then usually stimulated first at the *Igk* locus, and then at the *Igl* locus if a functional κ L chain is not produced. Productive rearrangement of at least one H and one L chain must occur in order for the B cell to survive.

- Much of the diversity in the antibody repertoire is derived from multiple germline V, D and J gene segments and their combinatorial joining, junctional diversity and random H–L chain pairing.

- Specificity of antigen–antibody binding is defined by the unique surface created by the combined V_H and V_L CDRs. The strength of antigen–antibody binding depends on the additive effect of several types of non-covalent binding forces.

- The affinity of antibody binding to a given antigen is the strength of the antigen–antibody interaction at one antigen-binding site. Avidity is a measurement of the total strength of binding where multivalent antibody binds multivalent antigen.

- Cross-reactivity of a given antibody may be due to recognition of the same or a similar epitope on different antigens.

Section A.I–II

1) Can you define these terms? gamma globulin, Fab, Fc, Ig domain, hinge region

2) What is the basic structure of an Ig molecule, and how do its C and V regions differ?

3) What is the Ig fold? Give two reasons why this feature is useful to a protein.

Section A.III

1) Can you define these terms? CDR, framework region

2) What are the hypervariable regions and how do they contribute to antibody diversity?

3) What is the function of the framework regions?

Section A.IV

1) Can you define these terms? mIg, sIg, SIg, tailpiece, J chain, secretory component, pIgR, transcytosis, ITAM

2) Describe the L chain isotypes. Explain why they don't contribute to antibody effector functions.

3) What are the five major H chain isotypes and how are they derived?

4) What is isotype switching and when does it occur?

5) Compare the production and functions of the three structural Ig isoforms.

6) Describe the components of the BCR complex and their functions.

Section B.I–III

1) Can you define these terms? gene segment, V exon, C exon, V(D)J recombination

2) Why isn't the antibody repertoire encoded in the germline of B cells?

3) Describe the three Ig loci and their complements of gene segments and C exons.

4) Does RNA splicing contribute to V exon assembly? If not, why not?

5) What C_H exons are first included in the primary transcripts of a rearranged *Igh* locus?

6) How does the production of a functional μ chain affect somatic recombination in the L chain loci? How does it affect cell survival?

Section B.IV

1) Can you define these terms? RAG, RSS

2) Why is V(D)J recombination of the Ig genes necessary for their expression?

3) How many hematopoietic cell types can carry out V(D)J recombination of the Ig genes?

4) In what order are V, D and J gene segments joined together, and what does the 12/23 rule have to do with this?

Section B.V

1) Can you define these terms? P nucleotide, N nucleotide

2) Describe four sources of antibody diversity associated with somatic recombination.

3) Describe two types of junctional diversity.

4) Give two reasons why the total diversity of the antibody repertoire is likely to be less than the theoretical estimate of 10^{11}.

Section C.I

1) Can you define these terms? epitope, determinant, induced fit

2) Which CDR plays the most important role in antigen binding and why?

Section C.II

1) Name four types of intermolecular forces that contribute to the antigen–antibody bond.

2) How can the antigen–antibody bond be reversed in the laboratory?

Section C.III

1) Can you define these terms? polyclonal, polymeric antibody

2) Distinguish between antibody affinity and antibody avidity.

Section C.IV

1) Can you define these terms? cognate antigen, cross-reaction

2) Describe two ways in which cross-reactivity can arise.

5

B Cell Development, Activation and Effector Functions

Never give advice unless asked.
German Proverb

B cell development takes place in a series of well-defined stages that can be grouped into two phases: the *maturation* phase and the *differentiation* phase. In the maturation phase, an HSC divides and eventually generates mature naïve B cells through a process that is tightly controlled by cytokines but <u>independent</u> of foreign antigen. The major developmental stages of the maturation phase include the HSC, the MPP, the CLP, the pro-B cell (progenitor B cell), the pre-B cell (precursor B cell), the immature naïve B cell, the transitional B cell, and the mature naïve B cell. Some stages are subdivided, as in "early" and "late" pro-B cells. The maturation phase begins in the bone marrow and ends with mature naïve B cells taking up residence in the secondary lymphoid tissues in the body's periphery. There are two stages in the differentiation phase: the activation of a mature naïve B cell by its specific antigen, and the generation by that cell of antigen-specific plasma cells and memory B cells. The antibodies produced by the plasma cells then carry out a range of effector functions that work to eliminate the original antigen. Which effector functions are deployed are a function of the isotype of antibody produced and the tissue in which the assault has occurred. The antibody response is crucial to immune defense, as exemplified by the plight of patients with primary immunodeficiencies caused by mutations that affect B lymphocytes exclusively (see Box 5-1).

A. B Cell Development: Maturation Phase

A general scheme of the maturation phase of B cell development is shown in Figure 5-1. For the pro-B and pre-B cell stages in particular, the stromal cells of the bone marrow are absolutely critical for continued development. The stromal cells secrete chemokines and cytokines and establish direct intercellular contacts with both pro-B and pre-B cells that are essential for their survival and progression.

I. PRO-B CELLS

Early pro-B cells are the first hematopoietic cells clearly recognizable as being of the B lineage. These cells are identified by their expression of certain B lineage markers and by the fact that all their Ig genes are still in germline configuration (i.e., the *Igh*, *Igk* and *Igl* loci have yet to undergo V(D)J recombination). In late pro-B cells, the first attempts at $D_H J_H$ joining in the *Igh* locus are initiated, such that the $D_H J_H$ sequences on at least one chromosome are clearly distinct from the germline *Igh* sequence.

II. PRE-B CELLS

It is in early pre-B cells that complete V(D)J rearrangement of the *Igh* locus can be detected; that is, a V_H gene segment is joined to $D_H J_H$ on one chromosome. Both the *Igk* and *Igl* loci remain in germline configuration and there is no expression yet of Ig L chain transcripts. These early pre-B cells transcribe the newly rearranged *Igh* locus to determine whether it can produce a functional Ig H chain of the μ isotype. Normally, an H chain must associate with an L chain in order to be conveyed to the cell surface, but, at this point, there are no L chains present. Instead, a protein homologous to the N-terminus of the L chain comes together with another protein homologous to the C-terminus of an L chain to form the *surrogate light chain* (SLC). Two SLC molecules associate with two copies of the candidate cytoplasmic μ chain plus the Igα/Igβ signaling heterodimer to form a *pre-B cell receptor complex* (pre-BCR). The pre-BCR is expressed transiently on the pre-B cell membrane to test the functionality of this particular $V_H D_H J_H$ combination (Fig. 5-2). A pre-BCR is not a true Ig and

Box 5-1 B Cell-Specific Primary Immunodeficiencies (B-PIs)

B-PIs account for 70% of all primary immuno-deficiencies and are classed by whether mature B cells are entirely absent, or present but non-functional. B-PI patients are usually free of infection at birth due to the presence of mater-nal antibodies. By age 7–9 months, however, these patients show abnormally low levels of one or more serum Igs and an increased sus-ceptibility to pyogenic organisms. Boys with **Bruton's X-linked agammaglobulinemia (XLA)** lack mature B cells due to an X-linked mutation of an intracellular signaling enzyme called Bruton's tyrosine kinase (Btk). Btk is

required for pre-B cell maturation. No antibody isotypes are detectable, leaving these patients highly vulnerable to infection. Similar clinical features result from **autosomal mutations** in the μ **H chain** or Igα genes that block the pro-B to pre-B cell transition. **Common variable immunodeficiency (CVID)** is a family of rela-tively prevalent, heterogeneous diseases char-acterized by the presence of circulating mature B cells but profoundly decreased levels of IgA, IgG and/or IgM. These deficits can be due to a variety of mutations, including defects in isotype switching, somatic hypermutation or

B–T cooperation (see later). In **selective Ig deficiencies**, antibodies of a particular isotype or directed against a particular pathogen are missing due to deletions or mutations in the *Igh* locus. Children with **anti-polysaccharide antibody deficiency** have normal levels of serum Igs but lack the ability to make antibod-ies to bacterial capsule polysaccharides. Patients with **hyper-IgE syndrome** have very high levels of serum IgE and suffer from severe abscesses. The underlying defects in these last two disorders are unknown.

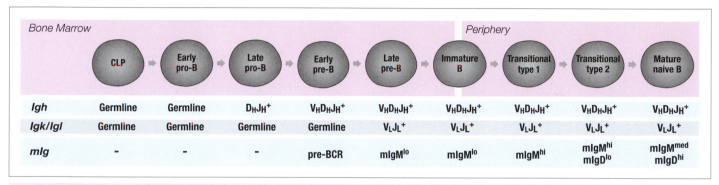

Fig. 5-1 **Major Stages of B Cell Development**

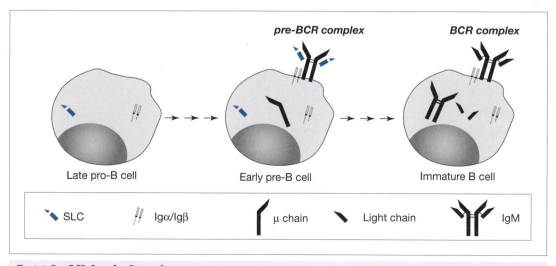

Fig. 5-2 **Pre-BCR Complex Expression**

cannot respond to antigen, but it is thought to bind to ligands on bone marrow stromal cells such that an intracellular signal is delivered telling the cell that a functional H chain has been successfully synthesized. If the particular $V_HD_HJ_H$ sequence selected during recombination is not functional, sequential $D_H \rightarrow J_H$ and $V_H \rightarrow D_HJ_H$ rearrangements are attempted at the second *Igh* allele and the resulting candidate μ chain is also

tested in pre-BCR form. Failure on both chromosomes occurs in about half of pre-B cells, pre-empting further maturation and leading to the apoptotic death of most of these cells in the bone marrow.

Late pre-B cells develop from those early pre-B cells that do produce a functional pre-BCR. Further *Igh* locus rearrange-ments are terminated and V(D)J recombination is stimulated at the *Igk* and *Igl* loci as described in Chapter 4. Only late pre-B cells that successfully express a complete Ig molecule containing fully functional H and L chains survive to become immature B cells. The presence of the complete BCR complex (membrane-bound Ig plus Igα/Igβ) on the B cell surface delivers an intracellular signal that terminates all Ig locus rearrangements.

III. IMMATURE B CELLS IN THE BONE MARROW: RECEPTOR EDITING

Despite their functional BCRs, immature B cells in the bone marrow do not proliferate or respond to foreign antigen. It is this population of cells that is screened for central tolerance; that is, immature B cells that recognize self antigens are removed. Bone marrow stromal cells express "housekeeping" molecules that are produced by all body cells. A B cell whose BCR binds to one of these molecules with high affinity is potentially autoreactive, and the release of such a B cell from the bone marrow into the periphery could lead to an attack on self tissues. The immature B cell therefore determines <u>before</u> it is released from the bone marrow whether its BCR recognizes the housekeeping molecules present on the surrounding stromal cells. If so, the autoreactive B cell receives an intracellular signal to halt development. The cell is then given a brief window of opportunity to try to further rearrange its Ig loci and stave off apoptosis by altering its antigenic specificity. This secondary gene rearrangement is called *receptor editing*.

Receptor editing occurs primarily in the L chain because, after $V_H D_H J_H$ recombination in the *Igh* locus is completed, there are rarely any D_H gene segments available to permit further rearrangements satisfying the structural rules of V(D)J recombination. In contrast, the light chain loci usually still have available many upstream V_L gene segments and a few downstream J_L segments after a productive round of *Igk* or *Igl* rearrangement. If receptor editing fails to achieve success, the B cell dies by apoptosis and is said to have been *negatively selected* (Fig. 5-3). However, if receptor editing is successful, the BCR no longer recognizes self antigen and the cell appears to receive a *positive selection* signal that sustains survival. On a daily basis, about 2–5% of the entire pool of immature B cells survives the selection processes that establish central B cell tolerance.

IV. THE TRANSITION TO MATURITY: IgM AND IgD COEXPRESSION

Immature B cells remain in the bone marrow for 1–3 days before commencing the expression of new adhesion molecules and homing receptors that allow them to leave the bone marrow and travel in the blood to the secondary lymphoid tissues. At this point, the developing B cell is known as a *transitional type 1 B cell* or T1 B cell. T1 B cells extravasate from the blood first into the red pulp of the spleen and then into its PALS. After about 24 hours in the PALS, T1 B cells become *transitional type 2 B cells*, or T2 B cells. Local production of

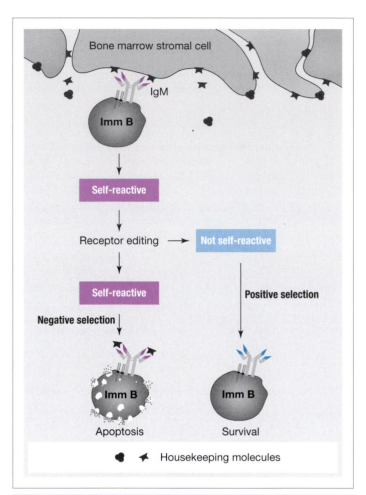

Fig. 5-3 **Receptor Editing in B Cell Central Tolerance**

a cytokine called BAFF (B lymphocyte activating factor belonging to the TNF family) is essential for this transition and T2 B cell survival. T2 B cells start to colonize the B cell-rich areas of the spleen and acquire the ability to emigrate to other secondary lymphoid tissues, particularly the lymph nodes. T2 B cells also commence the surface expression of IgD as well as IgM. Both Igs are derived from the same rearranged H and L chain genes containing the same $V_H D_H J_H$ and $V_L J_L$ exons, respectively, and therefore have the same V domains in their antigen-binding sites and the same antigenic specificity. It is only the C region of the H chain that differs. Coexpression of IgM and IgD occurs because the primary RNA transcripts that are synthesized from the rearranged H chain gene contain the sequences encoded by the $V_H D_H J_H$ exon and <u>both</u> the Cμ and Cδ exons. Differential processing of this transcript at alternative polyA addition sites followed by RNA splicing results in the removal of either the Cμ or the Cδ exon from the RNA, leaving an mRNA specifying either IgD or IgM, respectively (Fig. 5-4). Both pathways operate simultaneously to process the population of primary transcripts generated in the B cell, such that a single T2 B cell expresses two different isotypes of membrane-bound Igs.

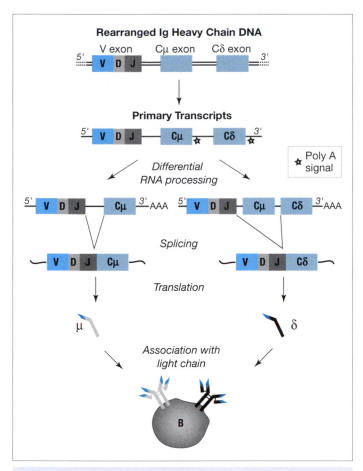

Fig. 5-4 IgM and IgD Coexpression

spans, the chances of these B cells being stimulated by specific antigen are extremely limited: only 1 in 10^5 peripheral B cells encounters specific antigen and avoids death by apoptosis. It is the binding of specific antigen to a B cell's BCRs that triggers the differentiation phase.

I. THE NATURE OF B CELL IMMUNOGENS

As was introduced in Chapter 1, an antigen is any substance that binds specifically to the antigen receptor of a T or B cell. However, some antigens that bind to a TCR or BCR fail to induce the activation of the lymphocyte even under optimal conditions. Immunologists have therefore coined the term *immunogen* to refer to any substance that binds to a BCR or TCR <u>and</u> elicits an adaptive response. Thus, all immunogens are antigens but not all antigens are immunogens. (Despite this very real distinction, the immunological community regularly uses "antigen" when "immunogen" would be more accurate.) Almost every kind of organic molecule (including proteins, carbohydrates, lipids and nucleic acids) can be an antigen, but only macromolecules (primarily proteins and polysaccharides) have the size and properties necessary to be physiological immunogens. In a natural infection, invading bacteria and viruses appear as collections of proteins, polysaccharides and other macromolecules to the host's immune system. Within this collection, there are many immunogens, each inducing its own adaptive immune response.

i) Responses to B Cell Immunogens

Three major classes of B cell antigens (or more properly, immunogens) exist: *T-independent-1* (Ti-1), *T-independent-2* (Ti-2) and *T-dependent* (Td) antigens. Both types of Ti antigens can activate B cells to produce antibodies without interacting directly with T cells. Td antigens can bind to the BCRs of B cells to initiate activation but cannot induce plasma cell differentiation and antibody production unless the B cell interacts directly with a Th effector cell activated by the same antigen. The Th effector cell is said to be supplying *T cell help* to the B cell. T cell help takes the form of cytokines and intercellular contacts mediated by *costimulatory molecules*. Costimulatory molecules are proteins on the surfaces of lymphocytes whose engagement by specific ligand is necessary for complete activation. It is T cell help that allows activated B cells to undergo somatic hypermutation, isotype switching and memory B cell production. As a result, responses to Td antigens are dominated by highly diverse IgG, IgE or IgA antibodies and generate memory B cells capable of mounting a faster, stronger secondary response in a subsequent exposure to that antigen.

B cell responses to Ti-1 and Ti-2 antigens do not involve direct interaction with effector Th cells and thus are very rapid and provide frontline defense against invaders. However, this absence of T cell interaction means that B cells responding to Ti antigens do not carry out extensive isotype switching or somatic hypermutation and therefore produce mainly IgM antibodies of limited diversity. In addition, memory B cells are not generated so that the intensity of the response in a subse-

V. MATURE NAÏVE B CELLS IN THE PERIPHERY

Once T2 B cells establish themselves in the lymphoid follicles, they become mature naïve B cells in the periphery (also known as *follicular B cells*). The vast majority of mature naïve B cells remain in the follicles and do not recirculate among the secondary lymphoid tissues (unlike mature naïve T cells). Mature naïve B cells show slightly lower levels of mIgM than do T2 B cells but higher levels of mIgD. Although mIgM is clearly essential for the response of the B cell to antigen, the function of mIgD has been more difficult to ascertain (see later). Mature naïve B cells also permanently lose expression of the recombination activating genes RAG-1 and RAG-2, so that no further changes in V(D)J gene segment usage can occur in either the mature B cell itself or in its memory and plasma cell progeny. These B cells are now poised to encounter antigen.

B. B Cell Development: Differentiation Phase

The adult human bone marrow churns out about 10^9 mature naïve B cells every day. However, over the course of their life

Table 5-1 Features of Responses to Td and Ti Antigens

Property	Td Antigen	Ti-1 Antigen	Ti-2 Antigen
Requires direct interaction with a T cell for B cell activation	Yes	No	No
Requires T cell cytokines	Yes	No	Yes
Epitope structure	Unique	Mitogen	Repetitive
Protein	Yes	Could be	Could be
Polysaccharide	No	Could be	Could be
Relative response time	Slow	Fast	Fast
Dominant antibody isotypes	IgG, IgE, IgA	IgM IgG rarely	IgM IgG sometimes
Diversity of antibodies	High	Low	Low
Stimulates immature and neonatal B cells	No	Yes	No
Polyclonal B cell activator	No	Yes	No
Memory B cells generated	Yes	No	No
Magnitude of response upon a second exposure to antigen	Secondary response level	Primary response level	Primary response level
Examples	Diphtheria toxin Purified Mycobacterium protein	Bacterial LPS	Pneumococcal polysaccharide

quent exposure to the antigen is stuck at the primary level rather than progressing to the secondary level. The principal features of responses to Td, Ti-1 and Ti-2 antigens are discussed next and summarized in Table 5-1.

ii) Properties of Ti Antigens

Ti-1 antigens contain a molecular region that acts as a *mitogen*. A mitogen is a molecule that non-specifically stimulates cells to initiate mitosis. In the case of a Ti-1 antigen, the mitogen portion binds to a surface receptor that is expressed by many cells, including by B cells. Simultaneously, the BCRs of the B cell bind to epitopes outside the mitogen region of the Ti-1 antigen. The mitogen receptor sends a strong signal to the nucleus of the B cell to proliferate. Because the expression of the mitogen receptor is not restricted to one B cell, many clones of naïve B cells can be activated at once by Ti-1 antigen molecules. *Polyclonal* B cell activation is said to have taken place. Because mitogenic activation does not depend on the BCR, Ti-1 antigens can also activate immature B cells and the B cells of newborn children.

Ti-2 antigens, which are found in many bacterial and viral structures and products, are generally large polymeric proteins or polysaccharides (and sometimes lipids or nucleic acids) that contain many repetitions of a structural element. This repetitive structure acts as a multivalent antigen that can bind with high avidity to the mIg molecules in neighboring BCR complexes on the surface of a B cell. The BCRs are said to be *cross-linked* because the antigen-binding sites of two (or more) mIg molecules are indirectly linked together by virtue of their binding to the same very large immunogen. This extensive cross-linking triggers intracellular signaling that leads to B cell activation, proliferation and differentiation in a way that bypasses the need for costimulatory contacts with an activated Th cell. However, these antigens cannot activate naïve B cells in the absence of cytokines produced (mainly) by activated T cells.

iii) Properties of Td Antigens

The TCR of a T cell recognizes a complex composed of an MHC molecule bound to a peptide of an antigenic protein; that is, the epitope for a TCR is pMHC. For a Th effector cell to supply T cell help to a B cell that has encountered an antigen, the original naïve Th cell must also have been activated by a peptide–MHC class II complex derived from the same antigen. Thus, as well as supplying a B cell epitope, a Td antigen must contain protein, since it must supply at least one peptide that can bind to MHC and form the T cell epitope. Other physical properties of a protein such as its foreignness, conformation and molecular complexity also affect its ability to be a Td antigen (Table 5-2).

iiia) Foreignness.

To be immunogenic, an antigen must be perceived as non-self by the host's immune system. Whether an antigen is perceived as non-self depends on whether there is a lymphocyte in the host's repertoire that can recognize the antigen. In the case of proteins that are similar to self proteins, many of the lymphocyte clones whose antigen receptors might have recognized that antigen should have already been removed

Table 5-2 Factors Affecting Td Immunogenicity

	More Immunogenic	Less Immunogenic
Foreignness	Very different from self	Very similar to self
Molecular complexity		
Size	Large	Small
Subunit composition	Many	Few
Conformation	Denatured, particulate	Native, soluble
Charge	Intermediate charge	Highly charged
Processing potential	High	Low
Dose	Intermediate	High or low
Route of entry	Subcutaneous > intraperitoneal > intravenous or gastric	
Host genetics Mutations to MHC and other antigen processing molecules resulting in:	Efficient presentation and peptide binding	Inefficient presentation and peptide binding

during the establishment of central tolerance. Thus, a protein that is widely conserved among different species will not be very immunogenic. In contrast, the more an antigen deviates structurally from self proteins, the more easily it is recognized as non-self. Accordingly, the introduction of such a protein into the host will usually trigger an intense immune response.

iiib) Complexity. The molecular complexity of a Td antigen encompasses the properties of *molecular size*, *subunit composition*, *conformation*, *charge* and *processing potential*. A protein of large size and/or composed of multiple subunits will likely contain both the T and B epitopes needed for a Td response. In general, Td antigens are at least 4 kDa and often close to 100 kDa in molecular size. However, large molecules are not good Td antigens if they are too simple in structure. For example, homopolymers (composed of a single type of amino acid) are not effective Td antigens but copolymers of two different amino acids are because a copolymer provides more possibilities for T and B epitopes. For the same reason, macromolecules with several different subunits are generally more effective Td antigens than are macromolecules containing a single subunit type.

The conformation of a molecule can affect immunogenicity. Most B cell epitopes are structures that project from the external surfaces of macromolecules or pathogens. The majority of these B cell epitopes are *conformational determinants*, a term used by immunologists to refer to epitopes in which the contributing amino acids may be located far apart in their linear sequence but which become juxtaposed when the protein is folded in its natural or *native* shape (Fig. 5-5). These epitopes depend on native state folding, so that they disappear when the protein is denatured. Other epitopes are linear, in that they are defined by a particular stretch of consecutive amino acids. Some linear determinants are present on the surface of a protein and are accessible to antibody whether the protein is in native or denatured form (external linear determinants).

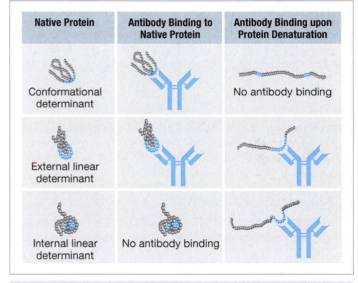

Fig. 5-5 Effect of Protein Conformation on Antibody Binding to Antigenic Determinants
[Adapted from Abbas A.K. *et al.* (1997). *Cellular and Molecular Immunology*, 3rd ed. W.B. Saunders Company, Philadelphia.]

Other linear determinants are buried deep within a molecule and are accessible to antibody only when the protein is denatured (internal linear determinants). Some determinants are *immunodominant*; that is, even though the antigen contains numerous epitopes, the majority of antibodies are raised to only a few of them. Immunodominant epitopes for B cells tend to be conformational determinants on the surfaces of macromolecules because these sites are more available for binding to BCRs than are internal epitopes. Structural considerations that affect the ease of binding between the epitope and the antigen receptor binding site may make one surface epitope immunodominant over another.

The charge or electronegativity of a molecule is also important for immunogenicity. Since B cell epitopes are most often

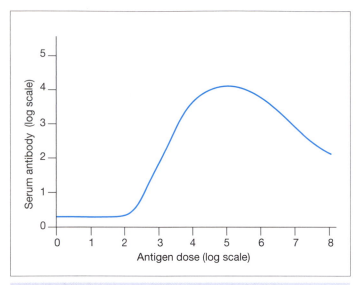

Fig. 5-6 Primary Antibody Response to Different Antigen Doses
[Adapted from Janeway C.A. *et al.* (1997). *Immunobiology*, 3rd ed. Current Biology Ltd., London.]

Table 5-3 **Modes of Immunogen Administration**			
	Abbr.	**Description**	**Immunogen Channeled to**
Oral	p.o.	By mouth	MALT
Parenteral			
Intravenous	i.v.	Into a blood vessel	Spleen
Intraperitoneal	i.p.	Into the peritoneal cavity	Spleen
Intramuscular	i.m.	Into a muscle	Regional lymph node
Intranasal	i.n.	Into the nose	MALT
Subcutaneous	s.c.	Into the fatty hypodermis layer beneath the skin	Regional lymph node
Intradermal	i.d.	Into the dermis layer of the skin	SALT

those exposed on the surface of a macromolecule in the aqueous environment of the tissues, those areas of globular proteins acting as B cell epitopes are often hydrophilic in nature. Proteins containing a high ratio of aromatic to non-aromatic amino acid residues often have numerous hydrophilic epitopes and are very effective Td antigens. However, very highly charged molecules elicit poor antibody production, possibly due to electrostatic interference between the surface of the B cell and the antigen.

iiic) Dosage. When a pathogen invades the body, it is able to replicate and so delivers substantial doses of a plethora of immunogenic molecules to the host. In the case of experimental immunizations with a single antigen, immunologists have determined that there is a threshold dose below which no adaptive response is detected because insufficient numbers of lymphocytes are activated. As the dose of antigen increases above the threshold, the production of antibodies rises until a broad plateau is reached that represents the optimal range of doses for immunization with this antigen (Fig. 5-6). At very high antigen doses, antibody production again declines, possibly due to mechanisms responsible for tolerance to self proteins (which are naturally present at high concentrations). Memory lymphocytes generally respond to a lower dose of antigen than naïve cells, but very high or very low doses of antigen can induce memory cells to adopt non-responsive states known as *high zone tolerance* or *low zone tolerance*, respectively (see Ch. 10). Antibody production is minimal at these dosages.

iiid) Route of entry or administration. The immunogenicity of a molecule can depend heavily on its route of administration. The food we eat contains many antigens, but because these antigens are "dead" (cannot replicate) and are introduced via the digestive tract, enzymatic degradation usually destroys them before an immune response is provoked. For this reason, antigens used for immunizations are usually introduced *parenterally*; that is, by a route other than the digestive tract (Table 5-3). The antigen is channeled to a nearby secondary lymphoid tissue to trigger lymphocyte activation. Interestingly, the dose of a molecule that provokes a strong immune response when injected by one route of administration may induce only a minimal response when introduced by another. More on the administration of immunogenic molecules appears in Chapter 14 on vaccines.

iiie) Potential for antigen processing and presentation. A protein's potential for uptake and intracellular processing by APCs to generate the peptides for pMHC is also important for Td immunogenicity. Macromolecules that are readily phagocytosed by APCs and proteins that are easily digested by the APC's enzymes tend to be more immunogenic. Subtle sequence differences in the genes encoding MHC molecules or other proteins involved in antigen processing can alter antigen processing and presentation, affecting the type, as well as the intensity, of the immune response to a given immunogen (see Ch. 6).

II. B CELL ACTIVATION BY Td IMMUNOGENS

i) Rationale for T Cell Help

A large proportion of the body's B cells will need T cell help to respond because Ti immunogens are only a small fraction of the antigens that a host encounters. Many of the molecules making up a pathogen are proteins of unique amino acid sequences that lack the large repetitive structures needed to cross-link BCRs and trigger B cell activation directly. Even when a protein is present in thousands of copies on the surface of an intact pathogen, steric hindrance may forestall extensive BCR cross-linking and prevent complete B cell activation. Enter T cell help supplied by a Th effector cell specific for the same antigen (see Ch. 9 for a description of Th effector cell development). The B cell receives both costimulation and cytokines that deliver signals completing the intracellular signaling pathway leading to complete B cell activation, proliferation, and antibody production. Without the signals provided by this "B–T cooperation", activation events such as protein kinase activation and increases in intracellular Ca²⁺ may be initiated in the B cell but neither cellular proliferation nor antibody production can occur.

ii) The Three Signal Model of B Cell Activation

The activation of a resting mature naïve B cell by a Td antigen occurs in three steps: (1) the binding of antigen molecules to the BCRs; (2) the establishment of intercellular contacts with a CD4⁺ Th effector cell; and (3) the binding of Th-secreted cytokines to cytokine receptors on the B cell surface. Each of these steps delivers a stimulatory signal to the B cell, but complete activation does not occur unless all three signals are received (Fig. 5-7).

iia) Signal 1. A resting, mature naïve B cell in the periphery is said to be in a *cognitive state* because it is capable of recognizing and reacting to its specific antigen. However, the binding of one antigen molecule to one antigen-binding site of an mIg is not sufficient to activate a B cell. Rather, multiple (in the range of 10–12) antigen–BCR pairs are required to reach the response threshold (equivalent to the cross-linking of Fabs imposed by a Ti antigen of repetitive structure). Within a minute of reaching the threshold of BCR stimulation, a signal that is conveyed down the length of the transmembrane Igα/Igβ proteins induces the expression of nuclear transcription factors. These factors then activate and regulate the gene transcription necessary to prepare the B cell to receive T cell help. Within 12 hours of antigenic stimulation, the B cell expands in size, increases its RNA content, and moves from the resting phase of the cell cycle to the phase preceding cell division. MHC class II molecules are upregulated on the B cell surface, as well as costimulatory molecules that are required for the B cell to receive "signal 2" from the Th cell. The expression of cytokine receptors necessary for the receipt of "signal 3" also commences. This B cell is said to be in a *receptive* state.

iib) Signal 2. Signal 2 is delivered by the engagement of costimulatory molecules on the B cell surface. If a B cell's BCRs are engaged by a Td antigen in the absence of costimulation, the lymphocyte eventually undergoes apoptosis or becomes *anergic* (unresponsive to its specific antigen). Only after receiving both signal 1 and signal 2 can a B cell accept signal 3 in the form of Th cell-secreted cytokines. The most important costimulatory event for B cell activation is the interaction between CD40 on B cells and CD40L (CD40 ligand) on activated Th effector cells. CD40 is constitutively expressed by mature naïve B cells, whereas CD40L is upregulated on a T cell's surface in response to activation by pMHC (see Ch. 9).

In order for the antigen-stimulated B cell to receive T cell help, it must come together with an activated Th effector cell to form a bicellular structure called a *B–T conjugate*. The formation of this conjugate is possible because the recognition of antigen by the BCR does more than just deliver signal 1: it also

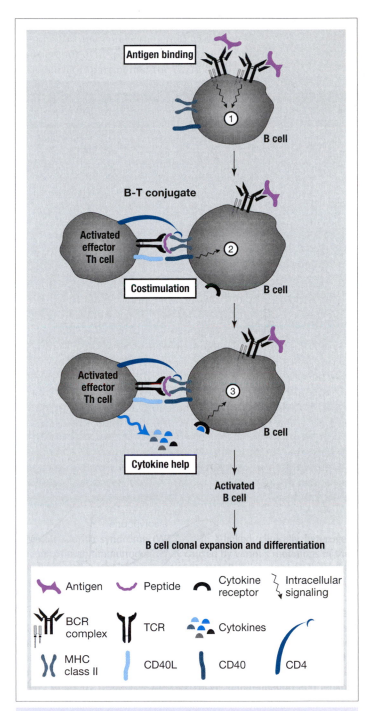

Fig. 5-7 Three Signal Model of B Cell Activation by Td Antigen

allows the B cell to internalize the antigen via receptor-mediated endocytosis, process it into peptides, and present T cell epitopes from the same antigen on its MHC class II molecules. Th effectors specific for this antigen can then recognize the pMHCs displayed by the B cell and are able to form pMHC-TCR "bridges" that mediate B–T conjugate formation. In addition, the CD4 coreceptor of the Th effector cell binds to a distal site on the MHC class II molecule of the pMHC bound by the TCR. These intercellular contacts (plus the interactions of several ICAM-related adhesion molecule pairs) ensure that

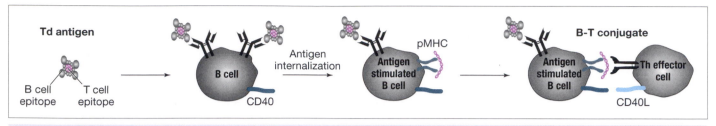

Fig. 5-8 **Linked Recognition**

the B–T conjugate is held together long enough for signal 2 to be delivered.

The fact that a B–T conjugate must be formed for the complete activation of a B cell by a Td antigen means that the B and T cell epitopes involved must be present <u>on the same macromolecular structure or in the same pathogen</u>. In other words, the B cell and the Th cell involved in a Td response are specific for different epitopes of the same antigen, a phenomenon called *linked recognition* (Fig. 5-8). It is linked recognition that results in B cells and Th cells mounting an efficient, coordinated response against a foreign entity. This requirement for linked recognition also means that only proteins of a certain size and complexity are effective Td immunogens, as they must contain both B and T cell epitopes. In addition, linked recognition reduces the likelihood that an autoreactive B cell (that has somehow escaped tolerance mechanisms) will be stimulated, since the Th help required can only come from an antigen-specific Th cell and there is a relatively small chance that an autoreactive Th cell specific for the same host protein will also have escaped tolerance and be present in the periphery.

iic) Signal 3. The binding of antigen to sufficient BCRs on a B cell coupled with CD40/CD40L costimulation induces the expression of new cytokine receptors (particularly the receptors for IL-1 and IL-4) on the B cell surface. However, without the binding of the relevant cytokines to these receptors, the B cell usually undergoes only limited proliferation. While some of the pertinent cytokines required for B cell activation can be secreted by a nearby activated macrophage or DC, most are

derived from the antigen-activated Th effector cell that has made contact with the B cell. Shortly after adhesive, pMHC-TCR and costimulatory contacts are established between the B cell and the Th effector cell, the Golgi apparatus is reorganized within the Th cell such that this secretory structure is closer to the site of contact. The release of the cytokines is then directed more precisely toward the B cell, increasing the efficiency of the delivery of "signal 3". A B cell that has already received signals 1 and 2 and binds IL-4 starts to proliferate vigorously. Other cytokines, including IL-2, IL-5 and IL-10, support this proliferation. As the progeny B cells continue to divide, IL-2, IL-4, IL-5, IL-6, IL-10, IFNs and TGFβ become important for inducing the differentiation of antibody-secreting plasma cells and memory B cells (see later in this chapter).

III. CELLULAR INTERACTIONS DURING B CELL ACTIVATION

Naïve B cell activation and the subsequent maturation of progeny cells into plasma cells and memory B cells depends on cellular interactions that occur in various anatomical structures within the secondary lymphoid tissues. These interactions have been best studied in the lymph nodes.

i) Lymph Node Paracortex and Primary Follicles
One of the most common places for antigen and lymphocytes to meet is in the lymph node. Prior to a B cell's encounter with antigen, the follicles within the cortex of a lymph node are known as *primary follicles* (Fig. 5-9). A primary follicle is filled

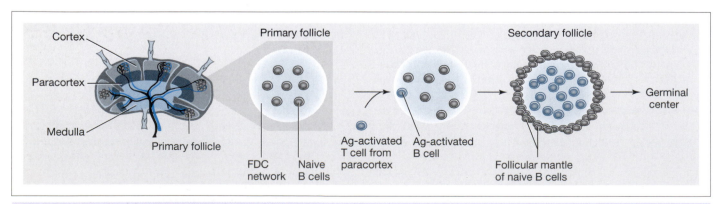

Fig. 5-9 **Primary and Secondary Follicle Structure in the Lymph Node**

with resting naïve B cells and contains a network of FDCs that can trap whole antigens on their surfaces. Let us suppose that an antigen X has penetrated the skin or mucosae and has been conveyed by the lymphatic system to the cortex and paracortex of the regional lymph node. A resting naïve B cell expressing a BCR specific for an epitope on antigen X may be present in a primary follicle of this node. If molecules of antigen X bind in sufficient numbers to the BCRs of this B cell, activation signal 1 is delivered. The now-receptive B cell displays more pMHCs derived from the antigen in anticipation of conjugate formation, and maintains its expression of CD40 in anticipation of signal 2. Meanwhile, the proteins of antigen X are digested and processed by the lymph node's various APCs, including DCs in the paracortex. These cells present pMHC complexes to naïve Th cells located in the paracortex, activating them and causing them to proliferate and generate Th effector cells that express CD40L (see Ch. 9). Thus, at this point, the two partners necessary for B–T conjugate formation are activated but localized in different parts of the lymph node. The antigen (Ag)-activated B cell follows a chemokine gradient emanating from the T cell-rich paracortex to the edge of the primary follicle. On the edge of the follicle, the antigen-activated B cell meets the Th effector cell specific for the same antigen. The B–T conjugate is formed that ensures delivery of signals 2 and 3 to the B cell and its full activation.

About 4–6 days after antigen contact, the activated B cell undergoes one of two fates. In some cases, the B cell on the edge of the follicle immediately proliferates and terminally differentiates into a population of *short-lived plasma cells* without carrying out isotype switching or somatic hypermutation. In other cases, the B cell partner of the B–T conjugate drags the Th effector cell back into the center of the primary follicle where the Th cell dissociates from the B cell to allow the latter's proliferation. The majority of the progeny of this B cell will become *long-lived plasma cells* while the remainder will become memory B cells.

ii) Secondary Lymphoid Follicles

An activated B cell destined to produce long-lived plasma cells and memory B cells undergoes rapid clonal expansion within the primary follicle that converts it into a *secondary follicle*. This expansion requires the presence of antigen (displayed on the surfaces of FDCs), continued engagement of CD40 by the CD40L of an activated antigen-specific Th effector cell, and stimulatory cytokines secreted by the Th effector. By 6–9 days after antigen contact, the uninvolved naïve B cells that filled the primary follicle are displaced by the proliferating B cell clone and are compressed at the edges of the follicle to form the *follicular mantle*.

iii) Germinal Centers

By 9–12 days after antigen contact, the secondary follicle polarizes into two distinct areas, a *dark zone* and a *light zone*, and becomes a *germinal center* (GC) (Plate 5-1). This process of polarization is called the *germinal center reaction*. Activated B cells are found first in the dark zone, where they continue to proliferate rapidly and become known as *centroblasts* (Fig. 5-10, #1). The dark zone of the GC is where the antibody

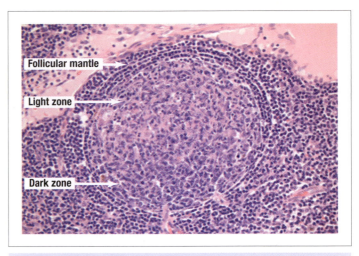

Plate 5-1 **Germinal Center**
[Reproduced by permission of David Hwang, Department of Pathology, University Health Network, Toronto General Hospital.]

repertoire undergoes its final diversification by somatic hypermutation of the V_H and V_L exons (#2). (Somatic hypermutation, affinity maturation and isotype switching are all discussed in detail in the following sections.) As centroblasts mature and differentiate further, they migrate into the light zone where they become known as *centrocytes* (#3). The light zone of the GC is where centrocytes bearing the newly generated somatic mutations interact with antigen on the FDCs and are either negatively or positively selected (#4). In the first step toward establishing *peripheral B cell tolerance* (see Ch. 10), negative selection in the GC induces B cells that no longer recognize the antigen (and could thus be autoreactive) to undergo apoptosis; the dead B cells are removed by macrophages (Mac). Positive selection ensures the survival of B cells that continue to recognize the FDC-displayed antigen with the same or increased affinity (affinity maturation). The light zone is also where the Ig C_H exons undergo isotype switching in progeny B cells to increase functional diversity. At the end of all these processes, the surviving centrocytes either return to the beginning of the GC cycle for further expansion, diversification and selection (#5a), or continue their differentiation (#5b) into long-lived plasma cells or memory B cells that exit the GC and enter the circulation and tissues (#6). The tremendous proliferation and differentiation in GCs throughout the lymph node can persist for up to 21 days after antigen encounter, after which the number and size of the GCs decrease unless there is a fresh assault by antigen.

IV. GERMINAL CENTER PROCESSES THAT DIVERSIFY ANTIBODIES

In Chapter 4, processes that increased the diversity of the antibody repertoire before encounter with antigen were discussed. Here we describe two processes, somatic hypermutation and affinity maturation, that take place in the GC and contribute to antibody diversity after the antigen encounter. Isotype switching, which contributes to functional diversity rather than antigen-binding diversity, also occurs in the GC.

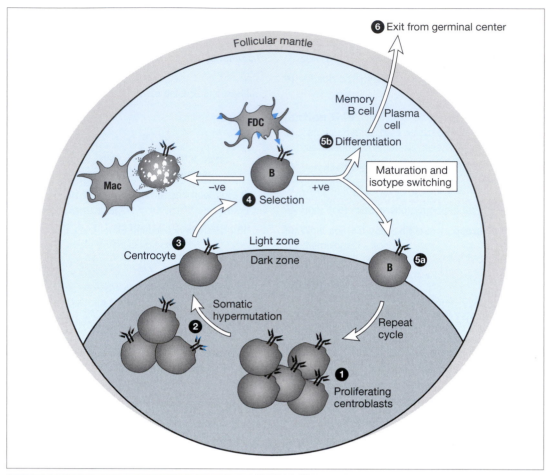

Fig. 5-10 **Germinal Center Formation and Function**
[Adapted from McHeyzer-Williams M.G. and Ahmed R. (1999). B cell memory and the long-lived plasma cell. *Current Opinion in Immunology* **11**, 172–179.]

i) Somatic Hypermutation

The V regions of the IgM antibodies produced by the first plasma cells generated by an activated B cell clone differ slightly in sequence from the V regions of the IgG antibodies produced by later progeny. In the early primary response, the proliferating centroblasts in the GC dark zone undergo a cycle of DNA replication in which their $V_HD_HJ_H$ and occasionally their V_LJ_L exons are particularly subject to the occurrence of random point mutations, a process called *somatic hypermutation* (Fig. 5-11). These mutations continue to accumulate until late in the primary response, resulting in sequence alterations mainly in CDR1 and CDR2 that do not usually destroy the ability of the antibody (Ab) to bind to the antigen (Ag) but often increase its binding affinity. In subsequent exposures to the same antigen (secondary and tertiary responses), the Ig genes of the progeny of activated memory B cells may undergo additional rounds of somatic hypermutation. The rate of point mutation in the V region is estimated to be about 1 change per 1000 base pairs per cell division, about 1000 times the rate of somatic mutation for non-Ig genes (which is why the phenomenon is called somatic hyper̲mutation). Somatic hypermutation requires the activity of an enzyme called *activation-induced cytidine deaminase* (AID) that converts cytidine to uridine.

AID is expressed exclusively in GC B cells in the presence of CD40 signaling.

ii) Affinity Maturation

Somatic hypermutation is a random process so that, theoretically, a set of mutations in the antigen-binding site of an antibody could increase, decrease or have no effect on its binding affinity. In reality, the supply of antigen trapped on the FDCs in the GC is limited, so that competition for binding to a particular epitope on an antigen occurs between centrocytes with different somatic mutations in their BCRs. After somatic hypermutation, centrocytes expressing a BCR of very low affinity for the epitope do not succeed in binding to it and thus do not receive a survival signal that rescues them from death by apoptosis. Centrocytes whose somatic mutations confer a BCR with a higher affinity for the epitope are more likely to bind to it and thus receive a survival signal than are centrocytes with BCRs of only moderate affinity for the epitope. In addition, centrocytes with high affinity BCRs are more efficient at internalizing the antigen and presenting pMHC to Th effector cells. This increased antigen presentation means that these B cells also preferentially receive growth stimulatory signals from the Th cells. These factors combine to ensure the survival and proliferation of B cell clones with increased affinity for the epitope. The higher affinity antibodies produced by these successful clones thus predominate later in the response.

Centrocytes with a significant increase in affinity for an epitope after somatic hypermutation preferentially become memory B cells. When these memory B cells are activated in a secondary response, their progeny plasma cells produce antibodies that recognize the same epitope as primary response antibodies but bind to it with increased affinity (hence the name, affinity "maturation"). Those centrocytes with more modest affinities for the epitope tend to become plasma cells.

iii) Isotype Switching

Centrocytes with a high affinity for antigen are also those cells that undergo isotype switching, replacing the original production of IgM antibodies with that of IgG, IgA or IgE antibodies of the same antigenic specificity. Although isotype switching is independent of somatic hypermutation and can occur without it, most B cells that express Igs of new isotypes have already undergone somatic hypermutation.

During isotype switching, the C_H region of the centrocyte undergoes a series of DNA cutting/rejoining events that can bring any of the downstream C_H exons next to the $V_H D_H J_H$ exon previously established by V(D)J recombination. The antigenic specificity of subsequent progeny B cells is the same (because the V exon is unchanged) but this specificity is now linked to a C region that may confer a different effector function. The actual mechanism of isotype switching, called *switch recombination*, is not yet fully understood but requires the same AID enzyme involved in somatic hypermutation. Switch recombination depends on the pairing of highly conserved *switch regions* ($S\mu$, $S\gamma3$, $S\gamma1$, $S\gamma2b$, $S\gamma2a$, $S\varepsilon$ and $S\alpha$ in the mouse) that lie just upstream of each C_H exon (except $C\delta$) (Fig. 5-12). Once the signal to switch is received, the DNA is likely looped out such that the selected C_H exon is juxtaposed next to the rearranged $V_H D_H J_H$ exon, the intervening sequences (including unused C_H exons) are deleted, and the DNA is repaired to

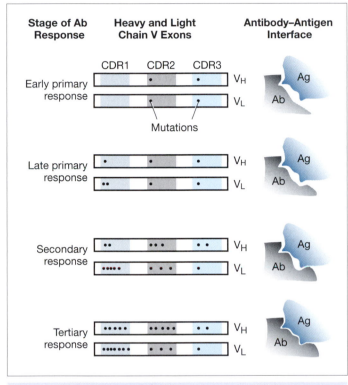

Fig. 5-11 **Somatic Hypermutation**

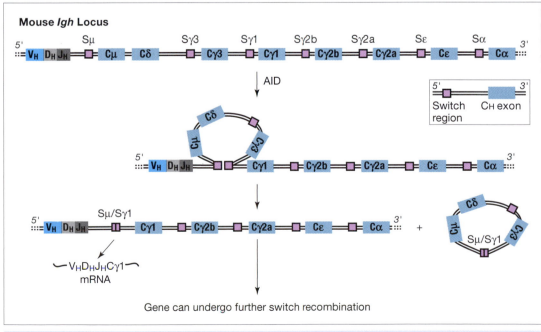

Fig. 5-12 **Mechanism of Switch Recombination**

restore the Ig gene. During the synthesis of any Ig, the C_H exon closest to the $V_H D_H J_H$ exon is preferentially transcribed, spliced and translated. Thus, for example, where switch recombination in a murine B cell has deleted the $C\mu$, $C\delta$ and $C\gamma3$ exons, only the $C\gamma1$ exon immediately 3' of the switch site is transcribed and spliced to the $V_H D_H J_H$ exon, and only IgG1 antibodies are synthesized. However, each daughter of this IgG1-producing B cell may subsequently undergo its own switch, delete one or more of the remaining 3' C_H exons, and start to produce the corresponding IgG2a, IgG2b, IgE or IgA antibodies.

Isotype switching cannot occur unless there is prior production of short transcripts (which are not translated) of the selected C_H exon *in its germline configuration*. These short transcripts may function as components of the switch recombination machinery. As shown in Table 5-4 for mice, cytokines (particularly TGFβ, IL-4 and IFNγ) can promote or inhibit the transcription of the selected C_H exon and thus the production of Igs of a particular isotype. Isotype switching is therefore heavily influenced by cytokines in the immediate microenvironment of the activated B cell. Box 5-2 describes primary immunodeficiencies that are largely associated with defects in isotype switching.

V. PLASMA CELL DIFFERENTIATION

i) Short-Lived Plasma Cells

In a response to a Td antigen, a minority of activated B cells do not experience the GC reaction. When these cells are first activated and positioned on the edge of the primary follicle, they fail to upregulate a transcriptional repressor called Bcl-6 that blocks access to the plasma cell terminal differentiation pathway. As a result, these B cells do not re-enter the primary follicle and instead immediately differentiate into short-lived plasma cells without undergoing isotype switching or somatic hypermutation. No memory B cells are produced via this pathway. Short-lived plasma cells have a half-life of 3–5 days and secrete only low affinity IgM antibodies. An encounter of a B cell with a Ti antigen also gives rise to short-lived plasma cells because, without the involvement of a Th cell, the activated B cells cannot upregulate Bcl-6. Short-lived plasma cells produced in the spleen are particularly important for the very early stages of the adaptive response against blood-borne Ti antigens.

Table 5-4 **Cytokine Effects on Isotype Switching in Mice**			
Cytokine	**Inhibits Isotype Switching to C_H Exon**	**Promotes Isotype Switching to C_H Exon**	**Ig Produced**
IL-4	$C\gamma2a$	$C\gamma1$ $C\epsilon$	IgG1 IgE
IFNγ	$C\gamma1$, $C\epsilon$, $C\alpha$	$C\gamma2a$ $C\gamma3$	IgG2a IgG3
TGFβ	$C\epsilon$	$C\alpha$ $C\gamma2b$	IgA IgG2b

ii) Long-Lived Plasma Cells

Most of the B cells responding to a Td antigen undergo the GC reaction and make the decision to become either a long-lived plasma cell or a memory cell. A centrocyte in a GC that has been positively selected and undergone isotype switching and somatic hypermutation but later experiences a loss of Bcl-6 function is directed to the plasma cell terminal differentiation path. In the presence of IL-2 and IL-10, these cells first become *plasmablasts* and then long-lived plasma cells that can secrete IgG, IgA and/or IgE antibodies. These mature plasma cells have enlarged rER and Golgi compartments and an increased number of ribosomes. Long-lived plasma cells express little or no CD40, MHC class II or mIg on their cell surfaces, can no longer receive T cell help, and are incapable of cell division. Once generated, these plasma cells migrate from the GCs primarily to the bone marrow but can also take up residence in the lymph node medulla or in the splenic red pulp. In these sites, the plasma cells can produce high affinity antibody for several months in the absence of any cell division or re-exposure to the original Td antigen. Up to 40% of the total protein synthesized by these mature plasma cells is immunoglobulin, most of which is released into the blood or tissues as secreted antibody.

iii) Mechanism of Antibody Synthesis

Prior to its differentiation, an activated B cell clone can make both the membrane-bound form of its Ig protein to serve in its

---

Box 5-2 Hyper IgM (HIGM) Syndromes

HIGM patients have normal or very high levels of circulating IgM but very low or absent levels of all other isotypes. Infections with opportunistic fungal and parasitic organisms are frequent. Patients may also suffer from recurrent bacterial infections. There are six HIGM syndromes, five of which are B cell-specific primary immunodeficiencies. HIGM1 is a T cell-specific primary immunodeficiency and is discussed in Chapter 9. **HIGM2** is caused by mutation of the AID enzyme critical for isotype switching and somatic hypermutation. **HIGM3** is due to CD40

mutations that impair isotype switching, memory B cell generation and somatic hypermutation as well as CTL responses. **HIGM4** patients have an unknown defect that acts downstream of CD40 and AID and results in a mild version of the HIGM2 phenotype. While isotype switching is impaired in HIGM4 B cells, somatic hypermutation is not. **HIGM5** patients have a defect in the gene encoding *uracil DNA glycosylase* (UNG) that may act downstream of AID to promote isotype switching. HIGM5 B cells show a partial defect in somatic hypermu-

tation. A mild form of HIGM sometimes called **HIGM-NEMO** occurs in association with an X-linked disorder in which patients have sparse hair and abnormal or missing teeth, and lack sweat glands. Cell-mediated and humoral responses (particularly to polysaccharide antigens) are impaired. The mutation lies in the X-linked gene encoding *NEMO* (NF-κB essential modulator), a regulator of the crucial transcription factor NF-κB.

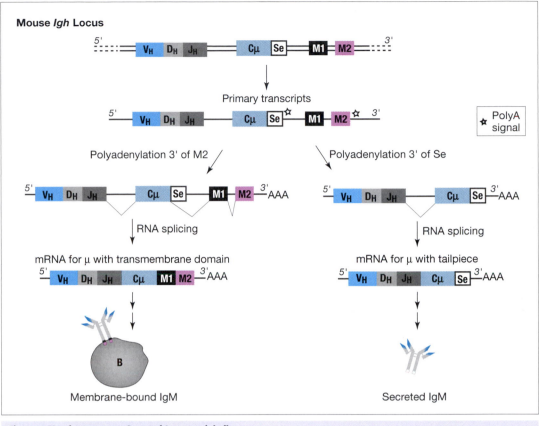

Fig. 5-13 **Membrane versus Secreted Immunoglobulin**

BCR, or the secreted form of its Ig protein to serve as circulating antibody. Varying proportions of these forms are produced as progeny of the original activated B cell differentiate into plasmablasts and then mature plasma cells. In the genome of any murine B cell, each C_H exon is followed by three small exons: *Se*, which encodes the Ig tailpiece of a secreted Ig; *M1*, which encodes the transmembrane domain of an mIg; and *M2*, which encodes the cytoplasmic domain of an mIg (Fig. 5-13). When the polyadenylation site 3′ of M2 is used during transcription, the processed transcripts retain the Se, M1 and M2 sequences. RNA splicing removes the Se sequence, leading to H chains with transmembrane domains (mIg). When the polyadenylation site 3′ of Se is used, only the Se exon is retained at the 3′ end of the processed transcript and sIg containing only the tailpiece is produced. In a resting B cell, the vast majority of H chain transcripts are polyadenylated and processed to produce mIg. However, as the progeny cells of an activated B cell differentiate into plasmablasts, signals are received that cause these cells to preferentially use the other polyadenylation site. The frequency of transcripts containing the transmembrane domain drops precipitously such that the level of mIgM on the cell surface decreases and sIgM is produced in large quantities. Because Se, M1 and M2 exons are associated with each C_H exon, sIgs of all isotypes can be produced after isotype switching. However, the tailpiece sequences that occur in the Cδ exon appear to be seldom used, since IgD is rarely secreted. A very similar mechanism produces mIgs and sIgs in humans.

VI. MEMORY B CELL DIFFERENTIATION

A GC centrocyte that has been positively selected and undergone isotype switching and somatic hypermutation is directed to the memory cell differentiation path if it receives sustained Bcl-6 signaling. This transcriptional repressor blocks the plasma cell differentiation pathway and forces the centrocyte to become a memory B cell.

i) General Characteristics

Memory B cells resemble naïve B cells in their small size and general morphology but carry different surface markers and have a longer life span. As the primary response terminates, memory B cells often take up residence in a body location where the antigen might next be expected to attack. For example, in response to an antigen first encountered in a lymph node, some of the memory B cells produced remain in the follicular mantle and are ready to react rapidly when a fresh dose of the antigen is conveyed to the lymph node. However, other memory B cells may leave the original lymph node and enter the blood, circulating among the body's chain of lymph nodes and maintaining peripheral surveillance for the antigen. In the case of an antigen first encountered in the spleen, the memory B cells produced during the primary response tend to congregate in the splenic marginal zones, precisely where blood-borne antigens collect. For many different antigens, at least some of the memory B cells produced localize preferentially among the epithelial cells of the skin and mucosae, thereby contributing to SALT and MALT.

ii) Secondary Responses

In a secondary response, the activation of memory B cells by antigen occurs in much the same way as in the primary response but is much more efficient for several reasons. Firstly, because of their increased numbers and types of adhesion molecules, memory B cells in the periphery can home to the primary follicles of a lymph node more rapidly than naïve B cells can migrate from the bone marrow. Secondly, due to affinity maturation, the BCR of a memory B cell has an increased affinity for antigen so that the cell is stimulated more easily and efficiently. Thirdly, memory B cells are both present in expanded numbers and can act as APCs for memory Th cells, removing the requirement of having to wait for a DC to activate a naïve Th cell (see Ch. 9). Fourthly, antigen presentation by a memory B cell is associated with faster upregulation of the costimulatory molecules B7-1 and B7-2 needed for complete activation of Th cells (see Ch. 9). Fifthly, the progeny of activated memory B cells differentiate into second generation plasma cells that produce antibodies that are already of greater affinity and diversified isotype.

Although no further somatic hypermutation can occur in mature plasma cells, this process can continue in the progeny of the first generation of memory B cell clones in the GCs. Positive selection then favors the survival of second generation memory B cell clones that display even greater affinity for the antigen. As the secondary response succeeds in clearing the antigen, the second generation plasma cells die off, leaving the second generation memory B cells to maintain peripheral surveillance. In a tertiary response, these second generation memory B cells may undergo additional somatic hypermutation and give rise to third generation plasma cells secreting antibodies with even higher affinities, better-suited effector functions, and/or more appropriate physiological localizations. The effectiveness of the immune response will thus be further improved.

iii) Memory Cell Life Span

Some immunologists believe that, after each round of antigenic stimulation, the progeny of memory B cells are more likely to become plasma cells (which die) than new memory cells (which survive). *In vivo*, this could translate into an increased number of effector cells in the secondary and subsequent responses, and a control on the possible overexpansion of one particular memory B cell clone. However, it also means that the host might one day no longer have memory cells of this clone to call upon when the relevant pathogen strikes. In this "decreasing potential hypothesis", immunological memory is ultimately limited. In addition, for reasons that are not yet understood, memory cells specific for different antigens have different life spans. These variations have implications for how frequently a booster shot must be given to ensure complete vaccination against a particular pathogen (see Ch. 14).

C. Effector Functions of Antibodies

The binding of antibody to antigen leads to clearance or destruction of the antigen and protection of the host. Since different antibody isotypes can trigger different effector functions, how an antigen is eliminated often depends on the isotype of the antibodies to which it is bound. Four types of effector functions can be ascribed to antibodies: *neutralization, classical complement activation, opsonization*, and *antibody-dependent cell-mediated cytotoxicity* (ADCC) (Table 5-5). Neutralization depends solely on the Fab region of the Ig molecule and so is isotype-independent. Classical complement activation depends on isotype because complement component C1q binds to the Fc regions of only certain Ig isotypes. For opsonization and ADCC, the Fc region of the antigen-bound antibody must interact with *Fc receptors* (FcRs) on the surfaces of innate leukocytes that are capable of carrying out these effector functions. These interactions are isotype-dependent because different FcR subtypes are expressed only on particular leukocytes and bind only to specific Ig isotypes. FcRs are described in Box 5-3.

Table 5-5 General Characteristics of Antibody Effector Functions

Effector Function	Isotype-dependent?	FcR-mediated?
Neutralization	No	No
Complement activation	Yes	No
Opsonization	Yes	Yes
ADCC	Yes	Yes

I. NEUTRALIZATION

Neutralization of an antigen is carried out by secreted or secretory antibodies. Certain viruses, bacterial toxins, and the venom of insects or snakes cause disease by binding to proteins on the host cell surface and using them to enter host cells. A neutralizing antibody that can recognize and bind to the virus, toxin or venom can physically prevent it from entering, thereby protecting the cell (Fig. 5-14). If preformed neutralizing antibodies exist in a host (due to a previous exposure to the antigen), the initial spread of the virus can be averted. Once an infection is entrenched, however, neutralizing antibodies are no longer sufficient and the action of CTLs is usually required for successful virus elimination (see Ch. 9 and 13).

II. CLASSICAL COMPLEMENT ACTIVATION

A pathogen coated with antibody can initiate the classical pathway of complement activation. In humans, IgM, IgG1, IgG2 and IgG3 are the antibodies best suited for activating complement in this way and so are sometimes called "complement-fixing" antibodies. The binding of antigen to these antibodies opens a site in the Fc region that allows the binding of C1q. However, the C1q molecule must bind simultaneously to <u>two</u> C1q-binding sites (and thus two separate Fc regions) for it to activate the cascade (refer to Fig. 3-5). A pathogen must

Box 5-3 **Fc Receptors**

The physical removal of an antibody–antigen complex often depends on the recognition of the Fc region of the Ig molecule by an FcR. FcRs are expressed on the surfaces of various leukocytes (including neutrophils, macrophages, NK cells and eosinophils) that can act to eliminate the antigen. There are FcRs for all antibody isotypes except IgD. The FcRs binding to IgG, IgA and IgE are the best characterized.

An FcR is named first for the Ig isotype to which it binds, indicated by a Greek letter. For example, the FcγRs bind to the Fc region of IgG antibodies. If major subtypes of an FcR exist, they are denoted by Roman numerals. For example, the three major FcR subtypes that bind to IgG are FcγRI, FcγRII and FcγRIII. These subtypes differ in amino acid sequence, affinity for the various IgG subtypes, cell type distribution, and effector functions. Within each FcR subtype, distinct but related receptors are indicated with a Roman capital letter. For example, FcγRIII occurs in two slightly different forms:

the FcγRIIIA receptor, which is expressed on NK cells, and the FcγRIIIB receptor, expressed exclusively on neutrophils.

Most FcRs display several Ig-like extracellular domains, a transmembrane domain, and a cytoplasmic domain that often contains ITAMs. Many membrane-bound FcRs are multisubunit complexes, whereas others are single polypeptide chains (see Figure). Several FcRs are composed of one subunit that confers Fc region binding specificity plus two other subunits involved in either transport to the cell surface or intracellular signal transduction in response to Ig binding. This intracellular signaling triggers the initiation of phagocytosis or antibody-dependent cell-mediated cytotoxicity (ADCC; see main text) and thus destruction of the antigen–antibody complex.

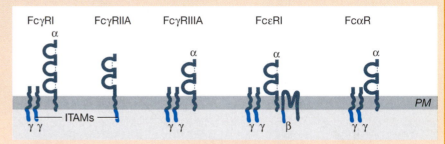

[Figure adapted from Daeron M. (1997). Fc receptor biology. *Annual Review of Immunology* **15,** 203–234.]

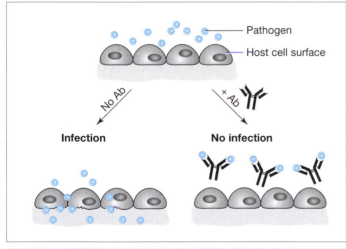

Fig. 5-14 Neutralization of Pathogen by Antibody

therefore usually be well coated with Ig molecules supplying Fc regions in close proximity before the cascade can commence and a MAC is assembled on the pathogen surface.

III. OPSONIZATION

Opsonization is the process by which an antigen is coated with a host protein (the opsonin) in order to enhance recognition by phagocytic cells such as neutrophils and macrophages (refer to Fig. 3-10). Antibodies are powerful opsonins because phagocytes express FcRs that bind strongly to the Fc regions of particular antibody isotypes. Clathrin-mediated endocytosis of the antigen–antibody complexes is then greatly stimulated. In

humans, it is antigen-bound IgG1 and IgG3 that best mediate opsonization.

IV. ANTIBODY-DEPENDENT CELL-MEDIATED CYTOTOXICITY (ADCC)

When a pathogen is antibody-coated but too large to be internalized by a phagocyte, antibody-dependent cell-mediated cytotoxicity can eliminate the invader. ADCC is carried out by certain leukocytes that express FcRs and have cytolytic capability, such as NK cells, eosinophils, and, to a lesser extent, neutrophils, monocytes and macrophages. Once a target entity (which could be a large bacterium, parasite, virus-infected cell or tumor cell) is coated by antibodies, the Fc regions of these antibodies bind to the FcRs of the lytic cell and trigger its degranulation (Fig. 5-15). The release of the hydrolytic contents of the granules in close proximity to the target damages the target's membrane such that its internal salt balance is disrupted and it lyses. NK cells and activated monocytes and macrophages whose FcγRs are engaged also synthesize and secrete TNF and IFNγ, which hasten the demise of the target.

NK cells (discussed in detail in Ch. 11) are the most important mediators of ADCC. In humans, these cells express FcγR molecules that bind to monomeric IgG1 and IgG3 molecules. In general, only targets that are well coated with IgG will trigger the release of the NK cell's damaging contents. Eosinophils express FcεR and FcαR molecules that can bind to IgE- or IgA-coated parasitic targets, particularly helminth worms. These pathogens are resistant to the cytotoxic mediators released by activated neutrophils and NK cells but are susceptible to the granule contents of eosinophils.

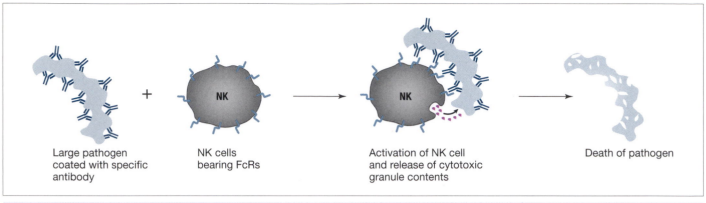

Large pathogen coated with specific antibody

NK cells bearing FcRs

Activation of NK cell and release of cytotoxic granule contents

Death of pathogen

Fig. 5-15 **Antibody-Dependent Cell-Mediated Cytotoxicity (ADCC)**

D. Immunoglobulin Isotypes in Biological Context

Each of the five major Ig isotypes has distinct physical properties and biological effector functions that depend on its C-terminal amino acid sequence and carbohydrate content. These isotype-specific properties influence where a given antibody may travel in the body and how it is involved in host defense. The physical and functional properties of human antibody isotypes are described here and summarized in Table 5-6. Antibodies are also used extensively in the laboratory and the clinic; some of these uses are described in Appendix F.

I. NATURAL DISTRIBUTION OF ANTIBODIES IN THE BODY

The bulk of Ig proteins in the body are present in the form of secretory IgA in the external secretions. These antibodies guard the mucosae where pathogens are likely to attempt entry. Next in relative abundance are the secreted antibodies that circulate throughout the body in the blood. Size considerations dictate that the pentameric sIgM molecule remains primarily in the blood vessels, but that the smaller sIgG, sIgA and sIgE molecules can diffuse freely from the blood into the tissues. No Ig is normally detected in the brain.

II. MORE ABOUT IgM

Monomeric mIgM is always the first Ig produced by naïve B cells. Early in a primary response to antigen, plasma cell progeny of an activated B cell secrete pentameric sIgM exclusively. Because all other antibody isotypes (except IgD) are generated by isotype switching that commences only late in a primary response or not until the secondary response, it is sIgM antibodies that are expressed first in any primary immune response, and those that are synthesized first in a newborn mammal. In an adult human, sIgM antibodies normally com-

Table 5-6 Major Physical and Functional Properties of Human Ig Isotypes

	IgG	IgM	IgA	IgE
Serum concentration (mg/ml)	3–20	0.1–1.0	1–3	0.0001–0.0011
Half-life in serum (days)	2–4	1	1	<1
Intravascular distribution (%)	45	80	42	50
Tailpiece	No	Yes	Yes	No
Secreted form	Monomer	Pentamer	Monomer (IgA1) Monomer, dimer, trimer (IgA2)	Monomer
Classical complement activation	IgG1 ++ IgG2 ++ IgG3 +++ IgG4 −	+++	IgA1 − IgA2 −	−
Placental crossing	+	−	−	−
Involved in allergy	−	−	−	+++

prise only about 5–10% of normal total serum Igs. The detection of increased sIgM levels in an adult indicates exposure to a novel antigen. sIgM antibodies are also produced in response to Ti antigens that can activate a B cell to secrete sIgM but do not induce isotype switching. An example of such Ti antigens are the foreign blood group proteins that might be encountered during a blood transfusion (see Ch. 17).

Because of its pentameric nature, the IgM antibody displays 10 Fab sites that can theoretically bind to a pathogen. In practice, however, steric hindrance usually prevents the IgM molecule from binding to more than five antigenic epitopes at once. Nevertheless, this number is sufficient for the IgM antibody to bind with high avidity to a large antigen or pathogen displaying multiple copies of the same antigenic determinant. IgM is thus able to reduce the infectivity of the pathogen and increase its clearance much more efficiently (using fewer molecules) than a monomeric Ig molecule can. In addition, because the individual binding sites of an IgM molecule are of relatively low affinity, they exhibit correspondingly higher levels of cross-reactivity to related epitopes. This property allows the host to "cast a broad net", maximizing the number of antigens recognized by each IgM-secreting B cell clone.

Pentameric sIgM bound to a pathogen surface is ideally suited for classical complement activation, since multiple Fc regions are already juxtaposed in the pentamer and provide the necessary two C1q-binding sites in close proximity. The classical cascade can thus be triggered by a single molecule of antigen-bound IgM. IgM molecules are also very effective neutralizers and easily prevent pathogens from binding to host receptors on epithelial cells. However, IgM is not an isotype prominent in either opsonization or ADCC because FcRs able to bind to IgM occur only rarely on the surfaces of the appropriate leukocytes.

Although the bulk of IgM is found in the blood, if vascular permeability has been increased during an inflammatory response, sIgM antibodies can exit the blood and enter the tissues to reach sites of infection. In addition, because of the presence of the J chain in the pentameric form of sIgM, these antibodies can occasionally acquire the secretory component by passage through epithelial cells and thus enter the external secretions as SIgM. Although their concentration in the external secretions is very low compared to that of SIgA, SIgM antibodies do make a valuable contribution to mucosal humoral immunity (see Ch. 12).

III. MORE ABOUT IgD

IgD is a monomeric, richly glycosylated Ig that is barely detectable in the blood (0.001% of total serum Ig). mIgD is observed mainly on the surfaces of mature, peripheral B cells that already express mIgM. The precise function of mIgD is unknown but this molecule is capable of sending signals to the nucleus via its associated Igα/Igβ heterodimer. It is thus possible that mIgD either regulates B cell maturation or prolongs the life span of mature B cells in the periphery. Some immunologists speculate that the inherent flexibility of mIgD (due to an extended hinge region) may allow this Ig to bind to antigens featuring epitopes that are widely spaced and could not be bound by the more

rigid mIgM molecule. In any case, mIgD disappears after stimulation of the B cell by antigen.

IV. MORE ABOUT IgG

IgG is the "workhorse" of systemic humoral immunity since it is the isotype most commonly found in the circulation and tissues. In the blood of normal adult humans, 70–75% of serum Ig is monomeric sIgG. The approximate proportions (which vary by individual) of its subclass molecules are: IgG1, 67%; IgG2, 22%; IgG3, 7%; IgG4, 4%. IgG is a key opsonin and important for phagocytosis and ADCC exerted by FcγR-bearing phagocytes and lytic cells, respectively. The IgG1 and IgG3 subclasses are particularly good opsonins and mediators of ADCC because FcγRs bind IgG1 and IgG3 antibodies with high affinity. FcγRs generally bind less well to IgG4, and hardly at all to IgG2.

Although not as efficient as sIgM, sIgG is also an important activator of complement. Free IgG molecules display readily accessible Fc regions but because this Ig is monomeric, it supplies only one C1-binding site per antibody. Two sIgG molecules must be brought together by mutual binding to antigen in order to furnish two C1q-binding sites in close enough proximity to trigger complement activation. IgG3 is the most efficient complement-fixing IgG subclass, while IgG1 is somewhat less efficient, and IgG2 is even less so. IgG4 is unable to bind C1q and so cannot activate complement at all.

IgG antibodies are unique in their ability to cross the mammalian placenta. The immune system of mammals is not fully developed at birth and is limited in its ability to eliminate microbes. In human infants, although independent IgM synthesis starts at birth, it may take as long as 6–12 months for adequate levels of serum and secretory Igs to be produced. To compensate, evolution has provided protection to fetuses and neonates through *passive immunity*, which is defined as "protection by preformed antibodies transferred to a recipient". In humans, maternal IgG1, IgG3 and IgG4 (but not IgG2) antibodies efficiently cross the placenta and enter the fetal circulation, preparing it for birth when it enters a pathogen-filled environment. The maternal sIgG is detectable in a human infant's blood for up until about 9 months after birth.

V. MORE ABOUT IgA

More IgA is produced per day in an adult human than all other Ig isotypes combined. In humans, 85–90% of IgA antibodies are produced mainly by plasma cells in the MALT and are found in the secretory form in the body's external secretions, including tears, saliva, mucous secretions of the gastrointestinal, urogenital and respiratory tracts, breast milk, and prostatic fluid, among others. The remaining 10–15% of IgA antibodies circulate in the blood and are produced by plasma cells in the lymph nodes, bone marrow and spleen. Most sIgA antibodies in the blood are monomeric, but 20% occur as multimers (up to hexamers). Memory B cells located in the diffuse submucosal lymphoid tissues, the Peyer's patches, and the tonsils produce large quantities of IgA antibodies because cytokines secreted by Th cell subsets in these MALT microenvironments promote successive switching to the IgA isotype.

Secretory IgA antibodies are of enormous importance because they facilitate antigen removal right at the mucosal surface, the most common site of initial pathogen attack. Neutralization is the predominant effector mechanism used by SIgA antibodies. Accordingly, these Igs are often said to have "antiviral" activity, since the polymeric nature of this antibody allows it to easily bind repeating epitopes on virus particles, impeding their attachment to mucosal cell surfaces. The expulsion of these particles from the body is then carried out by mechanical means, such as the movement of mucus by the undulating cilia of the respiratory tract. Because IgA antibodies are poor complement fixers and opsonins, the delicate mucosae are protected from potential damage caused by the inflammation associated with the activation of phagocytes, lytic cells or complement. Secretory IgA also contributes to the passive immunity that protects mammalian neonates, since this antibody is passed along in breast milk to defend the neonatal gut mucosa. Secreted IgA is useful for the elimination of helminth worms, because sIgA-coated parasites can be dispatched by ADCC carried out by eosinophils bearing FcαR.

VI. MORE ABOUT IgE

Monomeric secreted IgE is present in the serum at the lowest concentration of all isotypes, a mere 0.000003% of the total Igs in a healthy human. IgE antibodies do not cross the placenta, cannot fix complement, and do not function as opsonins. Nevertheless, IgE antibodies have a clinical impact that far outweighs their actual numbers and limited effector options. Firstly, IgE is essential for combating large parasitic worms. Th cells responding to these invaders secrete cytokines that influence activated B cells to undergo isotype switching to IgE. The serum concentration of IgE rises dramatically, and antigen-specific IgE molecules coat the surface of the worm. The presence of the IgE permits the recognition of the invader by FcεR-bearing eosinophils, the only lytic cell type competent to destroy these pathogens. Secondly, sIgE antibodies are responsible for the symptoms experienced in allergic reactions such as hay fever, and more severe conditions such as asthma and anaphylactic shock. These disorders are all manifestations of a type of immune reaction called *immediate hypersensitivity*. The symptoms and causes of this and other types of immune hypersensitivity are discussed in Chapter 18.

This concludes our discussion of B cells and the humoral response. The next two chapters deal with the MHC and antigen processing and presentation. An appreciation of these elements of the T cell adaptive response is necessary for a complete understanding of T cell activation, the lynch pin of both humoral and cell-mediated adaptive immunity.

CHAPTER 5 TAKE-HOME MESSAGE

- **B cell maturation proceeds from HSCs through the MPP, CLP, pro-B, pre-B, immature B, transitional B and mature B cell stages. Most of this development occurs in the bone marrow and is independent of antigen.**

- **Negative selection in the bone marrow removes B cells expressing potentially autoreactive BCRs and establishes central B cell tolerance.**

- **B cell differentiation takes place in secondary lymphoid tissues and involves the activation of mature B cells by antigen and the generation of memory B cells and antibody-secreting plasma cells.**

- **Ti-1 antigens contain mitogenic regions and are polyclonal activators. Ti-2 antigens are large polymeric molecules with repetitive structures or subunits capable of cross-linking mIg. Ti antigens activate B cells in the absence of T cell help.**

- **Td antigens are protein-containing macromolecules that supply both B and T cell epitopes. Complete B cell activation by a Td antigen requires: signal 1, antigen binding to BCRs; signal 2, costimulation supplied by an activated Th effector cell specific for the same antigen; and signal 3, Th-derived cytokines. Signals 2 and 3 constitute T cell help.**

- **T cell help is required for somatic hypermutation, affinity maturation, isotype switching and memory B cell differentiation.**

- **Somatic hypermutation, affinity maturation and isotype switching occur in the germinal centers and are responsible for antibody diversification after antigen encounter.**

- **The major effector functions of antibodies are neutralization, classical complement activation, opsonization and ADCC.**

- **Different Ig isotypes are best suited for different effector functions because isotype defines the structure of the antibody. Antibody structure determines if an Fc region can activate complement via C1q interaction, or engage FcRs expressed by cells mediating opsonized phagocytosis or ADCC.**

Section A.I–II

1) Can you define these terms? stromal cell, pre-BCR, B-PIs, CVID

2) Give two major differences between the maturation and differentiation phases of B cell development.

3) At what B cell stage does V(D)J recombination commence?

4) What is the surrogate light chain and why is it necessary?

Section A.III–V

1) What is receptor editing and why is it useful?

2) Distinguish between negative and positive B cell selection in the bone marrow.

3) Describe at the DNA level how IgM and IgD can be coexpressed by a B cell.

4) To which tissues do immature naïve B cells migrate upon exiting the bone marrow?

Section B.I

1) Can you define these terms? T cell help, BCR cross-linking, molecular complexity, conformational determinant, immunodominant, high zone tolerance, parenteral

2) Distinguish between antigens and immunogens.

3) Distinguish between Ti-1 and Ti-2 antigens.

4) How are Ti antigens physically different from Td antigens and what effect does this have on their immunogenicity?

5) What role do costimulatory molecules play in B cell activation?

6) How do the antibodies produced in response to Ti and Td antigens differ?

7) Describe five properties of a protein that affect its ability to be a Td immunogen.

Section B.II

1) Can you define these terms? B–T cooperation, B–T conjugate, cognitive B cell, receptive B cell, anergic B cell

2) Why does a B cell need a Th cell specific for the same antigen for complete activation?

3) Describe the three signal model of B cell activation.

4) How does linked recognition increase the efficiency of B cell activation?

Section B.III

1) Can you define these terms? centroblast, centrocyte, follicular mantle, GC light zone

2) How does a primary lymphoid follicle become a secondary follicle?

3) What is the germinal center reaction?

Section B.IV

1) Can you define these terms? AID, switch recombination, HIGM syndrome

2) How does somatic hypermutation diversify antibody specificity?

3) Why is affinity maturation useful to the immune response?

4) Distinguish between positive and negative B cell selection in the GC.

5) How does isotype switching diversify antibodies?

6) Give two examples of how cytokines influence isotype switching.

Section B.V–VII

1) Give three differences between short-lived and long-lived plasma cells.

2) Describe at the DNA level how Ig synthesis switches from mIg to sIg production.

3) Why is Bcl-6 important for memory B cell differentiation?

4) Give three reasons why secondary responses are faster and stronger than primary responses.

Section C.I–IV

1) How do FcRs function to facilitate opsonized phagocytosis and ADCC?

2) How does neutralization by an antibody protect a cell?

3) What human antibodies are best suited for classical complement fixation and why?

4) What human antibodies are best suited for opsonized phagocytosis and why?

5) Describe how ADCC is important for defense against helminth worms.

Section D.I–VI

1) Which antibody isotype occurs in the largest amount in the body and in what location?

2) What isotypes are the first antibodies produced in a primary response? In a newborn?

3) Why is IgM well suited for classical complement activation?

4) What are some possible functions of IgD?

5) Why is IgG well suited for opsonization and ADCC?

6) What is passive immunity and how can it protect an infant?

7) Why have IgA antibodies evolved to be poor complement fixers?

8) Which antibody isotype is prominent in allergy?

6

The Major Histocompatibility Complex

WHAT'S IN THIS CHAPTER?

Near or far, hiddenly,
To each other linked are,
That thou canst not stir a flower
Without troubling a star.
Francis Thompson

As was introduced in Chapter 2, recognition of antigen by αβ T cells is more complex than antigen recognition by B cells. While the BCR binds directly to a unitary epitope on a pathogen or foreign macromolecule, the TCR binds to a bipartite epitope (pMHC) composed of an antigen-derived peptide bound to an MHC protein encoded by a gene in the *major histocompatibility complex* (MHC).

A. Overview of the Major Histocompatibility Complex

Proteins encoded by the MHC were originally discovered in the 1930s during studies of tissue rejection in transplantation experiments. These proteins were therefore named for their association with *histocompatibility* (*histo*, meaning "tissue," and *compatibility*, meaning "getting along"). The genes controlling the histocompatibility of tissue transplantation were localized to a large genetic region containing multiple loci; hence, the term "complex". Moreover, the proteins encoded by these genes were found to have dramatic effects on histocompatibility. To distinguish these proteins from other molecules (encoded elsewhere in the genome) that had relatively minor effects on histocompatibility, these molecules were called the "major" histocompatibility molecules. Thus, the genes encoding these proteins were dubbed the "major histocompatibility complex" (MHC) genes. Soon after, it was discovered that MHC-controlled rejection of transplanted tissue was due to the mounting by the transplant recipient of an immune response against the donated cells (see Ch. 17). Although this finding implied that MHC gene products were directly involved in immune responses, it took several more decades for immunologists to define the normal physiological role of MHC-encoded proteins in presenting antigenic peptides to T cells.

The MHC-encoded proteins that are involved in most instances of antigen recognition by T cells are the *MHC class I* and *MHC class II* molecules. The TCRs of CD8+ T cells recognize peptides bound to MHC class I, while the TCRs of CD4+ T cells recognize peptides bound to MHC class II (Fig. 6-1). As is described in detail in Chapter 8, the CD8 coreceptor of CD8+ T cells also binds to MHC class I, while the CD4 coreceptor of CD4+ T cells binds to MHC class II. The MHC class I protein is a heterodimer consisting of a large transmembrane α chain non-covalently linked to a small non-transmembrane chain called *β2-microglobulin* (β2m). The MHC class I α chain is encoded within the MHC but β2m is not (Table 6-1). The MHC class II protein is composed of an α chain and a slightly smaller β chain, both of which are transmembrane proteins encoded by genes in the MHC. Despite this difference in composition, the tertiary structures of MHC class I and class II molecules are highly similar, apart from the peptide-binding groove. While almost all nucleated cells express MHC class I, only the few cell types that function as APCs (including DCs, macrophages and B cells) express MHC class II. Thus, almost any cell can serve as a target cell and present antigen to CTLs derived from CD8+ Tc cells, but only APCs can activate CD4+ Th cells.

I. HLA COMPLEX

In the human genome, the MHC is called the *HLA complex* (for *h*uman *l*eukocyte *a*ntigen complex). The HLA complex covers about 3500 kb on chromosome 6 and contains 12 major regions, as shown in Figure 6-2A. Each region contains dozens of genes, only some of which are functional and many of which do not appear to be involved in antigen presentation.

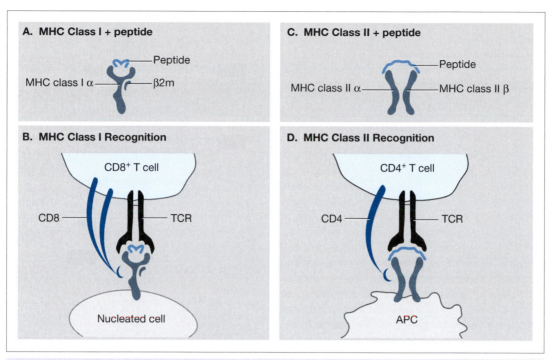

Fig. 6-1 **Recognition of MHC Class I and II Molecules by T cells**

Table 6-1 **Chromosomal Location of MHC Class I and II Genes**

Protein Encoded	Chromosome	
	Human	**Mouse**
MHC class I α chain	6	17
β2-microglobulin	15	2
MHC class II α chain	6	17
MHC class II β chain	6	17

The HLA-A, HLA-B and HLA-C regions are all MHC class I regions. Each contains a single functional gene encoding a human MHC class I α chain. The DP, DQ and DR regions are all MHC class II regions. Each contains multiple functional genes encoding both MHC class II α and β chains. The single genes within each of the HLA-E, -F and -G regions encode MHC class Ib proteins, while several genes in the DM and DO regions encode MHC class IIb proteins. MHC class Ib and IIb proteins structurally resemble MHC class I and II proteins, respectively, but are not directly involved in routine antigen presentation to T cells. MHC class Ib and IIb proteins are therefore considered to be "non-classical" MHC molecules (see Box 6-1). The MHC class III region is not known to encode any peptide-binding presentation molecules but contains many genes relevant to immune responses, including those encoding complement components, HSPs and the cytokines TNF and *lymphotoxin* (LT).

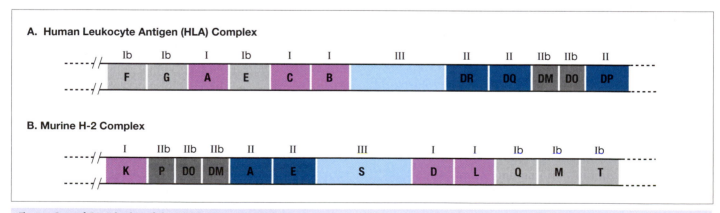

Fig. 6-2 **General Organization of the MHC in Humans and Mice**
[Source: http://imgt.cines.fr/.]

Box 6-1   Human Non-classical and MHC-like Genes and Proteins

The non-classical MHC class Ib and IIb molecules resemble the classical MHC proteins in structure but generally do not present peptides to αβ T cells (see Table). The class Ib genes in humans are located in the HLA-E, -F and -G regions. Some class Ib gene products are secreted (unlike the products of classical MHC class I genes), while others are membrane-bound. Two MHC class Ib proteins called HLA-E and HLA-F may function in antigen presentation to γδ T cells (see Ch. 11). HLA-G is expressed in placental cells during fetal development and may contribute to the prevention of maternal immune responses against the fetus (see Ch. 9). A gene called HFE, which is located in the HLA-E region, encodes an MHC class I-like protein that associates with β2m but does not have a peptide-binding groove. HFE appears to be involved in iron absorption, such that when HFE is defective, excessive iron is deposited in various organs. Two MHC class Ib proteins called MICA and MICB closely resemble MHC class I molecules in structure but are stress-induced molecules. The binding of MICA to a particular receptor on NK and γδ T cells can stimulate these cells (see Ch. 11). The human MHC class IIb proteins are located in the DO and DM regions. The HLA-DM protein regulates the loading of antigenic peptides onto MHC class II molecules (see Ch. 7), while the HLA-DO protein negatively regulates HLA-DM activity.

The CD1 proteins are encoded <u>outside</u> the MHC loci but share certain structural similarities and functions with classical MHC molecules. The MHC-like CD1 proteins associate with β2m but have binding grooves that are very hydrophobic. This type of groove preferentially binds fragments of lipid and glycolipid antigens. Certain T lineage cells can be activated by this non-peptidic form of antigen presentation (see Ch. 7).

	Class I	Class II	Class Ib	Class IIb	Class III	MHC-like
Gene encoded in MHC region	Yes	Yes	Yes	Yes	Yes	No
Polypeptides	Class I α plus β2m	Class II α plus Class II β	Class I α-like plus β2m	Class II α-like plus Class II β-like	Neither class I nor class II chains	Non-MHC chains plus β2m
Tissue expression	Almost ubiquitous	APCs	Restricted	APCs	Almost ubiquitous	Restricted
Soluble form?	Very rare	No	Some	Some	Yes	Some
Polymorphism	Extreme	Extreme	Limited	None	None	None
Function	Peptide presentation to $CD8^+$ T cells	Peptide presentation to $CD4^+$ T cells	Stimulation of γδ T or NK cells; fetus protection; iron absorption	Peptide loading of MHC class II	Complement components, inflammatory cytokines, heat shock and stress proteins	Lipid antigen presentation to T lineage cells
Examples	HLA-A HLA-B HLA-C	HLA-DP HLA-DQ HLA-DR	HLA-E HLA-F HLA-G HFE MICA MICB	HLA-DM HLA-DO	C4 TNF HSP70	CD1 isoforms

II. H-2 COMPLEX

In the mouse genome, the MHC is known as the *H-2 complex*. The H-2 complex is spread over 3000 kb on chromosome 17 and contains 12 major regions, as shown in Figure 6-2B. The K, D and L regions contain single functional genes that encode mouse MHC class I α chains. The A and E regions each contain a single functional gene encoding an MHC class II α chain, and one or more functional genes encoding an MHC class II β chain. The S region of the H-2 complex contains genes encoding the MHC class III proteins, again including complement proteins, HSPs, TNF and LT. The Q, T and M regions of the H-2 complex contain genes encoding class Ib proteins, whereas class IIb proteins are encoded by genes in the P, DO and DM regions.

B. MHC Class I and Class II Proteins

The MHC class I and class II proteins are heterodimeric molecules comprising an extracellular N-terminal peptide-binding region, an Ig domain-containing extracellular region, a hydrophobic transmembrane region, and a short C-terminal cytoplasmic region. The structure of the peptide-binding region in both MHC class I and class II molecules is such that the affinity of an MHC molecule for peptide is much lower than that of an antibody for its cognate antigen. This relaxed binding is a necessity if a given MHC molecule is to carry out its task of presenting a wide range of peptides for T cell perusal. It should also be noted that a given peptide may be capable of binding to different MHC class I or class II molecules, a phenomenon known as "promiscuous binding".

I. MHC CLASS I PROTEINS

Most nucleated cells sport a mixed population of MHC class I proteins. The peptides bound by these MHC class I molecules are generally of *endogenous* origin; that is, they are derived from the degradation of proteins synthesized within the cell (see Ch. 7). The vast majority of these peptides will be "self" in nature, because most proteins routinely produced within a cell at any one time are of host origin (as opposed to proteins of non-self origin, such as those generated during a viral infection). The MHC class I molecule does not discriminate among "self" and "non-self" peptides; that job is left to the TCRs of CD8⁺ T cells. Self peptide–MHC complexes do not trigger an immune response because T cells with the corresponding specificity are generally absent from the T cell repertoire due to the establishment of central tolerance (see Ch. 9). In contrast, non-self peptides complexed to MHC class I are recognized and trigger CD8⁺ T cell activation.

i) MHC Class I Component Polypeptides

In both mice and humans, MHC class I α chains are glycoproteins of about 44 kDa and composed of three extracellular globular domains (Fig. 6-3). Domains α1 and α2 at the N-terminal end of the chain non-covalently pair with each other to form the peptide-binding site, while the Ig-like α3 domain associates non-covalently with the β2m polypeptide. The α chain also supplies the transmembrane domain and the cytoplasmic domain. The α1 domain maintains its shape without disulfide linkage, but the α2 and α3 domains each have an internal disulfide bond. The other partner in the MHC class I molecule, the β2m protein, is a non-transmembrane polypeptide of about 12 kDa. β2m resembles a single Ig-like domain and associates non-covalently with the α3 domain of the MHC class I α chain to maintain the overall conformation of the MHC class I molecule. The association of the β2m chain with the α chain in the ER soon after protein synthesis is essential for the transportation of the complete heterodimer to the cell surface.

ii) MHC Class I Peptide-Binding Site

The groove-like peptide binding site of the MHC class I molecule is relatively small. As a result, MHC molecules cannot recognize large native antigens. Rather, antigens must be processed to small peptides before they can fit into the MHC groove and be presented to T cells. Each MHC molecule can bind a spectrum of small peptides with moderately high affinity but binds to only one peptide at a time. It has been estimated that each MHC class I molecule has the ability to bind several hundred different peptides (Fig. 6-4).

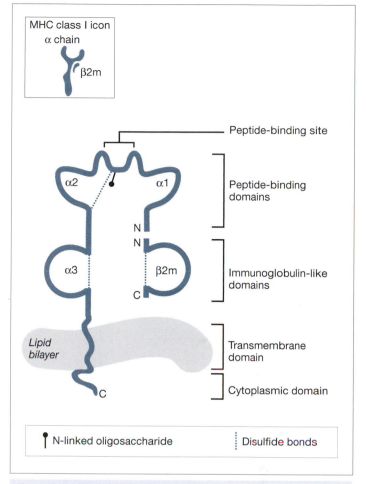

Fig. 6-3 **Structure of the MHC Class I Protein**

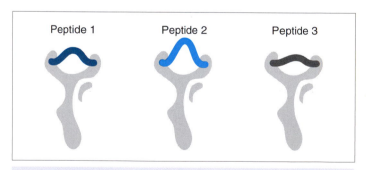

Fig. 6-4 **Accommodation of Different Peptides in the MHC Class I Binding Site**

The MHC class I peptide-binding groove is formed by the juxtaposition and interaction of the α1 and α2 domains of the α chain. The β2m chain contributes by interacting with the amino acids in α1 and α2 that form the floor of the groove. These interactions are strengthened and the entire MHC class I structure is stabilized when the groove is occupied by a peptide of 8–10 amino acids. The peptide is held in place in the groove by interactions between specific amino acids of the α1 and α2 domains and conserved "anchor residues" located

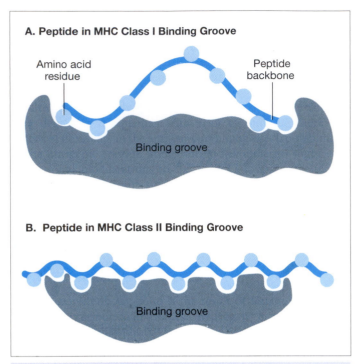

Fig. 6-5 **MHC Peptide-Binding Sites**

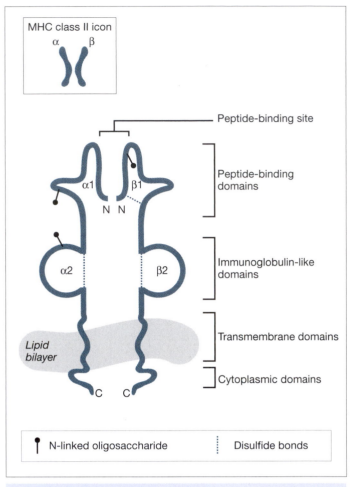

Fig. 6-6 **Structure of the MHC Class II Protein**

in the N- and C-termini of the peptide. The peptide anchor residues point "down" into the groove, while the central peptide residues project "up" toward the TCR (Fig. 6-5A). A sufficient degree of conformational flexibility exists such that peptides of widely varying amino acid sequence in the region between the anchor residues can occupy the groove. The ends of the MHC class I groove are closed, which means peptides larger than 8–10 amino acids can fit in only if their central residues can bulge upward out of the groove.

II. MHC CLASS II PROTEINS

As mentioned earlier, MHC class II molecules are found mostly on APCs. The peptides bound by MHC class II are of *exogenous* origin; that is, derived from the degradation of proteins that have entered the cell from the exterior via either phagocytosis or receptor-mediated endocytosis (see Ch. 7). Because APCs also capture and digest spent host proteins, the vast majority of peptides presented on MHC class II molecules are "self" and do not trigger CD4+ T cell activation because these specificities have been removed from the Th cell repertoire by the establishment of central tolerance. A Th response is induced when an APC presents a non-self peptide bound to MHC class II.

i) MHC Class II Component Polypeptides

In both humans and mice, the α and β chains of MHC class II proteins are glycoproteins of similar size and structure (24–32 and 29–31 kDa, respectively). Both chains contain an N-terminal extracellular domain, an extracellular Ig-like domain, a hydrophobic transmembrane domain and a short cytoplasmic tail (Fig. 6-6). The peptide-binding region is made up of the N-terminal α1 and β1 domains of the α and β chains, respectively. The α2 and β2 domains form globular loops that are homologous to the Ig fold but are not involved in peptide binding.

ii) MHC Class II Peptide-Binding Site

The MHC class II peptide-binding groove is similar in overall structure to that of MHC class I molecules (Fig. 6-5B). However, the ends of the MHC class II groove are open, permitting the binding of much longer peptides (up to 30 amino acids). Nevertheless, the majority of peptides found in MHC class II grooves are 13–18 amino acids long. The open ends of the MHC class II groove also mean that binding does not depend on conserved anchor residues at the ends of the peptides but is instead mediated by hydrogen bonding between the peptide backbone and the side chains of certain MHC amino acids. Researchers have found that antigenic peptides that are successfully bound to the floor of the MHC class II groove possess a particular conserved secondary structure (resembling a polyproline chain) in the portion of the peptide that aligns with critical acidic MHC residues located in the middle of the groove. As a result of this conformational requirement, MHC class II proteins generally bind a narrower range of proteins than do MHC class I proteins.

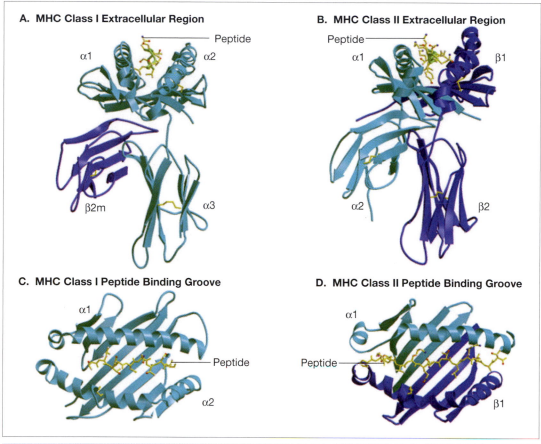

A. MHC Class I Extracellular Region

B. MHC Class II Extracellular Region

C. MHC Class I Peptide Binding Groove

D. MHC Class II Peptide Binding Groove

Plate 6-1 **X-Ray Crystal Structures of MHC Class I and II in the Mouse**
[Reproduced by permission of Bjorkman, P.J. (1977). MHC restriction in three dimensions: A view of T cell receptor/ligand interactions. *Cell* **9**, 167–170.]

III. X-RAY CRYSTALLOGRAPHY OF MHC CLASS I AND II MOLECULES

Much of the information on how MHC class I and II molecules bind to peptides has come from X-ray studies of crystallized pMHC complexes. Plate 6-1 shows side-view crystal structures of the carbon backbones of the extracellular regions of murine MHC class I (A; β2m shown on the left) and MHC class II (B). The structures of their respective peptide-binding grooves viewed from above are shown in C and D. The similarity of the tertiary structures of MHC class I and class II molecules can be clearly seen. Analyses of MHC crystal structures have shown that water plays an important role in peptide–MHC binding. The fit of the peptide in the groove is tightened when water molecules fill any gaps in the complex.

C. MHC Class I and Class II Genes

I. POLYGENICITY OF MHC CLASS I AND II GENES

Most proteins in our bodies are unique; that is, there is only one functional gene in the genome that encodes a protein carrying out that particular function. The MHC genes are unusual in that, due to gene duplication during evolution, two to three separate, functional genes encoding the same type of MHC class I or II polypeptide exist. This phenomenon is called *polygenicity*. These genes are named for their region of location and the chain they specify. For example, the HLA includes three loci, HLA-A, -B and -C, that all encode the same type of polypeptide: an MHC class I α chain. Similarly, genes giving rise to MHC class II α chains can be found in the HLA-DP, -DQ and -DR regions; these genes are called DPA, DQA and DRA genes, respectively. Also in each of the HLA-DP, -DQ and -DR regions are separate genes that encode MHC class II β chains; these are called DPB, DQB and DRB genes, respectively.

The MHC class II loci show further polygenicity in that each of the HLA-DP, -DQ and -DR regions may have more than one alpha chain gene and more than one beta chain gene. For example, within the HLA-DP region, there are two genes that could encode MHC class II α chains, DPA1 and DPA2, and two genes that could encode MHC class II β chains, DPB1 and DPB2. However, only the DPA1 and DPB1 genes are functional. Similarly, the HLA-DQ region contains the DQA1 and DQA2 genes which could encode MHC class II α chains, and the DQB1, DQB2 and DQB3 genes which could encode MHC class II β chains. However, only DQA1 and DQB1 are functional. In the HLA-DR region, a single gene designated DRA encodes the DR α chain and is functional in all humans.

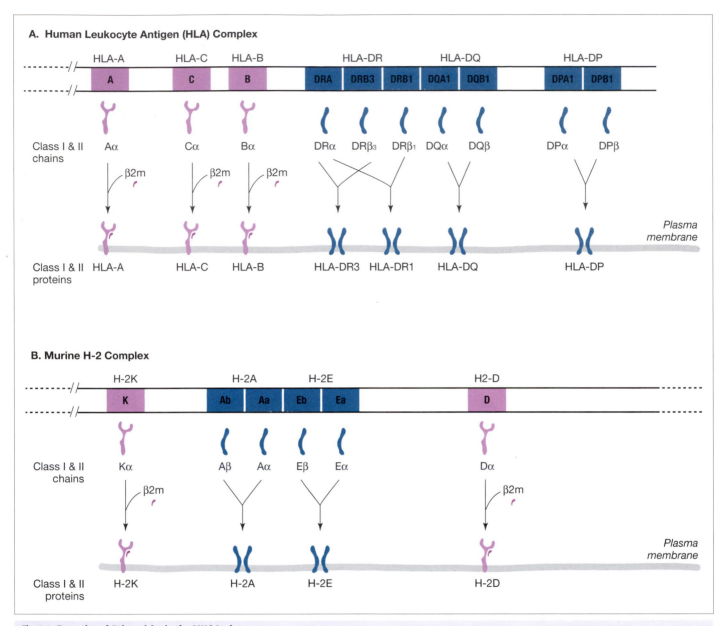

Fig. 6-7 **Examples of Polygenicity in the MHC Loci**

In contrast, not every individual carries the same number of DRB loci on his/her chromosomes. Nine different DRB genes have been identified, designated DRB1 to DRB9. While every individual has the DRB1 and DRB9 loci, different individuals may also have one or more DRB loci selected from among the DRB2 to DRB8 genes. However, only DRB1, DRB3, DRB4 and DRB5 are functional and encode DR β chains.

For unknown reasons, an α chain derived from a DP region gene almost always combines with a β chain derived from the DP region (and not from the DQ or DR regions) to form a complete MHC class II molecule. Similarly, a DQ α chain combines with a DQ β chain and a DR α chain with a DR β chain. Only very rarely do mixed MHC class II molecules such as HLA-DRA/DQB occur. Figure 6-7A illustrates how the products of the HLA loci can come together to form complete human MHC molecules.

Polygenicity also occurs in the mouse H-2 complex. Two loci, H-2K and H-2D, contain single genes encoding MHC class I α chains. MHC class II α chains are encoded by one functional gene called Aa (or Aα) within the A region of the mouse H-2, as well as by the Ea (or Eα) gene in the E region. Similarly, MHC class II β chains are encoded by the Ab (or Aβ) gene in the H-2A region and the Eb (or Eβ) gene in the H-2E region. Figure 6-7B shows examples of how products of the H-2 complex give rise to complete murine MHC class I and II molecules. Again, an α chain derived from an A region gene almost always combines with a β chain derived from the A region, and not with an E region β chain (and vice versa).

II. POLYMORPHISM OF MHC CLASS I AND II GENES

The vast majority (>90%) of vertebrate genes are *monomorphic*; that is, almost all individuals in the species share the same nucleotide sequence at that locus. In contrast, the MHC loci exhibit extreme polymorphism. *Polymorphism* is the existence in a species of several different alleles at one genetic locus. *Alleles* are slightly different nucleotide sequences of a gene; the protein products of alleles have the same function. For example, close to 600 alleles have been identified for the HLA-A gene, more than 900 for HLA-B and more than 300 for HLA-C (Table 6-2). A functional MHC class I molecule can consist of the protein product of any one of these HLA-A, -B or -C alleles associated with the invariant β2m chain. Multiple alleles also exist for the MHC class II genes, so that the product of any DPA allele can combine with the product of any DPB allele (and DQA with DQB, and DRA with DRB) to form a functional MHC class II protein. The degree of sequence variation among MHC alleles can be astonishing: differences of as many as 56 amino acids have been identified between individual alleles. Not surprisingly, this amino acid variation is concentrated in the peptide-binding site of the MHC protein. In MHC class I α chains, most of the polymorphism is localized in the α1 and α2 domains. The α3 domain is less polymorphic and more Ig-like, and the transmembrane and cytoplasmic domains are more conserved than any of the α domains. The β2m protein exhibits almost no polymorphism within a species or variation among species. In the case of MHC class II molecules, polymorphic variation among alleles is found in the α1 and β1 domains that constitute the peptide-binding site. Again, the transmembrane and cytoplasmic domains are highly conserved.

Due to the high level of polymorphism in the HLA, humans (which are outbred) are generally *heterozygous* at their MHC loci (have different alleles on the maternal and paternal chromosomes). In contrast, experimental mice (which have been repeatedly inbred to create pure strains) are *homozygous* for any given MHC gene (have the same allele on both the maternal and paternal chromosomes). In addition, in outbred populations, two individuals are very likely to have different nucleotide sequences at each HLA locus. These two individuals are said to be *allogeneic* to each other at their MHC loci (*allo*, meaning "other"). In an inbred population, not only is each individual homozygous at each MHC locus but all individuals in the population express the same MHC allele at a given locus. Such inbred animals express exactly the same spectrum of MHC molecules and are said to be *syngeneic* at their MHC loci (*syn*, meaning "same").

III. CODOMINANCE OF MHC EXPRESSION

The polygenicity and polymorphism of the MHC genes underlie the vast diversity of MHC molecules expressed within an outbred population. In an individual, the breadth of MHC diversity is increased by the fact that, at each MHC locus, the gene on both chromosomes is expressed independently, or *codominantly*. In other words, when a given MHC locus is expressed in an individual, the genes on both the maternal and paternal chromosomes produce the corresponding proteins. For example, in an individual heterozygous at the HLA-A locus, there are two MHC class I α chains produced (maternal and paternal) that can combine with β2m to form two different HLA-A proteins. Similarly, for an MHC class II molecule such as HLA-DP, two different α chains and two different β chains are produced that can combine to form four different HLA-DP proteins. For a locus such as HLA-DR, which comprises more than one DRB gene in most individuals, the number of possible HLA-DR heterodimers is much higher.

The net effect of polygenicity, polymorphism and codominant expression on the profile of MHC molecules expressed by an outbred individual is illustrated in Figure 6-8. In this example, panel A depicts the expression of a wide spectrum of

Table 6-2 **Numbers of HLA Alleles (as of August 2007)**	
	Number of Alleles
MHC Class I Genes	
HLA-A	580
HLA-B	921
HLA-C	312
MHC Class II Genes	
HLA-DPA1	23
HLA-DPB1	127
HLA-DQA1	34
HLA-DQB1	86
HLA-DRA	3
HLA-DRB	577

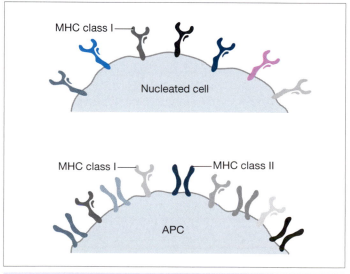

Fig. 6-8 **Spectrum of MHC Class I and II Expression in an Outbred Individual**

MHC class I molecules on the surface of a nucleated host cell. Panel B shows MHC expression on a typical APC, where multiple types of MHC class II molecules appear on the surface as well as multiple types of MHC class I molecules.

IV. MHC HAPLOTYPES

The MHC loci are closely linked, meaning that the specific set of alleles for all MHC loci on a single chromosome is usually passed on to the next generation as an intact block of DNA. This set of alleles is called a *haplotype*, and any individual inherits two haplotypes from his/her parents: the MHC block on the paternal chromosome and the MHC block on the maternal chromosome. In an outbred population, the high degree of polymorphism of the MHC can lead to great variation in haplotypes among unrelated individuals. Interestingly, within the human population, researchers have identified over 30 *ancestral haplotypes* that are shared not only within a single family but also among a large number of families. These ancestral haplotypes are thought to have originated in "founder" populations that settled in diverse geographic regions, so that a given ancestral haplotype is often associated with a particular ethnic background. For example, one ancestral haplotype is found predominantly among Basques and Sardinians, while another is specific to Eastern European Jews, and a third is exclusive to South East Asians.

In inbred mice, both parents have the same allele at each MHC locus, the maternal and paternal haplotypes are the same, and all offspring inherit the same single haplotype on both chromosomes. Immunologists frequently use a single term "short form" to indicate the haplotype of a particular strain. For example, the haplotype of the C57BL/6 mouse strain is denoted "H-2^b" where the "b" in H-2^b means that allele number 12 is present at the K locus, allele number 74 is at the Aβ locus, allele number 3 is at the Aα locus, allele number 18 is at the Eβ locus, and so on. In contrast, the CBA mouse strain has a haplotype of "H-2^k", where allele number 3 is present at the K locus, allele number 22 at Aβ, and so on. More detailed short forms can be used to indicate specific alleles in a haplotype. For example, the term "H-2D^b" means that the D allele being discussed is that which occurs in a mouse strain of the "b" haplotype. Researchers might also identify this allele as D^b, and verbalize it as "D of b".

V. EXPRESSION OF MHC GENES

The expression of MHC genes is tightly and differentially regulated, such that MHC class I is expressed on almost all healthy host cells but MHC class II expression is limited to APCs. As well, MHC protein expression may be upregulated or induced by cytokines and other stimuli released in a host cell's vicinity. Depending on the type of host cell and the tissue in which it resides, these stimuli may be either constitutively produced or induced during an immune response to injury, pathogens or tumors. For example, molecules in the walls of invading bacteria stimulate macrophages to produce TNF and LT, and viral infection induces the infected cells to synthesize

IFNs. The interaction of these cytokines with specific receptors on a host cell triggers intracellular signaling pathways that activate transcription factors. The activated transcription factors enter the host cell nucleus and bind to 5′ regulatory motifs in the DNA upstream of the MHC genes, altering their expression. An increase in MHC expression facilitates the amplification of an adaptive response by enhancing antigen presentation to T cells.

D. Physiology of the MHC

I. POLYMORPHISM AND THE BIOLOGICAL ROLE OF THE MHC

How did the MHC loci come to be so polymorphic? In an ancient, antigenically simple world, a primeval MHC molecule that displayed endogenous and exogenous protein fragments to T cells likely existed but was of very limited (or non-existent) variability. As the world became more antigenically complicated, it was individuals with multiple duplications of this primordial MHC gene that likely survived because they possessed more than one gene dedicated to presenting protein fragments. Perhaps concurrently, different alleles of each gene also evolved, each with a different sequence in the peptide-binding groove. A broader range of peptide binding and presentation molecules would have been generated. Today, the resulting polymorphism at multiple MHC loci ensures that each member of an outbred species is heterozygous at most if not all MHC loci, and thus has a very good chance of possessing at least one MHC allele capable of binding any given antigenic peptide. For the species as a whole, MHC polymorphism means a large catalog of MHC alleles is spread over the entire population. In the case of a devastating pathogen attack, a significant fraction of the population (but not all individuals) will be able to respond to the pathogen and survive to perpetuate the species.

II. MHC AND IMMUNE RESPONSIVENESS

Immunologists have long observed that some foreign proteins that provoke strong immune responses in some individuals fail to do so in others. Those individuals failing to mount a response were originally called *non-responders*, while those that did react were called *responders*. Among responders, there were subtle differences in the level of the response, leading to the description of individuals as either *low* or *high* responders. Immunologists soon mapped the genes behind immune responsiveness to the MHC, and showed that mice of different MHC haplotypes sometimes respond differently to a given peptide (Table 6-3). These variations in response levels to a given antigen can be interpreted as differences in the ability of particular MHC alleles to effectively present peptides from that antigen that can be recognized by T cells. In an inbred population, there is a greater possibility that an individual will be a non-responder; that is, an individual will lack an MHC allele that can lead to specific T cell activation during a challenge with a particular antigen. Two hypotheses, which may not be

Table 6-3 MHC Haplotype Correlated with Immune Responsiveness

Mouse Strain	H-2 Haplotype	Response to TGAL* Peptide
C3H	k	Low
C3H.SW	b	High
A	a	Low
A.BY	b	High
B10	b	High
B10BR	k	Low

*TGAL, synthetic peptide containing lysine, alanine, tyrosine and glutamic acid residues.

mutually exclusive, have been proposed to account for non-responsiveness: the *determinant selection* model and the *hole in the T cell repertoire* model.

i) Determinant Selection Model

For a T cell response to be mounted against a foreign protein, a host must possess at least one MHC allele with a groove that accommodates a peptide derived from that protein. Responsiveness then depends on the strength of binding between a given MHC allele and a given determinant (peptide), which in turn depends on structural compatibility. In other words, the MHC proteins in an individual "select" which determinants will be immunogenic in that individual as well as the extent of the response. Since a foreign protein is usually processed into three to four strongly immunogenic peptides, an outbred individual is very likely to possess an MHC allele capable of binding to at least one of these peptides and provoking an immune response. Such an individual is then a responder to this particular antigen, and their status as a high or low responder correlates with strong or weak binding of the peptide to MHC, respectively. On the other hand, if none of the individual's MHC molecules can bind to any of the peptides generated from the protein, the individual is a non-responder to this antigen.

The determinant selection model has been supported by experiments in which a given peptide is immunogenic only when it is bound to a particular MHC allele. For example, in inbred mice of the H-2^k haplotype, a certain peptide of an influenza virus protein readily provokes an immune response. More specifically, the determinant is recognized when presented to T cells on the MHC class I H-2K^k molecule. However, this same peptide fails to stimulate T cells in mice of the H-2^b haplotype. Instead, a peptide from a different part of the same influenza virus protein triggers the antiviral response in the H-2^b mice when presented on the MHC class I H-2D^b molecule.

ii) Hole in the T Cell Repertoire Model

Immune non-responsiveness may also result from tolerance mechanisms. It may be that, in non-responders, a particular foreign peptide–self MHC combination very closely resembles the structure of a self peptide–self MHC combination, such

that any T cell clones capable of recognizing the foreign peptide–self MHC combination were eliminated as autoreactive during the establishment of central tolerance. In a non-responder, this would result in a missing T cell specificity or a "hole" in the T cell repertoire relative to the repertoire of a responder.

III. MHC AND DISEASE PREDISPOSITION

An individual's MHC haplotype determines his/her responsiveness to immunogens. If an individual cannot mount an appropriate immune response to an immunogen associated with infection or cancer, the individual will likely suffer disease. If an immune response is mounted when it is inappropriate, disease in the form of autoimmunity or hypersensitivity (including allergy) can result. The direct link between immune responsiveness and particular MHC alleles means that certain MHC haplotypes may predispose individuals to particular susceptibilities or disorders.

In humans, many of the disorders linked to the possession of specific MHC alleles manifest as *autoimmune disease*. Autoimmune disease results when self-reactive T cell clones escape the tolerance mechanisms that would normally prevent these cells from entering or acting in the periphery. The individual may then possess T cells that can recognize self components and may attack tissues expressing these components. For example, type 1 (insulin-dependent) diabetes mellitus is thought to arise from an autoimmune attack on antigens expressed by the insulin-producing β cells of the pancreatic islets. Immune destruction of the β islet cells results in insulin deficiency and thus diabetes. For unknown reasons, the HLA-DQ8 allele is eight times more prevalent in groups of humans suffering from type 1 diabetes than it is in healthy populations. Similarly, 90% of Caucasian patients suffering from a degenerative disease of the spine called *ankylosing spondylitis* carry the HLA-B27 allele, whereas only 9% of healthy Caucasians do.

Several autoimmune diseases and their association with particular HLA alleles are given in Table 6-4. Note, however, that mere possession of a predisposing HLA allele is not usually

Table 6-4 Examples of HLA-Associated Disorders in Humans

Disease	Examples of Associated HLA Alleles
Ankylosing spondylitis	B27
Birdshot retinopathy	A29
Celiac disease	DR3, DR5, DR7
Graves' disease	DR3
Narcolepsy	DR2
Multiple sclerosis	DR2
Rheumatoid arthritis	DR4
Type 1 diabetes mellitus	DQ8, DQ2, DR3, DR4

sufficient to cause disease; other genetic and environmental factors are thought to be involved. A complete discussion of autoimmune disease is presented in Chapter 19.

This concludes our discussion of the structure and physiology of the MHC. In the next chapter, entitled "Antigen Processing and Presentation", we describe the derivation of antigenic peptides, and how MHC molecules associate with these peptides and present them to T cells.

CHAPTER 6 TAKE-HOME MESSAGE

- The MHC class I and II genes in the MHC encode cell surface proteins that present peptides to T cells.

- Non-classical MHC class I and class II genes as well as MHC class III genes also constitute part of the MHC.

- MHC class I and class II proteins are heterodimeric molecules with highly variant N-terminal domains that form a peptide-binding site.

- MHC class I is expressed on almost all cells in the body and generally presents peptides of endogenous origin.

- MHC class II is expressed by APCs and generally presents peptides of exogenous origin.

- MHC class I interacts with the CD8 coreceptor found on Tc cells and CTLs, while MHC class II interacts with the CD4 coreceptor on naïve and effector Th cells.

- The MHC genes are characterized by polygenicity, extreme polymorphism and codominant expression.

- Outbred populations are heterozygous at the MHC loci, while inbred mice are homozygous. Syngeneic individuals have the same MHC genotype, while allogeneic individuals have different MHC genotypes.

- Differences in MHC alleles are largely responsible for transplant rejection and variations in immune responsiveness to a given pathogen.

- Expression of particular MHC alleles is linked to autoimmune disease predisposition.

DID YOU GET IT? **A SELF-TEST QUIZ**

Section A

1) Describe the derivation of the term "major histocompatibility complex".

2) What polypeptide chains make up the MHC class I molecule? MHC class II?

3) What cell types express MHC class I? MHC class II?

4) To what type of cells does MHC class I present peptide? MHC class II?

5) Name five regions of the HLA complex and describe the nature of their gene products.

6) Name five regions of the H-2 complex and describe the nature of their gene products.

7) Name three non-classical MHC genes and describe the nature of their gene products.

8) What MHC-like proteins present non-peptidic antigens to T lineage cells?

Section B.I–II

1) Why do MHC proteins have relatively low binding affinity for peptides?

2) What is an anchor residue?

3) Do self peptides bound to MHC usually provoke immune responses? If not, why not?

4) What domains form the peptide-binding grooves of MHC class I molecules? Of MHC class II molecules? How do these grooves differ in structure?

5) Describe two ways in which the peptides binding to MHC class I differ from those binding to MHC class II.

Section C.I–III

1) Can you define these terms? polygenicity, monomorphic, polymorphic, allele, codominance

2) Which MHC molecule is more common in a human: DQA/DRB or DQA/DQB?

3) Where is amino acid variation concentrated in the MHC protein and why?

4) Distinguish between the terms "heterozygous" and "allogeneic".

5) How many different types of HLA-DQ proteins are likely present in an outbred individual?

Section C.IV–V

1) Can you define these terms? haplotype, ancestral haplotype

2) What does the term H-2K^b represent? How would you verbalize this term?

3) What effect does inflammation have on MHC expression?

Section D.I–II

1) Why does inbreeding sometimes put a species at risk for decimation by a pathogen?

2) Outline two theories accounting for variation in immune responsiveness to an antigen.

7

Antigen Processing and Presentation

Real generosity toward the future lies in giving all to the present.
Albert Camus

A. Overview of Antigen Processing and Presentation

Antigen processing is the complex process by which peptides are produced from antigenic proteins. *Antigen presentation* is the binding of these peptides to MHC molecules and the positioning of the resulting pMHC complexes on a host cell surface so that they can be inspected by T cells. Antigen processing and presentation are discussed in this chapter, whereas the recognition of pMHCs by the TCRs of T cells is discussed in Chapter 8.

Antigen processing provides the host with a means of scanning the proteins constantly being produced and turned over in the body. At any one time, almost every cell in the body displays several hundred thousand pMHCs on its surface. This population represents hundreds of distinct peptides, the vast majority of which are "self" in origin and elicit no T cell response in a healthy individual. In the case of an infection, a substantial proportion (up to 10%) of the peptides may be pathogen-derived, a number more than sufficient to trigger the activation of T cells specific for these pMHCs.

Antigenic peptides are produced by either the *exogenous* (or *endocytic*) processing pathway or the *endogenous* (or *cytosolic*) processing pathway. The exogenous processing pathway breaks down proteins that are synthesized or acquired from <u>outside</u> the host cell (extracellular proteins). This acquisition occurs when whole pathogens or their components and products are internalized by cells specialized for this purpose; that is, by APCs. The degradation of the pathogen proteins into peptides takes place within the endosomal compartment of the APC. In contrast, the endogenous processing pathway generates peptides from proteins that are synthesized <u>inside</u> any host cell (intracellular proteins). Such proteins include, among myriad self proteins, the components of pathogens that are able to replicate inside host cells, as well as proteins made by tumor cells. Protein degradation takes place within the

cytoplasm (or cytosol) of almost all host cells (including APCs).

Following their synthesis on the rER of a host cell, MHC class I and II molecules follow different intracellular trafficking pathways such that MHC class I molecules interact predominantly with the endogenous processing pathway, and MHC class II molecules interact with the exogenous processing pathway. As a result, MHC class I molecules usually display peptides from intracellular proteins, while MHC class II molecules display peptides from extracellular proteins. Because MHC class I molecules are expressed on all nucleated cells, and MHC class I molecules have a binding site for the CD8 coreceptor normally expressed by Tc cells and their CTL effectors, it is CD8+ T cells that respond to intracellular proteins produced by any infected cell in the body. In contrast, MHC class II molecules are expressed only by cells that can act as APCs. Because MHC class II molecules have a binding site for the CD4 coreceptor of Th cells, it is CD4+ Th cells that are activated by extracellular proteins taken up by APCs resident in lymphoid tissues or circulating past sites of pathogen entry into the body. These differences between MHC class I and II in cell type expression and coreceptor binding mean that the most appropriate subset of T cells is activated in response to the presence of a given threat. For example, in the case of an intracellular threat such as viral replication or tumorigenesis, Tc cells are activated whose daughter CTL effectors kill any host cell producing the intracellular antigen. In the case of an extracellular pathogen, Th cells are activated whose daughter Th effectors provide T cell help to antigen-specific B cells. The activated B cells produce the plasma cells and antibodies necessary to clear the pathogen from the host. In addition, IFNγ secreted by certain Th effectors hyperactivates macrophages so that they can more easily kill any extracellular pathogen they have engulfed. Thus, as summarized in Figure 7-1, it is the intracellular or extracellular origin of the antigen that determines the pathway used to process the protein and consequently the cells that respond to it.

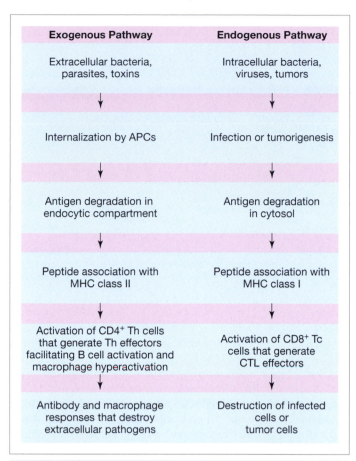

Fig. 7-1 **Overview of Exogenous and Endogenous Antigen Processing Pathways**

Exogenous Pathway	Endogenous Pathway
Extracellular bacteria, parasites, toxins	Intracellular bacteria, viruses, tumors
↓	↓
Internalization by APCs	Infection or tumorigenesis
↓	↓
Antigen degradation in endocytic compartment	Antigen degradation in cytosol
↓	↓
Peptide association with MHC class II	Peptide association with MHC class I
↓	↓
Activation of CD4+ Th cells that generate Th effectors facilitating B cell activation and macrophage hyperactivation	Activation of CD8+ Tc cells that generate CTL effectors
↓	↓
Antibody and macrophage responses that destroy extracellular pathogens	Destruction of infected cells or tumor cells

Table 7-1 **Comparison of APCs: DCs, Macrophages and B Cells**

	Mature DCs	Macrophages	B Cells
Receptor used to acquire antigen	PRRs	PRRs	BCR
Level of MHC class II	Very high	High	High
Level of constitutive costimulatory molecule expression	High	Moderate	Low
Capable of cross-presentation	+++	++	+/−
Activates naïve Th and Tc cells	Yes	No	No
Activates memory Th and Tc cells	Yes	Yes	Yes

B. Exogenous Antigen Processing

I. PROFESSIONAL APCs

{ Dendritic Cells
  Macrophages
  B cells. }

Peptides from antigens that are synthesized outside the cell, like those derived from extracellular bacteria, toxins or parasites, are captured by receptor-mediated endocytosis or phagocytosis and enter the exogenous processing pathway of APCs. Components of intracellular pathogens may also end up inside APCs when viable pathogens are captured during transit between cells or when components in the debris of dead cells are phagocytosed (see Ch. 9). "Professional" APCs are those few cell types that constitutively or inducibly express high levels of MHC class II and costimulatory molecules and so can activate CD4+ T cells. Professional APCs include mature DCs (which can activate both naïve and memory T cells) and macrophages and B cells (which cannot activate naïve T cells) (Table 7-1). "Non-professional" APCs are cell types (such as fibroblasts and epithelial cells) that can transiently express low levels of MHC class II if exposed to IFNγ during an inflammatory response. These cells play only a minor role in antigen processing and T cell activation.

i) Dendritic Cells as APCs

ia) DC subsets. Multiple DC subsets have been identified in both mice and humans. These subsets are distinguished by their marker expression, function, location in the body, and whether they were generated during "steady state" conditions or during the course of an infection. "Steady state" DCs are routinely present in an immature state in healthy individuals and are induced to undergo maturation upon infection. "Inflammatory" DCs are not normally present in a healthy individual but are induced to appear by the inflammation associated with infection. Inflammatory DCs are often derived from monocytes that extravasate into an inflamed tissue and encounter high levels of inflammatory cytokines and growth factors. In the context of total DC numbers, these cells are relatively uncommon.

Two classes of DCs are present under steady state conditions: *plasmacytoid DCs* and *conventional DCs* (Table 7-2). Plasmacytoid DCs are comparatively rare cells that are named for the morphology of their immediate precursor, which resembles a plasma cell and carries the B cell marker B220 but does not produce Igs. Upon activation, plasmacytoid DCs adopt the dendrite-like morphology of conventional DCs but show a response to antigen acquisition that is unique. Unlike conventional DCs, plasmacytoid DCs lack expression of most TLRs but express high amounts of TLR7 and TLR9 (refer to Ch. 3). These intracellular PRRs allow plasmacytoid DCs to sense RNA and DNA viruses. The stimulation of TLR7 and TLR9 induces the vigorous production of IFNα/β, which mediate innate defense against viruses and may activate other DC subsets. The production of these cytokines distinguishes plasmacytoid DCs from other DC subsets, as most activated DCs produce IFNγ but not IFNα/β. Plasmacytoid DCs are usually found in the blood, lymph nodes and thymus.

Conventional DCs can be further classified as *migratory DCs* or *lymphoid-resident DCs*. As mentioned in Chapter 2, all subtypes of DCs can be derived from early myeloid or

Table 7-2 Properties of Steady State DC Subsets

DC Subset	Function	Examples
Plasmacytoid	Secrete IFNα, IFNβ during early antiviral responses	Found in blood, lymph nodes, thymus
Conventional migratory	Front-line defense in peripheral tissues Acquire antigens in peripheral tissues and migrate through lymphatics to lymph nodes Deliver antigens to lymphoid-resident DCs or directly initiate T cell responses in local lymph nodes	Langerhans cells in epidermis Dermal DCs in dermis Mucosal DCs in mucosae of body tracts Interstitial DCs in non-lymphoid tissues
Conventional lymphoid-resident	Front-line defense in lymphoid tissues Do not migrate but acquire antigens from migratory DCs or antigens that have accumulated in lymphoid tissues; initiate T cell responses in lymph nodes Participate in central tolerance Initiate T cell responses to blood-borne antigens in the spleen	Lymphoid-resident DCs in lymph nodes Thymic DCs Splenic DCs

lymphoid precursors, although lymphoid-resident DCs are more likely than migratory DCs to be of lymphoid origin. Lymphoid-resident DCs include those present in the thymus and spleen and about half of those present in lymph nodes. Lymphoid-resident DCs do not travel the body in the lymph system and instead remain in one lymphoid site, collecting and presenting self and foreign antigens that enter that tissue. Several different subsets of lymphoid-resident DCs have been identified based on differential surface marker expression. Among these subsets are *thymic DCs*, most of which develop from intrathymic hematopoietic precursors and remain in this organ throughout their short life span. Thymic DCs most likely participate in the establishment of central tolerance, presenting peptides from self antigens to T cells developing within the thymus (see Ch. 9). Immature T cells that strongly recognize these pMHCs (and thus are self-reactive T cells) are then eliminated. *Splenic DCs* reside in the spleen and monitor blood-borne antigens. At least three subtly different subsets of splenic DCs have been identified in mouse spleen.

Migratory DCs are the classic DCs that collect antigens in the peripheral tissues and convey these molecules to the local lymph node via migration through the lymph system. Once in the lymph node, these DCs either deliver the antigen to lymphoid-resident DCs, or directly present pMHCs derived from the antigens to naïve T cells present in the node. The rate of migration of migratory DCs is increased in response to pathogen attack or inflammation. There are several subtypes of migratory DCs that differ slightly in their surface markers, tissue distribution, cytokines secreted upon activation, and effects on T cell activation. It is thought that most migratory DCs arise from monocytes that differentiate under non-inflammatory conditions. Examples of migratory DCs are *Langerhans cells* (LCs), which are relatively long-lived DCs present in the epidermis of the skin; *dermal DCs* present in the dermis of the skin; *mucosal DCs* in the mucosae lining the gastrointestinal, respiratory and urogenital tracts; and *interstitial DCs* present in almost all other non-lymphoid peripheral tissues.

Although conventional DCs are associated primarily with antigen presentation to T cells of the adaptive response, they have additional functions that extend their influence to the innate response, the B cell response, Th subset differentiation, and peripheral T cell tolerance. These properties of conventional DCs are outlined in Box 7-1 and explored in more detail in Chapters 9, 10 and 11.

ib) Immature conventional DCs. Immature conventional DCs that are either settled in their lymphoid tissue of residence or migrating through the lymph system initially express relatively low levels of MHC class II on their cell surfaces, potentially limiting their antigen display and thus their capacity to interact with T cells. However, migratory DCs (and their precursors) express large arrays of chemokine and cytokine receptors as well as adhesion molecules that allow them to migrate efficiently to non-lymphoid sites experiencing inflammation or infection. Once in the site of infection, immature migratory DCs expertly capture both whole pathogens and macromolecular antigens using the many PRRs on their cell surfaces (Fig. 7-2). Lymphoid-resident DCs can do the same for pathogens and antigens that have been brought into a lymphoid tissue either through the lymph or by a migratory DC. Among the most important PRRs expressed by immature DCs are the TLRs (including TLR-2, -3, -4, -5, -6, -7 and -8), FcγRII (a low affinity IgG receptor), and the *mannose receptor* (MR). The FcγRII molecule allows the efficient binding of any antigen complexed to IgG (through opsonization), whereas MR binds to a wide variety of glycoproteins and other antigens exhibiting exposed mannose residues. Immature DCs also express scavenger receptors (like CD91 and CD36) and complement receptors (like CR3) that allow the phagocytosis of whole pathogens and the uptake of antigenic proteins released by necrotic cells. All these sources of extracellular antigens supply an immature DC with large amounts of antigenic peptides to later display on MHC class II.

ic) Mature conventional DCs. As described in Chapter 3, "danger signals" are molecules whose presence alerts cells of the host's innate immune system that a threat is present. Bacterial and viral PAMPs, host stress molecules and complement components are among the most common danger signals

Box 7-1 The Multiple Talents of Dendritic Cells

DCs have several functions in addition to antigen processing and presentation. (1) Because immature DCs bear numerous TLRs and CRs, these cells contribute to the innate response by scanning the body for the presence of bacterial and viral products and complement components that serve as ligands for these receptors. DCs activated by TLR or CR engagement secrete cytokines (such as IFNγ, IL-12, IL-15 and IL-18) that promote the development and activation of NK and NKT cells. (2) Cytokines secreted by DCs also support B cell activation and promote plasma cell differentiation. (3) Different subsets of DCs can drive the differentiation of the Th cells they activate down one of two parallel paths described in Chapter 9. For example, a DC subset that preferentially produces IL-12 and the IFNs in response to engagement of its TLRs causes the Th cell it has activated to differentiate into a *Th1 cell*. Th1 cells secrete cytokines that assist in adaptive responses against intracellular threats. In contrast, a DC subset that preferentially produces IL-13 helps to influence the Th cell it has activated to differentiate into a *Th2 cell*. Th2 cells secrete cytokines that assist in adaptive responses against extracellular threats. Finally, a DC subset that preferentially produces IL-6 helps to influence the Th cell it has activated to differentiate into a *Th17 cell*. Th17 cells both participate in adaptive responses against extracellular threats and play a role in autoimmunity. All these Th subsets are discussed in Chapter 9. (4) Certain DCs contribute to peripheral T cell tolerance. A naïve T cell that binds to pMHCs displayed by a DC that does not express high levels of costimulatory molecules is killed or *anergized* (inactivated) rather than activated (see Ch. 9 and 10).

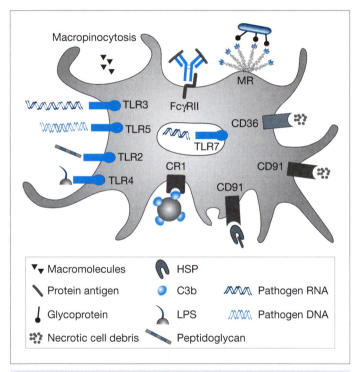

Fig. 7-2 **Examples of Antigen Capture Mechanisms in Conventional DCs**

Table 7-3 **Danger Signals**	
Danger Signal	**Examples**
Bacterial products	LPS, peptidoglycan, CpG, mannose-bearing carbohydrates
Viral products	Double-stranded and single-stranded RNA
Complement products	C3b, C4b, iC3b
Reactive oxygen intermediates	H_2O_2, $OH\cdot$, O_2^-
Stress molecules	HSPs, chaperone proteins

The precise molecular and cellular processes mediating DC maturation remain to be characterized. However, studies of migratory DCs have suggested that once maturation is initiated, the actin cytoskeleton of the DC is reorganized, the receptors used to internalize antigen are downregulated, and new patterns of chemokine receptors (particularly CCR1 and CCR7) are expressed. The antigen-loaded DCs migrate into the blood or lymph and travel to the secondary lymphoid organs where naïve T cells can be found. After reaching their lymphoid tissue destinations, DCs continue to mature, losing the capacity to capture antigen but gaining the capacity to activate naïve T cells (Table 7-4). Surface expression of MHC class II increases by 5- to 20-fold, allowing the now mature DC to rapidly present many copies of different antigenic pMHCs to Th cells. If the TCR expressed by a Th cell recognizes one of the pMHCs displayed by the mature DC, costimulatory molecules such as CD28 and CD40L are upregulated on the Th cell surface (see Ch. 9). The binding of CD40L on the Th cell to CD40 on the DC greatly upregulates DC expression of the B7 costimulatory molecules that bind to CD28. These high levels, which greatly exceed those expressed by other types of APCs, are necessary to supplement the signal delivered by TCR binding to pMHC. The DC is able to push the naïve Th cell over the activation threshold in a way that other APCs cannot.

(Table 7-3). An immature DC whose PRRs are engaged by danger signals is induced to commence the maturation process that allows it to present captured antigen to naïve T cells (Fig. 7-3). Immature DCs that are exposed to proinflammatory cytokines secreted by other activated innate leukocytes will also commence maturation. Thus, by contributing to the maturation of DCs, inflammation and danger signals caused by infection or injury set the stage for the subsequent activation of lymphocytes if an adaptive response becomes necessary. Importantly, the apoptotic cell death associated with normal development and cell turnover may furnish an immature DC with antigen but does not provoke maturation because danger signals and inflammation are absent in these cases.

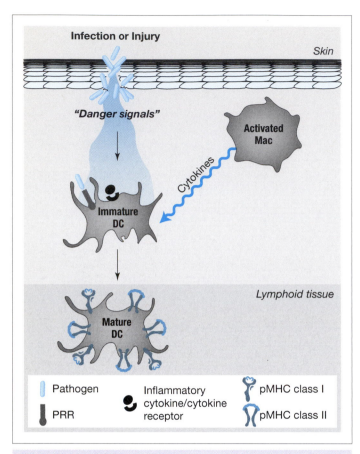

Fig. 7-3 DC Maturation

Table 7-4 **Comparison of Immature and Mature Conventional DCs**		
	Immature DCs	**Mature DCs**
Location	Peripheral tissues Secondary lymphoid tissues	Secondary lymphoid tissues
Surface MHC class II	Low	High
Antigen internalization capacity	High	Low
Costimulatory molecules	Low	High
Antigen presentation to T cells	Inefficient	Very efficient
Chemokine receptors	High CCR1, low CCR7	Low CCR1, high CCR7
Arrays of actin filaments	Present	Absent

As well as activating naïve Th cells, DCs are able to present peptides on MHC class I molecules and thus activate naïve Tc cells. Such presentation readily occurs if the DC is infected with a pathogen that replicates intracellularly. In addition, when a DC phagocytoses a whole pathogen or its components, some of the captured proteins may be diverted from the exogenous antigen processing pathway into the endogenous pathway. As a result, antigenic peptides from an intracellular pathogen may appear on MHC class I even if the pathogen has not infected the DC. This phenomenon is called *cross-presentation* and is described later in this chapter.

ii) Macrophages as APCs

Macrophages excel at ingesting whole bacteria or parasites or other large native antigens and quickly digesting them, producing a spectrum of antigenic peptides that are combined with MHC class II and presented to Th cells. However, unlike mature DCs, macrophages express only moderate levels of costimulatory molecules and so cannot activate naïve Th cells (see Ch. 9). Instead, macrophages make their contribution as APCs by activating memory or effector T cells. The IFNγ secreted by a Th effector interacting with a macrophage hyperactivates the macrophage, increasing antigen presentation and thus pathogen clearance. Activated macrophages also produce cytokines that promote Th effector cell differentiation (see Ch. 9) and upregulate MHC class II on other APCs (including DCs) in the immediate vicinity. Macrophages are thus important amplifiers of the adaptive response.

iii) B Cells as APCs

B cells are considered professional APCs because they constitutively express MHC class II. B cells use their aggregated BCRs to internalize protein antigens by receptor-mediated endocytosis. Unknown signals direct the internalized antigen into the exogenous processing pathway of the B cell such that peptide–MHC class II complexes appear on its surface. A B cell acting as an APC is one of the most efficient antigen presenters in the body, since the BCR binds specific antigen with high affinity and can thus capture antigens present at very low concentrations. However, B cells do not generally serve as APCs in the primary response to a Td antigen because antigen-specific B cells and the antigen-specific Th cells required to help them are very rare in an unimmunized individual. Thus, the chance that a naïve B cell recognizing antigen X and able to act as an APC is in close proximity to an equally rare naïve anti-X Th cell is exceedingly small. In addition, resting mature naïve B cells express only low levels of the costimulatory molecules required for full Th activation (see Ch. 9). However, once activated, B cells quickly upregulate B7 and become effective APCs. Moreover, in a secondary response, anti-X memory B and Th cells are present in significantly increased numbers. Memory B cells thus frequently serve as APCs in the secondary response and become increasingly prominent upon each subsequent encounter with antigen X.

II. GENERATION OF PEPTIDES VIA THE EXOGENOUS PATHWAY

An APC that has internalized a whole microbe or macromolecule encloses the entity in a membrane-bound transport vesicle. As described in Chapter 3, the transport vesicle enters the endocytic pathway and is successively fused to a series of protease-containing endosomes of increasingly acidic pH. Within the endolysosome, the proteins of the microbe or macromolecule are degraded into peptides of 10–30 amino acids. The peptides are then conveyed in a transport vesicle to specialized late endosomal compartments known as *MIICs* (MHC class II compartments; see later). It is in the MIICs where the peptides bind to newly synthesized MHC class II molecules. Peptides and MHC class II molecules must arrive in the MIICs in a synchronized fashion: if a peptide is not immediately bound by an MHC class II molecule, the peptide is rapidly degraded.

III. MHC CLASS II MOLECULES IN THE rER AND ENDOSOMES

During their synthesis on membrane-bound ribosomes, the α and β chains of the MHC class II molecule are cotranslationally inserted into the rER membrane. A third protein called the *invariant chain* (Ii) is coordinately expressed with the MHC class II chains and is also cotranslationally inserted into the rER. The function of Ii is to bind to newly assembled MHC class II heterodimers and protect the binding groove from being occupied by endogenous peptides present in the rER (see later). The MHC class II molecules are thus "preserved" for the exogenous peptides generated in the endocytic compartment. The part of the Ii molecule that sits in the MHC class II binding groove is called *CLIP* (class II associated invariant chain peptide) (Fig. 7-4A). Ii also contains a trimerization

domain that allows the formation of a nonameric complex consisting of three MHC class II heterodimers (three αβ pairs) and one Ii trimer (three Ii polypeptides) (Fig. 7-4B). Once complexed to Ii, the MHC class II molecules enter the Golgi complex but are then deflected away from the cell's secretory pathway and into its endocytic pathway by localization sequences in the Ii protein (Fig. 7-5, #1–3). In an endolysosome, most of the Ii protein is degraded by the sequential action of a family of *cathepsin* enzymes, leaving CLIP stuck in the grooves of the now monomeric MHC class II molecules. The MHC–CLIP complexes then enter the MIICs.

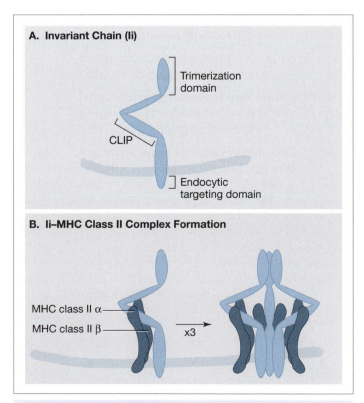

Fig. 7-4 **Interaction of MHC Class II and Invariant Chain**
[Adapted from Pieters J. (1997). MHC class II restricted presentation. *Current Opinion in Immunology* **9**, 89–96.]

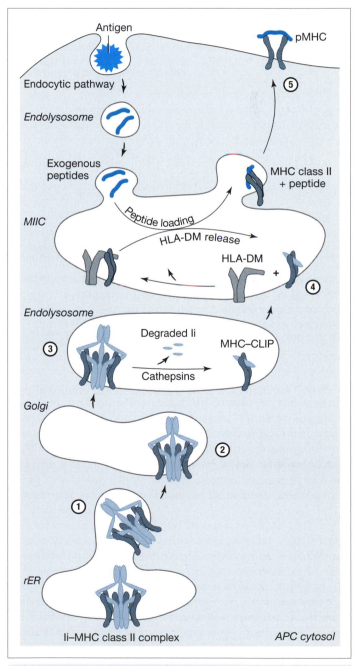

Fig. 7-5 **MHC Class II Antigen Presentation Pathway**

IV. PEPTIDE LOADING ONTO MHC CLASS II

The next step in exogenous antigen processing is the exchange of the CLIP peptide in the MHC class II binding groove for an exogenous peptide. A non-classical MHC molecule called HLA-DM in humans is essential for this process but the mechanism remains unclear. (The equivalent molecule in mice is called H-2DM.) Although HLA-DM closely resembles a conventional MHC class II molecule, it does not bind peptides. Instead, the association of HLA-DM with MHC–CLIP likely induces a conformational change that promotes the release of CLIP. HLA-DM then stabilizes the empty MHC class II heterodimer until the exogenous peptide is loaded (Fig. 7-5, #4). The actual mechanism of peptide loading has yet to be defined, but once an exogenous peptide binds stably in the MHC class II groove, the conformation of the MHC class II molecule alters again to force dissociation of HLA-DM. The pMHC is then transported out of the MIICs in a vesicle, inserted into the APC membrane by reverse vesicle fusion, and displayed to CD4$^+$ Th cells (Fig. 7-5, #5).

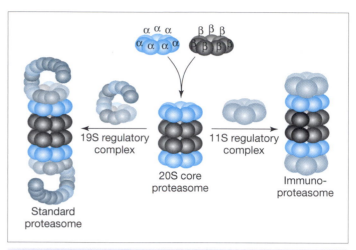

Fig. 7-6 **Proteasome Structure**
[Adapted from Rivett A.J. (1998). Intracellular distribution of proteasomes. *Currernt Opinion in Immunology* **10**, 110–114.]

C. Endogenous Antigen Processing

Endogenous antigen processing differs from exogenous processing in three ways: (1) Unlike the limited number of cell types that express MHC class II and so can function as professional APCs, almost any nucleated body cell expresses MHC class I and so can present peptides derived from intracellular antigens to CD8$^+$ CTL effectors. Thus, almost any body cell that has become aberrant due to old age, cancer or intracellular infection becomes a target cell for CTL-mediated cytolysis. (2) The processing of the protein antigen takes place in the cytosol rather than in the endocytic system. (3) Peptides generated in the cytosol meet newly synthesized MHC class I molecules in the rER rather than in the MIICs.

I. GENERATION OF PEPTIDES VIA THE ENDOGENOUS PATHWAY

Viruses or intracellular bacteria that have taken over a host cell force it to use its protein synthesis machinery to make viral or bacterial proteins. Antigenic proteins are thus derived from the translation of viral or bacterial mRNA on host cytosolic ribosomes. Similarly, abnormal proteins synthesized in the cytoplasm of tumor cells can give rise to peptides that appear to be "non-self" to the host's immune system and thus warrant an immune response. The process used to generate peptides from such foreign proteins is basically the same as that used to deal with misfolded or damaged host proteins. In the cytosol of every host cell are large numbers of huge, multisubunit protease complexes called *proteasomes*. The function of proteasomes is to degrade proteins into peptides. There are two types of proteasomes: the *standard proteasome* and the *immunoproteasome* (Fig. 7-6). Both types of proteasome

have at their core a structure called the *20S core proteasome*, a hollow cylinder of four stacked polypeptide rings made up of α and β subunits. The α subunits maintain the conformation of the proteasome core while the β subunits are its catalytically active components. The standard proteasome contains the 20S core plus two copies of a structure called the *19S regulatory complex*. The immunoproteasome contains a slightly modified 20S core and two copies of the *11S regulatory complex*.

Standard proteasomes are present in all host cells and routinely degrade spent and unwanted self proteins. The peptides produced by a standard proteasome are usually 8–18 residues in length. About 20% are the size (8–10 amino acids) that fits neatly into an MHC class I binding groove. Additional trimming of peptides that are initially too large can be carried out by peptidases in the cytosol. The degradation of host proteins in this way and subsequent loading of these peptides onto MHC class I allows constant scanning of self components and the monitoring of the cell's internal health. In contrast to standard proteasomes, immunoproteasomes are not present in most resting host cells but can be induced by exposure to the pro-inflammatory cytokines (such as IFNγ and TNF) that are present at high concentrations during a pathogen attack. Accordingly, immunoproteasomes most often produce peptides from foreign rather than self proteins and are responsible for most of the antigen processing associated with immune responses. The majority of peptides produced by the immunoproteasome are the optimal 8–10 amino acids in length.

II. TRANSPORT OF PEPTIDES INTO THE ENDOPLASMIC RETICULUM

To induce an immune response, the peptides generated by the proteasomes in a host cell's cytosol must access the binding site of an MHC class I molecule. However, the peptide and the peptide-binding site of the MHC class I molecule are on topologically opposite sides of the cell's membrane system. Unlike the case for MHC class II molecules and their peptides, no vesicle fusion event brings MHC class I molecules and their peptides together. Instead, peptides generated in the cytosol are transported directly into the rER where MHC class I molecules are synthesized.

In the membrane of the rER are positioned transporter structures known as *TAP (transporter associated with antigen processing)* (Fig. 7-7). TAP is a heterodimeric molecule composed of two subunits, TAP-1 and TAP-2, which are encoded by genes in the MHC. Structurally, TAP-1 and TAP-2 contain domains that project into the rER lumen, hydrophobic domains that span the rER membrane, and domains that extend into

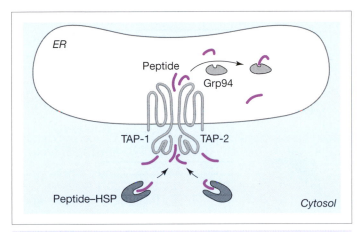

Fig. 7-7 **The TAP Transporter**

the cytosol and combine to form a single peptide-binding site. Peptides produced by the action of proteasomes are normally subject to very rapid degradation in the cytoplasm but can be saved from this fate by binding directly to TAP. Alternatively, the peptides can be bound by *chaperone* proteins that protect the peptides from degradation and escort them to TAP for transfer into the rER. Such chaperone proteins include certain HSPs, as described further in Box 7-2. Peptides that arrive independently or via chaperone delivery at the cytosolic TAP binding site are then transported through the rER membrane into the rER lumen. It has been estimated that TAP molecules can translocate 20,000 peptides/min/cell, more than enough to ensure a steady supply for loading onto nascent MHC class I molecules generated in the rER at a rate of 10–100/min. TAP preferentially imports peptides of 8–12 residues in length, although longer peptides can be transported with lower efficiency. Once in the ER lumen, the peptides meet one of four fates. Some peptides are bound immediately to MHC class I molecules, while others are temporarily taken up by chaperone proteins resident in the ER (such as *Grp94*; glucose-regulated protein 94) and protected from further degradation prior to loading onto MHC class I. Still other peptides are trimmed by ER-resident peptidases to achieve the correct length and C-terminus necessary for fitting into the MHC class I groove. Lastly, some peptides are rapidly re-translocated back into the cytosol where they are either degraded or re-imported back into the rER via TAP and subjected to further trimming.

III. MHC CLASS I MOLECULES IN THE rER

The α chain of an MHC class I molecule is synthesized on a membrane-bound ribosome and is cotranslationally inserted

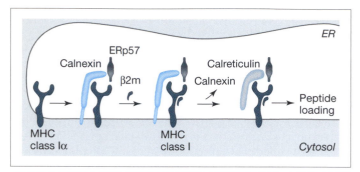

Fig. 7-8 Human MHC Class I Chaperone Proteins

into the membrane of the rER. As it enters the rER membrane, the MHC class I α chain associates with a transmembrane chaperone protein called *calnexin* and a non-transmembrane enzyme called *ERp57* (Fig. 7-8). Calnexin facilitates proper polypeptide folding and association of the α chain with the coordinately expressed β2m chain. In humans, calnexin is then replaced by a soluble chaperone protein called *calreticulin*. (In mice, either the original calnexin molecule or an incoming calreticulin molecule associates with the heterodimer after association with β2m.) ERp57 binds to both calnexin and calreticulin and works with these molecules to catalyze the formation of disulfide bonds in MHC class I α chains. ERp57 also promotes the loading of peptide into the MHC class I binding groove.

IV. PEPTIDE LOADING ONTO MHC CLASS I

To bring a newly synthesized MHC class I heterodimer (with its chaperones) into the vicinity of TAP and the peptides, the MHC class I molecule transiently interacts with a protein called *tapasin* that binds to ERp57, MHC class I and TAP (Fig. 7-9). Tapasin helps to stabilize the empty MHC class I heterodimer in a conformation suitable for peptide loading. Exactly how a peptide, either free or bound to a chaperone, accesses the MHC class I binding groove has yet to be determined but tapasin is important for this process. Tapasin also works with calreticulin to prevent improperly loaded peptide–MHC class I complexes from leaving the rER. Peptide loading is a crucial step in antigen presentation because an MHC class I heterodimer that is transported to the cell surface without a peptide in its groove is unstable and rapidly lost. With the insertion of a pMHC into the membrane of a host cell, it is ready for inspection by CD8⁺ T cells.

D. Cross-Presentation on MHC Class I

Cross-presentation refers to the presentation on MHC class I of peptides from extracellularly acquired antigens, and is thought to be a major means by which DCs can activate naïve Tc cells (see Ch. 9). Cross-presentation was discovered when

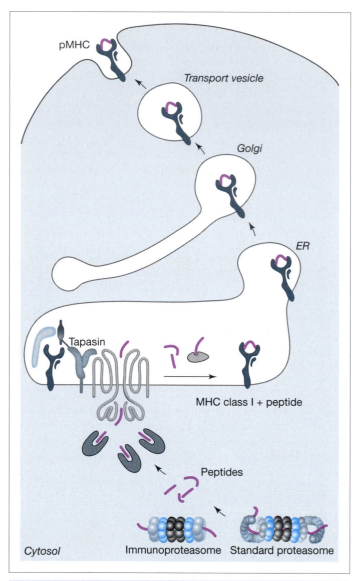

Fig. 7-9 MHC Class I Antigen Presentation Pathway

researchers found that CD8⁺ CTL responses could be mounted to certain antigens that were known to be extracellular, and that peptides from viral antigens could be presented on MHC class I even when the endogenous processing pathway was blocked. The phenomenon was called "cross-presentation" because the viral antigen appeared to physically "cross over" from an infected host cell to an uninfected APC that presented peptides from the antigen on MHC class I as if the antigen had originated in the interior of the APC. Although cross-presentation has been definitively demonstrated *in vitro*, its physiological significance remains a matter of debate among some immunologists.

How can exogenous peptides be loaded onto MHC class I? APCs (particularly DCs) most often acquire viral antigens by internalizing debris from infected cells that have undergone necrotic or apoptotic death. In rare cases, DCs may also

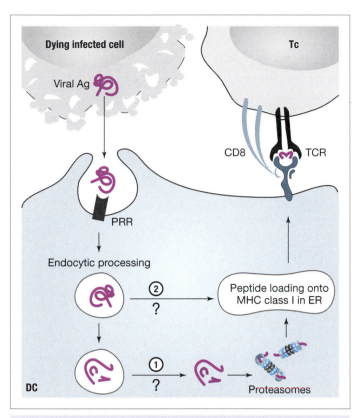

Fig. 7-10 **Model of Cross-Presentation on MHC Class I**

acquire portions of the membranes of live infected cells (by a process sometimes called "nibbling") but the mechanism is unclear. In all these situations, because the viral protein has entered the APC from the extracellular environment, it is initially directed to the endocytic system in the usual way. Thus, viral peptides appear on the APC surface associated with MHC class II and a Th response to the antigen can be initiated. However, during the initial processing of the viral proteins in the early endosomes, a fraction of the resulting polypeptides may be actively transported (by an unknown mechanism) from the endosomes into the cytosol, where the fragments are taken up by proteasomes and degraded to peptides (Fig. 7-10, #1). Alternatively, there is limited evidence suggesting that some protein-containing endosomes can fuse directly to the rER (Fig. 7-10, #2). In other cases, proteins may exit from the endosome, perhaps enter the Golgi and then the rER. Lastly, a very few proteins appear to be able to cross the plasma membrane directly, bypass the endocytic system and enter a proteasome. Regardless of the mechanism, the end result is that the viral peptides are loaded onto MHC class I just as if the viral protein had originated within the APC itself. The pMHCs are then displayed on the APC surface for the perusal of antiviral CD8+ T cells. As well as viral antigens, antigens from intracellular bacteria and parasites can also be processed and their peptides displayed on MHC class I via cross-presentation.

E. Other Methods of Antigen Presentation

I. ANTIGEN PRESENTATION BY MHC CLASS Ib MOLECULES

The glycoproteins encoded by the MHC class Ib genes (refer to Fig. 6-2) are closely related in structure to the classical MHC class I molecules. However, MHC class Ib molecules are less polymorphic, are expressed at lower levels, and have a more limited pattern of tissue distribution. Some MHC class Ib molecules occur in a secreted form and do not bind to antigenic peptides. Other MHC class Ib molecules are transmembrane proteins that can bind to certain subsets of foreign peptides and present them to subsets of αβ and γδ T cells in a TAP-dependent manner (see Ch. 11). However, the peptide-binding groove in these MHC class Ib molecules is partially occluded such that a narrower range of shorter peptides is presented.

II. NON-PEPTIDE ANTIGEN PRESENTATION BY CD1 MOLECULES

In Box 6-1, we described "MHC-like" molecules that are encoded outside the MHC but feature an MHC-like fold in their structures. Five MHC-like CD1 molecules have been identified: CD1a, CD1b, CD1c, CD1d and CD1e. Human APCs can express all five isoforms whereas mouse APCs express only the CD1d isoform. The CD1 proteins are of particular interest with respect to antigen presentation because these non-polymorphic proteins can present non-peptide lipid and glycolipid antigens to certain T cell subsets. The antigen-binding groove in CD1 molecules is much more hydrophobic than that of classical MHC molecules, suiting it to lipid binding. It is thought that a T cell "sees" a combined epitope composed of amino acids of the CD1 molecule plus a small portion of the carbohydrate head group of the lipid. T cells can be very discriminating in their recognition of CD1-presented antigens, failing to respond if the orientation of even a single hydroxyl group of the antigen is changed. In humans, CD1a, CD1b, CD1c and CD1e molecules present lipids from glycolipid and glycosphingolipid antigens to subsets of αβ Th and Tc cells. CD1c can also present an unknown ligand to a particular subset of γδ T cells (see Ch. 11). In both mice and humans, CD1d molecules present a very restricted collection of ceramide-based antigens to NKT cells and some T cell subsets (see Ch. 11). The consequences of CD1-mediated antigen presentation are virtually identical to those of peptide presentation; that is, activated NKT cells and Th effectors secrete cytokines while Tc cells generate CTLs that kill target cells by cytolysis.

Antigen processing and presentation by CD1 molecules appear to utilize elements of both the cytosolic and endocytic pathways. Like MHC class I, CD1 chains must be associated with β2m to be transported to the cell surface, but, unlike MHC class I, antigen loading of CD1 molecules does not take place in the rER. Like MHC class II, CD1 molecules are targeted to the endocytic system, but no association with either invariant chain or HLA-DM is required. Because of the presence or absence of particular amino acid motifs, CD1a molecules accumulate in early endosomes, while CD1c molecules tend to collect in intermediate endosomes, and CD1b molecules are directed to late endosomes. As illustrated in Figure 7-11, a mycobacterium phagocytosed by a human APC can be degraded in a phagolysosome to yield a collection of lipids (among other types of molecules). Depending on the structure of the lipid, it is sorted and confined to either an early endosome, an intermediate endosome or a late endosome. Once in an endocytic compartment, each lipid is loaded onto the type of CD1 molecule resident in that particular compartment. The lipid-loaded CD1 molecules are then transported to the plasma membrane for display on the APC surface. The actual mechanisms involved in the antigen loading of each type of CD1 molecule and their subsequent presentation on the APC surface remain unclear.

MHC molecules presenting foreign peptides are the body's signposts to the immune system that a T cell-mediated adaptive response is required. The next chapter discusses the genes and proteins of TCRs, the antigen receptor molecules that carry out pMHC recognition.

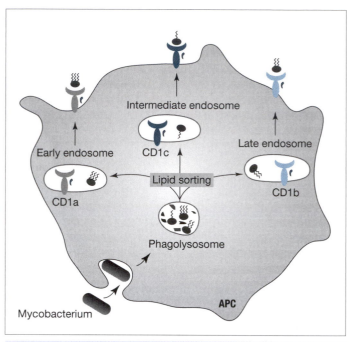

Fig. 7-11 **Lipid Antigen Presentation by Human CD1 Molecules**

- The pMHCs recognized by the TCRs of T cells are assembled by the exogenous or endogenous antigen processing and presentation pathways.

- In the exogenous pathway, extracellular antigens are internalized by APCs and degraded within endosomal compartments. The resulting peptides bind to MHC class II molecules to form pMHCs that are transported from the endosomes to the APC surface for recognition by CD4$^+$ Th cells.

- Professional APCS include DCs, macrophages and B cells. These cell types take up antigen efficiently and express MHC class II inducibly or constitutively. However, only mature DCs can activate naïve Tc and Th cells.

- In the endogenous pathway, antigens that are produced intracellularly as a result of host cell infection or transformation are degraded by cytoplasmic proteasomes. The resulting peptides are actively transported into the rER where they bind to MHC class I molecules to form pMHC complexes that are transported to the cell surface for recognition by CD8$^+$ CTLs.

- Cross-presentation refers to the presentation on MHC class I of peptides from extracellularly acquired antigens. DCs may use cross-presentation to activate naïve Tc cells.

- Antigens are often presented to $\gamma\delta$ T cells and NKT cells by non-classical MHC class Ib molecules and MHC-like molecules such as CD1.

DID YOU GET IT? A SELF-TEST QUIZ

Section A

1) Distinguish between antigen processing and antigen presentation.

2) Name the two most important antigen processing and presentation pathways.

3) Which processing pathway would be used to initiate an immune response against a liver cancer and why?

4) Why is it $CD4^+$ rather than $CD8^+$ T cells that respond to extracellular pathogens?

Section B.I

1) Can you define these terms? Langerhans cell, inflammatory DC, interstitial DC

2) What cell types can function as professional APCs and why?

3) Distinguish between the three main types of conventional DCs.

4) What is the function of thymic DCs?

5) Give three ways in which immature DCs differ from mature DCs.

6) How do danger signals promote DC maturation?

7) Does apoptotic cell death induce DC maturation? If not, why not?

8) What effect does interaction with a Th cell have on a mature DC? Why is it important?

9) Briefly outline how DCs influence Th differentiation.

10) Outline the functions of DCs in innate responses and in peripheral tolerance.

11) Why are macrophages considered amplifiers of the adaptive response?

12) Why are B cells efficient APCs for the secondary response?

Section B.II–IV

1) Can you define these terms? MIICs, CLIP, Ii

2) Where in the cell does the final degradation of extracellularly derived proteins most often occur?

3) Describe the structure and function of the invariant chain.

4) Where in the cell do exogenous peptides and MHC class II meet?

5) What is the role of HLA-DM during the loading of peptides onto MHC class II?

Section C.I

1) Describe three ways in which endogenous and exogenous antigen processing differ.

2) Distinguish between standard proteasomes and immunoproteasomes in terms of structure and function.

3) What is a heat shock protein and what does it do?

Section C.II–IV

1) Why are endogenous peptides transported into the rER?

2) Describe the structure and function of TAP.

3) What is a chaperone protein? Give two examples of such proteins and their functions.

4) How does tapasin facilitate pMHC formation?

Section D

1) How is cross-presentation thought to allow DCs to activate naïve Tc cells?

2) Describe three ways in which an uninfected APC might acquire antigens from intracellular pathogens.

Section E

1) How does antigen presentation by MHC class I and Ib molecules differ?

2) What types of antigens are presented by CD1 molecules?

3) Describe how elements of both the standard exogenous and endogenous antigen processing pathways are used in antigen processing and presentation by CD1 molecules.

8

The T Cell Receptor: Proteins and Genes

WHAT'S IN THIS CHAPTER?

The best way to have a good idea is to have lots of ideas.
Linus Pauling

A. TCR Proteins and Associated Molecules

As introduced in earlier chapters, the T cell receptor (TCR) is responsible for antigen recognition by T cells. TCRs are expressed by all T cells except their earliest precursors, so that most thymocytes and all mature T cells bear TCRs. Like B cells, the antigenic specificities of T cells are clonal in nature, meaning that (with rare exceptions) all members of a given T cell clone carry 10,000–30,000 identical copies of a receptor protein with a unique binding site. Like BCRs, TCRs possess V and C regions, and, like the Ig genes, the TCR genes undergo RAG-mediated recombination of V, D and J gene segments to produce a repertoire of receptors with considerable sequence diversity in the V region. However, TCRs differ from BCRs in two fundamental ways. Firstly, while the BCR repertoire can recognize and bind to virtually any structure, the spectrum of antigens recognized by TCRs is much more restricted. The vast majority of T cells bind to antigenic peptides that must be complexed to MHC molecules displayed on the surfaces of APCs or target cells. Only a small percentage of T cells recognize lipids or unprocessed antigens that may or may not be associated with MHC-related molecules. Secondly, while B cells secrete a form of their BCRs as antibody, T cells do not secrete their TCRs.

As introduced in Chapter 2, there are two types of TCRs defined by their component chains: TCRαβ and TCRγδ (Fig. 8-1). Mutually exclusive expression of these TCRs characterizes two distinct T cell subsets that develop independently: αβ T cells and γδ T cells. In humans, most mature T cells are αβ T cells, with only 5–10% being γδ T cells. While αβ T cells are concentrated in the secondary lymphoid tissues and function in adaptive responses, most γδ T cells are *intraepithelial* in location (tucked between the mucosal epithelial cells lining the body tracts) and participate in innate responses. Rather than pMHCs, γδ TCRs recognize a broad range of cell surface molecules that may be encountered in their natural, unprocessed forms. The biology of γδ T cells is discussed in more detail in Chapter 11.

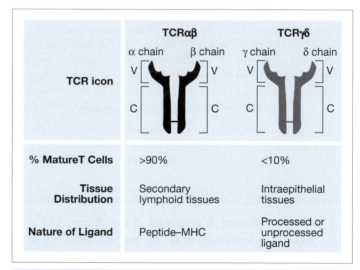

Fig. 8-1 **Basic Characteristics of TCRαβ and TCRγδ**

I. BASIC TCR STRUCTURE

Unlike mIg, which is made up of two light chains and two heavy chains and has two identical antigen-binding sites, a TCR is a heterodimeric glycoprotein with a single antigen-binding site. TCRαβ is composed of a TCRα chain (49 kDa) linked via a disulfide bond to a TCRβ chain (43 kDa), whereas TCRγδ consists of a TCRγ chain (40–55 kDa) linked via a disulfide bond to a TCRδ chain (45 kDa); TCRαδ or γβ structures have not been found in nature. Each TCR chain contains an Ig-like V domain, an Ig-like C domain, a cysteine-containing connecting sequence, a charged transmembrane portion, and a short cytoplasmic tail (Fig. 8-2). The V and C domains are arranged in Ig fold structures that are stabilized by intrachain disulfide bonds. The TCR V region is composed of the N-terminal ends of both TCR polypeptides and contains the antigen-binding site. Amino acids in the binding site establish contacts with both the antigenic peptide and the MHC molecule to which it is bound. Unlike the case for Igs (which

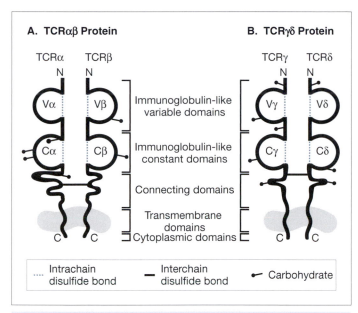

A. TCRαβ Protein

TCRα TCRβ
N N

Vα Vβ — Immunoglobulin-like variable domains

Cα Cβ — Immunoglobulin-like constant domains

— Connecting domains

— Transmembrane domains

C C — Cytoplasmic domains

B. TCRγδ Protein

TCRγ TCRδ
N N

Vγ Vδ

Cγ Cδ

C C

···· Intrachain disulfide bond   — Interchain disulfide bond   ⊸ Carbohydrate

Fig. 8-2 **Schematic Representations of TCRαβ and TCRγδ Proteins**
[With information from Klein J. and Horejsí V. (1997). *Immunology*, 2nd ed., Blackwell Science, Oxford.]

undergo isotype switching), a TCR's C region is fixed for the life of a given T cell clone. The short connecting sequence located between the TCR C domain and the transmembrane domain is analogous to the Ig hinge.

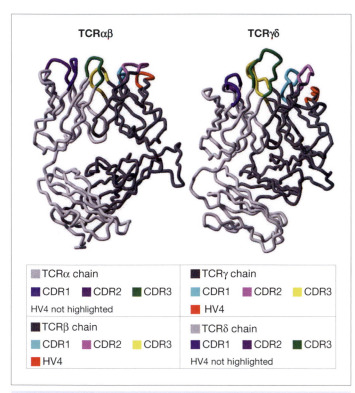

TCRαβ   **TCRγδ**

▨ TCRα chain			■ TCRγ chain		
■ CDR1	■ CDR2	■ CDR3	■ CDR1	■ CDR2	■ CDR3
HV4 not highlighted			■ HV4		
■ TCRβ chain			▨ TCRδ chain		
■ CDR1	■ CDR2	■ CDR3	■ CDR1	■ CDR2	■ CDR3
■ HV4			HV4 not highlighted		

Plate 8-1 **X-Ray Crystal Structures of TCRαβ and TCRγδ**
[Reproduced by permission of Ruldph M.G. and Wilson I.A. (2002). The specificity of TCR/pMHC interaction. *Current Opinion in Immunology* **14**, 52–65.]

As was true for the H and L chains of Igs, the V domain of each TCR polypeptide contains sites of increased amino acid variability. There are four such complementarity-determining or hypervariable (HV) regions in a TCR chain: CDR1, CDR2, CDR3 and HV4 (Plate 8-1). In TCRαβ, various CDR/HV regions of both the TCRα and TCRβ chains are involved in pMHC recognition, depending on the particular TCR and pMHC involved. In some cases, the CDR1, CDR2 and HV4 regions interact with residues on the MHC class I or class II protein itself, while the highly diverse CDR3 regions preferentially make contact with the antigenic peptide nestled in the MHC groove. In other cases, the CDR1 and/or CDR2 regions may bind to part of the peptide, while the CDR3 regions contact the MHC molecule but the HV4 regions fail to make a major contribution. The V domains of the TCRγ and δ chains also contain CDR1, CDR2, CDR3 and HV4 regions. However, because TCRγδ ligands are often non-peptides, the precise roles of the hypervariability regions in the binding of antigen to these receptors are thought to be slightly different. Antigen recognition by γδ TCRs is discussed in Chapter 11.

II. THE CD3 COMPLEX

i) Structure

The short cytoplasmic tails of TCR chains are too short for signal transduction. This type of problem is solved in the BCR complex by the association of mIg with the Igα/Igβ heterodimer. The ITAMs present in the cytoplasmic tails of Igα/Igβ intracellularly transduce the signal triggered by antigen binding to mIg. In T cells, while the extracellular V domains of TCRαβ or TCRγδ recognize antigen, the TCR heterodimer as a whole must associate non-covalently with a collection of invariant transmembrane proteins known as the *CD3 complex* to transduce the signal. The CD3 complex contains three heterodimeric proteins made up of variable combinations of five polypeptides designated CD3γ, CD3δ, CD3ε, CD3ζ (ζ, zeta), and CD3η (η, eta). The γ chain of the FcεRI receptor can occasionally participate in the CD3 complex because it shares certain structural similarities with the CD3ζ molecule. In general, the CD3 complex that clusters around a TCRαβ molecule is composed of a CD3εδ heterodimer, CD3εγ heterodimer, and either a CD3ζζ homodimer, a CD3ζη heterodimer or a CD3ζ–FcεRIγ heterodimer (Fig. 8–3). In the case of TCRγδ, the CD3εδ heterodimer is replaced by another CD3εγ heterodimer. Close to 90% of all T cells bear the CD3ζζ homodimer, with just under 10% carrying the CD3ζη heterodimer and a very small fraction expressing CD3ζ–FcεRIγ.

ii) Functions

The CD3 complex has two major functions. Firstly, as mentioned in the last section, the CD3 chains are required for intracellular signaling. Upon engagement of the TCR by pMHC, tyrosine residues in the CD3 ITAMs are phosphorylated by an intracellular signaling kinase called *Lck*. Additional signaling kinases can then be recruited to the receptor complex to propagate the signaling cascade. Secondly, the CD3 complex is required for TCR surface expression. In the

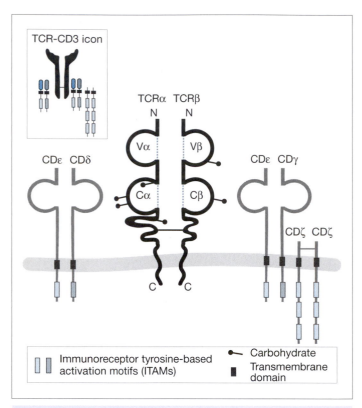

Fig. 8-3 **The TCRαβ–CD3 Complex**

Legend (Fig. 8-3):
Immunoreceptor tyrosine-based activation motifs (ITAMs)
Carbohydrate
Transmembrane domain

the peptide-binding groove, coreceptor binding does not depend on the identity of the antigenic peptide.

The binding by the coreceptors to MHC molecules stabilizes the interaction so that the TCR can determine whether the peptide of the pMHC fits into the TCR's binding site. It is still not known why the expression of CD4 and recognition of peptide–MHC class II are almost exclusively associated with T helper function, while CD8 expression and peptide–MHC class I recognition are features of T cells with cytotoxic powers.

ii) Structure

Despite their ostensibly equivalent functions, the CD4 and CD8 proteins show little similarity in either structure or amino acid sequence. In both mice and humans, CD4 is a transmembrane glycoprotein that is expressed as a single polypeptide on the cell surface. The CD4 protein contains four extracellular Ig-like domains that interact with the α2 and β2 domains of MHC class II; a transmembrane domain; and a cytoplasmic tail with sites that promote relatively strong association with Lck kinase (Fig. 8-4A). In both humans and mice, the majority

ER, the TCR heterodimer physically associates with the CD3 complex before moving to the Golgi for glycosylation and finally transport to the T cell surface. The invariant, Ig-like extracellular domains present in the CD3γ, CD3δ and CD3ε chains interact with the Ig-like extracellular domains in the TCR chains to help to keep the TCR and the CD3 complex together throughout transport and on the cell surface. In the absence of CD3 expression, the TCR remains stalled in the ER. The synthesis and incorporation of the CD3ζ chain into the CD3 complex controls the assembly and transport of the entire TCR–CD3 assembly.

III. THE CD4 AND CD8 CORECEPTORS

i) Nature

As introduced in Chapters 2 and 6, mature αβ T cells patrolling the body's periphery bear either the CD4 or CD8 coreceptor. In humans, about two-thirds of mature αβ T cells are CD4+ cells, whereas one-third are CD8+ cells. Most mature γδ T cells express neither CD4 nor CD8, although some γδ T cells in the gut are CD8+ (see Ch. 11). CD4 and CD8 are called "coreceptors" because a molecule of either one of these proteins colocalizes with a TCR on the T cell surface, and then binds to the same MHC molecule on the APC or target cell that is engaged by that particular TCR. CD4 binds to MHC class II molecules, whereas CD8 binds to MHC class I molecules. However, because CD4 and CD8 bind to sites on their respective MHC molecules that are in invariant regions outside

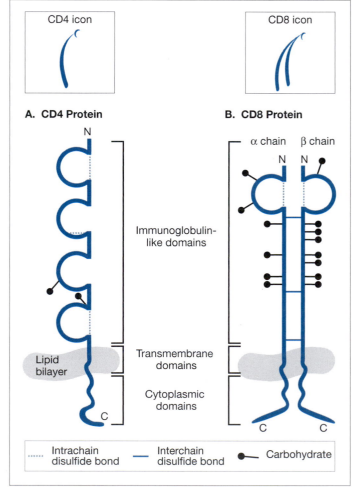

A. CD4 Protein

B. CD8 Protein

Fig. 8-4 **Structures of the CD4 and CD8 Coreceptors**
[With information from Klein J. and Horejsí V. (1997). *Immunology*, 2nd ed., Blackwell Science, Oxford.]

of CD8 molecules expressed on a T cell surface are CD8αβ heterodimers made up of CD8α and CD8β chains (Fig. 8-4B). Some intestinal intraepithelial T cells express a CD8αα homodimer. In both cases, the complete CD8 protein contains one Ig-like extracellular domain that binds to the α3 domain of MHC class I; a transmembrane domain; and a cytoplasmic tail that associates relatively weakly with Lck kinase.

iii) Functions

CD4 and CD8 have two major functions: (1) stabilization of TCR–pMHC binding by the interaction of CD4 with MHC class II and CD8 with MHC class I; and (2) recruitment of Lck to the TCR–CD3 complex. Although neither CD4 nor CD8 is absolutely required for the initial engagement of TCRαβ by pMHC, the adhesive contacts these molecules establish with the MHC molecule greatly enhance TCR–pMHC binding. With respect to recruitment, because of the positioning of the coreceptors near the TCR in the membrane, Lck that is physically associated with the cytoplasmic tail of CD4 or CD8 is brought into close proximity with the tails of the CD3 chains (Fig. 8-5). Lck then can phosphorylate the ITAMs in the CD3 tails, propagating the intracellular signaling cascade that leads to T cell activation (see Ch. 9).

B. TCR Genes

I. STRUCTURE OF THE TCR LOCI

The TCRα, β, γ and δ polypeptide chains are encoded by the *TCRA*, *TCRB*, *TCRG* and *TCRD* loci, respectively. The chromosomal locations of these loci in humans and mice are given in Table 8-1 and their exon/intron structures are shown in Figures 8-6 and 8-7. Like the genes encoding the Ig proteins, the genes encoding the TCR proteins are composed of V and C exons, with the V exon being made up of small V, D and J gene segments that are assembled at the DNA level by V(D)J recombination.

The *TCRA* and *TCRG* loci contain V and J gene segments but no D segments, making these loci analogous to the Ig light chain loci. Whereas there is one Cα exon in *TCRA* in mice and humans, there are two Cγ exons in human *TCRG* and three functional Cγ exons in mouse *TCRG*. The *TCRB* locus is like *Igh* in that it contains multiple V, D and J gene segments. Two Cβ exons are present in both species but it is unlikely

that these exons are functionally different because, in T cells, there is no mechanism analogous to the isotype switching that occurs in B cells. The sole function of the TCR C domains appears to be association with the CD3 complex, and Cβ1 and Cβ2 have equivalent roles in this respect. Although *TCRD* also has similarities to *Igh*, there are some startling differences. Firstly, in both mice and humans, *TCRD* is nested within the *TCRA* locus. *TCRD* contains its own Vδ, Dδ and Jδ gene segments and Cδ exon but also shares the use of some Vα gene segments. However, Vδ gene segments recombine only with Jδ and Cδ sequences and never with Jα or Cα sequences. The unconventional location of *TCRD* prevents the expression of

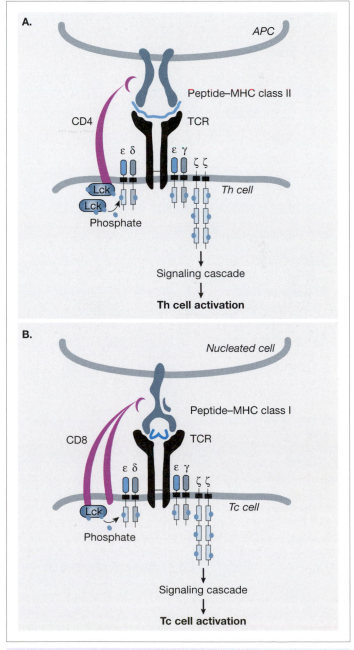

Fig. 8-5 **Role of Coreceptors in TCR Signaling**

Table 8-1 Chromosomal Localization of TCR Loci

Genetic Locus	Chromosome	
	Human	Mouse
TCRA	14	14
TCRB	7	6
TCRG	7	13
TCRD	14	14

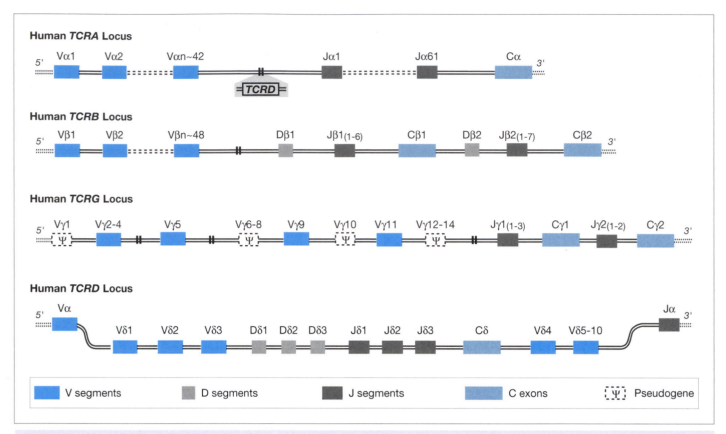

Fig. 8-6 **Genomic Organization of the Human TCR Loci** [With information from http://imgt.cines.fr/.]

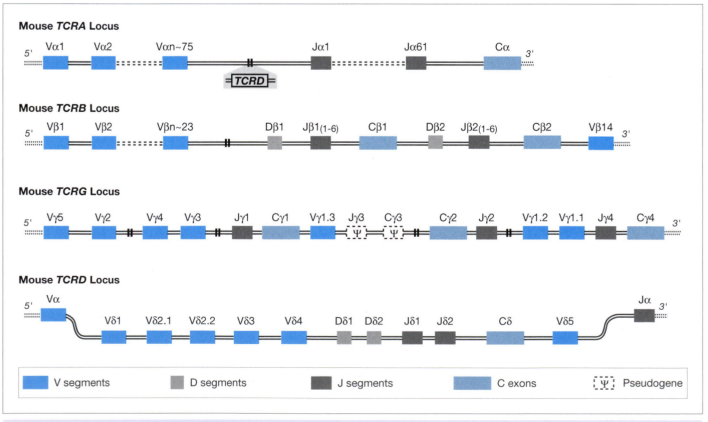

Fig. 8-7 **Genomic Organization of the Mouse TCR Loci** [With information from http://imgt.cines.fr/.]

both *TCRD* and *TCRA* on the same T cell, since the recombination of the Vα and Jα gene segments deletes the entire *TCRD* locus. Secondly, in addition to the use of one Dδ segment to create a VDJ exon, multiple Dδ gene segments can be used in tandem to create a VDDJ exon in mice or a VDDJ or even a VDDDJ exon in humans. The additional D–D joints present in these exons dramatically enhance the junctional diversity found in TCRδ chains.

II. ORDER OF REARRANGEMENT

When a T cell progenitor leaves the bone marrow and enters the thymus to become an immature thymocyte, its TCR genes are in the germline configuration. In the thymus, the immature thymocyte rearranges its TCR genes and eventually becomes either an αβ T cell or a γδ T cell. It is not yet clear whether a given thymocyte receives stimulation from the local environment that influences it to become either an αβ or γδ T cell, or whether commitment to one lineage or the other simply depends on the success of the gene rearrangements occurring in the relevant loci. It is also possible that these scenarios are not mutually exclusive. V(D)J recombination of the TCR loci is intimately tied to T cell development and is discussed in more detail in this context in Chapter 9.

i) *TCRA* and *TCRB* Rearrangement

Irrevocable commitment of a thymocyte to the TCRαβ lineage and the continued maturation of the clone depend on V(D)J recombination resulting in a functional TCRβ gene. The *TCRB* locus rearranges prior to the *TCRA* locus. Simultaneously in *TCRB* on the maternal and paternal chromosomes, V(D)J recombination first joins a Dβ gene segment to a Jβ segment, and then a Vβ segment to DβJβ. When a gene is completed, it is then tested for functionality via formation of the *pre-TCR*, a signaling complex composed of a newly produced candidate TCRβ chain combined with a surrogate TCRα chain called the *pre-T alpha chain* (plus the CD3 chains). Successful intracellular signaling initiated by the pre-TCR indicates that the candidate TCRβ protein is functional and thus that the rearrangement of the *TCRB* gene has been successful. Because signaling through the pre-TCR governs the continued differentiation of thymocytes, this molecule is discussed in more detail in Chapter 9.

The assembly of a functional TCRβ gene on one chromosome signals to the cell to suppress V(D)J rearrangement of *TCRB* on the other chromosome. In a thymocyte in which *TCRB* rearrangement has been unsuccessful on both chromosomes, the cell neither attempts to rearrange *TCRA* nor becomes a γδ T cell; instead, it dies by apoptosis. If a functional TCRβ gene is produced, V(D)J recombination of *TCRA* commences on both chromosomes. If *TCRA* rearranges productively on either chromosome, it can generate a TCRα chain that can combine with the newly synthesized TCRβ chain and appear on the cell surface as a functional TCRαβ. The *TCRD* locus is deleted by successful *TCRA* rearrangement. If *TCRA* rearrangement fails on both chromosomes, the cell dies by apoptosis.

ii) *TCRG* and *TCRD* Rearrangement

In thymocytes that eventually become γδ T cells, rearrangement commences simultaneously but independently in the *TCRG* and *TCRD* loci on both chromosomes. Despite the fact that the *TCRA* locus physically surrounds *TCRD*, *TCRA* does not undergo rearrangement. VJ joining of TCRγ gene segments occurs in the usual way, but the TCRδ D gene segments can be combined with each other to form tandem D–D or D–D–D units. The D, D–D or D–D–D entities are in turn joined to Jδ and then finally to Vδ to complete the V exon. More on gene rearrangement during the development of γδ T cells appears in Chapter 11.

III. V(D)J RECOMBINATION

The same RAG recombinases and DNA repair enzymes that execute V(D)J recombination in the Ig loci in developing B cells act on the TCR loci in thymocytes to produce functional TCR genes. Accordingly, the TCR gene segments are flanked by the same 12-RSS and 23-RSS discussed in Chapter 4. Moreover, as shown for *TCRB* in Figure 8-8, the RAG recombinase enzymes (purple ovals) follow the same 12/23 rule to juxtapose only those RSSs that are not of the same type. Importantly, D segments in the *TCRB* and *TCRD* loci are flanked on the 5′ side by a 12-RSS and on the 3′ side by a 23-RSS. In the *TCRD*

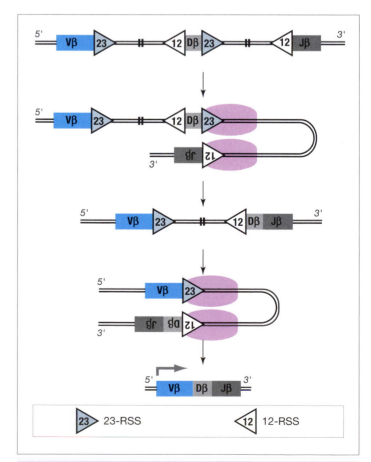

Fig. 8-8 RSS-Mediated V(D)J Recombination of *TCRB*

locus, where the Dδ segments are clustered together, this arrangement of the RSSs facilitates the tandem joining of Dδ segments prior to the addition of the J segment.

Despite the apparent duplication of the V(D)J recombination apparatus in B cells and T cells, the Ig genes are not rearranged in developing T cells and the TCR genes are not rearranged in developing B cells. It is thought that a mechanism that alters the chromatin of the Ig or TCR loci and thus their accessibility to the RAG recombinases must distinguish at an early stage between hematopoietic precursors that are destined to become B cells and those that become T cells.

IV. TCR GENE TRANSCRIPTION AND PROTEIN ASSEMBLY

Following V(D)J recombination to generate the V exon, the rearranged TCR gene undergoes conventional transcription from the promoter associated with the participating V gene segment. A single primary transcript that includes V and C exons is generated (Fig. 8-9). The primary transcript undergoes RNA splicing to bring the V and C exons together, thereby generating mature mRNAs that are translated into TCR polypeptides. In contrast to *Igh* genes, which contain separate exons specifying mIg or sIg, the TCR genes have only a TM-encoding exon (not shown). Thus, there is no production of alternative mRNA transcripts specifying membrane-bound versus secreted TCR proteins. After translation of the TCR transcripts in the ER, disulfide bonding links the TCRα and β chains, or the TCRγ and δ chains, to form membrane-bound TCRαβ or TCRγδ molecules, respectively. These heterodimers associate with the CD3 complex and are then transported to the plasma membrane where they appear on the cell surface as complete antigen receptor complexes.

V. TCR DIVERSITY

The mechanisms of isotype switching and somatic hypermutation that create diversity in antigen-activated B cells do not operate in T cells, so that the diversity in the T cell repertoire is established entirely by mechanisms that function prior to

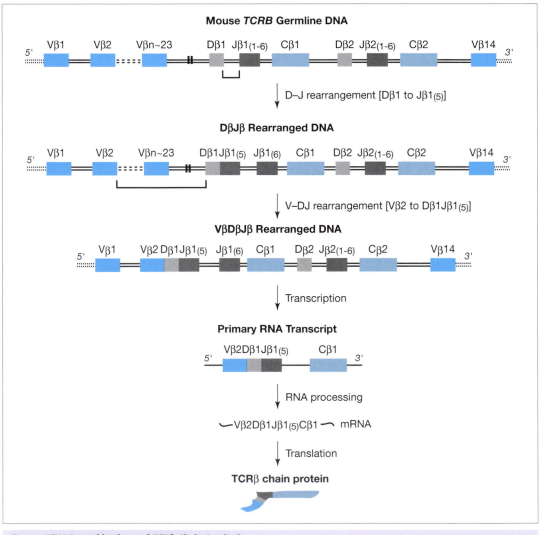

Fig. 8-9 **V(D)J Recombination and TCRβ Chain Synthesis**

antigenic stimulation. These are multiplicity of germline segments, combinatorial diversity, junctional diversity, and $\alpha\beta$ (or $\gamma\delta$) chain pairing.

i) Multiplicity and Combinatorial Joining of Germline Gene Segments

In mice and humans, the numbers of different V and D gene segments available for recombination in the TCR loci are much lower than the number of corresponding segments in the Ig loci, but the number of *TCRA* J segments is greater than the number of Ig J segments (Table 8-2). Overall, the contribution of this source of diversity to the maximum theoretical TCR repertoire is 10- to 30-fold lower than for the Ig repertoire. The random juxtaposition of TCR V, D and J segments during V(D)J recombination then contributes diversity that can be calculated just as for the Ig genes. For example, for the mouse TCRα chain, the number of possible combinations is 75 Vα × 61 Jα × 1 Cα = 4575, whereas that for mouse TCRβ is 23 Vβ × 2 Dβ × 12 Jβ × 2 Cβ = 1104. Using this methodology, one might also conclude that there are (considering functional segments only) 6 Vγ × 3 Jγ × 3 Cγ = 54 possible combinations for the mouse TCRγ chain, and 12 Vδ × 2 Dδ × 2 Jδ × 1 Cδ = 48 combinations for the mouse TCRδ chain. However, these theoretical calculations do not take into account certain joining preferences that occur in the TCR loci. For example, Cβ1 is found only in conjunction with Dβ1 and Jβ1, and Vγ segments tend to rearrange only with the closest DJγ. In addition, the gene segments that make up $\gamma\delta$ TCRs are not chosen entirely at random. Different $\gamma\delta$ TCRs appear to contain specific Vγ and Vδ gene segments depending on the cellular subset or anatomic location in which they are found (Table 8-3). Thus, the actual diversity derived from combinatorial sources is more limited than the theoretical diversity.

Fortunately, what is lost in combinatorial diversity is compensated for by variable D segment inclusion. Although the Ig loci contain higher numbers of D gene segments, only one D segment can join to an Ig J segment during a given rearrange-

Table 8-3 Use of TCRγ and δ Gene Segments in Murine Tissues

Tissue	Preferential γ Usage	Preferential δ Usage
Skin	Vγ3Jγ1Cγ1, Vγ5Jγ1Cγ1	Vδ1Dδ2Jδ2Cδ
Uterus	Vγ6Jγ1Cγ1	Vδ1Dδ2Jδ2Cδ
Tongue	Vγ6Jγ1Cγ1	Vδ1Dδ2Jδ2Cδ
Intestine	Vγ7JγnCγn	Vδ4DδnJδnCδ

ment. Diversity in the $\gamma\delta$ TCR repertoire is increased because TCR Dδ segments may join to each other as well as to Jδ segments to form VDJ, VDDJ or VDDDJ variable exons.

ii) Junctional Diversity

The mechanisms of generating junctional diversity that were discussed in the context of B cells in Chapter 4 also apply to T cells. Both P nucleotides and N nucleotides can be added to VD and DJ joints in TCR chains and give rise to amino acids that are not encoded in the germline. Because more than one Dδ segment may be included in tandem in a TCRδ chain, many more opportunities for P and N nucleotide addition occur at each D–D or D–J joint. It has been estimated that junctional diversity contributes billions of possible TCRδ chains to the TCR repertoire.

iii) Chain Pairing

The random pairing of TCRα and β chains (or TCRγ and δ chains) within a given $\alpha\beta$ (or $\gamma\delta$) T cell also contributes to TCR repertoire diversity. In the case of an $\alpha\beta$ T cell, the TCR's antigen-binding site is composed of the V domains of the one TCRα chain and the one TCRβ chain synthesized in that cell. However, since any one of the vast number of possible sequences for a TCRα chain gene can occur in the same T cell as any one of the even more numerous possibilities for a TCRβ chain, the total number of possible $\alpha\beta$ heterodimers approaches 10^{20}. Similarly, the repertoire of functional TCR$\gamma\delta$ heterodimers is about 10^{18}. These numbers compare very favorably to the 10^{11} specificities estimated for the Ig repertoire.

C. TCR–Antigen Interaction

The interaction between a TCR$\alpha\beta$ protein and its pMHC epitope underlies fundamental aspects of the cell-mediated adaptive immune response. Firstly, the strength of binding between a thymocyte's TCR and various pMHCs encountered in the thymus determines whether the thymocyte is negatively selected and dies, dies of "neglect", or is positively selected and survives to become a mature T cell (see Ch. 9). Secondly, the strength of binding between a mature $\alpha\beta$ T cell's TCR and pMHC presented by an APC in the periphery determines whether the T cell will be activated to proliferate and differentiate into effector cells, or will become non-responsive.

Table 8-2 Estimates of Numbers of Gene Segments in Mouse and Human TCR Loci

	Number of Gene Segments in Germline*			
	TCRA	TCRB	TCRG	TCRD
Mouse Gene Segments				
V	75	23	7 (1)	12
D	0	2	0	2
J	61	12	4 (1)	2
Human Gene Segments				
V	42	48	14 (8)	10 (5)
D	0	2	0	3
J	61	13	5	3

*Numbers in parentheses are the numbers of these gene segments that are pseudogenes.

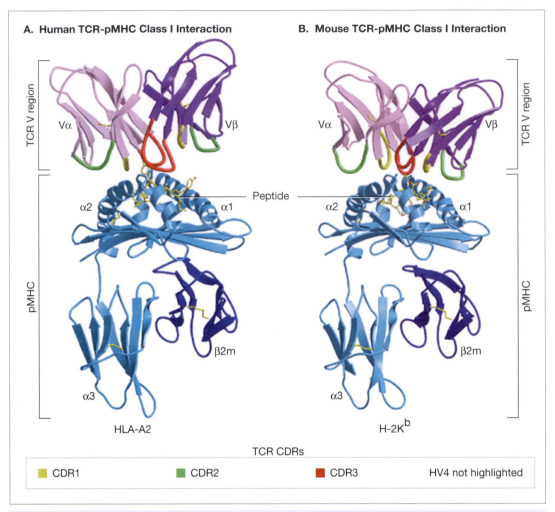

A. Human TCR-pMHC Class I Interaction

TCR V region
Vα — Vβ

Peptide
α2 — α1

pMHC

β2m

α3

HLA-A2

B. Mouse TCR-pMHC Class I Interaction

TCR V region
Vα — Vβ

α2 — α1
Peptide

pMHC

β2m

α3

H-2K^b

TCR CDRs

■ CDR1   ■ CDR2   ■ CDR3   HV4 not highlighted

Plate 8-2 **X-Ray Crystal Structures of Human and Mouse TCR–pMHC Interaction**
[Reproduced by permission of Bjorkman P.J. (1997). MHC restriction in three dimensions: a view of T cell receptor/ligand interactions. *Cell* **89**, 167–170.]

Immunologists still do not fully understand the molecular pathways governing these cell fate decisions. The structural aspects of TCR binding to pMHC are presented here, whereas issues associated with T cell activation/differentiation and peripheral T cell tolerance are discussed in Chapters 9 and 10, respectively.

Studies of TCR X-ray crystal structures have shown that the V domains of the TCRαβ heterodimer resemble the V domains of the Ig molecule, but that the interdomain pairing of the Cα and Cβ regions differs from that in the Ig C regions. In addition, in contrast to the relative independence of the Ig V and C domains, TCR Vβ and TCR Cβ are closely associated within the crystals. This association may confer a degree of inflexibility to that region of the TCR that is analogous to the Ig Fab region. Plate 8-2A depicts the V domains of a human αβ TCR interacting with peptide bound to the extracellular region of MHC class I, while Plate 8-2B shows the corresponding murine molecules. Comparable analyses of TCRγδ crystal structures have shown that TCRγδ differs physically from TCRαβ. In particular, the structure of the TCR Vδ domain looks more like Ig V_H than TCR Vα or Vβ. This finding is

consistent with the results of functional studies showing that γδ T cells recognize antigenic structures other than pMHCs (see Ch. 11).

In many TCRs, the TCR peptide-binding site itself is relatively flat except for a deep hydrophilic cavity between the TCRα CDR3 and TCRβ CDR3. When bound, the TCR is oriented in a diagonal position over the pMHC such that the flat region can interact with the peptide. Both the TCRα and β chains are usually involved in binding to both the MHC molecule and the peptide, and this binding occurs virtually simultaneously. In general, the highly variable CDR3 regions of the TCRα and β chains bind to the middle of a peptide lodged in the MHC binding groove as well as to points on the MHC protein backbone. The less variable CDR1 and CDR2 regions tend to bind to the ends of the peptide and to conserved sites on the MHC backbone.

The area of contact between the TCR and the pMHC is relatively small such that only a few of the residues in the peptide generally make contact with a TCR chain. This limited opportunity for intermolecular bonding means that the binding affinity of a TCR for pMHC ($K = ~5 \times 10^5$ M^{-1}) is significantly lower than that of an antibody for its antigen ($K = 10^7–10^{11}$ M^{-1}). This relatively modest affinity of TCR binding has two implications. Firstly, the initial contact between T cells and APCs or target cells is established not by TCR–pMHC interaction but rather by the binding of complementary pairs of adhesion molecules. Specific TCR–pMHC contacts are made only after the cells are held in close enough proximity by the adhesion molecules to permit the T cell to scan the pMHCs in the APC or target cell membrane. At this point, contacts between CD4 or CD8 and the MHC class II or I molecule, respectively, also become important in holding the cells together. Secondly, because of their modest affinity for their cognate ligands, TCRs can bind (with varying strength) to a surprisingly broad range of pMHCs. Such flexibility facilitates thymic selection because one peptide can positively select several thymocyte clones, amplifying the T cell repertoire. Thymic selection is discussed in more detail in Chapter 9.

The flexibility of TCR binding is largely due to the CDR3 regions. Whereas the CDR1, CDR2 and HV4 hypervariable regions are encoded by the V gene segment of a variable exon, the DNA sequences encoding the CDR3 regions span the VJ joint in the rearranged TCRα gene and the VD and DJ joints in the rearranged TCRβ gene (Fig. 8-10). Junctional diversity thus imparts extreme variability in the amino acid sequence, length and conformation of this region. Comparisons of the conformations of TCRs that have not bound to pMHC versus TCRs bound to pMHC have demonstrated that the CDR3 regions are capable of undergoing an enormous conformational shift in order to achieve the diagonal orientation favored for binding to pMHC. The adoption of this "induced fit" affects only the CDR3 regions and does not alter the conformation of either the rest of the TCR molecule or the MHC molecule. The sequence of binding events is variable: sometimes CDR3 initiates interaction with peptide first, and sometimes CDR1 or CDR2 binding to pMHC is established first.

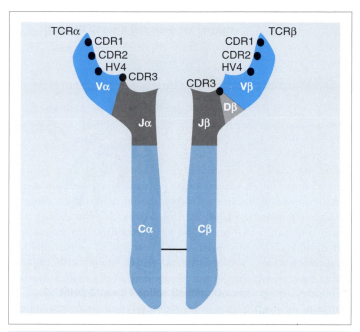

Fig. 8-10 **Correspondence of TCR Hypervariable Regions to TCR Gene Segments**

In any case, once a TCR finalizes its contacts with a given pMHC, the entire complex is stabilized and the flexibility of both the TCR and pMHC binding surfaces is lost.

This marks the end of our discussion of the TCR proteins and genes. In Chapter 9, we examine the development of T cells and the crucial role that TCRs play in the positive and negative selection processes that shape the T cell repertoire. Chapter 9 also describes T cell activation by antigen and the differentiation and functions of effector and memory T cells.

- There are two types of TCRs, TCRαβ and TCRγδ, which are expressed by αβ T cells and γδ T cells, respectively.

- TCRαβ molecules recognize peptides bound to either MHC class I or class II, whereas γδ TCRs can recognize antigens in their natural, unprocessed forms.

- TCR chains are incapable of signal transduction. This function is carried out by the ITAM-containing CD3 complex that associates with TCRαβ or TCRγδ.

- The TCRαβ molecule has four hypervariable regions that promote binding to a small collection of highly similar pMHCs.

- The *TCRA*, *TCRB*, *TCRG* and *TCRD* loci contain multiple V, D and J gene segments and one or two C exons. *TCRD* is nested within *TCRA*. V(D)J recombination assembles functional TCR genes in a strict order tied to T cell development.

- Although isotype switching and somatic hypermutation do not occur in T cells, the overall diversity of the T cell repertoire is greater than that of the Ig repertoire because of increased junctional diversity.

- αβ T cells express either the CD4 coreceptor that binds to a non-polymorphic region of MHC class II, or the CD8 coreceptor that binds to a non-polymorphic region of MHC class I.

- Coreceptor binding to MHC increases the adhesion between T cells and APCs or target cells, and facilitates Lck recruitment.

- The physical flexibility of the TCR binding site is largely due to the presence of four hypervariable regions in each TCR chain. CDR3 is particularly important for peptide binding by a TCR.

Section Introduction–A.I

1) What are intraepithelial cells?

2) Give two differences between TCRs and BCRs.

3) Give three differences between αβ T cells and γδ T cells.

4) What protein chains come together to form TCRs?

5) How do the hypervariability sites in the TCR chains differ from those in the Ig chains?

Section A.II

1) Describe the composition of the CD3 complex.

2) Why does the TCR need the CD3 complex?

3) What is Lck, what does it do and why is this important?

Section A.III

1) Why are CD4 and CD8 called "coreceptors"?

2) How do CD4 and CD8 differ in structure? Does this difference affect their function?

3) Give three functions of the coreceptors.

Section B.I

1) What loci encode the TCR chains? Is there anything unusual about the structure of these loci?

2) T cells do not undergo isotype switching. How is this reflected in their gene structure?

3) How do the D segments in the TCR loci differ from those in the Ig loci?

Section B.II–V

1) Which TCR locus is the first to rearrange in a thymocyte that will become an αβ T cell?

2) Which TCR locus is the first to rearrange in a thymocyte that will become a γδ T cell?

3) Give one hypothesis accounting for why the Ig genes are not expressed in developing T cells.

4) Why aren't TCR proteins secreted?

5) Give two reasons why the actual diversity in the TCR repertoire is less than the theoretical diversity.

6) Why does diversity in the TCR repertoire exceed that in the Ig repertoire?

Section C

1) Describe two cellular events governed by the affinity of binding between a TCRαβ and its pMHC epitope.

2) How does the structure of the TCRαβ V domain differ from that of the TCRγδ V domain, and what effect does this have on antigen recognition?

3) Describe how the hypervariable regions of the TCRαβ chain interact with peptide in the MHC binding groove.

4) How does the binding affinity of TCRαβ for pMHC differ from that of Ig for its antigen, and what are two implications of this difference?

5) Why is the flexibility of TCRαβ binding largely due to the CDR3 region?

9

T Cell Development, Activation and Effector Functions

Don't fall before you're pushed.
English Proverb

A. T Cell Development

I. COMPARISON OF B AND T CELL DEVELOPMENT

B and T cells are both lymphocytes and are derived from the same very early hematopoietic progenitors. However, their development differs in several important ways: (1) The thymus is required for the generation of the vast majority of mature peripheral T cells but not for that of mature B cells. (2) Naïve B cells are freshly produced at a virtually constant rate for the life of the individual. However, once the involution of the thymus commences around puberty, new naïve T cell production is sharply reduced and the maturing adult becomes increasingly dependent on the existing repertoire of T cells. (3) MHC molecules are involved in the establishment of central tolerance of T cells but not B cells. The TCR on a thymocyte must not only be functional (be derived from productive rearrangements of the TCR genes) but must also recognize the host's MHC molecules (so that it can "see" pMHC). This requirement imposes an additional layer of selection on T cells that developing B cells do not experience. (4) The TCR expressed on a T cell's surface is fixed for the life of the clone and cannot undergo the somatic hypermutation in response to activation that occurs following B cell activation. (5) Whereas the vast majority of functional B cells result from a single developmental program, functional T cells can result from several different paths. A developing thymocyte may give rise to γδ T cells or αβ T cells, and among the latter, Th or Tc cells. Once activated by antigen, a Th clone can further differentiate into subsets that differ slightly in their effector functions. In addition, certain thymocytes have the capacity to differentiate into *regulatory T cells* that can control the responses of activated Th and Tc cells. Regulatory T cells are discussed in Chapter 10.

The complex pathways mediating lymphocyte development, activation and differentiation involve many genes that can become targets for mutations leading to primary immunodeficiencies. Some of these disorders affect T cells primarily and

are described in Box 9-1. Other primary immunodeficiencies, which affect both T and B cells, are discussed in Box 9-2.

II. COLONIZATION OF THE THYMUS

During early embryonic development of a mammalian fetus, the thymus is empty of hematopoietic progenitors. The fetal thymus must be "colonized" or seeded with hematopoietic progenitors that subsequently proliferate and mature in the thymus into functional naïve T cells. As introduced in Chapter 2, T cells (like all hematopoietic cells) are derived from HSCs. Although HSCs are located in the bone marrow in an adolescent or adult individual, they are generated in the liver in a fetus. In either location, a proliferating HSC can differentiate into early progenitors (including MPPs and NK/T precursors), some of which leave the bone marrow and enter the blood circulation. NK/T precursors that express high levels of the chemokine receptor CCR9 exit the blood and enter the thymus. Depending on the cytokines and stromal cell ligands encountered by a given NK/T precursor, it may eventually differentiate into T cells, NK cells, NKT cells or lymphoid DCs.

The fetal thymus is colonized by NK/T precursors in a limited number of distinct waves that occur both before and after birth. The earliest prenatal waves migrate from the fetal liver, enter the fetal thymus, and give rise only to γδ thymocytes. Subsequent waves of NK/T precursors entering the thymus just before birth and thereafter give rise to both αβ and γδ thymocytes. However, after birth, thymocyte development is more and more biased toward αβ T cells such that the γδ T cells become a minor population. In addition, the bone marrow becomes the dominant site of generation of the NK/T precursors needed to replenish the thymus.

The developmental path of HSCs to mature T cells is generally the same in fetal, neonatal, adolescent and adult mice but displays slower kinetics in adolescent and adult individuals (Table 9-1). Importantly, the TdT enzyme responsible for

Box 9-1 Primary Immunodeficiencies Affecting T Cells

DiGeorge syndrome is a complex disorder in which thymus development is impaired to a variable degree. Patients suffer from recurrent infections, and have characteristically abnormal facial features, abundant hair, and cardiovascular and renal anomalies. Cleft palate is common, as is parahypothyroidism. However, not all patients show all symptoms. T cells that are present may show enhanced spontaneous apoptosis, or fail to proliferate in response to antigen. B cells are normal but humoral responses are often compromised due to the lack of T cell help. The precise molecular defect underlying DiGeorge syndrome is unknown but is thought to affect that stage of embryogenesis during which the primordial thymus, parathyroid glands, heart and face develop. In most (but not all) DiGeorge syndrome patients, translocations or large deletions involving chromosome 22 are observed.

X-linked lymphoproliferation (XLP) is a very curious primary immunodeficiency of young boys in which the main clinical manifestation is an inappropriate response to *Epstein-Barr virus* (EBV) infection. No other pathogen is linked to XLP. Affected boys develop potentially fatal infectious mononucleosis and show excessive numbers of anti-EBV lymphocytes in the liver. Serum IFNγ levels are high, T cell responses are skewed toward Th1, and NK cells cannot kill EBV-infected cells. XLP is caused by mutations of an X-linked gene called SAP that controls Th1 differentiation and IFNγ production.

Another group of T cell-specific primary immunodeficiencies causes a shared phenotype of potentially fatal susceptibility to infections with viruses, intracellular bacteria and fungi. Such consequences can result from **MHC deficiencies**, in which the MHC molecules necessary for antigen presentation to T cells are not expressed. In addition, mutations in the TAP transporter and in transcription factors and regulatory proteins required for MHC expression can block pMHC expression on APCs. These diseases can also result from defects in the **TCR signaling pathway**, such as mutations affecting the CD3 chains, Lck kinase or Ca²⁺ flux. In **hemophagocytic syndrome (HPS)**, mutations of proteins involved in CTL granule exocytosis lead (for unknown reasons) to uncontrolled and sometimes fatal CTL infiltration into various organs. In patients with **hyper IgM 1 syndrome (HIGM1)**, serum IgM is elevated while IgA and IgG are abnormally low due to a defect in T cell help. B cells are normal in HIGM1 patients but isotype switching is stymied by a mutation in the X-linked CD40L gene in T cells. The CD40L defect also impairs DC "licensing" (see later) such that the Tc cell-mediated response fails. HIGM1 patients thus suffer from recurrent infections with intracellular bacteria as well as opportunistic fungal and parasitic infections.

Box 9-2 Severe Combined Immunodeficiency Diseases (SCIDs)

The best known primary immunodeficiencies are the SCIDs. In all forms of SCID, T cell development and/or function are compromised such that both cell-mediated and humoral immune responses are impaired. Sometimes the B cells themselves are also faulty but often the B cell defects are secondary to the lack of T cell help. In addition, NK development and function may be impaired. No matter what the genetic defect, SCID patients generally present with chronic diarrhea, failure to thrive, and severe opportunistic infections. Without treatment, SCID is inevitably fatal in early childhood or adolescence.

X-linked SCID (XSCID) is caused by a mutation of the gene encoding a protein (the "common γ chain") that is required for signaling by the IL-2, IL-4, IL-7, IL-9, IL-15 and IL-21 receptors. Because both the IL-7 and IL-15 signaling pathways are disrupted, T and NK cells fail to develop. B cells are present but nonfunctional. Autosomal SCID diseases can arise from mutations in the signaling kinase **Jak3** (T and NK cells missing; B cells non-functional) or the IL-7 binding subunit of **IL-7R** (T cells missing, B cells non-functional; NK cells normal). T, B and NK cells are missing in patients with **ADA SCID** and **PNP SCID** due to adenosine deaminase (ADA) or *purine nucleoside phosphorylase* (PNP) deficiency, respectively. ADA and PNP are enzymes that prevent the buildup of toxic metabolites during nucleotide biosynthesis. These metabolites are selectively lethal to lymphoid cells. **RAG SCID** results from impaired V(D)J recombination due to a null mutation in either of the recombination activating genes RAG1 and RAG2. T and B cells are absent but NK cells are normal. **Omenn syndrome** is a milder disease caused by amino acid substitution mutations in the RAG genes that reduce the activity of the recombinases to 1–25% of normal.

Table 9-1 Comparison of Murine Thymocyte Development at Different Life Stages

Property	Fetus	Neonate	Post-Neonate
Origin of NK/T precursors	Fetal liver	Fetal liver	Bone marrow
TCRs in the periphery	Only TCRγδ No TCRαβ	Majority TCRγδ Minority TCRαβ	Minority TCRγδ Majority TCRαβ
Kinetics of progression from NK/T precursors to mature T cells	Fast	Fast	Slow
TdT expression	None	Initiated	Fully active
T cell repertoire diversity	Limited	Limited	Fully diversified
Generation of thymocytes	Continuous	Continuous	Minimal after involution of thymus

much of the junctional diversity generated during TCR gene rearrangement is not expressed until shortly after birth. Thus, the repertoire of T cell specificities available in the neonate is significantly less diverse than in older individuals.

III. THYMOCYTE MATURATION IN THE THYMUS

Shortly after their arrival in the thymus, NK/T precursors generate *thymocytes*. Thymocytes at different developmental stages are morphologically very similar and so are usually distinguished by either their patterns of surface marker expression or by their TCR gene rearrangement status. These parameters have been used to divide thymocyte maturation into three broad phases. These phases are the *double negative* phase (DN, stages DN1–DN4), in which thymocytes express neither CD4 nor CD8; the *double positive* phase (DP), in which thymocytes express both CD4 and CD8; and the *single positive* phase (SP), in which thymocytes express either CD4 or CD8 but not both (Fig. 9-1). Most DN3 thymocytes become αβ T cells but some become γδ T cells. Most DP thymocytes (that survive thymic selection; see later) become SP thymocytes and ultimately mature αβ T cells, but some DP thymocytes develop into NKT cells (see Ch. 11). Once SP thymocytes emerge from the thymus and enter the circulation and secondary lymphoid tissues, they are considered to be mature naïve CD4⁺ or CD8⁺ peripheral T cells.

i) The Thymic Environment

The development of thymocytes through the DN, DP and SP phases is totally dependent on the stromal cells that make up the thymic architecture (Fig. 9-2). (In mice, this dependence is clearly evidenced by the absence of T cells in the *nude* strain, which has a mutation in the development of thymic stromal cells.) As normal thymocytes mature, they pass through a succession of thymic microenvironments characterized by different mixes of stromal cell types. Among the most important of these stromal cells are *cortical thymic epithelial cells* (cTECs), *medullary thymic epithelial cells* (mTECs), *thymic DCs* and *thymic fibroblasts*. As is discussed later, thymic DCs, cTECs and mTECs are vital for the establishment of T cell central

tolerance. cTECs and mTECs also express cell surface ligands for *Notch1*, a cell fate protein expressed on the surface of thymocytes. Once Notch1 has bound to its ligand, the cytoplasmic domain of Notch1 interacts with transcription factors to promote T cell development while suppressing B cell development. Thymic fibroblasts secrete components of the extracellular matrix (such as collagen) that create a scaffolding used to concentrate the cytokines crucial for thymocyte development. Other components secreted by thymic fibroblasts are involved in controlling the adhesion of thymocytes to stromal cells and thus may direct thymocyte migration through the thymus.

ii) DN Phase

As already mentioned, the earliest thymocytes are said to be in the double negative or *DN phase* because they express neither CD4 nor CD8. These cells are also negative for TCR expression, cannot bind pMHC, and do not carry out effector functions. Within the DN phase are four subsets of thymocytes labeled DN1–4. In mice, these subsets are distinguished from each other by their expression of the surface markers c-kit (a cytokine receptor), CD44 (an adhesion protein) and CD25 (the α chain of the IL-2 receptor). (IL-2 is not required for thymocyte development and the function of CD25 in thymocytes is unknown.) In humans, expression of the markers CD34 (a putative adhesion protein), CD38 (an adhesion protein) and CD1a (an MHC-like protein) distinguishes the DN1–4 thymocyte subsets. The major markers expressed by developing human and murine T cells are compared in Table 9-2, and additional important molecules expressed during murine thymocyte development appear in Figure 9-3. The discussion here focuses on the better studied developmental path of murine thymocytes.

iia) DN1 subset. Murine DN1 thymocytes express both c-kit and CD44 (but not CD25) and reside in the thymic cortex near its junction with the medulla (refer to Fig. 9-2). The TCR genes remain in germline configuration. DN1 thymocytes are small and closely packed together among the cTECs. cTECs supply *stem cell factor* (SCF) that binds to c-kit on DN1 cells and

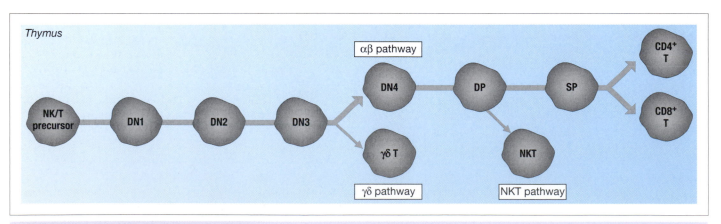

Fig. 9-1 **Model of Thymocyte Maturation in the Thymus**

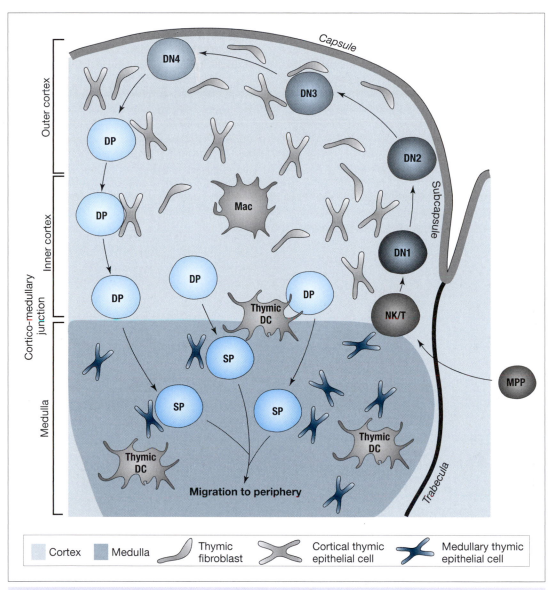

Fig. 9-2 Thymic Microenvironment and Location of Developing Thymocytes
[Adapted from Blackburn C.C. and Manley N.R. (2004). *Nature Reviews Immunology* **4**, 278–289.]

Table 9-2 Comparison of Murine and Human Thymocyte Markers

Stage	Murine	Human
DN1	CD44$^+$CD25$^-$	CD34$^+$CD38$^-$CD1a$^-$
DN2	CD44$^+$CD25$^+$	CD34$^+$CD38$^+$CD1a$^-$
DN3	CD44$^-$CD25$^+$	CD34$^+$CD38$^+$CD1a$^+$
DN4	CD44$^-$CD25^{lo}	CD34$^-$CD38$^-$CD1a$^+$CD4$^+$
DP-TCRβ	CD4$^+$CD8$^+$	CD4$^+$CD8$^+$
DP-TCRαβ	CD4$^+$CD8$^+$	CD4$^+$CD8$^+$
SP-CD4	CD4$^+$CD8$^-$	CD4$^+$CD8$^-$
SP-CD8	CD4$^-$CD8$^+$	CD4$^-$CD8$^+$

delivers a survival signal. Without c-kit signaling, the maturation process ceases and the DN1 cells die.

iib) DN2 subset. Murine DN2 thymocytes express CD25 as well as c-kit and CD44 and are sometimes known as *pro-T cells* (progenitor T cells). These thymocytes start to migrate toward the subcapsule of the thymus and thus are present primarily in the outer cortex. The TCR genes remain in germ-line configuration. DN2 thymocytes commence expression of the CD3 chains but it is unclear whether these proteins have a signaling function at this stage. Under the influence of IL-7 and SCF, DN2 thymocytes start to proliferate rapidly.

iic) DN3 subset. Murine DN3 thymocytes lose their expression of c-kit and CD44 but continue to express CD25. These cells stop proliferating and remain in the outer thymic cortex. The DN3 stage is critical in T cell development because five

Subset / Marker	DN Phase				DP Phase		SP Phase	
	DN1	DN2	DN3	DN4	TCRβ	TCRαβ	CD4+	CD8+
c-kit	+	+	–	–	–	–	–	–
CD44	+	+	–	–	–	–	–	–
CD25	–	+	+	Low	–	–	+	+
CD3	–	+	+	+	+	+	+	+
Rearranging TCR genes	–	–	*TCRB TCRG TCRD*	*TCRB*	*TCRA*	–	–	–
pTα	–	–	+	+	Low	–	–	–
RAG	–	–	+	Low	+	+	–	–
TdT	–	–	+	Low	+	+	–	–
TCR	–	–	–	–	–	+	+	+
CD4	–	–	–	Low	Med	+	+	–
CD8	–	–	–	Low	Med	+	–	+

Fig. 9-3 **Markers Characterizing the Phases of Murine Thymocyte Development (αβ T cells)**

key events occur: (1) DN3 thymocytes become restricted to the T lineage and eventually generate mature αβ and γδ T cells. (The development of the γδ T cell lineage is discussed in Chapter 11.) (2) The *TCRG*, *TCRD* and *TCRB* loci commence V(D)J recombination with concomitant upregulation of RAG and TdT. (3) DN3 thymocytes that eventually generate mature αβ T cells express a functional pre-TCR complex that allows them to determine if a functional TCRβ chain has been produced. (4) Successful rearrangement at the *TCRB* locus induces the cessation of further rearrangements at the *TCRG* and *TCRD* loci in these cells. (5) These DN3 thymocytes become *early pre-T cells* that are fully committed to the αβ T cell lineage and express a diverse repertoire of TCRβ chains.

In DN3 thymocytes that eventually generate αβ T cells, the *TCRB* locus is the first to undergo V(D)J recombination that leads to protein synthesis. Some rearrangement of the *TCRG* loci may also occur but functional chains are not produced. As introduced in Chapter 8, the productivity of a given rearranged TCRβ gene in a DN3 thymocyte is tested by the formation of a pre-TCR analogous to the pre-BCR structure in pre-B cells. The pre-TCR counterpart of the surrogate light chain in the pre-BCR is the *pre-T alpha chain* (pTα). pTα is an invariant protein first expressed in DN3 thymocytes that develop into αβ T cells; there is no equivalent in DN3 thymocytes that generate γδ T cells. pTα functions as a "surrogate TCRα chain" and brings a newly synthesized TCRβ chain (and the CD3 signaling chains) to the thymocyte membrane to form a pre-TCR complex (Fig. 9-4). The pre-TCR complex acts as a sensor so that if the TCRβ chain is functional, the cell receives a survival signal and proliferates vigorously, generating a clone of DN4 thymocytes that will eventually become αβ T cells. The process of testing newly produced TCRβ chains is called *β-selection*, and cells that survive β-selection are said to have passed the *pre-TCR checkpoint* (Fig. 9-5). The successful rear-

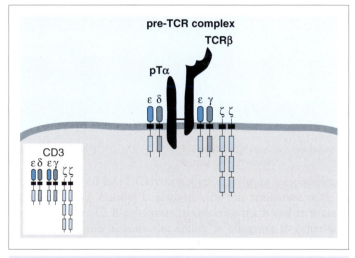

Fig. 9-4 **Pre-TCR Complex in DN3 Cells**

rangement of the *TCRB* gene on one chromosome signals to the cell to suppress both V(D)J rearrangement of the *TCRB* locus on the other chromosome and rearrangement of both *TCRG* loci. If *TCRB* rearrangement on both chromosomes has been unsuccessful, the cell neither attempts to rearrange its *TCRA* genes nor becomes a γδ T cell; instead, it dies by apoptosis. Indeed, only 10% of DN3 thymocytes successfully rearrange their TCRβ genes, are β-selected and enter the cell cycle. β-selection is thus directly linked to the proliferation of thymocytes that can proceed further in maturation.

iid) DN4 subset. Murine DN4 thymocytes, also called *late pre-T cells*, are slightly larger in size than DN3 cells and are concentrated in the subcapsular region of the thymic cortex (refer to Fig. 9-2). DN4 cells contain a functionally rearranged

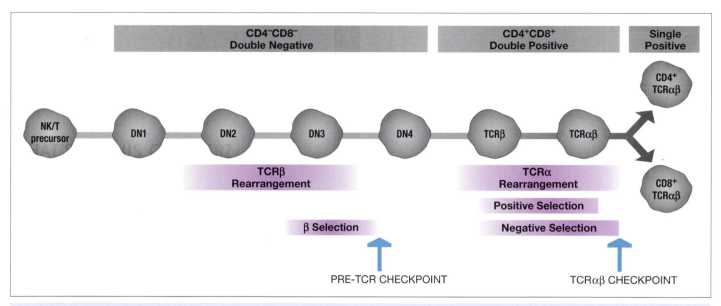

Fig. 9-5 Checkpoints of T Cell Development
[Adapted from Yeung, R.S., Ohashi, P. and Mak, T.W. (2007). T cell development. In Ochs, H.D., Smith, C.I. and Puck, J.M., Eds., *Primary Immunodeficiency Diseases. A Molecular and Genetic Approach*, 2nd ed. Oxford University Press Inc., New York.]

TCRβ gene, downregulate their expression of CD25, RAG and TdT, and start to express very low levels of CD4 and CD8.

iii) The DP Phase

In both humans and mice, the DP phase of αβ T cell development is dominated by the thymic selection processes that shape the mature αβ T cell repertoire. CD4 and CD8 expression levels are steadily upregulated and these coreceptors play increasingly important roles in directing thymocyte development. As they mature, DP thymocytes move from the subcapsular region through the outer cortex and back through the inner cortex toward the medulla (refer to Fig. 9-2).

iiia) TCRαβ pool expansion and *TCRA* locus rearrangement.

Early DP thymocytes receive signals through their pre-TCRs that drive their rapid proliferation. These signals appear to depend upon the assembly of pre-TCR itself and not upon interaction with a specific ligand. RAG and TdT expression resume and V(D)J recombination in both *TCRA* loci commences, resulting in the deletion of the *TCRD* loci (refer to Ch. 8). With the production of the first TCRα chains, pTα expression is gradually downregulated. Newly synthesized (but untested) TCRα chains combine with the proven TCRβ chains to form complete TCRαβ heterodimers (which may or may not be functional). *TCRA* rearrangement continues on both chromosomes until positive selection (see later) delivers a survival signal to those thymocytes with functional TCRs that recognize the host's MHC molecules with moderate affinity.

iiib) Thymic selection and the establishment of central T cell tolerance.

The establishment of central T cell tolerance requires that thymocytes with TCRs that recognize self antigen be eliminated before they leave the thymus. Central T cell tolerance is established by *thymic selection*, which includes the processes of *non-selection* (or "*neglect*"), *positive selection* and

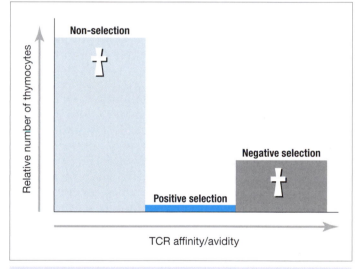

Fig. 9-6 Affinity/Avidity Model of Thymic Selection
[Adapted from Yeung, R.S., Ohashi, P. and Mak, T.W. (2007). T cell development. In Ochs, H.D., Smith, C.I. and Puck, J.M., Eds., *Primary Immunodeficiency Diseases. A Molecular and Genetic Approach,* 2nd ed. Oxford University Press Inc., New York.]

negative selection. Mature T cells must recognize both MHC and peptide simultaneously to mount an immune response, so that the TCRs of a DP thymocyte must bind to the host's MHC molecules (self-MHC) with at least moderate affinity. T cells that fail to produce functional TCRs, or produce TCRs that have no affinity for self MHC, cannot be activated in the periphery and thus are useless with respect to defending the host. These cells, which can number as high as 80% of developing thymocytes, are therefore "non-selected" and undergo apoptosis (Fig. 9-6). *Negative selection* removes from the T cell repertoire DP cells whose TCRs bind strongly to pMHCs made up of self peptides bound to self MHC. The activation of such T cells could spark a damaging autoimmune response in the host. The strong binding of the TCR to self pMHC triggers signaling in the thymocyte that induces its apoptosis. It is estimated that almost 20% of developing thymocytes are "deleted" from the thymus in this way. *Positive selection* preserves the 1–2% of developing thymocytes whose TCRs recognize self pMHCs neither too strongly nor too weakly. This type of binding appears to deliver a survival signal that allows the thymocytes to proliferate and mature further. Once released to the periphery, it is this small population of new T cells that is most likely to bear TCRs recognizing non-self peptide on self MHC, which is exactly what an APC or target cell will present when an immune response is required. The few remaining potentially autoreactive clones that escape thymic deletion, and clones that recognize self antigens which emerge only later in life (after positive selection has been completed), are neutralized by the mechanisms of peripheral tolerance (see Ch. 10).

Immunologists continue to debate exactly when and where in the thymus each type of selection occurs. Some maintain that positive selection takes place primarily in the cortex, whereas negative selection occurs later when the DP cells approach the medulla. Others believe that both positive and negative selection can occur in either the cortex or the medulla, and that these processes are temporally independent. What is now clear is that mTECs have a specialized role to play during negative selection, in the establishment of central tolerance to *tissue-specific antigens*. Tissue-specific antigens are self antigens that are normally expressed only in particular host tissues. mTECs express a transcription factor called AIRE (*autoimmune regulator*) that allows the expression of many of these tissue-specific antigens in the thymus, so that thymocytes bearing TCRs that recognize these antigens can be deleted. In the absence of AIRE, these tissue-specific antigens are not expressed in the thymus and thymocytes that are capable of responding to them escape to the periphery where they can mount autoreactive responses against host tissues expressing these antigens. Both humans and mice with a genetic deficiency of AIRE show symptoms of autoimmune disease (see Chapter 19).

iiic) Nature of signaling during thymic selection. Intracellular signals received by thymocytes during non-selection, positive selection and negative selection depend on the overall affinity/avidity of the interaction between the TCRs of a DP thymocyte and the pMHCs presented by the thymic APC it encounters. The level of intracellular signaling triggered by this interaction is also influenced by the level of aggregation of the TCRs and coreceptor molecules, by the type of thymic APC presenting the pMHC (thymic DC, mTEC or cTEC), and by the costimulatory molecules expressed by the thymic APC.

As noted earlier, the majority of DP thymocytes are non-selected because their TCRs cannot interact at all with the pMHCs presented by thymic APCs. Some of these thymocytes have out-of-frame rearrangements of the Vα and Jα TCR gene segments such that no TCRα protein can be produced. Other thymocytes have undergone successful V(D)J recombination but the TCR produced has a conformation that simply cannot bind to self MHC with any level of affinity. In both cases, TCR signaling is not triggered and these DP cells proceed down a default path of apoptosis and die in the cortex. With respect to positive selection, the moderate signaling triggered by weak binding of a DP thymocyte's TCR to the self pMHC of a thymic APC (particularly a cTEC) appears to be sufficient to induce the transcription of genes that rescue these clones from apoptosis. Only clones expressing these *anti-apoptotic* genes can subsequently proliferate and continue to mature into naïve T cells. With respect to negative selection, DP thymocytes with TCRs that bind strongly to self pMHCs presented by thymic APCs (particularly mTECs) experience intracellular signaling that drives the expression of genes promoting apoptosis. The effects of these *pro-apoptotic* genes appear to overwhelm any countersignaling by anti-apoptotic genes.

iiid) The TCRαβ checkpoint. DP thymocytes that have survived the gauntlet of thymic selection have passed the second developmental checkpoint, the *TCRαβ checkpoint* (refer to Fig. 9-5). These positively selected DP cells express a fully functional TCRαβ and both CD4 and CD8, and can thus interact with both MHC class I and class II. It is these thymocytes that proceed to the final phase of αβ T cell development. It is also these cells that will give rise to NKT cells (refer to Fig. 9-1 and see Ch. 11).

iv) The SP Phase

At this point, the class of MHC recognized by a positively selected DP thymocyte becomes fixed by the loss of expression of either CD4 or CD8. The progeny of this cell are then either CD4+ or CD8+ SP thymocytes. Which coreceptor is lost and which is retained on the surface of the thymocyte is determined by complex intracellular signaling pathways and the class of MHC molecule involved in the thymic selection of that thymocyte. As described in Chapter 8, CD4 and CD8 molecules bind to sites on MHC class II and MHC class I molecules, respectively. If the TCR of the thymocyte is specific for a selecting peptide presented on MHC class I by the cTEC or mTEC, there is an interaction between the MHC molecule and the CD8 coreceptor such that CD8 expression is retained while CD4 expression is lost. Conversely, if the thymocyte TCR interacts with a selecting peptide presented on MHC class II, an interaction occurs between the MHC molecule and CD4 coreceptor such that CD4 expression is retained while CD8 expression is lost. Thus, the TCRs of the descendants of SP CD4+ cells will bind to pMHCs involving MHC class II, and those of SP CD8+ cells to pMHCs involving MHC class I. We stress that this discrimination is not due to differences in the type of TCR expressed by CD4+ and CD8+ T cells; all TCRs expressed by all αβ T cells are derived from the same pool of TCRα genes and TCRβ genes.

Both CD4+ and CD8+ SP thymocytes loiter in the medulla of the thymus for a short time (2–3 days in the mouse) before they receive a final proliferative signal and expand their numbers. These progeny exit the thymus into the blood and travel to the secondary lymphoid organs, taking up residence as fully functional mature T cells. They survive in these locations (in the apparent absence of significant antigenic stimulation) for at least 5–7 weeks, and are poised to react upon meeting their specific antigens. For reasons that are not yet understood, the eventual effector function acquired by a T cell clone is largely dependent on which coreceptor it expresses, such that the vast majority of CD4+ SP thymocytes differentiate into mature naïve Th cells and CD8+ SP thymocytes usually differentiate into mature naïve Tc cells.

B. T Cell Activation

Like B cells, the complete activation of naïve T cells generally requires three signals: (1) the engagement of the antigen receptor by antigen; (2) costimulation; and (3) the receipt of cytokines. However, these signals differ slightly between B and T cell activation, and between naïve Th and Tc cell activation. Additional differences in the activation of effector and memory T cells also exist and are addressed toward the end of this chapter.

I. MEETING OF NAÏVE T CELLS AND DCs

The activation of most naïve T cells takes place in the paracortex of the lymph nodes, the secondary lymphoid organs where antigen-loaded mature DCs congregate and through which naïve T cells recirculate. As described in Chapter 7, immature migratory and lymphoid-resident DCs are experts at capturing antigens present as a result of infection or inflammation. The immature DCs use their panels of PRRs to initiate internalization of foreign entities. In the presence of pro-inflammatory cytokines and "danger signals" derived from these entities, the DCs commence maturation. A maturing migratory DC enters a lymph node via an afferent lymphatic vessel and settles in the paracortical region surrounding the HEVs penetrating the node. A lymphoid-resident DC is already in this location. In both cases, the DCs process their captured antigens and display antigenic peptides derived from them on MHC class II via exogenous processing, and on MHC class I (most likely) via cross-presentation. If a DC has become infected by the pathogen, intracellular antigens may also be processed via the endogenous pathway and presented on MHC class I.

Meanwhile, as described in Chapter 2, naïve Th and Tc cells are recirculating in the blood and throughout the secondary lymphoid tissues. In most cases, a naïve T cell enters the node via its HEVs and inspects the pMHCs displayed by the mature DCs congregated in the immediate vicinity of these vessels. The T cell "crawls" slowly over the surface of a DC in a process facilitated by several adhesion molecule pairs (Fig. 9-7). These molecules loosely hold the T cell and DC together so that the fit of a particular pMHC in the TCR's binding site can be checked. pMHCs that are bound with sufficient affinity/avidity by the TCR have the potential to activate the T cell.

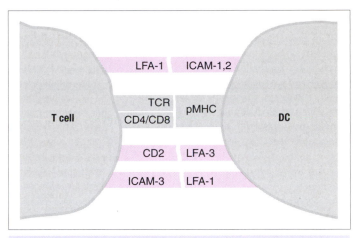

Fig. 9-7 Important Adhesion Contacts between Human T Cells and DCs

II. SIGNAL 1

Signal 1 is delivered when specific pMHCs displayed on a DC surface bind to multiple copies of a TCR expressed on a naïve Th or Tc cell surface (Fig. 9-8). The engagement of a TCR by pMHC leads to phosphorylation of its associated CD3 chains on their ITAMs by Lck kinase. Additional intracellular signaling enzymes are then recruited that associate with the cytoplasmic tails of the CD4 or CD8 coreceptor and the CD3 chains. Together, these enzymes mediate a cascade of chemical reactions that leads to the activation of several other enzymes. When this activation cascade occurs for multiple TCRs, the T cell receives signal 1.

Because the affinity of binding between a given pMHC and a TCR is relatively low (has a high "off" rate), a single pMHC does not engage a single TCR long enough to achieve complete activation of a naïve T cell. Neither is a transient interaction between a few pMHC–TCR pairs sufficient. Sustained interaction between the T cell and DC for several hours is needed to properly trigger the intracellular signaling pathways within the T cell that lead to the activation of the nuclear transcription factors necessary for new gene transcription. The TCRs required for sustained signaling are gathered together by the formation of an *immunological synapse* at the interface between the T cell and DC (Plate 9-1; green = points of contact). This process is driven by rearrangements of the T cell's actin cytoskeleton that are induced by the initial TCR–pMHC binding. These alterations result in the formation of rings of adhesion and signaling molecules that cluster around the TCR–pMHC pairs.

III. SIGNAL 2

In most cases, the engagement of TCRs by pMHCs is not sufficient to fully activate a naïve Th or Tc cell and signal 2 in the form of *costimulatory signaling* is required. Occasionally a Tc cell will encounter a pMHC (usually derived from a virus) that delivers such a strong signal 1 that costimulation is not required: this response is then independent of both costimulation and Th cell help.

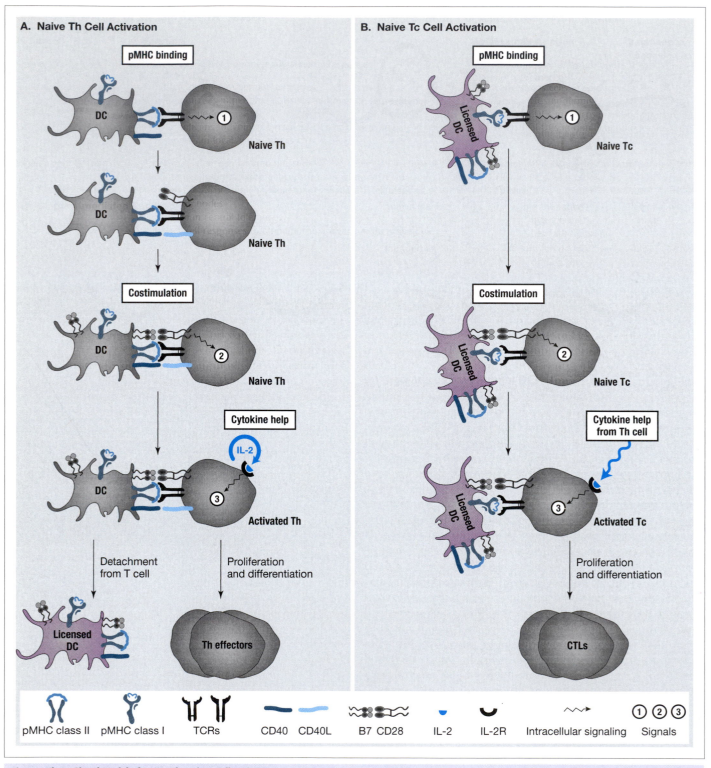

Fig. 9-8 **Three Signal Model of Naïve Th and Tc Cell Activation**

In the case of Th cells, the receipt of signal 1 leads to the upregulation of an important costimulatory molecule called *CD28* on the Th cell's surface (refer to Fig. 9-8A). However, in order for CD28 to convey signal 2 to the Th cell nucleus, it must bind to its ligand B7 on the surface of the DC presenting the activating pMHC. When the immunological synapse first starts to form, the mature DC involved does not express optimal levels of B7. A critical consequence of the delivery of signal 1 to a Th cell and the initial binding of CD28 to B7 is the upregulation of the transmembrane protein *CD40 ligand* (CD40L) on the Th cell surface. Once CD40L on the Th cell engages CD40 expressed by the DC, the DC greatly increases

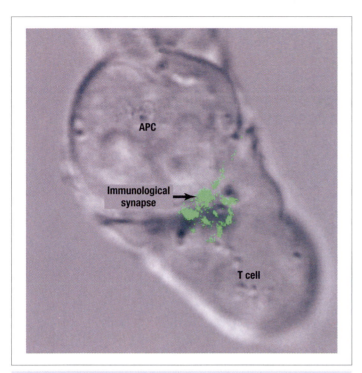

Plate 9-1 The Immunological Synapse
[Reproduced by permission of Vincent Das and Andres Alcover, Institut Pasteur.]

its expression of B7 and thus its binding to CD28 on the Th cell. As a result, a vigorous signal 2 is delivered that enhances the activatory intracellular signaling that is occurring within the T cell as a result of signal 1.

Although they upregulate CD28, most Tc cells do not express CD40L even after receiving signal 1. Thus, Tc cells cannot induce a DC to initiate CD40 signaling and upregulate B7 expression. Instead, Tc cells rely on CD28 engagement resulting from interaction with a DC that already expresses B7 due to a previous interaction with an antigen-activated Th cell. Some immunologists say that these DCs have been "licensed" for Tc activation (refer to Fig. 9-8B). This licensing of DCs by Th cells is one component of the T cell help provided by Th cells for Tc responses.

For both Tc and Th cells, CD28 signaling lowers the T cell activation threshold necessary to activate new gene transcription in the T cell and push it to proliferate and differentiate. In the absence of CD28 costimulation, naïve T cells are anergized instead of activated and fail to respond to pMHC (see Ch. 10). Costimulation via CD28 has several important molecular effects: (1) IL-2R expression is induced on the T cell surface, allowing the cell to receive signal 3. (2) Th cells start to secrete large quantities of IL-2, as well as other important cytokines and chemokines. (3) The expression or upregulation of additional costimulatory and regulatory molecules is induced in both Th and Tc cells.

CD28 signaling is controlled by the action of the negative regulator *cytotoxic T lymphocyte associated molecule 4* (CTLA-4). Expressed by both Tc and Th cells, CTLA-4 competes with CD28 for binding to the B7 costimulatory ligand. Because CTLA-4 has a much higher affinity for B7 than does

CD28, CTLA-4 displaces CD28 and recruits molecules to the TCR complex that shut down the intracellular signaling driving T cell activation. CTLA-4 does not appear on the T cell surface until 2–3 days after T cell activation is initiated, giving the adaptive response time to eliminate the threat before T cell activation is damped down.

IV. SIGNAL 3

A naïve Th or Tc cell that has received signals 1 and 2 upregulates the receptors (particularly IL-2R) that permit it to receive signal 3 in the form of cytokines, chemokines and growth factors (refer to Fig. 9-8). Activated Th cells produce many of the cytokines that can bind to these newly expressed receptors, the chief among them being IL-2. IL-2 is believed to deliver the most important signal for the proliferation of newly activated T cells. Although a Th cell on its own can make sufficient IL-2 to meet its requirements (*autocrine* IL-2 production), a Tc cell usually cannot. Thus, another component of T cell help provided to Tc cells by Th cells is the production of IL-2 (and possibly other cytokines) necessary for Tc proliferation.

C. Th Cell Differentiation and Effector Function

I. OVERVIEW

Once a naïve Th cell is fully activated, it starts to produce copious amounts of IL-2 and proliferates vigorously. The progeny generated are called *Th0 cells*. Depending on cytokines and other factors present in the immediate microenvironment, Th0 cells then terminally differentiate into two major types of effector T cells: *Th1 cells* and *Th2 cells*. In the presence of antigen presented by an APC (which can be a DC, macrophage or B cell at this stage), these effector T cells take action to eliminate the threat. Th1 cells facilitate effector functions specialized for the disposal of intracellular invaders, whereas Th2 cells promote effector functions designed to counter extracellular threats. A third type of effector T cell known as the *Th17 cell* has also been identified. This subset appears to differ from Th1 and Th2 cells in its differentiation and cytokine secretion. Some of the emerging information about this new defender is summarized in Box 9-3.

II. DIFFERENTIATION OF Th CELLS INTO Th1 AND Th2 EFFECTORS

About 48–72 hours after antigenic stimulation leading to activation and proliferation, most Th0 cells begin to differentiate into either Th1 or Th2 cells (Fig. 9-9, top). Which subset emerges depends on the cytokines in the local microenvironment, which are in turn determined by which innate response cells have been activated by the invading entity. Intracellular pathogens such as viruses and intracellular bacteria induce IFNγ production and trigger macrophages and DCs to produce IL-12 and IL-27. A Th0 cell that has its TCR engaged by specific pMHC and encounters IL-12, IL-27 and IFNγ becomes

Box 9-3 Th17 Cells

Th17 cells were first identified through studies of autoimmune disease in mice. Although these disorders had previously been blamed on the actions of dysregulated Th1 cells, it turned out that the inflammatory lesions in many of the affected animals contained a previously unidentified T effector cell subset that secreted, not IFNγ and IL-2 as Th1 cells do, but an unusual combination of the pro-inflammatory cytokines IL-17 and IL-6. These cells were thus dubbed "Th17 cells" because the production of IL-17 is unique to this subset. Subsequent experiments have confirmed that when mice are rendered IL-17- or IL-17R-deficient, these animals become resistant to the induction of autoimmune diseases. In humans, elevated levels of IL-17 have been detected in the blood, cerebrospinal fluid and tissues of patients with the autoimmune diseases multiple sclerosis, rheumatoid arthritis or inflammatory bowel disease (see Ch. 19).

Much is still unknown about Th17 cells, including the precise pathway of their differentiation. *In vitro* at least, a Th0 cell exposed to a cytokine milieu of TGFβ and IL-6 differentiates into a population of Th17 cells (see Figure). The involvement of these cytokines marks Th17 development as distinct from that of Th1 cells (IL-12-dependent) and Th2 cells (IL-4-dependent). It is theorized that TGFβ, considered to be an immunosuppressive cytokine, inhibits the production of IL-12 and/or IL-4 needed for Th1 or Th2 differentiation, respectively. Conversely, the IFNγ that bolsters Th1 production, as well as the IL-4 required for Th2 differentiation, suppress Th17 differentiation. Once differentiated, Th17 cells require the IL-12-related cytokine IL-23 to support their survival and proliferation.

The role of Th17 cells in normal adaptive immune responses remains to be clarified. In wild-type mice, Th17 cells are found in their greatest numbers in the intestinal lamina propria and mesenteric lymph nodes, and appear to be important for efficient immune responses against certain bacterial and fungal infections. The IL-17 produced by Th17 cells may induce endothelial cells and monocytes to secrete inflammatory mediators such as TNF that in turn promote the mobilization of neutrophils. Th17 cells also appear to be involved in autoimmune diseases and immune responses to tumor cells but their exact roles are unknown.

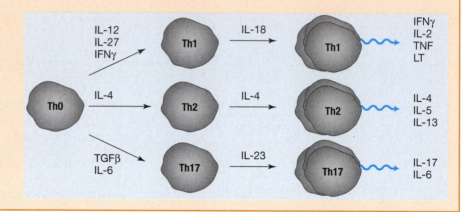

irreversibly committed to the Th1 subset 5–7 days after antigen stimulation. IL-18 then supports the survival and proliferation of the newly produced Th1 effector cells. On the other hand, extracellular pathogens such as certain bacteria and parasites do not induce IL-12 production by macrophages and DCs. Instead, these invaders stimulate an unknown cell type (which might be a mast cell or NKT cell) to secrete IL-4. In the absence of IL-12 and IFNγ but in the presence of IL-4, a Th0 cell generates Th2 effectors. IL-4 then sustains the survival and proliferation of the newly produced Th2 cells. As well as cytokines, some immunologists believe that subtly different DC subsets may deliver intercellular signals that play a role in determining Th differentiation. For example, there is some evidence in mice that certain DCs express Notch ligands capable of supplying signals to Th0 cells that promote Th2 differentiation.

The bottom part of Figure 9-9 gives an overview of the activation of newly differentiated Th effectors, their effector functions, and the types of pathogens they target. These matters are discussed in the following sections.

III. ACTIVATION OF Th1 AND Th2 CELLS

i) Localization

Once differentiated, Th1 and Th2 cells may remain in the lymph node to supply T cell help to naïve Tc cells in the para-cortex and naïve B cells in the primary follicles. Alternatively, the Th1 and Th2 cells may leave the node and follow chemokine gradients to sites of inflammation. Both subsets initially express CCR7, which acts within a lymph node to permit migration of the effector cells from the paracortex to the edges of the primary lymphoid follicles (where naïve B cells are located). However, as the response progresses, fully differentiated Th1 and Th2 cells express different panels of chemokine receptors and thus exhibit differential trafficking patterns. Later members of Th1 effector clones express CCR1, CCR5 and CXCR3 that draw the Th1 cells to sites of inflammation in the peripheral tissues where defense against intracellular pathogens is usually required. In contrast, Th2 effectors begin to preferentially express CCR3, CCR4 and CCR8. These receptors direct Th2 cells to sites such as the mucosae where responses against extracellular pathogens and toxins are needed.

ii) Interaction with APCs

Effector Th cells encountering pMHCs presented by APCs either in the lymph node or in the site of attack are activated essentially in the same way as naïve Th cells but with some important differences. Compared to naïve Th cells, effector Th cells express higher levels of adhesion molecules that stabilize the immunological synapse more rapidly, facilitating TCR

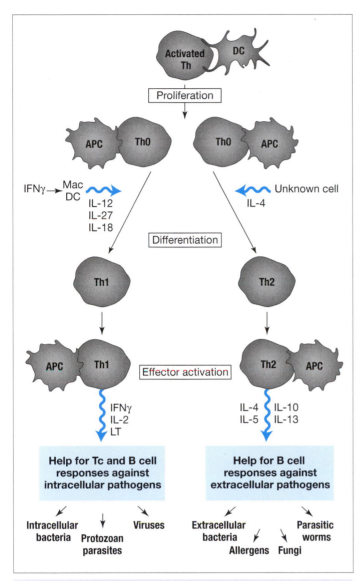

Fig. 9-9 **Th1/Th2 Differentiation and Effector Functions**

naling downregulates the expression of chemokine receptors, preventing the effector cell from migrating away from the site where antigen has been encountered. Supplementary costimulation for Th1 responses is delivered by *OX40*. OX40 expressed on the surface of a Th1 cell binds to *OX40 ligand* (OX40L) expressed on APCs. In contrast, supplementary costimulation in Th2 responses is delivered by the *inducible costimulatory molecule* (ICOS). ICOS, which binds to *ICOS ligand* (ICOSL) expressed on APCs, is highly expressed by Th2 effector cells but rarely by Th1 effectors.

IV. EFFECTOR FUNCTIONS OF Th1 AND Th2 CELLS

A brief comparison of the properties of Th1 and Th2 effectors is given Table 9-3.

i) Th1 Effector Functions

Th1 effectors supply T cell help to Tc and B cells providing cell-mediated and humoral defense against intracellular pathogens. Th1 cells secrete a panel of cytokines that is dominated by IL-2, IFNγ and LT ("Th1 cytokines"). IL-2 drives T and B cell proliferation and enhances ROI production by macrophages (Fig. 9-10). IFNγ and LT hyperactivate macrophages and spur them to secrete additional cytokines, undertake vigorous phagocytosis, and upregulate NO production. IFNγ also increases NK cell and macrophage expression of high affinity FcγR molecules that promote ADCC, and influences B cells to switch to the production of the Ig isotypes most effective against intracellular pathogens (IgG1 and IgG3 in humans). IgG1 and IgG3 are the antibodies best suited for opsonization, phagocytosis and complement activation, and also bind with high affinity to FcR on NK cells, macrophages and other phagocytes, further increasing ADCC. In addition, Th1 cytokines increase the antigen-presenting potential of macrophages by upregulating MHC class II and TAP. Th1 cells support the activation of Tc cells by producing IL-2 and by providing CD40/CD40L contacts for DC licensing.

triggering. While an estimated 20–30 hours of sustained TCR signaling is required for naïve T cell activation, only 1 hour is required for effector Th cell activation. Effector Th cells are thus activated by significantly lower quantities of pMHC. In addition, far less costimulation by the APC is required. As a result, effector Th cells respond efficiently to pMHC presented by DCs, macrophages or B cells or even by non-hematopoietic cells such as keratinocytes. In general, B cells are the principal APCs presenting antigen to Th2 cells, whereas macrophages predominate as APCs in interactions with Th1 cells.

iii) Differential Costimulatory Requirements

Although both Th1 and Th2 effector cells require a low level of CD28/B7 costimulation to stave off anergy and commence cytokine production, Th2 cells appear to need less of such signaling for activation than do Th1 cells. In both cases, CD28 engagement reduces the time required to achieve activation, thus avoiding prolonged stimulation. In addition, CD28 sig-

Table 9-3 **Comparison of Th1 and Th2 Effector Cells**

Property	Th1 Effectors	Th2 Effectors
Cytokines important for differentiation	IL-2, IL-12, IFNγ, IL-27	IL-4
Distinguishing surface markers	IL-12R CXCR3, CCR5, CCR1	IFNγR CCR3, CCR4, CCR8
Preferred APCs	Macrophages	B cells
Costimulation	CD28/B7 (low) OX40/OX40L	CD28/B7 (very low) ICOS/ICOSL
Cytokines secreted	IFNγ, IL-2, LT	IL-4, IL-5, IL-13, IL-10, IL-6, IL-3, IL-1
Type of immune response promoted	Humoral, cell-mediated	Humoral
Pathogens combated	Intracellular	Extracellular

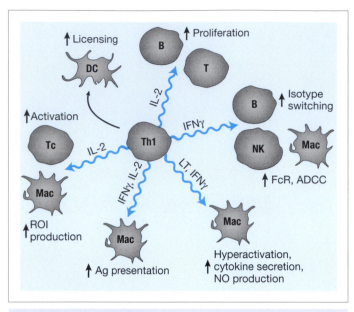

Fig. 9-10 **Th1 Effector Functions**

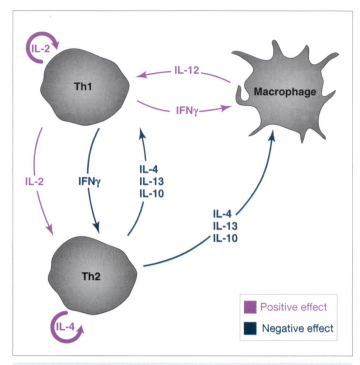

Fig. 9-12 **Th1/Th2 Cross-Regulation**

ii) Th2 Effector Functions

Th2 differentiation is usually induced upon invasion by extracellular pathogens. Th2 cells tend to promote humoral responses because these cells secrete IL-3, IL-4, IL-5, IL-6, IL-10 and IL-13 ("Th2 cytokines"). A major function of Th2 cells is to establish CD40–CD40L contacts with B cells and to secrete IL-4 and IL-5, cytokines that induce switching to the Ig isotypes most effective against extracellular pathogens (Fig. 9-11). Such isotypes, which include IgG4 in humans, are those best suited for neutralization. IgG4 is not very proficient at

complement activation or ADCC, which is an advantage in combating pathogens in mucosal sites where the inflammation induced by these effector functions could be damaging (see Ch. 12). IL-4 and IL-13 inhibit pro-inflammatory cytokine production, downregulate NO production, and decrease FcγR expression on macrophages, blocking ADCC. However, IL-4 upregulates MHC class II expression on APCs such as macrophages, DCs and B cells and thereby contributes to Th stimulation. IL-4 and IL-13 also enhance the humoral response by stimulating B cell proliferation. IL-5 promotes the growth, differentiation and activation of eosinophils crucial for the elimination of large parasites such as helminth worms. IL-3, IL-4 and IL-10 combine to promote the activation and proliferation of mast cells, also effective against large parasites. In general, however, IL-10 acts as a brake on immune responses and balances the driving forces exerted by other cytokines. For example, IL-10 inhibits the pro-inflammatory functions of macrophages and abrogates their production of IL-12 and MHC class II. IL-10 also downregulates B7 expression on macrophages and DCs.

iii) Th1/Th2 Cross-Regulation

Because of the cytokines they produce, the Th1 and Th2 subsets can cross-regulate each other's differentiation and activities, either positively or negatively (Fig. 9-12). For example, Th1 cells produce large amounts of IL-2 that can promote the proliferation of both Th1 and Th2 cells. However, the IFNγ produced by Th1 cells has a direct antiproliferative effect on Th2 cells and inhibits further Th2 differentiation. On the other hand, IFNγ stimulates macrophages to produce IL-12 which promotes Th1 differentiation. Th2 cells do not make substantial amounts of IL-2 or IFNγ and instead secrete IL-4,

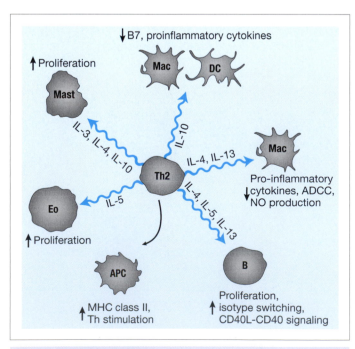

Fig. 9-11 **Th2 Effector Functions**

IL-13 and IL-10. These cytokines suppress IFNγ and IL-2 secretion by Th1 cells, inhibit further Th1 differentiation, and downregulate macrophage production of IL-12.

V. NATURE OF Th RESPONSES

Among immunologists, it is said that an immune response has either a Th1 or Th2 character or *phenotype*, depending on the predominant Th subset and cytokines observed in the host during that response. An attack on a host by intracellular pathogens stimulates the production by DCs and macrophages of cytokines favoring Th1 development, leading to the mounting of a *Th1 response*. Conversely, invasion by extracellular pathogens most often promotes the development of a *Th2 response*. Immunological disease states also tend to have either a Th1 or Th2 phenotype. For example, allergies are associated with a prevalence of Th2 cells, whereas Th1 cells dominate in transplant rejection and in many autoimmune disorders.

Because Th1 cells cannot later become Th2 cells (or vice versa), it was originally thought that Th1 and Th2 responses were mutually exclusive and that the presence of one subset might preclude the development of the other. However, unlike the phenotype of an individual Th clone, the overall phenotype of an immune response to a given pathogen can change with time. For example, mice infected with the parasite causing malaria first develop a Th1 response in which Th1 cells secrete IFNγ. This cytokine activates macrophages to secrete cytotoxic cytokines and produce large quantities of NO, and induces B cells to switch to the production of parasite-specific IgG2a antibodies. This Th1 phase is designed to deal with the intracellular stage of the infection. However, 10 days after the initial attack, the parasite adopts an extracellular phase that triggers a Th2 response characterized by high serum levels of IL-4, IL-10 and parasite-specific IgG1 antibodies.

D. Tc Cell Differentiation and Effector Function

I. OVERVIEW

Cytotoxic T cell responses can be thought of as occurring in five stages: (1) activation of the naïve Tc cell by a licensed DC in a secondary lymphoid tissue; (2) proliferation and differentiation of the activated Tc cell into daughter cells called *pre-CTLs*; (3) differentiation of a pre-CTL in an inflammatory site into an "armed" CTL; (4) activation of the armed CTL by encounter with specific non-self peptide presented by MHC class I on a target cell; and (5) CTL-mediated destruction of the target cell as well as other cells displaying the identical pMHC. Target cells of CTLs include cells infected with intracellularly replicating pathogens, tumor cells, and foreign cells entering the body as part of a tissue transplant. We emphasize that an activated Tc cell has no lytic powers at all: only its mature CTL progeny develop cytotoxicity.

II. GENERATION AND ACTIVATION OF CTLs

i) Differentiation of CTLs

In the presence of IL-2 and other cytokines usually secreted by an activated Th1 cell, an activated Tc cell proliferates and generates pre-CTL precursor cells (Fig. 9-13). These pre-CTLs leave the lymph node and travel to the site of pathogen attack. In the presence of IL-12, IFNγ and IL-6 produced by activated macrophages and DCs, these pre-CTLs differentiate into

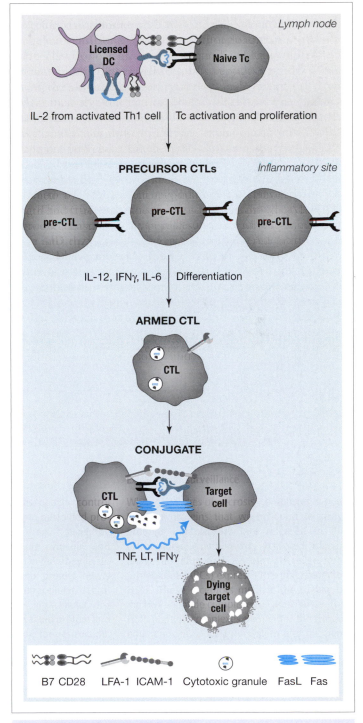

Fig. 9-13 **CTL Generation and Cytotoxicity**

mature CTLs whose cytoplasm contains cytotoxic granules. These mature CTLs (that have yet to encounter antigen) are said to be "armed" and do not need to carry out any additional protein synthesis to be effective killers. This effector generation process is completed within 24–48 hours of TCR stimulation of the original Tc cell. Importantly, because pre-CTL differentiation into armed CTLs requires inflammatory cytokines, the development of cytotoxicity and the power of the CTL response are reserved for situations in which a threat is actually present.

ii) Activation of Armed CTLs and Conjugate Formation

Within an inflammatory site, an armed CTL binds weakly to one host cell after another in search of its specific pMHC. The armed CTL detaches without incident if the affinity of TCR–pMHC binding is too low. However, should the armed CTL encounter a pMHC for which it is specific (the TCR binds with sufficient affinity), the host cell becomes a target. Stimulation of the TCR of an armed CTL rapidly increases the binding affinity of adhesion molecule pairs between the CTL and the target cell, forming a bicellular conjugate (refer to Fig. 9-13). The CTL then delivers a "lethal hit" of chemical mediators that rapidly causes target cell death. This speed is important to ensure the killing of an infected cell before too many of the progeny of the replicating pathogen can escape to new cells. Much lower concentrations of specific pMHC and the engagement of far fewer TCRs are required to activate armed CTLs compared to naïve Tc cells: only the engagement of a single TCR by a single specific pMHC is needed and no costimulation is required.

III. MECHANISMS OF TARGET CELL DESTRUCTION

Target cell destruction by CTLs can occur via the *granule exocytosis pathway,* the *Fas pathway,* and/or the release of *cytotoxic cytokines* such as TNF and LT (Fig. 9-14). The pathway used depends on the nature of the attacking intracellular pathogen. Granule exocytosis accounts for the majority of target cell killing by CTLs.

i) Granule Exocytosis

The granule exocytosis pathway refers to the release of the contents of the CTL's cytotoxic granules. Soon after conjugate formation, the cytoskeleton of the CTL reorganizes so its cytotoxic granules are brought to the site of CTL–target cell contact. The granules fuse with the CTL membrane and the cytotoxic contents of the granules are directionally exocytosed toward the target cell membrane. Among these granule contents are *perforin* and the *granzymes.* Perforin is a pore-forming protein and granzymes are serine proteases. It is not clear how these proteins actually enter the target cell, but after they do, they are immediately confined to the endocytic system. Perforin then facilitates the release of the granzymes from the endolysosomal vesicles into the cytoplasm of the target cell. The granzymes then initiate an internal disintegration pathway that leads to the degradation of important intracellular substrates, including DNA. Upon the destruction of these substrates, the target cell dies.

ii) Fas Pathway

Fas is a transmembrane "death receptor" that is widely expressed on mammalian cells. Engagement of Fas on a target cell by *Fas ligand* (FasL) expressed by an armed CTL results in the death of the target cell. Naïve Tc cells do not express FasL, but after activation by encounter with antigen, FasL is synthesized and stored in specialized transport vesicles in differentiating CTLs. Upon conjugate formation, the FasL-containing transport vesicles fuse with the CTL plasma membrane and anchor FasL on the CTL surface. The FasL engages Fas on the target cell and induces its apoptosis.

iii) Cytotoxic Cytokines

CTLs also produce cytotoxic cytokines, particularly TNF and LT. Apoptosis of a target cell can be induced by the binding of TNF produced by a CTL to TNF receptor 1 (TNFR1) on the target cell surface. LT, which also binds to TNFR1, has a similar effect. CTLs also secrete IFNγ, whose action in this context is more indirect. IFNγ stimulates B cells to produce antibodies that facilitate killing via ADCC or complement activation. As well, IFNγ upregulates MHC class I on nearby host cells, enhancing antigen display and making target cells more visible to scanning CTLs.

IV. DISSOCIATION

About 5–10 minutes after delivery of a lethal hit, the adhesion molecules on the CTL resume a low affinity conformation that allows the CTL to dissociate from the damaged target cell. The target cell succumbs to apoptosis within 3 hours of dissociation, while the CTL commences synthesis of new cytotoxic granules and moves off to inspect other host cells. A single armed (and re-armed) CTL can attach to many host cells in succession, delivering lethal hits without sustaining any damage itself. How the CTL avoids self-destruction by its granule contents is a mystery.

E. Control of Effector T Cells

After effector T cells have removed the antigenic stimulus sparking a primary immune response, there is no further need for their presence. Three mechanisms act in concert to bias the pro-apoptotic/anti-apoptotic balance and induce the death of Th effector cells and CTLs: *activation-induced cell death* (AICD), *cytokine "withdrawal"* and *T cell exhaustion.* AICD is a form of apoptosis induced when the intracellular signaling triggered by TCR engagement by antigen becomes prolonged. This extended signaling is thought to induce the transcription of pro-apoptotic genes such as FasL and TNFR1 within the effector T cell, and to decrease its expression of anti-apoptotic molecules. These changes set the effector T cell up to be killed by contact with Fas or TNF expressed by a neighboring cell. In the early stages of a primary response, effector T cells are protected from AICD by the low levels of CD28–B7 signaling they receive. This minimal costimulation ensures that pro-

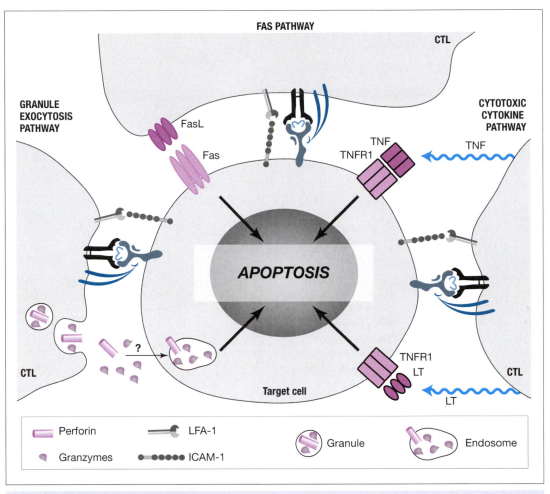

Fig. 9-14 **Mechanisms of CTL Cytotoxicity**

apoptotic genes are not transcribed to a level that overwhelms the effects of anti-apoptotic genes.

Cytokine withdrawal is a mechanism that was first observed for IL-2 *in vitro*. Clones of T cells die if all IL-2 is removed from the culture medium after activation. It seems that this phenomenon also happens *in vivo*, when antigen is being mopped up and less of it is around to stimulate DCs and other APCs to secrete cytokines. In the absence of antigen and these cytokines, T cell activation cannot be sustained, IL-2 production falls, and effector T cells die. In this case, the induction of T cell apoptosis does not involve Fas or TNF. Rather, the lack of cytokines prevents cytokine receptor engagement and thus blocks the delivery of a vital survival signal to the T cell nucleus. In the absence of this survival signal, certain pro-apoptotic genes are expressed that kill the cell.

Although most effector T cells eventually succumb to AICD or cytokine withdrawal, entire activated T cell clones are sometimes eliminated by "exhaustion". In this case, continuous exposure to antigen causes the T cells to divide so relentlessly into effectors that they burn out metabolically without generating memory cells. Effector cells, as well as the memory cells that normally would have dealt with a subsequent assault, are completely absent from the host.

F. Memory T Cells

The few antigen-specific progeny T cells generated in a primary response that survive AICD or IL-2 withdrawal are, or give rise to, memory T cells. Upon a second assault by the same pathogen, memory T cells mount a secondary response that is faster and stronger than the primary response. These differences are attributable to the localization, increased numbers and enhanced capacities of memory T cells. Memory T cells recognize the same pMHC as naïve and effector cells but have properties intermediate between them (Table 9-4).

I. MEMORY T CELL LOCALIZATION

Like naïve T cells, the vast majority of memory T cells appear to adopt a resting state. However, like effector T cells, memory T cells are not confined to the secondary lymphoid tissues and express homing receptors that allow them to recirculate widely throughout the peripheral tissues of the body. The cytokine milieu in which a naïve T cell was activated can influence which adhesion molecules and homing receptors are expressed by its progeny memory T cells. These molecules in turn determine the access of the memory T cells to a particular tissue. For example, suppose an antigen enters the body via a mucosal

Table 9-4 Comparison of Properties of Naïve, Effector and Memory T Cells

Property	Naïve T Cell	Effector T Cell	Memory T Cell
Preferred tissue	Secondary lymphoid tissues	Peripheral tissues, inflammatory sites, secondary lymphoid tissues (Th effectors)	Peripheral tissues, diffuse lymphoid tissues, inflammatory sites
Preferred APCs	DCs	Macrophages, B cells, DCs	DCs, B cells, macrophages
Costimulation	CD28/B7 (high)	CD28/B7 (low to none)	Minimal to none
Duration of TCR signaling for activation	20–30 hours	<1 hour	<1 hour
Cell division	Slow	Rapid	Moderate
Sensitivity to AICD	Low	High	High
Life span	5–7 weeks	2–3 days	Up to 50 years

route (say, through the upper respiratory tract) and activates naïve Tc cells in this location. Memory CD8[+] T cells are generated that express proteins (including chemokine receptors) that allow them to concentrate preferentially in multiple mucosal sites, including the linings of the digestive, respiratory and reproductive tracts (see Ch. 12). These sites are locations where a subsequent attack by a mucosal antigen might be anticipated.

II. MEMORY T CELL ACTIVATION

Whereas naïve Th and Tc cells are activated exclusively by pMHCs presented by DCs in the secondary lymphoid tissues, memory Th and Tc cells have less stringent requirements. Memory Th cells are dispersed in a broader range of anatomical sites than are naïve Th cells and can respond to pMHC presented by DCs, B cells or macrophages located almost anywhere in the periphery. Similarly, memory Tc cells can respond to infected host cells located almost anywhere in the body provided that these cells display the appropriate pMHC on their surfaces. In terms of signaling, the activation of both memory Th and memory Tc cells more closely resembles that of an effector cell than a naïve cell. Activation can occur at very low concentrations of antigen with only minimal costimulation (if any) and the duration of TCR signaling required is much shorter. Although naïve Tc cells usually require T cell help from antigen-specific Th cells to become activated, memory Tc cells do not. Once activated, memory T cells of both subsets proliferate more readily and for longer periods than their naïve counterparts. Memory T cells die quite readily due to AICD.

III. MEMORY T CELL EFFECTOR FUNCTIONS

Activated memory Th and Tc cells follow much the same differentiation pathways as naïve Th and Tc cells but complete them more quickly (within 24 hours as opposed to 4–5 days). Some immunologists maintain that many memory T cells are not really resting but instead are maintained in a type of "pre-activation" state (which may correlate with their intermediate marker phenotype). This theory holds that pre-activation may make it easier for the cells to immediately differentiate into new effectors capable of quickly combating an aggressive pathogen. It remains unclear whether activated memory T cells can give rise to a new generation of memory cells as well as effector cells.

IV. MEMORY T CELL LIFE SPAN

Most memory T cells persist for at least several months and often years, greatly exceeding the longevity of both naïve and effector T cells. The length of life span of a memory T cell clone varies with the nature of the antigen that provoked the primary response. We see evidence of this variability in the immunization schedules of different vaccines (see Ch. 14). Just one dose of some vaccines (e.g., against the polio virus) provides immunity for life, whereas "booster" doses of other vaccines (e.g., against the bacterium causing tetanus) must be given every few years to maintain protection. Immunologists are still divided over whether the persistence of memory lymphocytes depends on a periodic low level of stimulation by tiny amounts of residual antigen. Such stimulation might induce the expression of base levels of anti-apoptotic molecules that would permit the memory T cell to survive.

Can memory T cells protect a host forever? Studies of the aging of the immune system indicate that memory T cells arising from a given clone can be stimulated only so many times before they fail to proliferate in response to antigen. As well, the production of new naïve T cells by the thymus declines precipitously after involution, curtailing the generation of new memory T cells. Thus, as an individual ages, the numbers of both naïve and memory T cells that can be activated to generate effector T cells is ultimately limited, and the host becomes increasingly susceptible to pathogens toward the end of his/her life.

We have now described all the cellular components of an adaptive immune response and have discussed how that response removes non-self entities. In the next chapter, we examine peripheral tolerance, a collection of mechanisms that controls those mature naïve lymphocytes in the periphery which escaped the establishment of central tolerance and whose antigen receptors are directed against self antigens.

- HSCs give rise to NK/T precursors that colonize the thymus and generate thymocytes. Thymocytes mature through the DN, DP and SP phases defined by CD4/CD8 expression.

- Development of $\alpha\beta$ T cells is controlled by two checkpoints: the pre-TCR checkpoint and the TCR$\alpha\beta$ checkpoint.

- Thymic selection establishing central T cell tolerance involves the non-selection, positive selection or negative selection of DP thymocytes. Selection is determined by the affinity/avidity of TCR binding to pMHCs presented by TECs.

- Thymocytes that survive selection lose expression of one coreceptor to become SP thymocytes, eventually exiting to the periphery as mature CD4$^+$ Th and CD8$^+$ Tc cells.

- Naïve Th and Tc cells are activated by engagement of multiple copies of their TCRs by pMHCs presented on mature DCs in the lymph node.

- Th cell activation involves the formation of the immunological synapse between the naïve Th cell and the DC. The synapse allows the sustained triggering of multiple TCRs that delivers "signal 1."

- "Signal 2" is delivered by costimulatory contacts such as CD28–B7, whereas "signal 3" is delivered by the binding of cytokines such as IL-2.

- Signal 1 induces the phosphorylation of the ITAMs in the CD3 chains, and signals 2 and 3 trigger subsequent intracellular signaling that leads to the new gene transcription necessary to support proliferation and effector cell differentiation.

- Naïve Tc cells are activated by interaction with DCs that have been "licensed" by Th cells to express B7.

- An activated naïve T cell proliferates and differentiates into effector T cells that eliminate antigens, and into memory T cells that mediate secondary immune responses.

- Effector and memory T cells require lower levels of TCR engagement and costimulation than do naïve cells, and express different adhesion molecules and chemokine receptors.

- Depending on the cytokines in the microenvironment, activated Th cells generate Th1 or Th2 effector cells that secrete different panels of cytokines. These cytokines can either act against pathogens or support B cell and Tc cell activation. Th17 effector cells secrete inflammatory cytokines and mediate autoimmune disease.

- Activated Tc cells generate armed CTL effectors that kill tumor cells and infected target cells by perforin- and granzyme-mediated cytotoxicity, Fas ligation, and/or secretion of cytotoxic cytokines.

- The duration of Th1, Th2 and CTL responses is controlled by activation-induced cell death (AICD), T cell exhaustion and IL-2 withdrawal.

- Memory T cells persisting in the host take up residence in peripheral tissues and rapidly differentiate into new effectors upon a second exposure to an antigen.

Section A.I–II

1) Give four ways in which B cell and T cell development differ.

2) What cell types arise from NK/T precursors?

3) Which waves of NK/T precursors entering the thymus give rise to αβ T cells?

4) Why is T cell diversity in the neonatal repertoire less than in the adult repertoire?

Section A.III.i

1) What are cTECs and mTECs and why are they important?

2) Why is Notch1 described as a "cell fate protein"?

3) Give three ways in which the thymus is vital for T cell development.

Section A.III.ii

1) What are the four stages of the DN phase of T cell development? How are these phases similar in surface marker expression? How do they differ?

2) What is SCF and what is its function in the thymus?

3) What DN stage is also known as a pro-T cell? An early pre-T cell? A late pre-T cell?

4) What five key events occur during the DN3 stage?

5) Describe the composition and function of the pre-TCR complex.

6) What is β-selection and why is it important?

Section A.III.iii–iv

1) Name the three components of thymic selection and describe how affinity/avidity of TCR engagement defines each of them.

2) Why is negative selection essential for the establishment of central T cell tolerance?

3) What is AIRE and why is it important for thymic selection?

4) What factors influence the intracellular signaling triggered by TCR engagement during thymic selection?

5) Give two reasons why a thymocyte might be non-selected.

6) What are the effects on thymic selection of the expression of anti- and pro-apoptotic genes?

7) What is the TCRαβ checkpoint?

8) How does coreceptor expression in SP thymocytes determine the effector function of mature T cells?

Section B.I–II

1) What are the three signals of naïve T cell activation?

2) Briefly outline how naïve T cells and antigen-bearing DCs meet in the lymph node.

3) Why is sustained TCR triggering necessary to activate naïve T cells?

4) What is the immunological synapse?

Section B.III–IV

1) What is the most important costimulatory interaction for naïve Th cells and why is it necessary?

2) What is DC licensing and why is it necessary?

3) Give three effects of CD28-mediated costimulation.

4) What is the function of CTLA-4?

Section C.I–III

1) Briefly define the two major types of effectors derived from Th0 cells and outline their functions.

2) Which cytokines drive Th1 differentiation and what is their source? Th2 differentiation?

3) How do Th1 responses differ from Th2 responses?

4) What is a Th17 cell?

Section C.IV–V

1) Give three ways in which the activation of Th effectors differs from that of naïve cells.

2) What molecules provide supplementary costimulation for Th1 cells? Th2 cells?

3) Give two ways in which Th1 cells support cell-mediated immunity, and two ways in which Th2 cells support humoral immunity.

4) Why is IL-10 considered to act as a "brake" on immune responses?

5) Describe how Th1 and Th2 cells can cross-regulate each other.

Section D

1) Can you define these terms? pre-CTL, armed CTL, granule exocytosis, death receptor, dissociation.

2) Describe the five stages of cytotoxic T cell responses.

3) How does the activation of the armed CTL differ from that of a naïve Tc cell?

4) What are perforin and the granzymes and what do they do?

5) Name three cytotoxic cytokines and describe how they contribute to target cell elimination.

Section E

1) Distinguish between AICD and IL-2 withdrawal.

2) What happens if IL-2 is withdrawn from a culture of activated T cells?

Section F

1) Give three reasons why memory T cells react faster and stronger than naïve T cells.

2) How is the phenotype of memory T cells intermediate between naïve and effector T cell phenotypes?

3) Give two reasons why older individuals have less effective immune systems.

WHAT'S IN THIS CHAPTER?

If you wish to be brothers, drop your weapons.
Pope John Paul II

The adaptive immune responses that act in the peripheral tissues of the body are very powerful and, in a healthy individual, tightly regulated. This regulation is exerted in two ways: *tolerance* and *control*. Tolerance mechanisms prevent lymphocyte activation whereas control mechanisms rein in the actions of effector cells.

At the cellular level, peripheral tolerance is manifested when the interaction between a mature peripheral lymphocyte and a cognate antigen does <u>not</u> result in activation of that lymphocyte. The lymphocyte is said to be *tolerized*. Such a lack of response occurs when any one of the three signals required for lymphocyte activation is blocked. A major role of such lymphocyte inhibition is the maintenance of tolerance to self tissues. This peripheral "self tolerance" is vital because it prevents the activation of autoreactive lymphocytes that have escaped central tolerance mechanisms and persist in a repertoire of lymphocytes that has been otherwise shaped to recognize non-self. Tolerance to non-self entities in the periphery is not physiological but can be induced experimentally. "Experimental tolerance" has helped immunologists to understand much about tolerance mechanisms in general.

Whether a lymphocyte recognizes a non-self or a self antigen, if it becomes activated, its response ultimately needs to be damped down. In the case of a lymphocyte activated by a non-self antigen, regulation minimizes collateral damage to nearby tissues and eventually returns the body to a steady state. If an autoreactive lymphocyte becomes activated despite attempts to tolerize it, control measures normally prevent or minimize the damage that might otherwise be sustained by the healthy self tissues that are the target of the response. Success in implementing self tolerance and control mechanisms ensures that the host focuses the power of the immune response on non-self antigens; failure paves the way for uncontrolled tissue damage and the potential development of autoimmune disease.

In this chapter, we describe the mechanisms necessary to implement peripheral self tolerance and control lymphocyte responses. We also discuss two special peripheral tolerance situations: *maternal–fetal tolerance* and *neonatal tolerance*. Finally, we examine *experimental tolerance*, the situation in which a non-self antigen is introduced artificially into the body under circumstances that suppress the induction of an immune response to the antigen.

A. Self Tolerance of Lymphocytes in the Periphery

Because the antigen receptors of B and T cells are randomly generated, a certain proportion of developing lymphocytes will inevitably bear receptors directed against self antigens. Most B cells with such potentially autoreactive BCRs undergo receptor editing in the bone marrow during the establishment of B cell central tolerance such that these B cells no longer recognize self antigen (refer to Ch. 5). Most T cells with potentially autoreactive TCRs are eliminated by deletion during the establishment of central T cell tolerance in the thymus (refer to Ch. 9). However, if an autoreactive lymphocyte is released into the periphery because receptor editing fails, or a relevant self antigen is not expressed at sufficient levels in the bone marrow or thymus to induce negative selection, mechanisms of peripheral self tolerance will attempt to ensure that the autoreactive cell cannot be activated to attack self tissues.

I. T CELL SELF TOLERANCE

i) DC-Mediated Tolerization

As noted in Chapter 9, a resting immature DC expresses only low levels of MHC class II. Upon the receipt of an external "danger signal", the antigen-capturing immature DC commences maturation and upregulates the MHC class II and costimulatory molecules necessary to activate naïve Th cells (Fig. 10-1A). Once licensed, the DC may also cross-present

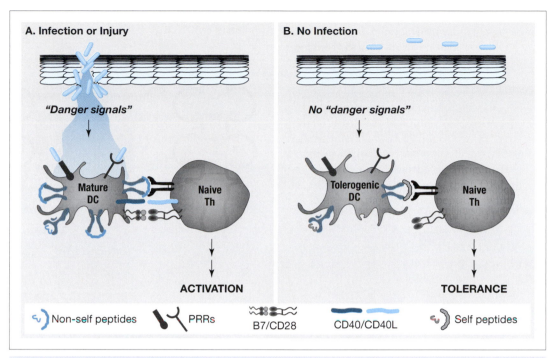

A. Infection or Injury

"Danger signals"

Mature DC

Naive Th

ACTIVATION

B. No Infection

No "danger signals"

Tolerogenic DC

Naive Th

TOLERANCE

| Non-self peptides | PRRs | B7/CD28 | CD40/CD40L | Self peptides |

Fig. 10-1 **Model of Th Cell Tolerization in Absence of Danger Signals**

antigenic peptide on MHC class I and thus initiate Tc cell activation. However, DCs mediate not only naïve T cell activation but also naïve T cell tolerization.

As mentioned previously, immature migratory DCs are broadly distributed in the peripheral tissues where these cells constantly take up antigens from the local microenvironment. In the absence of pathogen attack or injury, the only antigens encountered by these DCs are self antigens that are either shed by healthy tissues, or acquired by the engulfment of apoptotic cells during normal turnover. Danger signals are not present and an immature DC is not induced to mature. Under these conditions, the DC does not activate naive T cells but it may tolerize them. Originally, it was thought that a T cell was tolerized by interaction with an immature DC simply because these DCs do not express costimulatory molecules. However, there is evidence that, when a DC acquires antigen in the absence of danger signals and then encounters a T cell specific for a pMHC that it is presenting, the DC develops a *tolerogenic* capacity relative to that T cell. A "tolerogenic" DC is incapable of delivering signal 2 due to properties of the cell that are not fully understood but can be independent of its costimulatory molecule expression (Fig. 10-1B). Thus, if an immature migratory DC presents self antigen to an autoreactive Th cell that escaped central tolerance, and if the TCRs of this Th cell bind firmly to pMHCs presented by such a DC, the DC acts in a tolerogenic manner. Instead of activating the autoreactive Th cell, the DC induces either its *clonal deletion* or *anergization*. These events also result in Tc tolerization in many cases, as a Tc cell is only rarely fully activated in the absence of a licensed DC and the signal 3 cytokines produced by an activated Th cell.

ia) Clonal deletion. The most important mechanism by which peripheral Th cell tolerance is maintained is the clonal deletion of autoreactive Th cells. Naïve autoreactive Th cells that interact with pMHC presented by a tolerogenic DC are usually induced to undergo apoptosis. The precise mechanism has yet to be fully elucidated but appears to be independent of the apoptosis induced by Fas or TNFR engagement.

ib) Anergization. Those autoreactive Th cells that receive signal 1 in the absence of signal 2 but do not undergo apoptosis are *anergized*. An anergized Th cell survives but is inactivated and cannot produce effector cells (Fig. 10-2A). Moreover, if the antigenic pMHC is encountered a second time, the anergic Th cell still cannot respond even if the pMHC is presented by a fully mature DC expressing the appropriate costimulatory molecules (Fig. 10-2B). The anergic Th cell can maintain this unresponsive state for up to several months. The intracellular signaling pathways that induce anergy have yet to be fully defined.

ii) Clonal Exhaustion

As well as by DC-mediated mechanisms, tolerance can be invoked by the elimination of an entire T cell clone due to clonal exhaustion (refer to Ch. 9). In this situation, continuous exposure to an antigen forces the responding T cells to proliferate and generate effectors so rapidly that they burn out without generating memory T cells. Some immunologists believe that tolerance to many self antigens that are present in the body in high abundance may be established very early in life in this way. That is, during embryogenesis, the presence of large

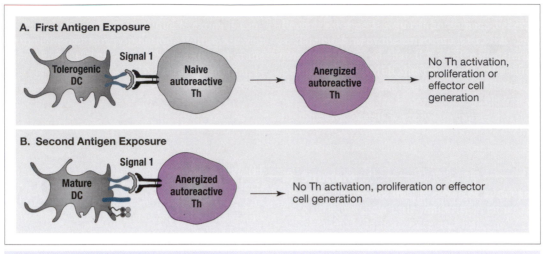

Fig. 10-2 **Anergization of Th Cells**

amounts of self antigen causes the exhaustion of autoreactive clones that escaped central T cell tolerance, ensuring peripheral tolerance to these self elements.

II. B CELL SELF TOLERANCE

Autoreactive B cells that escape central B cell tolerance are controlled by peripheral tolerance mechanisms that differ slightly from those just discussed for autoreactive T cells. If an autoreactive B cell encounters self antigen in the periphery, it receives signal 1 but still depends on an antigen-specific Th effector cell to deliver signals 2 and 3. Thus, even if "danger" is present, if the required Th cell has already been deleted by central tolerance mechanisms, the B cell cannot be activated. Instead, the B cell is anergized and then succumbs to apoptosis within 3–4 days (Fig. 10-3A). This reliance of the B cell on the Th cell for activation allows the host to benefit, without undue risk of increased autoreactivity, from the somatic hypermutation that occurs in the Ig genes. Although somatic hypermutation might produce a mutation that causes a B cell to become autoreactive, this B cell is unlikely to encounter a Th cell able to supply signal 2 because most autoreactive T cells are eliminated during the establishment of central T cell tolerance. Moreover, the TCR genes do not undergo somatic hypermutation, so that a non-autoreactive T cell cannot suddenly become autoreactive.

Sometimes an appropriate autoreactive Th cell happens to be present in the peripheral repertoire so that the potential for activation of an autoreactive B cell exists. However, in these cases, the absence of danger signals will work to maintain B cell self tolerance. As illustrated in Figure 10-3B, an autoreactive B cell may have had its BCRs engaged by self antigen such that it receives signal 1. In the absence of danger, any Th cell of the appropriate specificity will have been anergized so that there can be no delivery of signal 2 to the B cell. Again, the B cell is anergized and forced into apoptotic death.

B. Control of Lymphocyte Responses in the Periphery

Once lymphocytes have been activated, various mechanisms in the periphery control the quality, intensity and duration of the resulting adaptive response. These mechanisms include *regulatory T cells*, *immunosuppressive cytokines*, *immune deviation* and *immune privilege*.

I. REGULATORY T CELLS

Certain subpopulations of T cells, collectively known as *regulatory T cells*, can control the responses of activated conventional T cells, including autoreactive T cells. The existence of CD4+ regulatory T cells has been firmly established but there is less consensus on whether CD8+ regulatory T cells also contribute to the control of immune responses.

i) CD4+ Regulatory T Cells
As summarized in Table 10-1, there are various subsets of regulatory CD4+ T cells that differ with respect to their surface markers, precursor cells, suppressive mechanisms and regulatory effects. A brief description of each follows.

ia) T_{reg} cells. The best-documented regulatory T cells are commonly called T_{reg} *cells* and are recognized by their high surface levels of CD4 and CD25 (the IL-2Rα chain). The T_{reg} population comprises about 10% of all normal mouse or human peripheral CD4+ T cells. Most T_{reg} cells develop from precursors in the thymus, like conventional T cells. Some immunologists call these thymus-derived cells "natural T_{reg}" or nT_{reg} cells. Upon their release into the periphery, the newly produced nT_{reg} cells interact with tolerogenic DCs that render the nT_{reg} cells anergic. After anergization, nT_{reg} cells gain the ability to block the proliferation of conventional CD4+ T cells of <u>any</u> antigenic specificity (including autoreactive T cells) and to inhibit IL-2

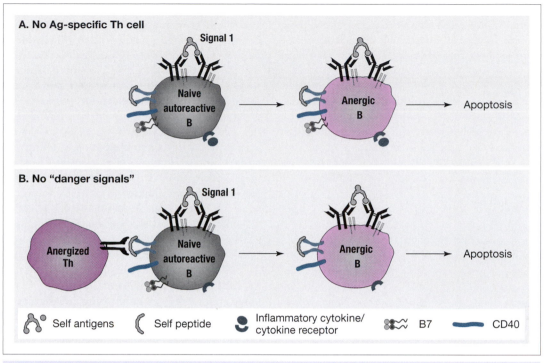

Fig. 10-3 **Anergization of B Cells**

Table 10-1 **Types of CD4$^+$ Regulatory T Cells**

	nT$_{reg}$	iT$_{reg}$	Th3	Tr1
Characteristic markers	CD4$^+$CD25^{hi} CTLA-4^{hi} Foxp3$^+$	CD4$^+$CD25^{hi} CTLA-4^{hi} Foxp3$^+$	CD4$^+$CD25^{lo} CTLA-4^{med} Foxp3$^-$	CD4$^+$CD25^{lo} CTLA-4^{lo} Foxp3$^-$
Derivation	Thymic precursor	Th0 cell + mature DC + TGFβ	Th0 cell + modulated DC + TGFβ	Th0 cell + modulated DC + IL-10
Suppressive mechanism	Intercellular contact	Secretion of TGFβ	Secretion of TGFβ	Secretion of IL-10, TGFβ
Regulatory effects	Suppresses activated T cells May induce Th0 cells to produce Tr1 and Th3 cells	Suppresses activated T cells	Suppresses activated T cells	Suppresses activated T cells

production. These suppressive effects are <u>not</u> cytokine-mediated but rather require direct intercellular contact with the conventional T cell. In addition to nT$_{reg}$ cells, there is some evidence for the existence of a minor population of "induced T$_{reg}$" (iT$_{reg}$) cells that arise from Th0 cells that interact with a fully mature DC. The development of these iT$_{reg}$ cells appears to depend on the presence of high local concentrations of the immunosuppressive cytokine TGFβ (see later). iT$_{reg}$ cells are thought to suppress effector T cells mainly via TGFβ secretion.

Unlike conventional T cells, both nT$_{reg}$ and iT$_{reg}$ cells express a transcription factor called *Foxp3*. Foxp3 is essential for both the development and function of T$_{reg}$ cells. T$_{reg}$ cells also constitutively express high surface levels of the negative regulator CTLA-4. As described in Chapter 9, CTLA-4 plays a critical role in winding down conventional T cell activation toward the end of an adaptive response because this molecule competes with CD28 for binding to B7, eventually displacing CD28 and shutting down the receipt of signal 2. Although mice lacking CTLA-4 show uncontrolled lymphocyte proliferation and autoimmunity, it remains controversial whether CTLA-4 actually contributes to the suppressive functions of T$_{reg}$ cells. CTLA-4 does appear to be important for T$_{reg}$ generation.

Recent work has suggested that, as well as their effects on T cells, T$_{reg}$ cells may have effects on APCs that cause these cells to become tolerogenic, thereby further contributing to the control of T cell responses. Importantly, T$_{reg}$ cells have been found to express TLRs, and engagement of these TLRs by the appropriate PAMPs early in a pathogen infection stimulates the proliferation of T$_{reg}$ cells while diminishing their suppressive activity. The anti-pathogen response thus proceeds full force to eliminate the invader. Toward the end of the infection, however, the expanded T$_{reg}$ cell population regains its suppressive potency and shuts down the now-unnecessary anti-pathogen T cell response.

ib) Th3 and Tr1 cells. *Th3 cells* and *Tr1 cells* are two other regulatory T cell subsets that can suppress activated conventional T cells in a non-specific manner. However, rather than by intercellular contacts, Th3 and Tr1 cells shut down activated T cells by secreting immunosuppressive cytokines. Th3 cells preferentially secrete TGFβ whereas Tr1 cells secrete IL-10 plus low amounts of TGFβ. Th3 cells express low levels of CD25 and moderate levels of CTLA-4, whereas Tr1 cells express only low levels of both CD25 and CTLA-4. Neither subset expresses Foxp3.

Unlike T$_{reg}$ cells, Th3 and Tr1 cells are thought to be derived from conventional Th0 cells that interact with *modulated DCs*. A modulated DC is an immature DC that acquires a self antigen or a benign foreign antigen (such as food) in the absence of danger signals but in the presence of certain immunosuppressive molecules, particularly IL-10 and TGFβ. When a modulated migratory DC travels to the regional lymph node and interacts with a naïve conventional Th0 cell, the Th0 cell is not activated to produce Th1 or Th2 effectors as it would be if it interacted with a mature DC, nor is it anergized as it would be if it interacted with an tolerogenic DC. Rather, studies in cell cultures suggest that the Th0 cell is induced to proliferate and differentiate into either Th3 or Tr1 cells, depending on the cytokine milieu (Fig. 10-4). Once generated, the Th3 and Tr1 cells home back to the tissue where the immature DCs first acquired the self antigen or benign foreign antigen and downregulate any conventional T cell responses to that antigen in that site. As well as after interaction between Th0 cells and modulated DCs, Th3 and Tr1 cells may arise from a Th0 cell that establishes intercellular contacts with a T$_{reg}$ cell. The suppressive functions of Th3 and Tr1 cells derived in this way appear to be comparable to those of Th3 and Tr1 cells derived from modulated DC–Th0 interaction.

ii) CD8⁺ Regulatory T Cells

Some immunologists believe that regulatory subsets of CD8⁺ T cells also exist, but it has been difficult to clearly define these cell populations. Some CD8⁺ regulatory T cells are believed to secrete IL-10 and to act directly on naïve and effector T cells to block their proliferation. Other subsets appear to have their primary effects on APCs, in that an immature DC that interacts with such a CD8⁺ regulatory T cell becomes a form of modulated DC. This DC then anergizes, rather than activates, any antigen-specific CD4⁺ Th0 cell with which it subsequently interacts. Alternatively, the modulated DC may induce the

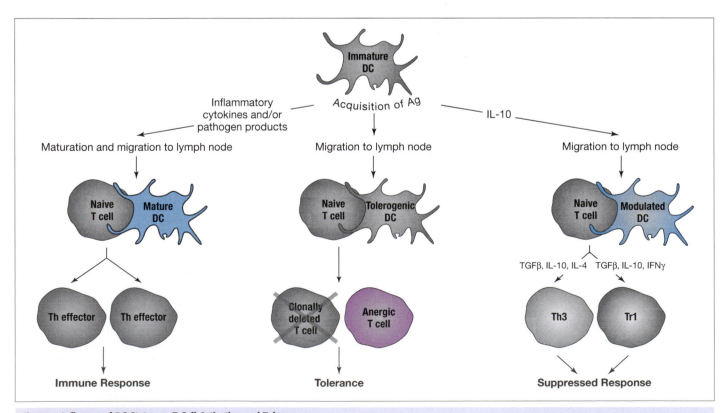

Fig. 10-4 **Influence of DC Status on T Cell Activation and Tolerance**

antigen-specific Th0 cell to differentiate into Th3 and Tr1 cells. The importance of CD8$^+$ regulatory T cells to the control of immune responses *in vivo* is under investigation.

II. IMMUNOSUPPRESSIVE CYTOKINES

Certain cytokines, in particular IL-10 and TGFβ, have immunosuppressive effects and act as brakes on innate and adaptive immune responses, including those initiated against self antigens. Although most often produced by innate response cells activated during a pathogen invasion, immunosuppressive cytokines may also be synthesized in significant amounts by iT$_{reg}$, Th3 and Tr1 cells. IL-10 downregulates TCR-induced intracellular signaling in a responding T cell, inhibits macrophage activation and inflammatory cytokine secretion, blocks APC function, prevents the proliferation of Th cells, and destabilizes the mRNAs of many cytokines, including IL-2. The ensuing lack of IL-2 compromises signal 3 delivery, thus promoting T cell anergization and thereby contributing to the establishment of T cell peripheral tolerance. TGFβ inhibits macrophage and NK cell activation, blocks the proliferation and IL-2 production of activated T cells, downregulates Ig synthesis by B cells, and interferes with the stimulatory effects of IL-2 on T and B cells.

III. IMMUNE DEVIATION

Immune deviation describes a phenomenon in which an adaptive response that has the potential to cause direct or indirect damage to a tissue appears to be "converted" to a less harmful response. This mitigation of damage was originally viewed as a form of tolerance. Immune deviation is caused by a bias toward the differentiation of either Th1 or Th2 effectors following the activation of a peripheral Th0 cell. Depending on the tissue involved, the differentiation of one subset can avoid the potential damage that might have been mediated by the other subset. For example, the tissues of the eye are very sensitive to the products and effects of Th1 responses, which include IFNγ production and macrophage hyperactivation. Factors within the microenvironment of the eye, including particular cytokines and/or presentation of pMHC by specific DC subsets, direct the differentiation of an activated Th0 cell toward Th2 effectors. Th2 cells then promote the mounting of relatively mild humoral responses rather than tissue-damaging cell-mediated responses. Thus, the immune response is still present but its effects have been blunted by "deviating" effector generation toward a Th2 phenotype.

IV. IMMUNE PRIVILEGE

Immune-privileged sites are anatomical sites that are naturally less subject to immune responses than most other areas of the body. Immune-privileged sites include the central nervous system and brain, the eyes and the testes. Even foreign antigens accessing these sites do not generally trigger immune responses. Originally, it was thought that self tolerance in immune-privileged sites stemmed from physical barriers that blocked lymphocyte access. However, it is now clear that immunosuppressive cytokines, the actions of regulatory T cells, and immune deviation are often involved. In addition, non-lymphoid cells in the eye and testis can eliminate activated T cells via Fas killing. Fas is widely expressed on non-lymphoid cells and is induced on T cells after activation, whereas FasL is expressed on only a very few cell types. Significantly, these cell types are often found in immune-privileged sites. For example, the Sertoli cells in the testis constitutively express high levels of FasL. Interaction of Fas on an activated T cell with FasL on a Sertoli cell induces the activated T cell to undergo apoptosis before it has a chance to differentiate into effector cells that could attack testicular tissue or secrete damaging inflammatory cytokines.

C. Special Tolerance Situations

I. MATERNAL–FETAL TOLERANCE

A mammalian mother does not normally reject her fetus, despite the fact that half of the histocompatibility molecules the fetus expresses are derived from the father and are therefore frequently "foreign" to the mother. Maternal tolerance to paternal histocompatibility molecules is maintained during fetal development but the potential for reactivity to these antigens eventually returns after the birth. Maternal–fetal tolerance is not primarily due to physical barriers, since the fetal and maternal tissues that come together in the placenta are not separated by basement membranes. Furthermore, although the maternal and fetal circulatory systems are physically distinct, the barrier is not absolute and fetal antigens and cells do enter the maternal circulation and reach the mother's secondary lymphoid organs.

Several mechanisms are thought to contribute to maternal–fetal tolerance. (1) The hormonal changes associated with reproduction may prepare the uterus to accept the "fetal graft" so that the fetus is not seen as a foreign and injurious entity. That is, once pregnancy is established, the unique hormonal and cytokine microenvironment of the placenta favors immunosuppression and immune deviation to Th2 responses. Indeed, Th1 responses are associated with early fetal loss. In addition, certain placental cells produce high levels of progesterone, which dampens intracellular signaling such that effector T cells are suppressed. (2) The tissues forming the maternal–fetal interface are populated with non-professional APCs that lack costimulatory molecules. Moreover, pregnancy does not usually represent a "danger" that would induce the production of inflammatory cytokines and the activation of professional APCs. (3) If a professional APC somehow becomes activated, local Tr1 and Th3 cells at the maternal–fetal interface usually suppress any incipient immune response. Tr1 and Th3 cells have been implicated in the prevention of miscarriage in animals. (4) The tissues of both the placenta and developing fetus are almost devoid of MHC class I and II molecules. In fact, the only MHC genes that an early human fetus is known to express are the non-polymorphic HLA-E and HLA-G genes (described in Ch. 6). The product of the HLA-E gene inhibits maternal uterine NK cells (see Ch. 11), whereas the product of the HLA-G gene is involved in the formation of new blood vessels required for embryonic development. Fetal cells also express the RCA proteins DAF and MCP that can interfere with maternal complement activation (refer to Ch. 3). (5) Some placental cells express FasL, so that Fas-expressing maternal T cells activated by an encounter with fetal antigens can be killed before fetal tissues are attacked.

II. NEONATAL TOLERANCE

Early studies of immunity and tolerance made use of skin graft rejection as an experimental readout for immune responsiveness. If the skin of a mouse of inbred strain A were transplanted onto an adult mouse of inbred strain B, the graft was immediately rejected due to MHC differences between strains A and B (see Ch. 17). However, if the spleen and bone marrow cells of a strain A mouse were injected into neonatal mice of strain B, the strain B mice were later (as adults) able to accept skin grafts from strain A mice. This tolerance proved to be lifelong. Immunologists concluded that there was a period shortly after birth when tolerance to foreign antigens could be "learned", before the animal became fully capable of mounting protective immune responses. This concept became known as "neonatal tolerance".

We now know that the decreased immune responsiveness observed in neonates is due to immune responses that are inherently harder to induce and tolerance mechanisms that are more easily triggered (Table 10-2). For example, in very young animals, FDCs are inefficient at trapping antigen, and APCs have a reduced antigen presentation capacity. B cells of early neonates, unlike those of adults, are difficult to provoke into mounting antibody responses and remain highly susceptible to tolerance induction for several days. When antigen exposure does result in a response, neonatal B cells do not differentiate into plasma cells as quickly as do adult B cells. Neonatal antibody responses to most (but not all) Ti antigens are strong, while responses to Td antigens are weak and often delayed, a state that persists for 2–8 weeks after birth in mice.

With respect to T cells, absolute numbers of mature naïve T cells are 10,000-fold lower in neonatal mice than in adults, although their expression of cytokine receptors and costimulatory molecules is comparable. These lower numbers effectively create a situation in which T cells in the neonate encounter more antigen on a per cell basis than do T cells in an adult. As is detailed in Section D of this chapter, very high doses of an antigen can be tolerogenic, an effect known as "high zone tolerance". Thus, neonatal T cells are tolerized at a dose that is immunogenic in the adult. Lymphocyte trafficking patterns are also different in neonates and adults, in that naïve T cells can access non-lymphoid peripheral tissues much more easily in the former and tend to accumulate in lung and skin rather than in the spleen and lymph nodes where antigen and mature DCs are concentrated. Finally, neonatal T cells tend to secrete higher levels of immunosuppressive cytokines than do adult T cells. Some or all of these factors could lead to subthreshold levels of signal 1 or 2 being delivered, resulting in T cell deletion or anergization and thus tolerance to the antigen.

Table 10-2 Comparison of Neonatal versus Adult Immunity

	Neonate	Adult
Original observations	Tolerant to foreign skin graft	Mounted immune response against foreign skin graft
FDC antigen trapping capacity	Low	High
APC presentation capacity	Low	High
Anergizing APCs	Many	Few
B cell differentiation into plasma cells	Delayed	Rapid
B cell responses to Ti antigen	Usually strong	Strong
B cell responses to Td antigen	Weak and delayed	Strong
Relative number of mature naïve T cells	1	10,000
Antigen dose inducing high zone tolerance	Relatively low	High
Localization of mature naïve T cells	Secondary lymphoid tissues; peripheral tissues such as skin and lung	Secondary lymphoid tissues
T cell cytokine secretion	IL-10, TGFβ	IL-2, IFNγ (Th1) IL-4, IL-5 (Th2)
Importance of regulatory T cells to peripheral tolerance	Low	High

Interestingly, neonatal tolerance depends very little on regulatory T cells. The role of regulatory T cells in peripheral tolerance in adult animals was first discovered during studies of the development of autoimmunity in mice. If the thymus was removed from a neonatal mouse between 2 and 4 days after birth, the adult animal later developed autoimmune gastritis due to an attack on the stomach lining by autoreactive peripheral T cells. However, if normal T cells were transferred to the thymectomized neonatal mouse, the autoimmunity in the adult was mitigated. The interpretation of these observations was that the regulatory T cells necessary to control autoreactive T cells in the adult periphery could not mature in the absence of a thymus. The fact that the thymectomized mouse did not develop autoimmunity during the neonatal period suggests that these regulatory T cells, which are vital for tolerance in the adult, do not make a major contribution to neonatal tolerance.

Table 10-3 Immunogens versus Tolerogens

	Immunogen	Tolerogen
Number of antigenic epitopes per molecule of antigen	Moderate	High
Size	Large, aggregated, polyvalent	Small, disaggregated
Solubility	Insoluble	Soluble
Dose	Moderate	Very high or very low
Schedule of Ag administration	Low number of moderate doses	Many small doses
Route of Ag administration	Subcutaneous	Intravenous, oral

D. Experimental Tolerance

Much of what is known about self tolerance has been derived from experimental models in which a lack of responsiveness to a <u>foreign</u> antigen is induced. That is, various experimental contrivances are used to convince the host animal's body that a foreign antigen is "self" such that an immune response is not mounted against it. Such an animal is said to be tolerized to the antigen. A *tolerogen* is an experimental foreign antigen that binds to the antigen receptors of lymphocytes but suppresses, rather than induces, activation.

I. CHARACTERISTICS OF EXPERIMENTAL TOLERANCE

Experimental tolerance is most easily induced in an animal with a lymphoid system that is not at mature full strength, such as in a neonatal animal or an animal with an immune system that has been damaged by treatment with drugs or irradiation. Under these conditions, these animals fail to mount an immune response to a normally immunogenic molecule. Even after the immunocompromising treatment is discontinued and the immune system is allowed to recover, a subsequent exposure to the same immunogen elicits no response. However, the immune system is not globally suppressed because responses to other, unrelated immunogens remain intact. In other words, experimental tolerance is antigen-specific. However, experimental tolerance is not usually permanent and its maintenance depends on continued exposure to the tolerogen in either a persistent or an intermittent fashion. Loss of the tolerogen for a prolonged period slowly restores normal responsiveness to the foreign antigen. This reversal can occur if the previously tolerized cells recover their immune reactivity, or if newly generated lymphocytes with normal reactivity to the antigen emerge. In general, experimental B cell tolerance usually lasts only a few weeks in the absence of antigen, while experimental T cell tolerance can persist for several months.

II. CHARACTERISTICS OF TOLEROGENS

The physical and behavioral characteristics of tolerogens are compared to those of immunogens in Table 10-3.

i) Nature of Tolerogenic Molecules

For reasons that are still unclear, some molecules are naturally more tolerogenic than others. For example, while L-amino acid polymers are immunogenic at almost any dose, D-amino acid polymers are tolerogenic at the same doses. For other molecules, a slight chemical modification is enough to turn an immunogen into a tolerogen for the same lymphocyte clone. Also important is the density of the antigenic epitope on the molecule. Having a moderate number of epitopes per molecule promotes immunogenicity, but a large number of epitopes per molecule favors tolerance induction.

ii) Molecular Size

Small, soluble molecules do not often make good immunogens but they can make good tolerogens. Immunologists speculate that small molecules may induce tolerance because their relatively innocuous presence does not trigger the danger signals necessary to activate the DCs processing them. A T cell whose TCR recognizes pMHCs derived from the small molecule therefore receives signal 1 but not signal 2, and is consequently anergized.

For B cells, a tolerogen must be capable of stimulating the BCRs such that signal 1 is delivered. However, the stimulation must not be so complete that the B cell is activated even in the absence of signal 2 (CD40L) delivered by an activated Th cell, as happens with many Ti antigens (refer to Ch. 5). Very small soluble molecules cannot cross-link BCRs at all and therefore do not tolerize (or activate) B cells.

iii) Dose

Although moderate doses of an antigen induce an immune response, very high and very low doses of the same antigen may induce tolerance. These types of tolerance are called *high zone* and *low zone* tolerance, respectively.

iiia) High zone tolerance. To illustrate high zone tolerance, consider the fate of mice that have been inoculated with either a very high dose or a moderate dose of a purified bacterial antigen. When these animals are later injected with live bacteria, the mice that received the moderate dose mount a humoral response that rescues them, but those that received the high dose do not make anti-bacterial antibodies and die. In some cases, this high zone tolerance is caused by a *B cell receptor blockade*, in which very large amounts of an antigen persistently occupy the BCRs without cross-linking them. This blockade affects signal 1 delivery because the ITAMs on the Igα and Igβ chains in the BCR complex are incompletely phosphorylated. An abnormal and possibly inhibitory signal is transduced that alters downstream signaling events such that B7 expression is not upregulated. Extensive downregulation of mIgM also occurs on these anergic B cells, further precluding signal 1 delivery. However, the failed antibody response could also be due to T cell tolerization, since B cells require T cell help to respond to Td antigens. High zone tolerance for T cells occurs at much lower antigen doses than those required for the same effect on B cells. In addition, whereas B cell tolerization is not observed until about 2 days after antigen administration, high zone tolerance for T cells is evident within just hours of exposure. T cell high zone tolerance often results from clonal exhaustion.

iiib) Low zone tolerance. Tolerization can also occur if an animal is exposed to a very low dose of antigen (too low to provoke an immune response) over a long period of time. Studies in mice have shown that repeated administration of very low doses of an antigen suppresses antigen-specific antibody production. Low zone tolerance has been demonstrated for a wide range of antigens in neonatal animals and immunocompromised adults, but for only a few antigens in healthy adults. The mechanism underlying low zone tolerance is unknown.

iv) Route of Administration

The route by which an antigen accesses the body determines how it is processed and affects whether it is immunogenic or tolerogenic. The outcome is likely related to the frequencies and types of APCs that take up the antigen, and the microenvironment in which they do so. The most immunogenic mode of delivery is the subcutaneous route. Langerhans cells resident in the skin are very effective APCs, meaning that antigen delivered in this way often triggers a robust immune response. In contrast, antigen administered intravenously often results in tolerance. Intravenous antigen is conveyed to the spleen, where it is presented primarily by naïve splenic B cells and non-professional APCs. In the absence of inflammatory cytokines, these cells express only low levels of costimulatory molecules (if any) and cannot deliver a robust signal 2. Responding T cells are therefore generally anergized and the immune response to the antigen is negligible.

Oral administration is a very effective way of inducing peripheral tolerance to immunogens, at least under laboratory conditions. Most dietary antigens are degraded in the stomach before ever reaching the small intestine, but some partially digested or even intact molecules do get absorbed into the circulation and thus are distributed systemically. In animal models, tolerance to such a systemic antigen can be observed within 5–7 days after the antigen is consumed. T cell tolerance after a single feeding can last for up to 18 months, whereas B cell oral tolerance lasts for 3–6 months. Despite intensive study, it is still not clear how oral tolerance is achieved. Several mechanisms are thought to make a contribution. (1) Presentation of gut antigens by non-professional APCs lacking costimulatory capacity may anergize antigen-specific T cells in the GALT (see Ch. 12). (2) Since food is usually harmless, there are no danger signals associated with the incoming antigen that could activate DCs. T cells interacting with these DCs should also be anergized. (3) Continual low doses of antigen may induce the differentiation of regulatory T cells in the gut, and the immunosuppressive cytokines secreted by these cells should suppress effector T cell responses. With respect to the human situation, attempts have been made in clinical trials to induce oral tolerance to particular antigens that are the targets of allergic, autoimmune or other detrimental immune responses. However, these approaches have yet to be successful, and the therapeutic potential of oral tolerance induction is therefore currently uncertain.

As we reach the end of this chapter, we have completed our study of two important cellular components of the adaptive immune response: B cells and αβ T cells. In the next chapter, we examine cells that are considered to bridge adaptive and innate immunity: NK cells, γδ T cells and NKT cells.

- Adaptive immune responses are tightly regulated by mechanisms of tolerance and control. Tolerance mechanisms prevent lymphocyte activation whereas control mechanisms rein in the actions of activated lymphocytes.

- The major physiological role of peripheral self tolerance is to ensure that any autoreactive lymphocytes that escaped negative selection during the establishment of central tolerance cannot initiate a response that damages host tissues.

- Tolerogenic DCs result from an encounter of immature DCs with antigen in the absence of "danger signals". Modulated DCs result from an encounter of immature DCs with antigen in the absence of danger and in the presence of immunosuppressive cytokines.

- If a naïve autoreactive T cell encounters its cognate pMHC presented by a tolerogenic DC, the T cell either dies by apoptosis due to clonal deletion or is inactivated due to anergization.

- An anergized T cell does not generate effector or memory cells and is unable to respond to cognate antigen the next time it meets the antigen under optimal conditions.

- T cell clonal exhaustion may establish tolerance to highly abundant self antigens.

- B cell anergy may occur if the BCRs are engaged by a Td antigen but the appropriate Th cell is not available, or if the Th cell fails to upregulate costimulatory molecules.

- Regulatory subsets of CD4$^+$ and CD8$^+$ T cells may anergize T lymphocytes via intercellular contacts or secretion of IL-10 and/or TGFβ, and may modulate DCs.

- Interaction of a Th0 cell with a modulated DC may generate regulatory T cells.

- The immunosuppressive cytokines IL-10 and TGFβ downregulate most immune responses, whereas immune deviation results in a less harmful response by influencing Th1 versus Th2 development.

- Immune-privileged sites show a decreased frequency of immune reactivity due the presence of immunosuppressive cytokines, immune deviation, death receptor upregulation, and/or the actions of regulatory T cells.

- Maternal–fetal tolerance is mediated by uterine hormonal status, lack of classical MHC class I expression, expression of non-classical MHC molecules, immunosuppressive cytokines, and presence of non-professional APCs and regulatory T cells.

- Neonatal tolerance refers to the fact that adaptive responses in neonates are weak and delayed compared to responses in adults. This deficit is due to differences in APCs, B and T cells rather than to differences in response and tolerance mechanisms.

- Experimental tolerance is tolerance induced to a foreign antigen and is often mediated by introducing the antigen via a non-standard route or in a tolerizing dose.

Introduction

1) Can you define these terms? autoreactive, tolerized.

2) Distinguish between tolerance and control mechanisms.

Section A

1) Distinguish between immature, tolerogenic and mature DCs.

2) Distinguish between clonal deletion and anergy.

3) What is the response of an anergized T cell to a second encounter with antigen presented by a mature DC?

4) What is clonal exhaustion and how might it contribute to self tolerance?

5) Describe two mechanisms by which a B cell can be tolerized.

Section B.I

1) Distinguish between the three main types of CD4$^+$ regulatory T cells (T$_{reg}$, Th3 and Tr1) in terms of surface markers and mechanisms of tolerization.

2) What is an iT$_{reg}$ cell and how does it differ from an nT$_{reg}$ cell?

3) Why is the expression of TLRs on T$_{reg}$ cells significant?

4) Describe two ways in which Th3 and Tr1 cells can be generated.

5) Describe two ways in which CD8$^+$ regulatory T cells might mediate tolerance.

Section B.II–IV

1) Why are IL-10 and TGFβ considered to be immunosuppressive? Give three examples for each.

2) What is immune deviation and why is it considered a tolerance mechanism?

3) Describe three mechanisms that can exert tolerance in an immune-privileged site.

Section C

1) Why is maternal–fetal tolerance required?

2) Describe four mechanisms that help to maintain maternal–fetal tolerance.

3) What was the original observation that led to the notion of a special type of "neonatal tolerance"?

4) Give four reasons why neonates are naturally more tolerant to antigens than are adults.

Section D

1) Describe two situations in which experimental tolerance is easily induced.

2) How can experimental tolerance to an antigen be maintained?

3) Give three characteristics of effective tolerogens.

4) Define "high zone" and "low zone" tolerance.

5) Why is the intravenous administration of a molecule usually tolerogenic?

6) Give three mechanisms thought to contribute to oral tolerance.

11

NK, γδ T and NKT Cells

If it's natural to kill, why do men have to go into training to learn how?
Joan Baez

In Chapter 3, we introduced NK cells, γδ T cells and NKT cells as cell types that bridge innate and adaptive immunity in both form and function. On one hand, the responses of NK, γδ T and NKT cells to infection or injury can be considered part of innate immunity because they are rapid and involve broadly specific recognition of antigen that is independent of classical pMHC complexes. This more generalized recognition is essential to the primary physiological function of these cells as defenders that clear pathogens as part of the induced innate response. On the other hand, NK cells, γδ T cells and NKT cells can be considered part of adaptive immunity because these cells are related to the T cell lineage, and γδ T cells and NKT cells express TCRs derived from V(D)J recombination. In addition, NK, γδ T and NKT cells can directly influence αβ T cells and B cells and their effector actions. In this chapter, we discuss the distribution, function and development of NK cells, γδ T cells and NKT cells. A schematic representation of the major cell surface markers distinguishing these cell types from αβ T cells is given in Figure 11-1. The overview in Figure 11-2 illustrates the development of these cell types from NK/T precursors that remain in the bone marrow to generate the majority of NK cells, or migrate to the thymus to generate γδ T cells, αβ T cells, NKT cells and thymic NK cells.

A. Natural Killer (NK) Cells

I. OVERVIEW

Histologically, NK cells are large, non-phagocytic lymphoid cells that possess cytoplasmic granules containing perforin and granzymes. At the molecular level, NK cells are distinguished from NKT, B and T cells by their lack of expression of TCRs or BCRs and the germline configuration of their TCR and BCR genes. Although NK cells are found at their highest frequency in spleen, liver, uterus and peripheral blood, they also congregate in the mucosal epithelium and so are important components of MALT (see Ch. 12). Moderate numbers of resting NK cells can be found in the bone marrow, lymph nodes and peritoneum. In situations of infection or inflammation, NK cells can be rapidly recruited to almost any tissue in the body. The life span of a mature NK cell (in the absence of activation) is about 7–10 days.

The primary functions of NK cells are to induce the cytolysis of tumor cells or virus-infected cells and to secrete cytokines (Fig. 11-3). Cytolysis may be induced by *natural cytotoxicity*, ADCC or cytotoxic cytokines. Some NK cell-secreted cytokines also regulate T and B cell function and differentiation. Importantly, unlike activated T and B cells, activated NK cells do not need to proliferate and differentiate into separate effec-

Fig. 11-1 **Characteristic Surface Markers of NK, γδ T, NKT and αβ T Cells**

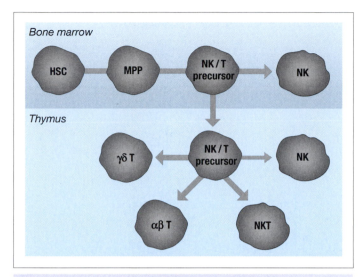

Fig. 11-2 **Overview of Development of NK, γδ T and NKT Cells**

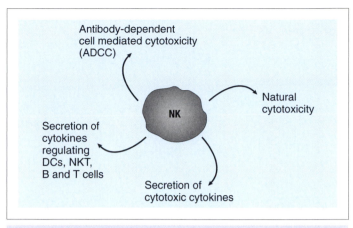

Fig. 11-3 **NK Cell Effector Functions**

tor cells in order to function. Thus, the peak NK cell response can be detected within hours of infection.

II. EFFECTOR FUNCTIONS

i) Natural Cytotoxicity
ia) "Self-deficit" model of NK-mediated natural cytotoxicity. Just like the granules of CTLs, the granules of NK cells contain perforin and granzymes that induce target cells to undergo apoptosis. However, these granules are preformed in an NK cell and do not have to be synthesized in response to activation, as occurs in CD8+ Tc cells. The triggering of NK-mediated natural cytotoxicity depends on a balance between competing signals initiated by two sets of surface receptors of broad

binding specificity: the NK *activatory* receptors and the NK *inhibitory* receptors. The NK activatory receptors are triggered by ligands that may be constitutively expressed on healthy cells, or ligands that may be induced or upregulated in response to viral infection, malignant transformation or other cellular stresses. Some non-classical MHC molecules have also been identified as activatory ligands. In contrast, most NK inhibitory receptors bind only to classical MHC class I molecules expressed by the host. When an NK cell scans the surface of a cell, the overall amount of activatory signaling that results must be counterbalanced by adequate inhibitory signaling or the NK cell is activated and releases the cytotoxic contents of its granules. In the case of normal host cells (almost all of which express MHC class I), the NK inhibitory receptors are engaged and send an inhibitory signal that dominates normal activatory signaling (Fig. 11-4A). The NK cell is not activated and the normal cell is spared. In contrast, infected

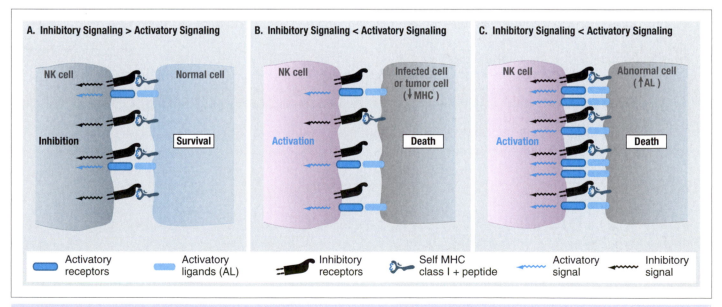

Fig. 11-4 **Outcomes of NK Activatory/Inhibitory Receptor Signaling**

and cancerous cells frequently downregulate their MHC class I expression. In these cases, an NK cell's inhibitory receptors are not engaged in sufficient numbers to prevent the activatory receptors from completing their intracellular signaling (Fig. 11-4B). The NK cell is activated and the abnormal cell is killed. Natural cytotoxicity will also result if relatively normal levels of self MHC class I expression are overwhelmed by an abnormal increase in the expression of activatory ligands by a host cell (Fig. 11-4C). Thus, NK cells are on the alert to eliminate targets that have a deficit of self MHC class I expression relative to the expression of various activatory ligands. Many immunologists refer to this as the "missing self" model of NK recognition, as many of the original studies involved situations in which self MHC class I was completely absent from the target cells.

Although no effector differentiation is required, resting NK cells don't usually acquire significant cytolytic competence until they are primed through exposure to cytokines such as IFNα/β, IL-2, IL-12 or IL-15. These cytokines are often present in the site of infection due to the actions of other activated innate response cells. The priming cytokines induce upregulation of the expression of multiple activatory and inhibitory receptors on NK cells as well as adhesion molecules that stabilize the binding of an NK cell to a potential target cell.

ib) Activatory and inhibitory receptors. Activatory and inhibitory receptors on NK cells are generally transmembrane proteins. The extracellular domains of both types of receptors are responsible for ligand recognition and often share similar molecular features. On this basis, they can be categorized into structural classes. The *natural cytotoxicity receptor* (NCR) class contains only activatory receptors, whereas the *natural killer group 2* (NKG2) and *killer Ig-like receptor* (KIR) classes include both activatory and inhibitory members. The KIR receptors are named according to the number of Ig domains present as well as the length of the cytoplasmic tail. For example, KIR2DS means "KIR, two Ig-like domains, short tail", and KIR3DL means "KIR, three Ig-like domains, long tail".

The opposing functions of activatory and inhibitory receptors can be attributed to differences in their intracellular domains. The NK activatory receptor proteins possess positively charged transmembrane residues and short cytoplasmic tails that contain few intracellular signaling domains. These chains are not expressed on their own and do not mediate any signal transduction in isolation. Instead, activatory receptor proteins associate with a homodimer composed of an accessory signaling molecule such as CD3ζ, the γc chain, or one of two adaptor proteins called DAP10 and DAP12. All of these molecules possess negatively charged transmembrane domains and all (except DAP10, which transduces signals in a slightly different way) contain ITAMs that facilitate signal transduction. Upon the binding of an activatory ligand to an activatory receptor complex, the ITAMs in the associated chain are phosphorylated and a signal that promotes natural cytotoxicity is conveyed to the interior of the NK cell. In contrast, inhibitory receptor proteins usually have long cytoplasmic tails containing *immunoreceptor tyrosine-based inhibition motifs* (ITIMs).

Table 11-1 Examples of Human NK Activatory and Inhibitory Receptors

Receptor Name	Receptor Class	Associated Signaling Chain	Example of Ligand
NK Activatory Receptors			
NKp46	NCR	FcεRIγ or CD3ζ	Viral hemagglutinin
NKp44	NCR	DAP12	Viral hemagglutinin
NKp30	NCR	CD3ζ	Unknown
CD94/NKG2C	NKG2	DAP12	HLA-E
NKG2D	NKG2	DAP10	MICA, MICB
KIR2DS1	KIR	DAP12	HLA-C
NK Inhibitory Receptors			
CD94/NKG2A	NKG2	None	HLA-E
CC94/NKG2B	NKG2	None	HLA-E
KIR3DL2	KIR	None	HLA-A
KIR3DL1	KIR	None	HLA-B
KIR2DL1	KIR	None	HLA-C

When an inhibitory receptor is stimulated by the binding of MHC class I, kinases and phosphatases are recruited to the receptor complex. The ITIMS are phosphorylated such that signal transduction within the NK cell is inhibited. It is the balance of these inhibitory and activatory signals that determines whether the NK cell is activated.

Some reasonably well-studied human NK activatory receptors appear in Table 11-1 (upper half) and in Figure 11-5. NKp46, NKp44 and NKp30 are members of the NCR class of activatory receptors and are expressed exclusively on NK cells. These receptors are largely responsible for direct NK cell killing of virus-infected and tumor cells. An example of an activatory receptor of the NKG2 class is the CD94/NKG2C receptor, composed of the NKG2C protein bonded to the CD94 protein, plus the ITAM-containing DAP12 protein. One ligand for CD94/NKG2C is HLA-E, a non-classical MHC class Ib molecule. Perhaps the most important NKG2 activatory receptor is NKG2D, in which a NKG2D homodimer associates with DAP10. (In mice, NKG2D associates with both DAP10 and DAP12). Human NKG2D binds to a number of MHC class I-related molecules, including the MICA and MICB molecules introduced in Chapter 6. NKG2D ligands tend to be transmembrane proteins that are induced on many types of epithelial cells in response to heat shock or other cellular stresses. NKG2D ligands are also frequently upregulated on virus-infected cells and cancer cells, making these cells NK targets. Some KIR receptors are also activatory. For example, KIR2DS1 binds to the conventional MHC class I molecule HLA-C and promotes NK cell activation.

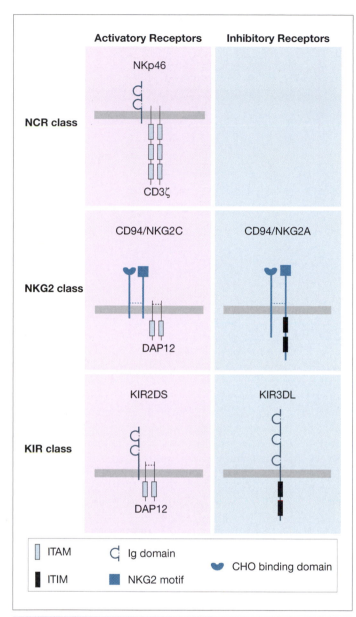

Fig. 11-5 Examples of NK Activatory and Inhibitory Receptors

blood vessels. This interaction between KIR-2DL4 and HLA-G thus promotes a successful pregnancy.

ii) NK-Mediated ADCC

NK cells express large amounts of FcγRIIIA (CD16). This FcR can trigger ADCC by binding to IgG molecules that have engaged epitopes on tumor cells or virus-infected cells. The engagement of FcγRIIIA activates the NK cell and causes it to release the contents of its cytotoxic granules (refer to Fig. 5-15).

iii) Cytokine Secretion

In response to infection, phagocytes produce IFNα/β, IL-12 and TNF that induce primed NK cells to synthesize copious quantities of IFNγ. Subsequent activation of primed NK cells by overwhelming activatory receptor engagement or FcγRIIIA stimulation leads to the production of a new battery of chemokines, growth factors and cytokines, including not only IFNγ but also TNF, IL-1, IL-3 and IL-6. These molecules have various effects on cells of both the innate and adaptive immune responses, as detailed in Appendix E.

III. DEVELOPMENT

i) Developmental Pathway

NK cells originate from the same bone marrow-derived HSCs, MPPs and NK/T precursors that give rise to T cells. Figure 11-6 outlines the later stages of NK cell development in mouse bone marrow, which has been better studied than the human situation. NK/T precursors that remain in the bone marrow become subject to signaling delivered by IL-15 as well as a molecule crucial for NK development called *Flt3 ligand* (*Flt3-L*). Bone marrow stromal contacts are also essential for NK development at this stage. Under the continued influence of IL-15, *pro-NK cells* are produced that eventually develop into *immature NK cells* that express only a few NK inhibitory receptors. Mature NK cells then emerge that express the full complement of activatory and inhibitory receptors. A secondary source of NK cells arises from NK/T precursors that have migrated from the bone marrow to the thymus (refer to Ch. 9). These precursors give rise mainly to T cells but also to a subset of thymic NK cells.

ii) NK inhibitory Receptor Repertoire

Within an individual, different NK cells express different combinations of inhibitory receptors on their cell surfaces. How is each NK cell's "repertoire" of inhibitory receptors assembled? A sequential expression model has been proposed in which developing NK cells gradually accumulate new types of self MHC-specific inhibitory receptors. It may be that certain activatory receptors, triggered by interaction with their ligands on normal host cells, build up a cascade of intracellular signaling inside the developing NK cell. At some point, the inhibitory receptors start to be expressed, and different classes accumulate on the cell surface until the signaling generated by the triggering of all the inhibitory receptors exceeds the level of signaling generated by all the activatory receptors. It is thought

Examples of some important NK inhibitory receptors are given in the lower half of Table 11-1 and in Figure 11-5. Inhibitory members of the NKG2 class include CD94/NKG2A and CD94/NKG2B. CD94/NKG2A binds specifically to HLA-E molecules, providing an inhibitory counterpart to the CD94/NKG2C activatory receptor. CD94/NKG2A function is particularly important for blocking maternal uterine NK cell function during pregnancy and protecting the trophoblast and fetus (whose cells express paternal "non-self" MHC molecules; refer to Ch. 10). KIR receptors that block NK cell activation include KIR3DL2, KIR3DL1 and KIR2DL1, which bind to HLA-A, B or C, respectively. Interestingly, engagement of KIR2DL4 on maternal NK cells by HLA-G expressed on the placenta does not trigger cytotoxicity but instead induces maternal uterine NK cells to secrete cytokines that induce the formation of new

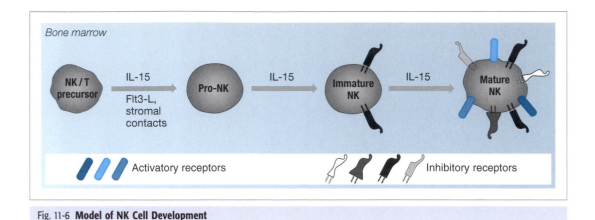

Fig. 11-6 **Model of NK Cell Development**

that the cell then receives a maturation signal that halts and fixes the expression of new inhibitory receptors.

iii) NK Cell Tolerance

NK cells are usually described as "self-tolerant" because they do not attack normal self cells. According to the self-deficit model, an NK cell with the potential to attack a self cell would be an effector whose collection of inhibitory receptors failed to adequately recognize self MHC. The inhibitory receptors of such an NK cell would not be engaged, allowing the activatory ligands on normal self cells to trigger the NK cytolytic program and precipitate the destruction of the self cells. Thus, during development, an NK cell that did not express sufficient inhibitory receptors recognizing self MHC would undergo some kind of tolerance process to either delete the cell or render it anergic or harmless. However, there are some aspects of NK cell tolerance that are difficult to explain solely on the basis of interactions between NK inhibitory receptors and MHC class I molecules. For example, in both humans and mice, individuals lacking MHC class I expression do not have a problem with NK cell "autoreactivity". Moreover, developing NK cells do not inflict tissue damage even though there appears to be a short period in which the maturing NK cells have effector function capability but still lack expression of inhibitory receptors. These observations may be explained by the recent discovery of ligands other than MHC class I molecules that can bind to NK inhibitory receptors. Thus, both MHC-dependent and MHC-independent mechanisms may play a role in establishing NK cell tolerance. The details of these mechanisms remain to be elucidated.

B. γδ T Cells

I. OVERVIEW

In Chapters 8 and 9, we described how two distinct types of T lymphocytes can be distinguished based on the polypeptide chains making up their TCRs. In both humans and mice, the majority of T cells in the body are αβ T cells bearing TCRs containing the TCRα and TCRβ chains. Mature αβ T cells carry either the CD4 or CD8αβ coreceptor. In contrast, γδ T cells express TCRs composed of the TCRγ and TCRδ chains and carry either CD8αα or no coreceptor at all. Whereas cells bearing TCRαβ recognize only pMHC complexes, γδ TCRs interact with ligands in a way that is similar to ligand recognition by PRRs. That is, γδ T cells can respond to antigens that are derived from a broad range of pathogens and abnormal or stressed host cells. In so doing, γδ T cells do not require the involvement of conventional MHC nor the processing and presentation of peptide antigens by professional APCs. Intact proteins or peptides from pathogens or stressed host cells, and non-protein antigens such as phosphorylated nucleotides, can serve as TCRγδ ligands. Furthermore, these ligands may be either soluble or cell surface-bound. Once activated, γδ T cells respond by proliferating and differentiating into γδ Th and CTL effectors much like αβ T cells, but do so much more rapidly, in smaller numbers, and in the apparent absence of conventional costimulation. A comparison of the properties of γδ and αβ T cells is given in Table 11-2.

II. ANATOMICAL DISTRIBUTION

The anatomical distribution of γδ T cells is strikingly different from that of αβ T cells. Only very low numbers of γδ T cells are found in the secondary lymphoid organs and thymus of mice and humans. Instead, γδ T cells are interspersed among the epithelial cells of the skin and in the mucosae of various body tracts. Because of this localization, these γδ T cells constitute one type of *intraepithelial lymphocyte* (IEL), a cell population that also includes small numbers of αβ T cells. When IELs are found in the gut, they are called iIELs, for "intestinal IELs". When IELs are resident in the top layers of the skin, they are sometimes called *dendritic epidermal T cells* (DETCs). The localization of γδ T cells in the skin or mucosae allows them to be among the first defenders to confront invading pathogens or injurious substances.

Unlike αβ T cells, which utilize highly random combinations of V gene segments for their TCRs regardless of their anatomical location, the γδ T cells resident in a particular tissue express a dominant or "canonical" TCR. For example, γδ T cells in mouse skin predominantly express TCRs contain-

Table 11-2 Comparison of the Properties of γδ and αβ T Cells

	γδ T cells	αβ T cells
Type of immunity	Induced innate	Adaptive
Frequency among T cells	0.5–5%	95–99%
Anatomical distribution	Epithelial layers in the skin and mucosae	Primary and secondary lymphoid organs and tissues
TCR	TCRγδ	TCRαβ
Repertoire diversity	Limited	Almost unlimited
Receptor specificity	Promiscuous	Specific
Distribution of epitope recognized by a given TCR	Broadly expressed on range of pathogens or stressed cells	Expressed on one pathogen
Ligand recognized	Intact host and pathogen proteins (e.g., phospholipids, glycolipids, phosphonucleotides, pyrophosphates, HSPs)	Peptide–MHC complexes
APCs required for antigen presentation	No	Yes
Mature T cell coreceptor expression	CD4−CD8− or CD4−CD8αα+	CD4+CD8− or CD4−CD8αβ+
CD28- or CD40-mediated costimulation required for activation	No	Yes
Days to effector cell generation	1–2 days	7 days
Types of effector cells generated*	γδ Th1, γδ Th2, γδ CTL, γδ regulatory T cells	Th1, Th2, CTL, regulatory T cells
Capable of long-lived memory response	No	Yes

*Not all types of effectors are found in all γδ T cell responses.

ing Vγ3Vδ1 or Vγ5Vδ1, whereas the genital epithelium and the tongue feature an abundance of Vγ6Vδ1-expressing cells. Vγ7 appears more often than not in the TCRs of iIELs, whereas Vγ2Vδ2 cells dominate in the tonsils, spleen and peripheral blood. In humans, Vγ1Vδ2 TCRs are prevalent on iIELs, whereas Vγ2Vδ2 and Vγ9Vδ2 TCRs occur on γδ T cells in the skin. Vγ2Vδ2 TCRs are also found on γδ T cells in human peripheral blood.

III. ANTIGEN RECOGNITION AND ACTIVATION

Unlike αβ TCRs, the TCRs of some subsets of γδ T cells can bind directly to low molecular weight non-peptide antigens, without the need for presentation by another molecule or cell. However, the TCRs of other γδ T cell subsets depend on non-peptide antigen presentation by non-classical MHC class Ib molecules. Still other γδ T cell subsets express TCRs that recognize antigens presented by members of the non-polymorphic CD1 family of MHC-like molecules that can present non-peptide antigens to αβ T cells (refer to Ch. 7). In general, γδ TCRs lack the fine antigen specificity of αβ TCRs and are often broadly cross-reactive.

γδ T cells have been shown to respond to a wide variety of bacterial, protozoan and viral molecules and products, includ-

ing peptides, proteins, pyrophosphates, phospholipids, lipoproteins, phosphorylated oligonucleotides and alkyl amines (Fig. 11-7A). Some γδ T cell subsets (as defined by V gene usage) appear to be specific for certain types of determinants and thus may counter a specific group of pathogens. The non-random anatomical distribution of such subsets may then optimize the chance that γδ T cells of the appropriate specificity will be in local abundance to defend against a given invader (Table 11-3). For example, the Vγ2Vδ2 subset is not only found in human skin and peripheral blood but also defines specificities effective at recognizing pathogens normally found in these locations, such as Epstein-Barr virus and mycobacteria. Other γδ T cells are specific for stress molecules that are expressed by host cells suffering injury, infection or cancerous transformation and that do not appear on the surfaces of healthy host cells. The expression of a single stress antigen in response to a myriad of different infections or injuries allows a γδ T cell population with a limited antigen receptor repertoire to monitor a wide range of assaults to the host epithelium. Whereas some stress molecules recognized by γδ T cells are small pyrophosphate-like molecules, others are peptides or whole proteins, such as the HSPs released in the debris of necrotic cells (Fig. 11-7B). Other human γδ T cell clones are activated by binding to the non-polymorphic, lipid antigen

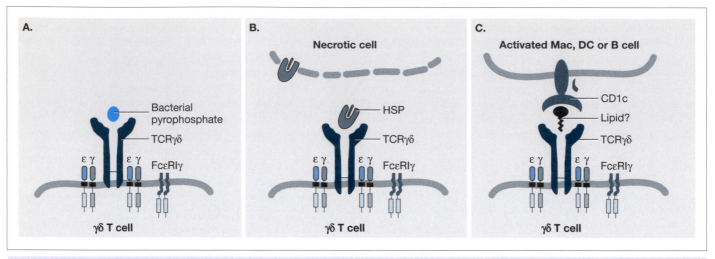

Fig. 11-7 Examples of Antigen Binding to γδ TCRs

Table 11-3 Examples of V Gene Usage and γδ T Cell Antigens Recognized

Species	γδ Subset	Anatomical Location	Antigen Recognized
Human	Vγ2Vδ2	Skin, peripheral blood	Small pyrophosphate epitopes derived from Epstein-Barr virus, *Mycobacterium*, *Plasmodium*, *Leishmania* or *Salmonella*
	Vγ9Vδ2	Skin	HSP 58
Murine	Vγ3	Skin	Stress antigen expressed by skin cells
	Vγ6	Uterus, tongue	Stress antigen expressed by lung epithelial cells

The proliferation of γδ T cells in response to injury or infection is generally very rapid, as befits cells participating in the induced innate response. Indeed, the release of cytokines by γδ T cells can precede the activation of αβ T cells by several days. However, the details of γδ T cell activation remain a mystery. At least *in vitro*, small phosphorylated metabolites are able to activate some γδ T cells without the need for CD28- or CD40-mediated costimulation. There is also evidence that, in sites of inflammation, APCs may be able to help activate γδ T cells by supplying stimulatory cytokines or (unknown) intercellular contacts. Some γδ T cell subsets express NK inhibitory receptors that may influence their activation. Indeed, host cells deficient in MHC class I expression have been shown to activate certain γδ T cell subsets.

IV. EFFECTOR FUNCTIONS

Once activated by antigen, γδ T cells generate effectors in a manner similar to αβ T cells, although the signaling pathways linking TCR stimulation to new gene transcription appear to be slightly different. Some γδ T cells are cytotoxic like αβ Tc cells, developing into CTLs that eliminate infected cells and tumor cells via Fas-mediated cytotoxicity or the secretion of perforin/granzymes or cytotoxic cytokines (Fig. 11-8). Other γδ T cells are like αβ Th cells and secrete cytokines and growth factors that affect other leukocytes. Indeed, Th1 and Th2 subtypes of γδ T cells have been identified based on their secretion of IL-2 and IFNγ, or IL-4, IL-5, IL-6 and IL-10, respectively, although the Th1 subtype predominates. Cytokines produced by activated γδ T cells may then support the differentiation of activated αβ Th0 cells into Th1 or Th2 effectors, and influence isotype switching in B cells.

Other molecules secreted by γδ T cells influence leukocyte trafficking and wound healing. For example, chemokines produced by activated γδ IELs promote the migration of neutrophils and macrophages toward damaged epithelium. Activated γδ IELs can also induce neighboring epithelial cells to produce

presentation molecule CD1c expressed by macrophages, DCs and B cells (Fig. 11-7C). It is unknown what ligand is present in the CD1c binding groove. Lastly, there is some evidence in mice suggesting that certain γδ TCRs can bind directly to at least some MHC class Ib molecules.

As well as stimulation through their γδ TCRs, at least some γδ T cells appear to experience some kind of costimulation through another receptor. Many γδ T cells express the NKG2D NK activatory receptor that binds to the stress antigens MICA and MICB, and MICA and MICB are known to stimulate the responses of certain human TCRγδ-bearing IELs. It remains unclear how signaling through the γδ TCR and NKG2D might cooperate and whether both signals are required for the activation of some γδ T cells.

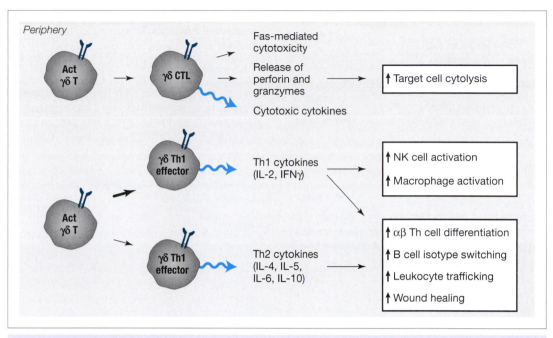

Fig. 11-8 γδ **T Cell Effector Functions**

the antimicrobial compound NO. Activated γδ T cells that have matured in the skin to become DETCs can secrete *keratinocyte growth factor* (KGF), a molecule that stimulates the growth and differentiation of epithelial cells.

Immunological memory does not appear to be a feature of γδ T cell responses although two phases of γδ T cell proliferation are observed in response to some infections. The first phase is a rapid response occurring immediately after infection and prior to the primary αβ T cell response. The second phase is a proliferative burst that occurs 2–3 days after the primary αβ T cell response concludes. However, the γδ T cells in this second phase do not appear to be fully functional memory cells and have only a restricted capacity to develop into second generation effectors. Thus, defense by γδ T cells appears to be stuck at the primary response level during a secondary challenge with antigen.

V. DEVELOPMENT

As introduced in Chapter 9, the first waves of T cells produced in a human or murine embryo are γδ T cells rather than αβ T

cells. Rearrangement of the γδ TCR genes can be detected in thymocytes as early as 8 weeks in the human fetus and by day 12.5 of gestation in the mouse. Under the influence of fetal thymic stromal cells, distinct waves of γδ T cells fan out to populate specific regions of the body. In cells of the very first wave exiting the murine thymus, the rearranged γδ TCR invariably contains Vγ5, and these cells populate the skin as DETCs. Subsequent waves include Vγ6[+] cells heading for the mucosae of the urogenital tract and then Vγ7[+] cells destined to be iIELs. These γδ T cells provide immune defense in the fetus and neonate before adaptive immunity mediated by the more powerful αβ T cells is fully established. In particular, αβ Th1 responses are weak in very young animals. Perhaps not coincidentally, higher frequencies of γδ T cells are present during the neonatal period than in later life, and many of these cells have a Th1 phenotype.

γδ T cells arise from the same NK/T precursor that generates αβ T cells and NK cells. A model of the developmental pathway of thymocytes destined to become γδ T cells in shown in Figure 11-9. At around the DN3 stage of thymocyte development, the *TCRB*, *TCRG* and *TCRD* loci begin to rearrange. At some

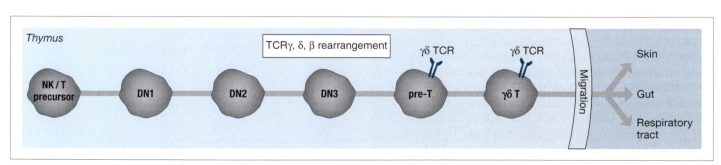

Fig. 11-9 γδ **T Cell Development and Dissemination**

unknown point prior to *TCRA* rearrangement, some developing DN3 cells express a complete γδ TCR that successfully initiates intracellular signaling precluding further *TCRB* rearrangement. Access to the αβ T cell developmental path is blocked, and the pre-TCR expression and β-selection characteristic of αβ T cell development do not occur. These cells eventually leave the thymus as mature γδ T cells and migrate mainly to the epithelial layers of the skin, gut and respiratory tract.

Some γδ T cells are thought to mature *extrathymically*. That is, rather than entering the thymus, certain NK/T precursors may migrate to the mesenteric lymph nodes or the *intestinal cryptopatches* (small regions of epithelial and lymphoid tissue in the small intestine) and may develop into γδ T cells in these sites. The evidence supporting this hypothesis comes from analyses of animals with an abnormal thymus structure, in that normal numbers of functional γδ T cells, but not normal numbers of αβ T cells, are able to develop in these mutants. Thus, it is possible that extrathymic development can compensate to some degree when the normal thymic pathway is compromised. At least in the laboratory, extrathymic γδ T cell development requires IL-7, which is conveniently produced by stromal cells, keratinocytes and gut epithelial cells that are known to colocalize and interact with γδ T cells.

Positive and negative thymic selection appear to occur for γδ T cells maturing in the thymus in a manner similar to the selection that shapes the αβ T cell repertoire, although the molecular details are not understood. At least some γδ T cell clones seem to be positively selected by host cell stress molecules and MHC class Ib molecules expressed on cTECs. Autoreactive γδ T cells are assumed to be negatively selected or anergized in the thymus but the mechanism remains undefined. There is some evidence that positive selection of extrathymic γδ T cell clones does occur but the peripheral host ligands involved are unknown. Similarly, γδ T cell peripheral tolerance appears to exist but the underlying mechanisms are unclear. More complete identification of the physiological ligands of γδ T cells should clarify how selection and tolerance operate for these cells.

C. NKT Cells

I. OVERVIEW

NKT cells are T lineage cells that share physical and functional characteristics with both T cells and NK cells. Morphologically, NKT cells closely resemble conventional T cells, and low numbers of NKT cells are found virtually everywhere T and NK cells are found: in peripheral blood, spleen, liver, thymus, bone marrow and lymph nodes. Following activation, NKT cells can immediately commence cytokine secretion without first having to differentiate into effector cells. The rapidity of their response makes NKT cells important players in the very first lines of innate defense against some types of bacterial and viral infections.

In contrast to the vast diversity of TCRαβ sequences expressed by the αβ T cell population, NKT cells carry a "semi-invariant" TCRαβ. "Semi-invariant" refers to the fact that the TCRα chain is essentially invariant among the NKT cells in a species, whereas the TCRβ chain is diversified. For example, in humans, the TCRs of almost all NKT cells contain an α chain made up of Vα24 plus Jα18. In mice, NKT cells overwhelmingly bear αβ TCRs containing a TCRα chain of the composition Vα14/Jα18. With respect to coreceptors, mouse NKT cells are either $CD4^+CD8^-$ or $CD4^-CD8^-$, whereas human NKT cells are $CD4^+CD8^-$, $CD4^-CD8^+$ or $CD4^-CD8^-$.

II. ANTIGEN RECOGNITION AND ACTIVATION

Rather than interacting with the polymorphic peptide-binding MHC molecules recognized by conventional αβ T cells, the TCRs of NKT cells recognize glycolipid or lipid structures presented on non-polymorphic CD1d molecules expressed by professional and non-professional APCs. The prototypic glycolipid antigen recognized by NKT cells is *α-galactosylceramide* but several other antigens of a similar structure have been identified. As described in Chapter 7, CD1d structurally resembles MHC class I but traffics through the endosomes of the exogenous antigen presentation pathway. The binding groove of the CD1d molecule tethers the lipid tail of a glycolipid antigen, while the carbohydrate head group of the antigen projects out of the groove for recognition by the TCR of the NKT cell. *In vivo*, experimental administration of α-galactosylceramide to a mouse potently activates its NKT cells, protecting against infection with various pathogens and promoting tumor rejection. Although pathogen-derived NKT ligands have been difficult to define, some bacterial glycolipids have been identified that can act as physiological ligands for at least some human and murine NKT cells. As well, there is *in vitro* evidence for the existence of NKT TCRs that recognize host stress glycolipid ligands. Again, however, very few natural mammalian ligands for NKT TCRs have been elucidated. The best known host ligand is a lysosomal glycolipid called *iGb3* that has been shown to stimulate murine NKT cells. Immunologists theorize that self glycolipid antigens such as iGb3 are somehow presented upon microbial infection or during stress, triggering an NKT response.

Human and murine NKT cells express inhibitory and activatory NK receptors, including NKG2D and CD94/NKG2A in mice and humans, and certain KIRs in humans. Thus, the activation of NKT cells, like that of NK cells, may be regulated to a degree by a balance of signaling by activatory and inhibitory receptors.

III. EFFECTOR FUNCTIONS

The characteristics of NKT activation are consistent with a cell type prominent in induced innate immunity. Mature NKT cells activated in the spleen, liver or bone marrow are stimulated to undergo rapid clonal expansion such that their numbers peak within 3 days of antigen encounter. However, these activated NKT cells can immediately carry out their effector functions without the need for differentiation. Thus, NKT cells supply timely and effective defense during the interval needed by

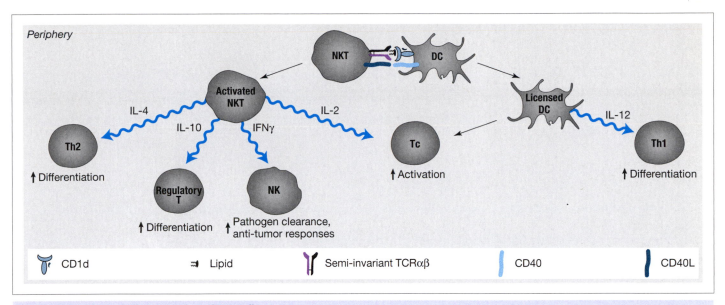

Fig. 11-10 Activation and Effector Functions of NKT Cells

conventional T cells for proliferation and differentiation into the effectors of the more finely tailored adaptive response. By 9–12 days after first encountering the antigen, NKT cell numbers return to their resting levels.

The most important *in vivo* effector function of activated NKT cells appears to be cytokine secretion (Fig. 11-10). NKT cells carry preformed mRNAs for IL-4 and IFNγ and can also rapidly synthesize IL-2 and IL-10, so that massive amounts of these cytokines are produced within 1–2 hours of NKT cell activation. This simultaneous production of arrays of pro- and anti-inflammatory cytokines means that NKT cells may promote or suppress immune responses in a manner that is hard to predict. The early burst of IL-4 produced by activated NKT cells may contribute to the initiation of Th2 differentiation by nearby activated αβ Th0 cells. Similarly, the IL-2 and IFNγ produced by activated NKT cells may help to prime some NK cells and supply help for Tc activation where Th help might be suboptimal. In contrast, the IL-10 secreted by NKT cells has been shown to promote regulatory T cell differentiation, perhaps helping to control immune reactivity. It is important to note that NKT cells are not always beneficial and can sometimes have deleterious effects on a host. For example, the TCRs of some NKT cells recognize the oxidized lipid antigens deposited in the arteries of patients with *atherosclerosis* ("hardening of the arteries"). The activation of these NKT cells and the ensuing inflammatory response can exacerbate the disease.

Activated NKT cells have important effects on DCs and NK cells. Upon interaction of its TCR with CD1d on a DC, the NKT cell upregulates its expression of CD40L. Interaction of this CD40L with CD40 on the DC activates the latter, both licensing it for Tc activation and spurring it to produce IL-12 to promote Th1 differentiation. Exposure to IL-12 also causes the activated NKT cell to selectively increase its secretion of IFNγ. This IFNγ helps to drive the differentiation of Th1 cells and amplifies the NK cell response. The effector actions of

these cells then contribute to the elimination of pathogens and prevent tumor growth and metastasis (see Ch. 16). *In vitro*, activated NKT cells can use FasL- or perforin/granzyme-dependent mechanisms to kill the same types of targets as NK cells (infected cells and tumor cells). However, it remains unclear how relevant NKT cytolytic capacity is *in vivo*.

NKT cells may also play a role in preventing autoimmune disease. Mice that are prone to the development of type 1 diabetes show abnormally low levels of NKT cells, and the NKT cells that are present have functional defects. Interestingly, if diabetes-prone mice are treated in advance with normal NKT cells, the experimental onset of diabetes can be prevented. Whether NKT cell defects occur in humans with type 1 diabetes is currently unclear. Impressive therapeutic results have also been obtained in mouse models of other autoimmune diseases where the animals were treated with NKT cells activated by glycolipid antigens.

IV. DEVELOPMENT

The reader will recall from Chapter 9 that the positive selection of conventional αβ T cells in the thymus involves the interaction of TCRαβ molecules on DP thymocytes with pMHCs on mTECs and cTECs. In contrast, the positive selection of NKT cells involves the interaction of semi-invariant TCRαβ molecules (that are randomly generated by TCR gene rearrangement) on certain DP thymocytes with CD1d on neighboring DP thymocytes (Fig. 11-11). It is currently unclear what self-glycolipid ligands are presented by CD1d to facilitate positive selection. Negative selection of NKT cells can be experimentally induced by engineering an encounter of developing NKT cells with CD1d bound to an experimental agonist ligand, or even with high levels of CD1d in the absence of ligand. It is not yet known if these findings are relevant *in vivo* but, given that NKT cells express a diverse repertoire of semi-invariant TCRs, it is highly likely that NKT cells must undergo negative

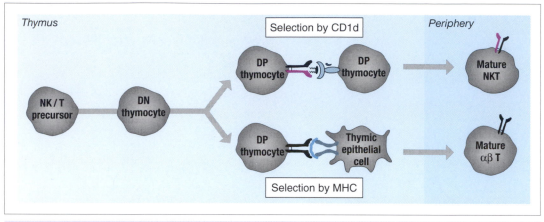

Fig. 11-11 **Selection of NKT Cells versus αβ T Cells during Development**

selection to avoid overt autoreactivity. In any case, once immature NKT cells are positively selected, IL-15 produced by a non-hematopoietic cell type plays an important role in continued NKT maturation. Immature NKT cells migrate from the thymus into the periphery, and particularly into the spleen and liver. In these locations, development continues until mature NKT cells are produced.

Embryologically speaking, mature NKT cells appear relatively late on the scene. In humans, immature NKT cells are present in the fetal and neonatal thymus and cord blood but may not yet be fully competent to secrete cytokines. Indeed, NKT cells do not appear to complete their maturation until they reach the periphery at about 3–5 days after birth. This situation stands in contrast to that of NK cells, which are fully mature and functional in fetal life. Interestingly, mature NKT cells appear to accumulate with age in human peripheral blood.

We have now come to the end of our study of the basic cell types of the immune system, the modes of antigen recognition of these cells, and their effector functions. These various cell types, some of which are elements of innate immunity, some of which strictly mediate adaptive immunity, and others that bridge the innate and the adaptive responses, combine to distinguish self from non-self, and danger from benign circumstance. In Chapter 12, we explore SALT and MALT, where many of these cell types combine to provide frontline defense at the most common ports of pathogen entry.

CHAPTER 11 TAKE-HOME MESSAGE

- NK, γδ T and NKT cells are considered to bridge the innate and adaptive responses because these cells are closely related to the T lymphocyte lineage, recognize antigens in a more general way than do αβ T cells, have direct and rapid effects on pathogens, and influence B cells and αβ T cells by the cytokines they secrete.

- A balance of signaling by activatory and inhibitory receptors expressed on the NK cell surface controls the activation of NK cells. NK activation occurs when the NK cell encounters a target cell exhibiting a relative deficit of self MHC class I.

- NK cells kill virus-infected cells and tumor cells by ADCC or perforin/granzyme-mediated natural cytotoxicity.

- γδ T cells are prominent in the mucosal and cutaneous epithelial layers of the body.

- γδ TCRs bind to non-peptide antigens derived from pathogen proteins and host stress molecules that are either recognized directly or presented on CD1c or non-classical MHC class Ib molecules.

- Activated γδ T cells generate CTLs that kill infected target cells by perforin/granzyme-mediated cytotoxicity, as well as Th effectors that secrete cytokines influencing αβ T cells and other leukocytes.

- NKT cells bear a semi-invariant αβ TCR that recognizes glycolipid and lipid ligands presented on CD1d by APCs.

- NKT cells rapidly secrete large amounts of pro-inflammatory and anti-inflammatory cytokines that allow these cells to enhance some immune responses (e.g., anti-pathogen, anti-tumor) while suppressing others (e.g., autoreactivity). NKT cells can promote Th1 and Th2 differentiation, enhance NK cell functions, and license DCs for Tc activation.

Introduction

1) Why are NK, γδ T and NKT cells considered to bridge the innate and adaptive responses?

Section A.I

1) How is an NK cell similar to a CTL? How is it different?

2) What are the primary functions of NK cells?

3) Name two ways by which NK cells carry out cytolysis of target cells.

4) Do activated NK cells generate separate effector cells? If not, why is this significant?

Section A.II

1) Describe how NK inhibitory and activatory receptors can control NK cell activation.

2) Define "deficit of self" and describe two situations in which it might occur.

3) Describe the structure and ligand of one member of each of three classes of NK activatory receptors.

4) Why do NK activatory receptors have to associate with a signaling chain?

5) Describe the structures and ligands of one member of each of two classes of NK inhibitory receptors.

6) How does an ITIM work?

7) Why are NK cells experts at ADCC?

8) What cytokines are secreted by NK cells?

Section A.III

1) What two molecules are essential for the development of NK cells from NK/T precursors?

2) Why are NK cells considered "self-tolerant"?

Section B.I–II

1) Give three ways in which γδ T cells differ from αβ T cells.

2) What kinds of molecules can serve as ligands for γδ TCRs?

3) How does the anatomical distribution of γδ T cells differ from that of αβ T cells?

4) What are iIELs? DETCs?

5) Give two examples of canonical γδ TCRs.

Section B.III–IV

1) What sorts of pathogens supply antigens for γδ T cells?

2) Give an example of a stress antigen recognized by γδ T cells.

3) What kinds of effector cells are generated by activated γδ T cells?

4) Describe two ways each in which γδ T cells regulate the innate response and the adaptive response.

5) Is memory a feature of γδ T cell responses? If not, what impact does this have?

Section B.V

1) Describe how the TCR loci rearrange in a thymocyte destined to generate mature γδ T cells.

2) What is extrathymic T cell development? Describe two sites in the body where it may occur.

3) Why is our knowledge of γδ T cell selection and tolerance incomplete?

Section C

1) What is a semi-invariant TCR?

2) What ligands are recognized by NKT cells and how are they presented?

3) Do activated NKT cells need to generate separate effector cells in order to function?

4) What are the primary functions of NKT cells *in vivo* and *in vitro*?

5) Give two effects each of NKT-secreted cytokines on the innate and adaptive responses.

6) From what type of thymocyte do NKT cells arise?

7) How are positive and negative selection of NKT cells thought to occur?

12

Mucosal and Cutaneous Immunity

WHAT'S IN THIS CHAPTER?

True realism consists in revealing the surprising things which habit
keeps covered and prevents us from seeing.
Jean Cocteau

Most human pathogens gain access to the body by penetrating its skin or mucosae. In Chapter 2, we introduced the concepts of the MALT (mucosa-associated lymphoid tissue) and SALT (skin-associated lymphoid tissue). The MALT and SALT are made up of collections of APCs and lymphocytes that are located just under the mucosae or skin (Table 12-1). These collections function as independent arms of the immune system that are responsible for *mucosal immune responses* and *cutaneous immune responses*, respectively. Mucosal and cutaneous responses are distinct from the *systemic immune responses* discussed so far in this book, i.e., those responses initiated by lymphocyte–APC interaction in a draining lymph node. Most mucosal and cutaneous immune responses are mounted locally where antigen is first encountered and without involving a draining lymph node. The effector cells that are generated then home specifically to mucosal and cutaneous sites to exert their effector functions. However, an accompanying systemic immune response in which effector cells attack the antigen in other locations throughout the body can be induced when antigen-bearing APCs of the MALT or SALT migrate to the local draining lymph node and activate naïve T and B cells in this location.

A. Mucosal Immunity

I. OVERVIEW

As introduced in earlier chapters, the mucosae are thin layers of epithelial cells that line a body passage such as the gut, respiratory tract or urogenital tract. The mucosae get their name from their capacity to produce *mucus*, a very viscous solution of polysaccharides in water that covers the *apical* (lumen-facing) membrane of an epithelial cell. The mucus contains various secretory antibodies (particularly SIgA and SIgM) and antimicrobial molecules that help protect the mucosae from pathogen invasion. In adult humans, the area of all the mucosae exceeds 400 m^2, a huge expanse that is constantly defended against pathogen assaults by immune responses mounted in the MALT. The MALT includes several subsystems of lymphoid elements associated with each of the body tracts. These subsystems, which differ in structure and cellular composition, are called the GALT, NALT and BALT based on their locations in the gut, nasopharynx and bronchi, respectively.

Scientists working in the field of mucosal immunity speak of *inductive sites* and *effector sites* (Fig. 12-1). An *inductive*

Table 12-1 The "ALTs"

System Name	Subsystem	Definition	Tissue Defended
MALT		Mucosa-associated lymphoid tissue	Mucosae of body tracts
	GALT	Gut-associated lymphoid tissue	Mucosae of small and large intestines
	NALT	Nasopharynx-associated lymphoid tissue	Mucosae of nose, tonsils, throat
	BALT	Bronchi-associated lymphoid tissue	Mucosae of bronchi and bronchioles in the lungs
SALT		Skin-associated lymphoid tissue	Skin on body surface

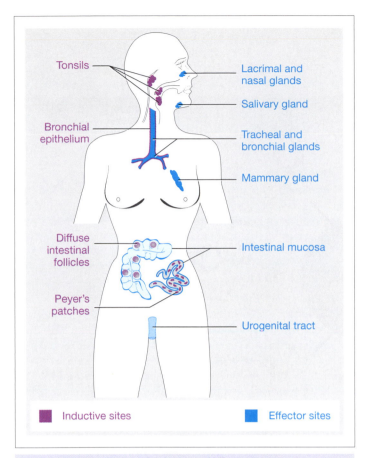

Fig. 12-1 Examples of Mucosal Inductive and Effector Sites

BALT and NALT, the main inductive sites are the bronchial epithelium and the collection of tonsils in the nasopharynx, respectively. A mucosal *effector site* is a specific area in the mucosae to which effector lymphocytes are dispatched after mucosal T and B cells are activated in a given inductive site. These activated lymphocytes migrate through the lymphatics and blood to various effector sites and complete their differentiation into Th effectors, CTLs and plasma cells in these locations. Thus, defense at multiple and widely separated mucosal effector sites may occur in response to lymphocyte activation in one inductive site. For example, when an antigen is captured in a PP (inductive site), an antibody response may be detected not only in the intestinal mucosa but also in the urogenital tract and in tissues as remote as the mammary glands (effector sites). Among the most important mucosal effector sites are the *exocrine glands*, such as the salivary and lacrimal glands. These glands produce protective external secretions (like saliva and tears) that contain antimicrobial molecules and secretory antibodies.

II. THE GUT-ASSOCIATED LYMPHOID TISSUE (GALT)

The integration of various components of the GALT provides immune defense against ingested pathogens and toxins (Fig. 12-2).

i) Basic Structure
ia) Gut epithelium. The gut is designed to both absorb nutrients from food and repel pathogens and noxious substances. These functions depend largely on the single layer of gut epithelial cells that acts as the gut mucosa. The gut epithelium is not flat but rather is folded into repeated *crypt* ("cave") and *villus* ("hill") structures.

A pathogen that arrives in the gut lumen encounters many cellular and soluble barriers that mediate innate and adaptive defense against penetration. The invader first has to compete for nutrients and living space with the billions of commensal

site is a specific area in the mucosae where an antigen is encountered and a primary adaptive response is initiated. Inductive sites in the GALT include organized lymphoid structures in the small intestine called *Peyer's patches* (PPs), the appendix, and diffuse collections of lymphocytes and APCs scattered within and just under the gut epithelium. In the

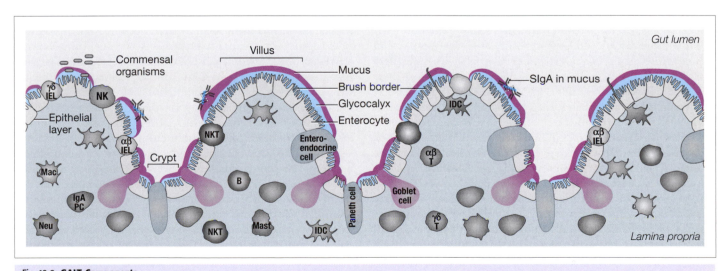

Fig. 12-2 GALT Components
[Adapted from Fagarasan S. and Honjo T. (2002). Intestinal IgA synthesis: regulation of front-line defences. *Nature Reviews Immunology* **3**, 63–72.]

organisms that normally live in the intestinal tract. Many commensal microbes secrete toxins that can seriously impede a pathogen. In addition, pathogens can be trapped by the sticky *mucus* coating the mucosal surface. The mucus is produced by *goblet cells* within the gut epithelium. The goblet cells also synthesize and secrete into the mucus antibacterial molecules such as *lysozyme*, which breaks down cell wall components, and *lactoferrin*, which sequesters iron needed for bacterial growth. Mucus also acquires high concentrations of secretory antibodies by pIgR-mediated transcytosis across epithelial cells (see later and refer to Ch. 4). These antibodies prevent a wide variety of pathogens from establishing a foothold on the epithelial layer.

Underneath the mucus, the exterior surfaces of gut epithelial cells form dense *microvilli* collectively known as the *brush border*. The brush border is coated with the *glycocalyx*, a thick layer of glue-like molecules anchored in the apical membrane. The negative charge of the glycocalyx repels many pathogens. The glycocalyx also contains several types of hydrolytic enzymes that degrade microbes or macromolecules attempting to make contact with the epithelial cells. *Paneth cells*, which are located at the bottom of intestinal crypts, produce antimicrobial proteins, whereas *enteroendocrine cells* in the villi secrete hormone-like molecules with stimulatory effects on surrounding cells. *Enterocytes*, which comprise the majority of cells lining the gut, have a primary function in nutrient absorption. However, enterocytes also express TLRs and can be activated to produce low levels of inflammatory cytokines in response to toxins or PAMPs. These cytokines recruit neutrophils and mast cells from the circulation to the site of infection and promote the activation and differentiation of αβ and γδ iIELs within or near the epithelial layer. Low numbers of NK cells as well as extremely low numbers of NKT cells are tucked into the gut epithelium and may also be stimulated by enterocyte-produced cytokines.

ib) Lamina propria. Beneath the *basolateral* (tissue-facing) surface of the gut epithelium is a loose connective tissue called the *lamina propria* (refer to Fig. 12-2). The lamina propria is home to numerous macrophages and neutrophils as well as low numbers of NKT cells, mast cells and immature DCs (IDCs). Only small numbers of γδ T cells are present and almost no NK cells. However, memory αβ T cells (both CD4[+] and CD8[+]) and memory B cells are abundant. Many of these lymphocytes are diffusely distributed in the lamina propria, while others are organized into *intestinal follicles* (see later). The constant attack by pathogens on the gut mucosa means that, at any one time, about 10–15% of lamina propria B cells have differentiated into plasma cells, the vast majority of which synthesize IgA. The total secretion of antibody by these IgA-producing plasma cells (IgA PCs) outstrips the combined output of antibodies synthesized by plasma cells in the spleen, bone marrow and lymph nodes.

ii) Antigen Sampling

iia) Intestinal follicles and the follicle-associated epithelium (FAE).

At inductive sites in the gut, lymphocytes are usually organized into *intestinal follicles*. Intestinal follicles may occur singly, as can be found scattered along the entire length of the intestine, or in small groups, or in larger groups of 30–40, as occur in the PPs and appendix. Where follicles are grouped, they lie directly under small flat sections of gut epithelium called *follicle-associated epithelium* (FAE) that are specialized for the capture of gut antigens (Fig. 12-3). The vast majority of cells in the FAE are enterocytes but 10–20% of FAE cells are large, odd-shaped cells called *M cells* ("membranous" or "microfold" cells) that are experts at antigen transcytosis. The region between the M cell and the underlying intestinal follicles is called the *dome*. Within the dome are populations of APCs that take delivery of antigen transcytosed by the M cell. The follicles beneath the FAE are separated from one another by *interfollicular regions* that contain high concentrations of mature αβ T cells surrounding an HEV.

The apical surface of an M cell lacks the glycocalyx and thick brush border present on enterocytes, allowing the M cell to easily internalize antigens either by macropinocytosis, clathrin-mediated endocytosis or phagocytosis (Fig. 12-4). Once the antigens are transcytosed across the cytoplasm of the M cell, they are released into an *intraepithelial pocket* created by the invagination of the M cell basolateral membrane. The antigens can then be taken up by APCs in the dome. As well as IDCs and macrophages, the dome contains CD4[+] and CD8[+] αβ T cells, resting B cells, and T$_{reg}$ and Th3 regulatory T cells. The intestinal follicle itself is made up of a GC containing activated B cells and FDCs.

iib) GALT DCs.

Several subsets of conventional IDCs have been defined in the murine GALT, and different subsets appear

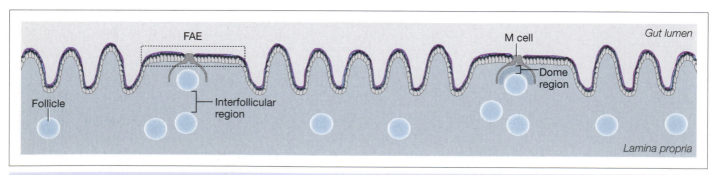

Fig. 12-3 Intestinal Follicles and the Follicle-Associated Epithelium (FAE)

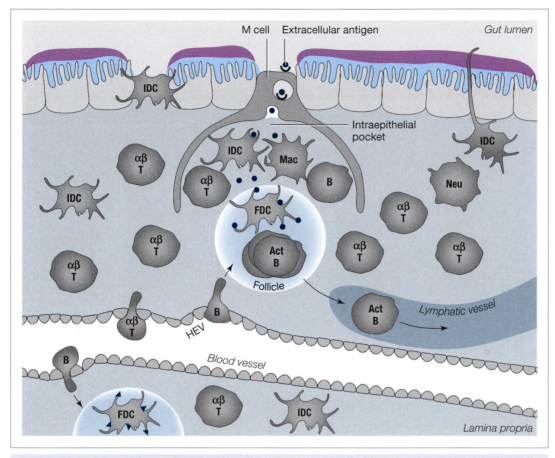

Fig. 12-4 **Uptake of Antigen by M Cells in the FAE**

to have slightly different functions. Among these are two IDC subsets distinguished by their expression of two particular chemokine receptors. IDCs that express CX_3CR1, which binds to the chemokine fractalkine ($Cx3CL1$), are found throughout the gut, including in the regular villi, the FAE regions, and the domes in the PPs. As well as receiving antigen by M cell-mediated transcytosis, CX_3CR1^+ DCs can participate directly in gut antigen sampling by extending cellular processes, called *transepithelial dendrites*, between epithelial cells into the gut lumen. These dendrites routinely capture soluble food antigens as well as commensal organisms and any pathogens present. IDCs that express CCR6, a chemokine receptor that binds to the chemokine CCL20, are found only in the domes of the PPs. These DCs acquire antigen solely by M cell-mediated transcytosis.

Under steady state (non-threatening) conditions, most GALT IDCs function to tolerize naïve T cells with the potential to respond to commensal organisms or food antigens, either by inducing the anergy or death of these cells, or by inducing regulatory T cell differentiation. When significant inflammation is present and the PRRs of GALT IDCs in the site of attack are engaged, these DCs commence maturation and activate mucosal T cells directed against the relevant antigens. Maturing DCs may also migrate from the dome in the GALT to the draining mesenteric lymph node, where the antigens they bear may be used to activate naïve T and B cells that can mount a

systemic response. Plasmacytoid DCs present in the PPs can contribute to defense of the gut by secreting IFNα/β.

III. THE NASOPHARYNX- AND BRONCHI-ASSOCIATED LYMPHOID TISSUES (NALT AND BALT)

i) Basic Structure

The NALT and BALT provide immune defense against dubious substances and pathogens in inhaled air (Fig. 12-5). The NALT comprises the nasal submucosal glands, the tonsils, and the epithelial layers lining the nasopharynx (upper airway). The components of BALT include the bronchial submucosal glands; the epithelial layers lining the trachea, the bronchi and the lungs; and the follicles and diffuse collections of lymphocytes underlying the epithelium of the lower airway. Depending on location, the epithelial layers of the respiratory tract show variation in their structure. Some regions are covered by a single layer of epithelial cells (like the gut), while others are covered in several epithelial layers arranged to form "stratified" epithelium.

The hairs in the nose constitute the first physical barrier against harmful entities attempting to breach the NALT. As well, the sweeping movement of cilia on epithelial cells in the upper respiratory tract and coughing in the throat can help to clear invaders attempting to gain a foothold. The mucus coating the nasopharyngeal epithelial cells not only physically

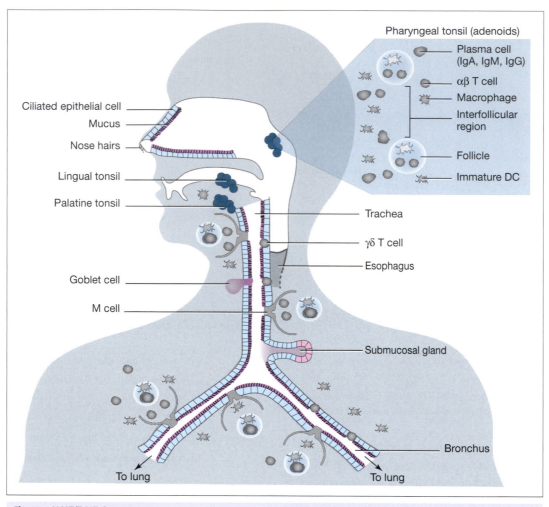

Fig. 12-5 **NALT/BALT Components**

traps microbes but also contains SIgA, lactoferrin, lysozyme and other antimicrobial molecules that can damage a pathogen. The nasopharyngeal epithelial cells themselves can secrete inflammatory cytokines (IL-1, TNF and IL-6), growth factors and chemokines that summon cells of the innate response to the site of assault. In the BALT, sweeping by cilia on epithelial cells in the lower respiratory tract pushes back invaders. Bronchial and bronchiolar epithelial cells are also important sources of IL-5, IL-6, IL-10 and TGFβ, cytokines that promote B cell isotype switching to IgA and plasma cell differentiation.

In the human NALT, the tonsils are the most important inductive sites. A tonsil is composed of a network of reticular cells that supports lymphoid follicles and interfollicular tissue. The follicles contain prominent GCs featuring FDCs and numerous B cells. The interfollicular tissue contains large numbers of αβ T cells and professional APCs. Interestingly, the tonsils also appear to serve as effector sites in the NALT, since high concentrations of fully mature plasma cells that express IgG as well as IgM and IgA can be found in these structures. In the BALT, many of the inductive sites are structurally similar to those in the gut. Groups of mucosal follicles that resemble PPs and underlie M cells can be found in the lamina propria underlying the bronchial airway epithelium.

ii) Antigen Sampling

Antigen uptake in the airway is thought to be easier than that in the GALT because the respiratory tract lacks the harsh degradative enzymes and low pH of the gut. In areas of respiratory tract lined by a single epithelial layer, antigen sampling can be carried out by M cells. In the tonsils and regions of the airway covered by stratified epithelium, antigens are conveyed to inductive sites by IDCs that can extend their processes through the epithelial layers to the luminal surface to acquire inhaled antigens. These IDCs commence maturation and return to the underlying diffuse lymphoid tissues in the lamina propria to initiate a mucosal response, or migrate to the more distant draining lymph nodes to initiate a systemic response.

IV. IMMUNE RESPONSES IN THE GALT, NALT AND BALT

i) Character of Mucosal Responses and the Influence of DC Subsets

The mucosae are inherently fragile structures that can easily be injured by the products and actions of cells activated during inflammation. For example, the constant mounting of vigorous inflammatory responses in the gut, with their accompanying production of TNF and IFNγ, could seriously damage the intestinal villi. For this reason, mucosal immunity depends to a large extent on SIgA, an antibody that coats mucosae throughout the body and can protect against attack without inducing inflammation. Potential invaders include commensal bacteria in the lumens of the body tracts, which do not normally harm the host but must be prevented from penetrating below the epithelium. The constant production of SIgA directed against these commensals protects the mucosa and maintains homeostasis under steady state conditions. As detailed in Chapters 4 and 5, the production of this SIgA depends on isotype switching to IgA in mucosal B cells, an event that requires the Th2 cytokines IL-4 and IL-10 as well as TGFβ.

Most immunologists now believe that the nature of mucosal DC subsets may inherently skew the differentiation of the Th0 cells that they encounter. For example, studies *in vitro* have shown that one subset of IDCs that resides in the dome region of the murine PPs secretes high levels of IL-10. Researchers believe that naïve mucosal Th0 cells interact most often with this dome population of IDCs and thus are induced to generate either regulatory Th3 and Tr1 cells producing TGFβ and IL-10, or mucosal Th2 effectors that produce IL-4, IL-5 and IL-10. In the presence of this particular cocktail of cytokines, antigen-activated B cells in the intestinal follicles undergo isotype switching to IgA and Th1 responses and IFNγ production are suppressed. Effective protection is provided without overt inflammation.

When an aggressive pathogen attacks the gut, a Th1 response may be needed to help contain the invasion. Under these circumstances, a second IDC subset found in the interfollicular areas of the murine PPs comes into play. Interfollicular PP DCs that acquire antigen in the presence of danger signals furnished by the invader preferentially produce IL-12, inducing locally activated mucosal T cells to differentiate into Th1 effectors. Some of the maturing interfollicular PP DCs may also migrate to the draining lymph node and activate naïve T cells in this location, inducing a systemic Th1 response against the pathogen. Interestingly, some mature interfollicular PP DCs may induce the expression of CCR9 on at least some of the T cells in the lymph node. The expression of CCR9 causes these activated T cells to home preferentially to mucosal lymphoid tissues and join mucosal T cells in the battle. In this way, Th1 and Th2 mucosal and systemic responses to the pathogen can all be mounted. A similar scenario is believed to hold for responses in the BALT and NALT.

ii) Mucosal Production and Function of Secretory Antibodies

Secretory antibodies are the principal weapons that prevent pathogens and toxins from penetrating the mucosae. These antibodies are key components of mucus and other body secretions such as saliva and tears. Although SIgM is present in some body secretions, the vast majority of secretory antibodies are SIgA. What is the source of this SIgA? Most of the body's activated B cells are located near the mucosae and the exocrine glands, and 80% of these B cells produce polymeric sIgA that is converted to SIgA (refer to Ch. 4). The body's production

of SIgA far exceeds that of any other isotype, with about 2–3 grams of SIgA synthesized in the average adult human gut every day.

Secretory IgA has several features and functions that make it ideal for mucosal defense. Firstly, SIgA is naturally localized on the surface of the mucosa, making this antibody a superb neutralizer of pathogens or toxins trying to make contact with epithelial cells. Secondly, independent of antigenic specificity, the carbohydrate moieties of SIgA molecules can bind to adhesion molecules expressed by many pathogens, trapping the invaders on the luminal surface. Thirdly, at least in the gut, about half of all SIgA antibodies are unusually cross-reactive, meaning that a broader range of threats can be countered with fewer antibodies. Fourthly, SIgA is not an efficient activator of complement, so that there is less chance of activating the cascade and initiating damaging inflammation. Fifthly, SIgA is highly resistant to a wide variety of host and microbial proteases, including those in the mammalian gut.

The series of events underlying the protection of the mucosae by SIgA is illustrated in Figure 12-6. Foreign antigen taken up across the epithelial barrier in an inductive site is processed so as to activate both antigen-specific B cells and αβ T cells in the local MALT. The surrounding cytokine milieu induces B cell proliferation and isotype switching to IgA, after which these B cells (IgA B) circulate in the lymph and blood, home to mucosal effector sites, and extravasate through local HEVs. In these locations, the IgA B cells complete their differentiation into mature IgA-producing plasma cells. The plasma cells secrete polymeric IgA in the vicinity of the basolateral surface of the epithelial cells lining the gut. As described in Chapter 4, the polymeric IgA is bound by pIgR expressed on the mucosal epithelial cells present in effector sites and is transcytosed across the cell (inset of Fig. 12-6). Upon exocytosis, the pIgR is cleaved such that the secretory component remains attached to the antibody, resulting in release of SIgA into the mucus.

iii) CTL Responses

If a pathogen avoids trapping by mucus or binding by secretory antibodies and penetrates the mucosa, it may be captured by APCs and its antigens processed to activate naïve Tc cells that are resident in mucosal inductive sites. In PPs (at least), Tc cells are found in the interfollicular areas surrounding the intestinal follicles. M cells in the FAE overlying the follicles may also capture pathogen antigens, or the pathogen may replicate and its progeny may infect nearby epithelial cells. Mucosal DCs in the dome can acquire pathogen antigens from M cells or from necrotic or apoptotic epithelial cells and cross-present pMHCs that activate antigen-specific Tc cells. Activated Tc cells leave the inductive site and antigen-specific CTLs appear in effector sites, although it is unclear precisely where CTL differentiation takes place and whether a draining lymph node is involved.

iv) A Common Mucosal Immune System

Pathogen invasion at one location in the intestine can lead to SIgA protection not only of mucosae along the entire length of the gut, but also in the respiratory tract, salivary glands, lacrimal glands, ocular tissue, middle ear mucosa and even the

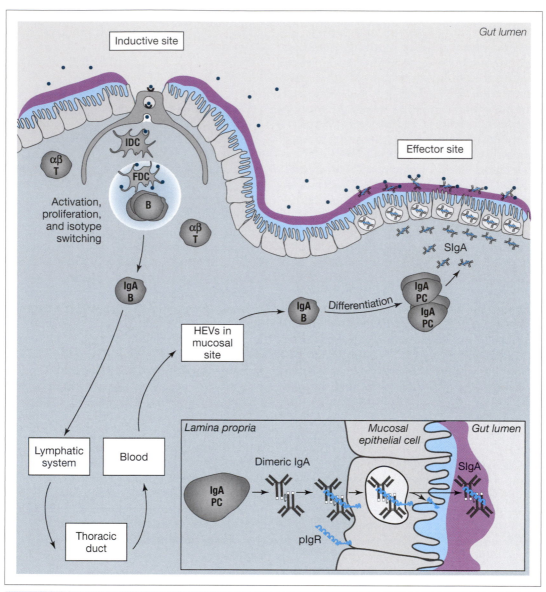

Fig. 12-6 **From Antigen Uptake to Secretory IgA Production**

lactating mammary glands. Similarly, antigen introduction intranasally can result in detectable antigen-specific SIgA in the saliva, tonsils, trachea, lung and gut. This disseminated protection has given rise to a concept called the *common mucosal immune system* (CMIS), in which the migration of mucosal B cells from an inductive site through the blood and lymphatic system to several effector sites is governed by shared expression of mucosal homing receptors. These receptors, which differ from those expressed by conventional B cells activated in the lymph nodes, bind to addressin proteins exclusively expressed in mucosal effector sites. For example, as shown in Figure 12-7, a conventional B cell bearing the α4β1 homing receptor circulates systemically and binds to the addressin VCAM-1 expressed by activated endothelial cells in sites of inflammation. In contrast, a mucosal B cell bearing the α4β7

receptor ignores these sites and instead binds to MAdCAM-1 expressed by endothelial cells in mucosal effector sites. In the case of mucosal T cells, the expression of the α4β7 receptor, as well as that of the chemokine receptor CCR9 (which binds to the chemokine TECK secreted by mucosal epithelial cells), is induced specifically by interaction with mucosal DCs. These T cells are then drawn to mucosal effector sites.

Interestingly, the immune responses induced at different mucosal effector sites are not of uniform strength, being strongest at those sites closest to the inductive site or in tissues sharing lymph drainage. For example, if the PPs in the GALT are the inductive site, a strong antibody response will be detected in effector sites in the nearby small intestinal mucosa (GALT) but only a weak response will be observed in the more distant tonsils (NALT).

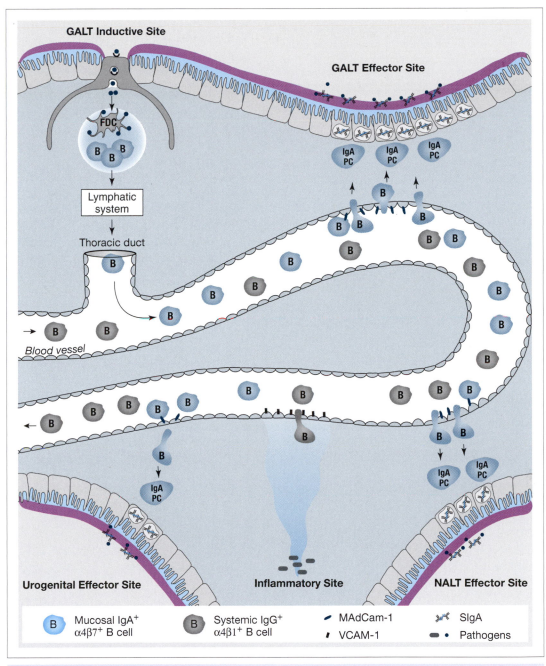

Fig. 12-7 **A Common Mucosal Immune System**

V. IMMUNE RESPONSES IN OTHER MALT

i) MALT in the Urogenital Tract

The vagina differs from the GALT and NALT/BALT in that it lacks the organized lymphoid structures typically present in MALT inductive sites. Intraepithelial DCs and macrophages occur in the cervical and vaginal epithelium but M cells and lymphoid follicles are absent. Thus, introduction of an antigen into the vagina promotes neither mucosal nor systemic immune responses. This lack of responsiveness is evolutionarily desirable because an immune response to incoming sperm could block reproduction and thus species survival. However, mucosal IgA responses can be detected in the murine vagina following immunization via the intranasal, intragastric or even intramuscular routes, so that the CMIS offers protection from pathogen attack. Secretory IgA can be found in vaginal secretions, confirming that the vagina is a mucosal effector site.

The penile urethra is both an inductive and effector site. Epithelial cells lining the penile urethra express pIgR and the lamina propria underlying the cuurethral mucosa contains many IgM- and IgA-secreting plasma cells. As a result, high concentrations of SIgA and SIgM can be found in the secretions coating the urethral mucosa. IgG-producing plasma cells may also be present in the lamina propria. Mucosal DCs reside among the urethral epithelial cells, whereas macrophages and large populations of CD4+ and CD8+ memory T cells are found in the lamina propria.

ii) MALT in the Ear

The middle ear cavity is lined with a thin covering of mucus that overlies a mucosa made up of several types of secretory, ciliated and non-ciliated epithelial cells. The mucus is constantly conveyed toward the Eustachian tube and nasopharynx by the beating of the cilia on the ciliated epithelial cells. This action helps to keep the middle ear cavity sterile because the tide of organisms trying to access the middle ear from the nasopharynx is continually swept backward. The antimicrobial molecules present in the mucus also take their toll on the invaders. Comparatively few pathogens access the middle ear from the exterior through the auditory canal, and those that attempt it are usually thwarted by the tough keratinized layer of squamous epithelium covering the exterior side of the tympanic membrane.

There are very few organized lymphoid structures or cells in a healthy middle ear cavity, meaning that it is not an inductive site. However, when infection occurs, the cavity becomes a mucosal effector site, complete with local production of SIgA. Antigen-specific SIgA, antibodies of other Ig isotypes, and pro-inflammatory cytokines such as IL-1, IL-6 and TNF can be detected in the middle ear fluid of infected individuals.

iii) MALT in the Eye

The conjunctiva and anterior ocular surface of the eye are particularly delicate tissues. Thus, as discussed in Chapter 10, the eye is an immune-privileged site in which immune responses and inflammation are generally discouraged. Cells and macromolecules cannot readily pass through the walls of the blood vessels supplying the eye, and the eye is not connected to a draining lymph node. Antigens that do access the eye are captured by intraocular APCs (including specialized subsets of DCs) that have been influenced by the high concentrations of TGFβ present in the aqueous humor of the eye to promote Th2 development. The intraocular APCs migrate from the eye into the blood and thence to the spleen, where lymphocyte activation occurs. Effector Th2 cells home back to the eye where they support non-inflammatory antibody responses.

B. Cutaneous Immunity

Cutaneous immunity protects the skin from damage caused by infection or injury. The immune system elements that underlie this defense are collectively known as the SALT (skin-associated lymphoid tissues).

I. COMPONENTS OF THE SALT

As introduced in Chapter 2, the skin is composed of the *epidermis*, the *dermis* and *hypodermis* (Plate 12-1 and Fig. 12-8). The epidermis is not vascularized and is separated from the underlying dermis by the *basement membrane*. The dermis contains both lymphatic and blood vessels. Beneath the dermis is the hypodermis, a fatty layer that provides passive barrier defense and support for lymphatics and blood vessels. However, the hypodermis functions chiefly as an energy source and so will not be discussed further here.

i) Epidermis

ia) Keratin layer. The tough outer layer of the skin that resists penetration by inert stimuli as well as by microbes is made up of filaments of a resilient, fibrous protein called *keratin*. Keratin is produced by specialized squamous epithelial cells called *keratinocytes*, which comprise over 90% of the cells in the epidermis. The epidermis is divided into several stratified layers, the outermost representing the oldest keratinocytes. New keratinocytes are constantly being produced from beneath in the lower layers of the epidermis such that the skin eternally renews itself from the inside out. Keratinocytes are generated in organized waves, with the cells in each wave being physically connected by specialized intracellular junctions known as *desmosomes*. The desmosomes ensure the formation of regimented horizontal layers of keratinocytes that divide and migrate upward as a unit. As they age and are pushed up to the skin surface by younger cells beneath them, the older keratinocytes increase their production of keratin fibrils. As

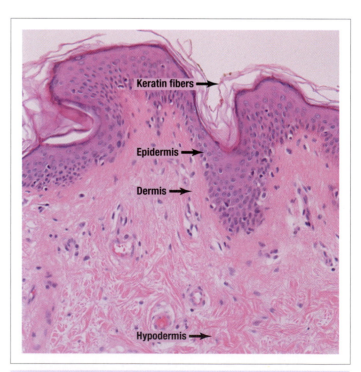

Plate 12-1 **The Skin**
[Reproduced by permission of Danny Ghazarian, Princess Margaret Hospital, University Health Network, Toronto.]

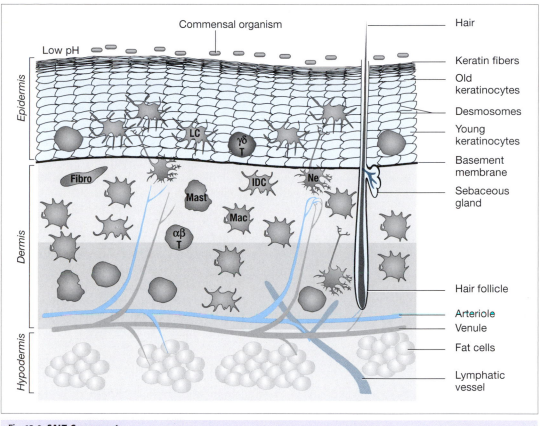

Fig. 12-8 **SALT Components**

the keratinocytes approach the skin surface, their nuclei disintegrate and their lysosomes burst, releasing contents that both kill the cell and polymerize the keratin into a thick, inanimate layer. The most exterior layers of keratinized shells are eventually lost as flakes of dead skin. This constant turnover of the keratinocytes prevents microbes from becoming entrenched.

In addition to the physical barrier thrown up by the keratin layer, billions of commensal organisms accumulate on the skin surface. These organisms compete with pathogens for both space and nutrients and secrete antimicrobial substances to which they themselves are resistant. Among these substances are lipases that break down fats in the skin into free fatty acids, thereby reducing the pH of the skin surface and discouraging pathogen replication. The acidity of the skin is also maintained by sebum produced by the *sebaceous glands* originating in the dermis.

ib) Lower epidermis. Below the keratin layer lie the differentiating strata of living keratinocytes. Below the keratinocytes, just above the basement membrane, are relatively small numbers of αβ and γδ epidermal T cells and immature skin DCs known as *Langerhans cells* (LCs). (B cells are not generally found in skin.) The LCs acquire antigen by infiltrating their long, slender processes between keratinocytes to capture antigens that have penetrated the keratin layer. As well as

synthesizing keratin, the keratinocytes constitutively secrete low levels of growth factors and cytokines that promote the survival and activation of the LCs and T cells. Keratinocytes also constitutively express several TLRs and rapidly release inflammatory cytokines and chemokines in response to TLR engagement. Innate cells responding to these cytokines produce IFNγ, which promotes LC maturation. Because LCs express CD1 proteins as well as MHC class I and II, they are ideal presenters of both peptide or glycolipid antigens to αβ or γδ epidermal T cells, respectively.

ii) Basement Membrane

The youngest keratinocyte layer of the epidermis is separated from the underlying dermis by the basement membrane. The basement membrane is composed of collagen and other molecules produced by epidermal keratinocytes in combination with fibronectin produced by dermal fibroblasts. Because there are no blood vessels in the epidermis, the nutrients required to sustain the keratinocytes must exit the circulation in the dermal blood vessels and diffuse across the basement membrane. Leukocytes that access the epidermis, including the T cell and LC populations, also migrate from the dermal blood vessels. The migrating cells secrete enzymes that dissolve small regions of the basement membrane, allowing passage of the leukocytes into the epidermis.

iii) Dermis

Compared to the tightly packed cells of the epidermis, the dermis is a much roomier mixture of structural fibers, blood vessels, lymphatics and low numbers of cells. Nerve fibers also crisscross the dermis, stretching up through the basement membrane. Collagen fibers provide structural support for the skin, whereas elastin gives skin its resilience and hyaluronic acid traps water molecules and keeps the skin taut and moist. Collagen, elastin and hyaluronic acid are synthesized by dermal fibroblasts. The dermis also contains neurons and leukocytes such as macrophages, mast cells, dermal IDCs (which are distinct from LCs), and αβ memory T cells. Leukocytes access the dermis by extravasating through the endothelial cell layer lining the dermal post-capillary venules. Macrophages are scattered throughout the dermis, whereas dermal T cells and mast cells cluster around the arterioles and venules penetrating the dermis. When a pathogen attacks, the endothelial cells of the dermal post-capillary venules secrete chemokines and express vascular addressins that promote the extravasation of additional leukocytes into the dermis.

II. IMMUNE RESPONSES IN THE SALT

The sequence of events leading to innate and adaptive responses in the SALT is illustrated in Figure 12-9. A pathogen that breaches the outer keratinized layer of the skin and penetrates

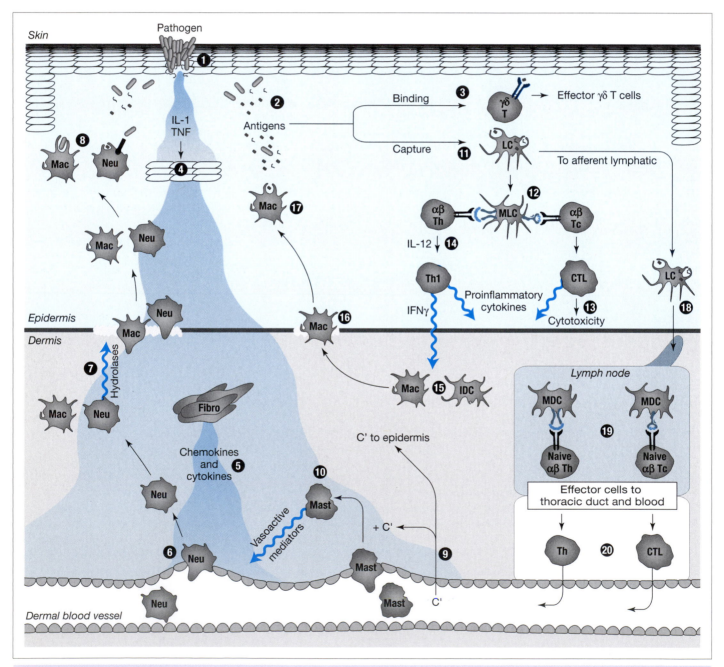

Fig. 12-9 **Immune Responses in SALT**

into the epidermis causes damage to living keratinocytes (#1). Antigens are supplied either from the keratinocytes themselves and/or from the microbes leaking through the breached keratinized layer (#2). γδ T cells immediately recognize common skin stress or bacterial antigens and are activated upon engagement of their TCRs (#3). Damage to a keratinocyte's membrane also induces that cell to release IL-1 and TNF. These inflammatory cytokines induce other keratinocytes to synthesize and release chemokines, growth factors and additional cytokines within the epidermis (#4). These molecules establish a chemokine gradient that diffuses down through the basement membrane into the dermis. Dermal fibroblasts respond with the synthesis of additional inflammatory cytokines and chemokines (#5). Some of these proteins reach the endothelial cells of the dermal blood vessels, promoting local vasodilation and selectin expression. The extravasation of leukocytes, particularly neutrophils, into the dermis is thus facilitated (#6). Once in the dermis, neutrophils produce hydrolases that degrade the basement membrane (#7), allowing leukocytes to penetrate into the epidermis. As neutrophils and macrophages enter the site of assault, they are activated by the presence of pro-inflammatory cytokines and upregulate their PRRs (particularly TLRs) and phagocytic receptors (#8). These cells then internalize any pathogens present and often deploy the respiratory burst to kill them. Phagocytosis may be enhanced if the pathogens are opsonized by complement components (C') that have diffused from local dermal blood vessels into the site of attack (#9). These complement components may also trigger the degranulation of mast cells (#10), which release substances that increase local blood vessel dilation and sustain inflammation. The importance of the cutaneous innate response to defense of the skin has recently been highlighted by the discovery of a new class of drugs that binds to certain TLRs and enhances cutaneous immunity to both viruses and tumors.

The adaptive response in the SALT is initiated when antigens released from dying keratinocytes and microbes are taken up by LCs (Fig. 12-9, #11). If the cytokine milieu is rich enough, the LCs mature within the epidermis (mature LC; MLC) and present peptides derived from stress or bacterial antigens to epidermal αβ Th and Tc cells (#12). Because these αβ T cells are primarily memory cells that have homed to the skin, the response is almost as rapid as that of the γδ T cells. Within 24 hours, the memory αβ T cells commence differentiation into CTLs and Th effectors. Cells that have internalized the pathogen and thus display its antigenic peptides on MHC

class I are destroyed by antigen-specific CTLs (#13). If there are high local concentrations of IL-12 in the site of assault, the differentiating Th effectors are biased toward Th1 development (#14). In contrast to the gut, Th1 responses are well tolerated by the skin due to its inherent toughness.

With the activation of the epidermal T cells, more and more pro-inflammatory cytokines are secreted. IFNγ and bacterial products diffuse from the epidermis into the dermis and activate dermal macrophages and IDCs (Fig. 12-9, #15). Activated dermal macrophages produce enzymes that degrade the basement membrane, making it easier for later waves of leukocytes to access the epidermis. The dermal macrophages themselves may cross into the epidermis (#16) under the influence of chemokines secreted by LCs, and undertake vigorous phagocytosis in this location (#17). If the response becomes prolonged, the IFNγ secreted by epidermal Th1 cells will drive the macrophages to become hyperactivated and gain enhanced microbicidal powers.

The traffic across the basement membrane can also go the other way. LCs bearing antigen may enter the dermis from the epidermis and access lymphatic channels leading to the local draining lymph node (Fig. 12-9, #18). Naïve T cells in the node may then be activated (#19) and mount a response to the antigen. The effector Th cells and CTLs generated in the node are induced to express the homing receptor *cutaneous lymphocyte antigen* (CLA). Thus, once the effector cells are released into the blood, they are directed back to inflammatory sites in the dermis (#20).

Should the CTL, Th1 effector and macrophage responses in a local skin site not prove sufficient to contain the pathogen, a switch may be made to Th2 conditions that promote a humoral response. Hyperactivated macrophages that fail to dispose of an invader start to produce more IL-10 than IL-12, and dermal mast cells contribute large amounts of IL-4. In the continuing presence of antigen, Th2 effectors that are generated from epidermal T cells migrate to the draining lymph node and interact with antigen-stimulated B cells in this location. Plasma cells are produced that secrete antibodies into the blood. The blood circulation eventually carries these antibodies back to the dermis and the site of attack.

We have now covered all the basic elements of the immune system and have described their roles in innate and adaptive immune responses. In Chapter 13, we present a discussion of the major classes of pathogens and how each is dealt with by the mammalian immune system.

- The MALT and SALT arms of immunity mount mucosal and cutaneous responses, respectively, to protect the linings of the body tracts and the skin.

- Most mucosal and cutaneous immune responses are initiated locally rather than in a draining lymph node and the effector cells produced home to mucosal and cutaneous sites. A systemic response can be induced when APCs of the MALT or SALT migrate to the draining lymph node and activate naïve T and B cells in this location.

- Passive anatomical barriers, SIgA-containing mucus, and intraepithelial leukocytes provide initial defense in the MALT.

- M cells in the FAE of inductive sites capture pathogens and convey them to mucosal APCs and T cells residing in the dome covering the B cell-containing lymphoid follicles. DCs may also capture antigen by extending processes through the epithelium into a tract lumen.

- Effector lymphocytes migrate to multiple mucosal effector sites, including the exocrine glands, to establish a common mucosal immune response.

- Immune responses in the MALT are generally biased toward SIgA production rather than inflammation so that damage to the relatively fragile mucosae is avoided. Aggressive pathogens requiring inflammation for control can trigger Th1 responses. Different DC subsets may determine the types of responses mounted.

- The SALT consists of diffuse collections of APCs and T cells in the epidermis and dermis. B cells are not prominent in the SALT.

- Keratinocytes provide a physical barrier and secrete pro-inflammatory cytokines and chemokines that mobilize and activate phagocytes and LCs in the lower epidermis.

- Pathogen antigens captured by LCs either activate epidermal memory T cells or are conveyed through the basement membrane and lymphatics to naïve T cells in the local lymph node. Effector T cells are generated that home back to the site of pathogen attack in the skin.

- The skin is tough enough to support Th1 responses, at least initially. However, inflammation and cellular activation triggered by a persistent pathogen may result in an ongoing and damaging response by hyperactivated macrophages.

Introduction, Section A.I

1) Can you define these terms? MALT, SALT, GALT, BALT, NALT, apical

2) How do mucosal and cutaneous immune responses differ from systemic immune responses? Are they mutually exclusive?

3) What is mucus?

4) Distinguish between mucosal inductive sites and effector sites. Give two examples of each.

Section A.II

1) Can you define these terms? villus, crypt, brush border, glycocalyx, basolateral, FAE.

2) Describe four elements of non-induced innate defense in the gut mucosae.

3) What are the functions of goblet cells? Paneth cells? Enteroendocrine cells? Enterocytes?

4) What cell types contribute to the IEL population in the gut?

5) Describe the structure and cellular components of the lamina propria.

6) Describe the structure and function of M cells.

7) What is a Peyer's patch?

8) What is a transepithelial dendrite and what does it do?

Section A.III

1) What body structures are major components of the NALT? The BALT?

2) Describe four elements of non-induced innate defense in the NALT.

3) What is a tonsil and what does it do?

4) Describe two methods of antigen sampling in the airway.

Section A.IV

1) Give four reasons why SIgA production is the major means of humoral defense in the MALT.

2) Describe how antigen uptake leads to the appearance of SIgA in the body's secretions.

3) What is the common mucosal immune system?

Section A.V

1) Why does the introduction of an antigen into the vagina generally promote neither a mucosal nor systemic immune response?

2) Why is the vagina considered a mucosal effector site?

3) Is the penile urethra an inductive site or an effector site?

4) Describe three elements of mucosal immunity in the middle ear.

5) How is the spleen involved in immune defense of the eye?

Section B.I

1) Can you define these terms? epidermis, dermis, hypodermis, keratin, desmosome, basement membrane.

2) How do dead keratinocytes contribute to immune defense?

3) Besides keratin, what are two other elements involved in non-induced innate defense of the skin?

4) Describe the localization and function of Langerhans cells.

5) How does the structure and cellular composition of the dermis differ from that of the epidermis?

Section B.II

1) Describe the roles of living keratinocytes and dermal fibroblasts during immune responses in the SALT.

2) Describe the localization and roles of $\gamma\delta$ T cells and LCs during immune responses in the SALT.

3) Why are Th1 responses tolerated well in the SALT but not the MALT?

4) Give two functions of dermal macrophages that may occur during immune responses in the SALT.

5) What is CLA and why is it important?

PART II
CLINICAL IMMUNOLOGY

13

Immunity to Infection

If one way be better than another, that you may be sure is Nature's way.
Aristotle

A. The Nature of Pathogens and Disease

Infectious diseases, which cause about 14 million human deaths annually, are caused by six types of pathogens: *extracellular bacteria, intracellular bacteria, viruses, parasites, fungi* and *prions*. Bacteria are microscopic, single-celled, prokaryotic organisms. Extracellular bacteria do not have to enter host cells to reproduce, whereas intracellular bacteria do. Viruses are submicroscopic, acellular particles that consist of a protein coat surrounding an RNA or DNA genome. To propagate, a virus must enter a host cell and exploit its protein synthesis machinery. Parasites are eukaryotic organisms that take advantage of a host for habitat and nutrition at some point in their life cycles. Parasites often damage a host but kill it only slowly. Parasites may be tiny, single-celled *protozoans* or large, multicellular *helminth worms*. Fungi are eukaryotic organisms that can exist comfortably outside a host but will invade and colonize that host if given the opportunity. Fungi may be single-celled or multicellular. Prions are infectious proteins that cause neurological disease by altering normal proteins in the brain of the infected host.

Infection occurs when an organism successfully avoids innate defense and colonizes a niche in the body. What follows is a biological "horse race" in which the pathogen tries to replicate and expand its niche, while the immune system tries to eliminate the pathogen (or at least confine it). Only if the replication of the pathogen causes detectable clinical damage does the host experience "disease". Microbial toxins released by a pathogen can cause disease even in the absence of widespread colonization. Immunopathic disease may occur if host tissues are unintentionally injured by the immune response as it strives to destroy a pathogen.

B. Innate Defense against Pathogens

Mechanisms of innate defense that operate against essentially all pathogens are summarized in Figure 13-1. The first obsta-cles encountered by any pathogen are the intact skin and mucosae (Fig. 13-1, #1, 2). In the skin, pathogens are prevented from gaining a firm foothold by the tough keratin layer protecting the epidermis, competition with commensal micro-organisms, and the routine shedding of skin layers. Pathogens ingested into the gut or inhaled into the respiratory tract are trapped by mucus or succumb to microbicidal molecules in the body secretions or to the low pH and hydrolases of the gut. However, a breach of the skin or mucosae may allow a pathogen access to subepithelial tissues. Barrier penetration may also occur in individuals whose immune systems have been compromised by either disease or therapeutic immunosuppression. These lapses in immune defense may allow *opportunistic pathogens* to cause disease. In contrast, *invasive pathogens* can enter the body even when surface defenses are intact. Invasive organisms assaulting the mucosae frequently gain access via the M cells of the FAE or by binding to host cell surface molecules that trigger receptor-mediated internalization.

A pathogen that penetrates the skin or mucosae triggers the flooding of the site with acute phase proteins, pro-inflammatory cytokines such as IL-1 and TNF, and complement components. Coating of the pathogen by C3b or MBL allows its elimination by the alternative or lectin complement cascades, respectively (Fig. 13-1, #3). At a cellular level, an invading pathogen first encounters PRR-mediated defense provided by resident macrophages, NK cells, $\gamma\delta$ T cells and NKT cells (#4–#7). These cells attempt to eliminate the pathogen or infected cells by clathrin-mediated endocytosis or phagocytosis, secretion of cytotoxic cytokines, or by natural cytotoxicity. These cells also contribute toxic NO and ROIs to the extracellular milieu. Chemokines produced in the ensuing inflammatory response draw neutrophils (#8) and other leukocytes from the circulation into the area of infection to assist in the fight. If a pathogen enters the blood, innate defense falls to monocytes and neutrophils in the circulation. Organisms that reach the liver or the spleen are confronted by resident macrophages.

As the innate response proceeds, DCs that have matured due to exposure to pathogen components such as LPS and CpG motifs migrate to the lymphoid follicles in the local lymph

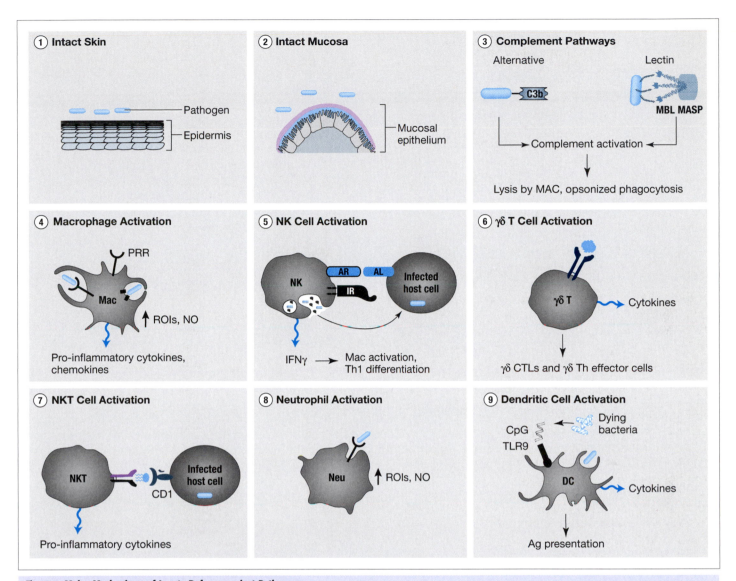

Fig. 13-1 **Major Mechanisms of Innate Defense against Pathogens**

node or the spleen and present pathogen-derived pMHCs to naïve T cells of the adaptive response (Fig. 13-1, #9). The effector lymphocytes are then drawn to the site of infection by chemokines secreted by activated DCs and macrophages. As detailed in Sections C–G that follow, the adaptive effector mechanisms best suited to countering a particular pathogen are determined by the invader's lifestyle and mode of replication.

C. Immunity to Extracellular Bacteria

I. DISEASE MECHANISMS

Extracellular bacteria attempting to establish an infection tend to accumulate in interstitial regions in connective tissue, in the lumens of the respiratory, urogenital and gastrointestinal tracts, and in the blood. These organisms often secrete proteins that penetrate or enzymatically cleave components of the mucosal epithelium, allowing access to underlying tissues. A wide variety of extracellular bacteria enter the M cells in the FAE, whereas others exploit host cell receptors on other host cell types. Examples of diseases caused by infections with extracellular bacteria are given in Table 13-1.

Many disease symptoms caused by extracellular bacteria can be attributed to their toxins. *Exotoxins* are toxic proteins actively secreted by both *gram-positive* and *gram-negative* bacteria. Gram-positive bacteria have cell walls containing a thick layer of peptidoglycan that is colored purple after Gram staining. Gram-negative bacteria have cell walls containing a thin layer of peptidoglycan plus LPS that is colored red after Gram staining. *Endotoxins* are the lipid portions of the LPS molecules embedded in the walls of gram-negative bacteria. Endotoxins are not secreted but rather are released only when the cell walls of gram-negative bacteria are damaged. A given gram-negative bacterial species may supply both exotoxins and endotoxins.

Table 13-1 Examples of Extracellular Bacteria and the Diseases They Cause

Pathogen	Disease
Bacillus anthracis	Anthrax
Borrelia burgdorferi	Lyme disease
Clostridium botulinum	Botulism
Clostridium tetani	Tetanus
Corynebacterium diphtheriae	Diphtheria
Escherichia coli O157:H7	Hemorrhagic colitis
Helicobacter pylori	Ulcers
Haemophilus influenzae	Bacterial meningitis
Neisseria meningitides	Bacterial meningitis
Neisseria gonorrhoeae	Gonorrhea
Staphylococcus aureus	Food poisoning, toxic shock
Streptococcus pyogenes	Strep throat, flesh-eating disease
Streptococcus pneumoniae	Pneumonia, otitis media
Treponema pallidum	Syphilis
Vibrio cholerae	Cholera
Yersinia enterocolitica	Severe diarrhea

Different exotoxins and endotoxins cause disease by different means and in different locations. For example, infection with *Vibrio cholerae* results in the local release of an exotoxin that binds to gut epithelial cells and induces the severe diarrhea that characterizes cholera. *Clostridium botulinum* produces a neuro-exotoxin that blocks the transmission of nerve impulses to the muscles, resulting in paralysis. In contrast, damage to a host caused by an endotoxin is always immunopathic. The LPS of gram-negative bacteria can activate macrophages and induce them to release pro-inflammatory cytokines, particularly TNF and IL-1. Although a little TNF and IL-1 is a good thing, the very high concentrations of these cytokines that are secreted in response to a significant gram-negative bacterial infection induce high fever and *endotoxic (septic) shock* (see Box 13-1).

II. IMMUNE EFFECTOR MECHANISMS

i) Humoral Defense

Because extracellular bacteria cannot routinely "hide" within host cells, antibodies are generally highly effective against these species. Polysaccharides present in bacterial cell walls make perfect Ti antigens for B cell activation (Fig. 13-2, #1), while other bacterial components supplying Td antigens induce primarily a Th2 response that provides T help for antibacterial B cells (#2). Neutralizing IgM antibodies dominate in the vascular system, while smaller IgG antibodies protect the tissues. These antibodies neutralize bacteria by physically preventing

---

Box 13-1 Endotoxic Shock

Severe infections with gram-negative bacteria can have catastrophic effects on a host, not because of the pathogen itself but because of the host's immune response to the invader. The cell walls of gram-negative bacteria contain LPS, and the lipid component of LPS is an endotoxin. Exposure to large amounts of an endotoxin induces activated macrophages to produce massive amounts of TNF. While a little TNF is necessary to combat infections, very high concentrations of this cytokine stimulate activated macrophages to overproduce the pro-inflammatory cytokine IL-1. This excess IL-1 in turn amplifies LPS-induced production of IL-6 and IL-8 by vascular endothelial cells and macrophages. The resulting systemic tidal wave of pro-inflammatory cytokines has devastating effects on the body (see Figure). The capillaries suffer from "disseminated intravascular coagulation", which is a blockage caused by neutrophil aggregation and excessive blood clot formation. Clotting factors are subsequently depleted, leading to hemorrhage. Cardiomyocytes in the heart begin to fail due to

the NO accumulation that follows the cytokine-induced overactivation of iNOS. Slowing of the blood circulation contributes to a drop in blood pressure followed by circulatory collapse. This collapse is promoted by the onset of "capillary leak syndrome", in which increased permeability of the capillaries (due to the cytokine storm) allows fluid and proteins to leak into the tissues. Glucose metabolism by the liver and the

muscles is disrupted, leading to metabolic failure. Fever and diarrhea are also present. The combination of circulatory collapse and metabolic failure may result in irreversible damage to host organ systems that can result in death within hours of infection. Close to 600,000 people a year experience endotoxic shock in North America, and more than 10% of cases have a fatal outcome.

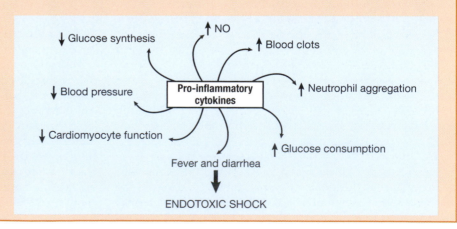

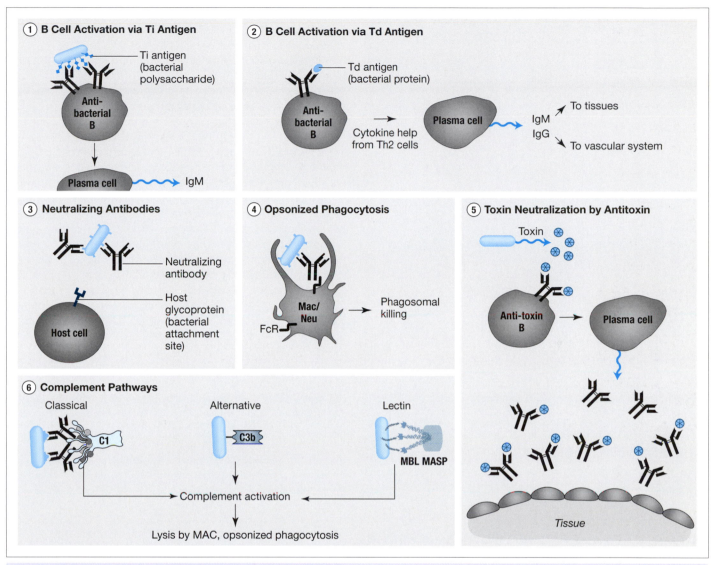

Fig. 13-2 **Major Mechanisms of Immune Defense against Extracellular Bacteria**

them from attaching to host cell surfaces (#3). Even though they do not need to enter host cells for replication, most extracellular bacteria try to adhere to host cells to avoid being swept off or out of the host by skin sloughing or movement of the intestinal contents. Antibodies can also serve as opsonins, coating the bacterium such that it is engulfed by phagocytic leukocytes expressing FcRs (#4). Once captured inside the phagocyte, extracellular bacteria are usually very vulnerable to killing via pH changes, defensins, and the ROIs and RNIs associated with the phagosomal respiratory burst. Antibodies made against bacterial exotoxins are called *antitoxins*. Antitoxins neutralize a toxin by preventing it from binding to the cells it would otherwise damage (#5). If the toxin is the sole element causing disease in the host, the production of the antitoxin alone will be enough to restore health. For example, human resistance to tetanus or diphtheria relies solely on anti-

toxins directed against the *Clostridium tetani* exotoxin or *Corynebacterium diphtheriae* exotoxin, respectively.

ii) Complement

All three pathways of complement activation can be brought to bear on extracellular bacteria (Fig. 13-2, #6). Antibodies of the appropriate isotype (and particularly IgM) will bind to complement component C1q to trigger the classical cascade. The alternative pathway can be activated by peptidoglycan in gram-positive cell walls or LPS in gram-negative cell walls. The lectin pathway is activated by the binding of MBL to distinctive sugars arrayed on bacterial cell surfaces. Almost all types of bacteria can be eliminated by phagocytosis facilitated by the binding of opsonins such as C3b that are produced during complement activation. In addition, bacteria possessing a membrane can be dispatched by MAC-mediated lysis. Comple-

Table 13-2 Evasion of the Immune System by Extracellular Bacteria

Immune System Element Thwarted	Bacterial Mechanism
Antibodies	Alter expression of surface molecules Secrete anti-Ig proteases
Phagocytosis	Block binding of phagocyte receptors to bacterial capsule Hide temporarily in non-phagocytes Inject bacterial protein that disrupts phagocyte function
Complement	Prevent C3b binding by lack of suitable surface protein, steric hindrance by surface proteins, C3b degradation Inactivate various steps of complement cascade Capture host RCA proteins Induce host production of antibody isotypes that are poor complement-fixers

ment is particularly crucial for defense against the *Neisseria* group of gram-negative extracellular bacteria.

III. EVASION STRATEGIES

Strategies used by extracellular bacteria to evade immune responses are summarized in Table 13-2.

i) Avoiding Antibodies

Some extracellular bacteria, such as the gonococci, ensure their adhesion to host tissues by routinely and spontaneously changing the amino acid sequence of the bacterial proteins used to stick to the host cell surface. Neutralizing antibodies directed against the original bacterial protein may not "see" the new version, allowing the bacteria to establish an infection. Other bacteria secrete proteases that cleave antibody proteins and render them nonfunctional. For example, *Haemophilus influenzae* expresses IgA-specific proteases that degrade sIgA in the blood and SIgA in the mucus.

ii) Avoiding Phagocytosis

The polysaccharide coating of encapsulated bacteria protects them from phagocytosis by conferring a charge on the bacterial surface that inhibits binding to phagocyte receptors. In addition, although C3b may still attach to the bacterial surface, the capsule sterically interferes with the binding of phagocyte receptors to the C3b so that opsonized phagocytosis of the bacterium is much less efficient. Some non-encapsulated extracellular bacteria avoid capture by phagocytes by temporarily entering non-phagocytes such as epithelial cells and fibroblasts. To gain access to these cells, the pathogens may inject bacterial proteins into the host cell that promote either macropinocytosis or cytoskeletal rearrangements facilitating bacterial uptake.

Extracellular bacteria may also inject bacterial proteins that have direct anti-phagocyte activity. For example, *Yersinia enterocolitica* injects a bacterial phosphatase into macrophages that binds to certain tyrosine-phosphorylated host proteins required for intracellular signaling and actin reorganization. When the bacterial phosphatase dephosphorylates these host proteins, phagocytosis of the bacterium is blocked.

iii) Avoiding Complement

Some extracellular bacteria can avoid complement by virtue of their basic structure. For example, the organism that causes syphilis, *Treponema pallidum*, has an outer membrane devoid of transmembrane proteins and so offers almost no place suitable for C3b deposition. Other bacteria have cell wall lipopolysaccharides that contain long, outwardly projecting chains that prevent the MAC from assembling directly on the bacterial surface. Many extracellular bacteria synthesize substances that inactivate various steps of the complement cascade. For example, group B streptococci contain sialic acid in their cell walls that degrades C3b and blocks alternative complement activation. Other streptococci produce proteins that bind to the normally fluid phase RCA protein Factor H and fix it onto the bacterial surface. In its hijacked site, the recruited Factor H makes any C3b that has attached susceptible to degradation. Certain salmonella species express proteins that interfere with the terminal steps of complement activation, while gonococci and meningococci induce the host to produce antibody isotypes (such as IgA) that are poor at fixing complement. These "blocking antibodies" compete with complement-fixing antibodies for binding to the bacterial surface, reducing MAC formation. Steric hindrance by the blocking antibodies also interferes with C3b deposition.

D. Immunity to Intracellular Bacteria

I. DISEASE MECHANISMS

Like extracellular bacteria, most intracellular bacteria access the host via breaches in the mucosae and skin, but some are introduced directly into the bloodstream by the bites of vectors such as ticks, mosquitoes and mites. (A *vector* is an intermediary organism that introduces the pathogen into the ultimate host.) Once inside the host, intracellular bacteria elude phagocytes, complement and antibodies by moving right inside host cells to reproduce. Epithelial and endothelial cells, hepatocytes and macrophages are popular targets. Because macrophages are mobile, bacteria that infect these cells are quickly disseminated all over the body.

Intracellular bacteria generally enter host cells by clathrin-mediated endocytosis and are thus first confined to a clathrin-coated vesicle. Some species remain in the vesicle whereas others leave it (by an ill-defined mechanism) to take up residence in the cytosol. Because of their desire to replicate within a host cell and keep it alive for this purpose, intracellular bacteria are generally not very toxic to the host cell and do not produce tissue-damaging bacterial toxins. However, their intracellular lifestyle makes these organisms difficult to

Table 13-3 Examples of Intracellular Bacteria and the Diseases They Cause

Pathogen	Disease
Brucella melitensis	High fevers, brucellosis
Chlamydia trachomatis	Eye and genital diseases
Legionella pneumophila	Legionnaire's disease
Listeria monocytogenes	Listeriosis
Mycobacterium leprae	Leprosy
Mycobacterium tuberculosis	Tuberculosis
Mycoplasma pneumoniae	Atypical pneumonia
Rickettsia rickettsii	Rocky Mountain spotted fever
Salmonella typhi	Typhoid fever
Salmonella typhimurium	Food poisoning
Shigella flexneri	Enteric disease

eradicate completely and chronic disease may result. Examples of diseases caused by intracellular bacteria appear in Table 13-3.

II. IMMUNE EFFECTOR MECHANISMS

i) Neutrophils and Macrophages

Early infections by intracellular bacteria are frequently controlled by the defensins secreted by neutrophils because these proteins can destroy the invaders before they can take refuge inside a host cell (Fig. 13-3, #1). Those bacteria that escape the defensins and are taken up by neutrophil phagocytosis find themselves, not in a haven for replication, but rather within a phagosome that can kill them via the powerful respiratory burst. Activated macrophages also play an important role in the phagocytosis and killing of intracellular bacteria (#2). In both neutrophils and macrophages, the killing of phagocytosed bacteria is frequently enhanced by certain host proteins present within the phagolysosomal membrane, and also by host enzymes in the ER or Golgi that regulate the maturation of pathogen-containing phagosomes. These enzymes are greatly upregulated in response to IFNs and LPS.

In addition to phagocytosis, macrophages can carry out TLR-mediated endocytosis of intracellular bacteria. For example, lipoprotein and lipoglycan components of mycobacteria are readily recognized by TLR2 and TLR4. Once activated by TLR engagement, the macrophages produce proinflammatory cytokines that promote NK cell activation and Th1 differentiation.

ii) NK Cells and γδ T Cells

NK cells stimulated by macrophage-derived IL-12 detect infected host cells by their deficit in MHC class I expression (which is typically downregulated by the infection) and destroy them by natural cytotoxicity (Fig. 13-3, #3). In addition, acti-

vated NK cells secrete copious amounts of IFNγ, which promotes macrophage activation directly and Th1 cell differentiation indirectly. γδ T cells are also important in combating at least some intracellular infections. Many species of intracellular bacteria (particularly the mycobacteria) release small phosphorylated molecules (including pyrophosphate) as they attempt to colonize the host. These metabolites trigger the generation of γδ T cell effectors that either carry out cytolysis or secrete IFNγ (#4).

iii) CD8+ T Cells

CTLs are critical for resolving many intracellular bacterial infections. If the bacterium replicates in the cytosol of the infected cell, some of its component proteins enter the endogenous antigen processing pathway and are presented on MHC class I, marking the cell as a target for CTL-mediated destruction (Fig. 13-3, #5). These CTLs will have been generated from pathogen-specific naïve Tc cells that were activated in the draining lymph node. This Tc activation will have been initiated by DCs that acquired antigens derived from the degradation of a phagocytosed bacterium or a dying host cell, followed by cross-presentation of peptides from these antigens on MHC class I. Interestingly, CTLs rarely use Fas-mediated apoptosis or perforin-mediated cytolysis to kill target cells infected with intracellular bacteria, in contrast to their destruction of virus-infected cells. Rather, CTLs eliminate these targets by relying on secreted TNF and IFNγ and/or granule components with direct antimicrobial activity. Indeed, individuals lacking the IFNγ receptor are highly susceptible to mycobacterial infections.

iv) CD4+ T Cells

CD4+ T cells make a significant contribution to defense against intracellular bacteria not only because of the IL-2 they secrete to support Tc differentiation but also because Th1 cells are required for macrophage hyperactivation. It is not unusual for intracellular bacteria phagocytosed by macrophages to be resistant to routine phagosomal killing, and the IFNγ produced by activated Th1 effectors hyperactivates the macrophages such that they gain enhanced microbicidal powers. The sequence of events starts when bacterial antigens either secreted by the bacteria themselves or released by necrotic infected cells are taken up by DCs. In the local lymph node, peptides from these antigens are bound to MHC class II and presented to CD4+ T cells (Fig. 13-3, #6). IL-12 produced by macrophages activated early in the infection favors the differentiation of Th1 effectors. Activated Th1 cells then supply the intercellular contacts (particularly CD40L) plus TNF and IFNγ that drive macrophage hyperactivation. A hyperactivated macrophage produces large quantities of ROIs and RNIs that efficiently kill almost all intracellular pathogens. If the bacterium is still resistant, however, a hyperactivated macrophage may go on to participate in *granuloma formation* (see later) to contain the threat.

The importance of the Th1 response to defense against intracellular pathogens is clearly illustrated in human immune responses to *M. leprae* infection. Patients who are predisposed

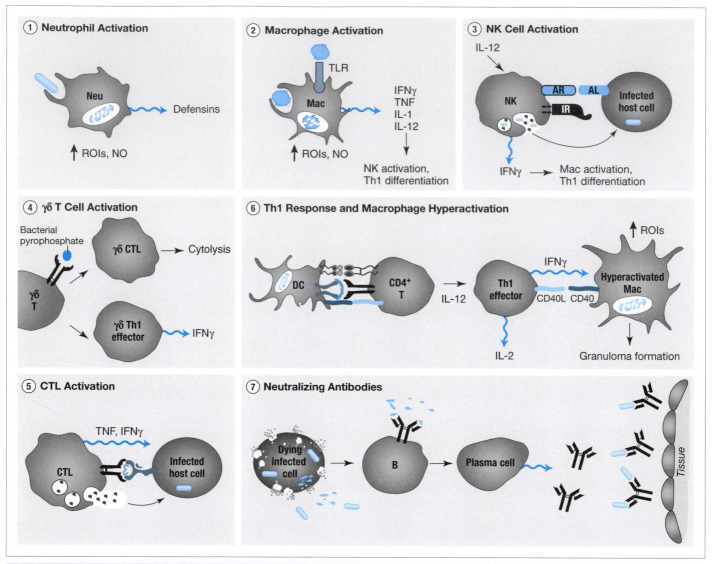

Fig. 13-3 **Major Mechanisms of Immune Defense against Intracellular Bacteria**

to mounting Th2 responses (i.e., their Th cells preferentially secrete IL-4 and IL-10) suffer from a devastating form of leprosy known as lepromatous leprosy. In contrast, patients who usually mount Th1 responses (i.e., their Th cells preferentially secrete IFNγ) present with tuberculoid leprosy, which is generally a less severe form of the disease. The cell-mediated immunity favored by a Th1 response is more effective against intracellular pathogens than the Th2 responses that promote humoral immunity.

v) Humoral Defense

Antibodies can make an important contribution to host defense against at least some intracellular bacteria. Bacterial components released from a dying infected cell may activate B cells to produce neutralizing antibodies (Fig. 13-3, #7). These antibodies may bind to newly arrived bacteria or to bacterial progeny that have been released into the extracellular milieu but have not yet infected a fresh host cell. The antibody-bound

bacteria are unable to enter host cells and are eliminated by opsonized phagocytosis or classical complement-mediated lysis, curbing pathogen spread.

vi) Granuloma Formation

When an intracellular pathogen like *Mycobacterium tuberculosis* is able to resist killing by CTLs and hyperactivated macrophages, the body attempts to wall off the pathogen in a cellular structure called a *granuloma* that forms around the infected macrophages (Fig. 13-4 and Plate 13-1). The inner layer of a granuloma contains macrophages and CD4+ T cells, whereas the exterior layer is composed of CD8+ T cells. Eventually the granuloma exterior becomes calcified and fibrotic, and cells in the center undergo necrosis. In some cases, all the pathogens trapped in the dying cells are killed and the infection is resolved. In other cases, a few pathogens remain viable but dormant within the granuloma, causing it to persist. Granuloma persistence is an overt sign that the disease is becoming

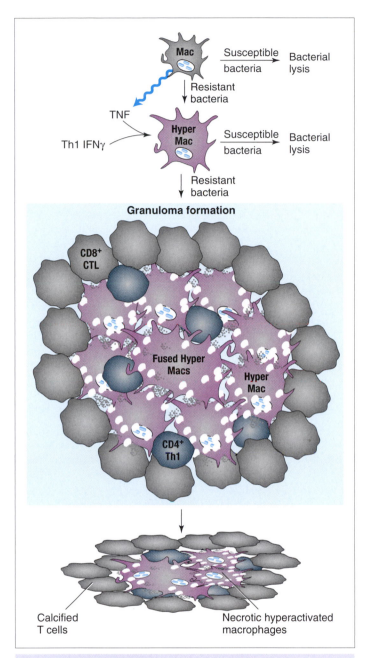

Fig. 13-4 **Granuloma Formation**

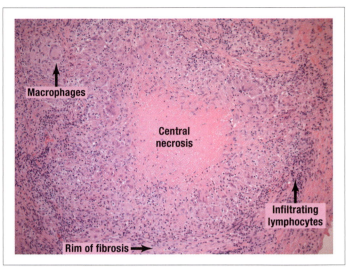

Plate 13-1 **Granuloma**
[Reproduced by permission of David Hwang, Department of Pathology, University Health Network, Toronto General Hospital.]

Table 13-4 **Evasion of the Immune System by Intracellular Bacteria**

Immune System Element Thwarted	Bacterial Mechanism
Phagosomal destruction	Infect a non-phagocyte Synthesize molecules blocking lysosomal fusion, phagosomal acidification, ROI/RNI killing Recruit host proteins blocking lysosome function
Hyperactivated macrophages	Block expression of host genes needed for macrophage hyperactivation
Antibodies	Spread to new host cell via pseudopod invasion
T cells	Reduce antigen presentation by APCs

chronic. If the granuloma breaks down, the trapped pathogens are released back into the body to resume replication. Should the host be immunosuppressed and unable to marshal the T cells and macrophages necessary to combat this fresh assault, the pathogen may reach the blood. As the bacteria travel in the circulation, they can infect organs throughout the body and even precipitate death.

Cytokines play a critical role in granuloma formation. Sustained IFNγ production by Th1 cells and CTLs is required to maintain macrophage hyperactivation. TNF production by hyperactivated macrophages is crucial not only for early chemokine synthesis (to recruit leukocytes to the incipient granuloma) but also for aggregating these cells and establishing the

"wall" around the invaders. IL-4 and IL-10 secreted by Th2 cells late in an adaptive response control granuloma formation, damping it down as the bacterial threat is contained.

III. EVASION STRATEGIES

Evasion strategies used by intracellular bacteria are summarized in Table 13-4.

i) Avoiding Phagosomal Destruction

Some intracellular bacteria avoid phagosomal killing by replicating in non-phagocytic cells. For example, *M. leprae* infects the Schwann cells of the human peripheral nervous system. Other intracellular bacteria deliberately enter phagocytes but then inactivate them or take steps to escape phagosomal killing. For example, *L. monocytogenes* accesses mouse phagocytes via host FcRs and CRs but then synthesizes a protein called *liste-riolysin O* (LLO) that induces pore formation in the phagolysosomal membrane. The bacterium escapes through the pore into the relative safety of the cytoplasm. When *M. tuberculosis* finds itself being engulfed in a macrophage phagosome, it recruits to the phagosome a host protein called TACO that inhibits the fusion of the phagosome to lysosomes. *M. tuberculosis* also produces NH_4^+, which reverses the acidification of phagolysosomes and promotes fusion with harmless endosomes. In addition, *M. tuberculosis* infection interferes with the expression of host genes needed for microbicidal action and macrophage hyperactivation. As a result of all these measures, mycobacteria can survive within host phagosomes for long periods. Certain salmonella species produce molecules that decrease the recruitment of NADPH oxidase to the phagolysosome, inhibiting ROI/RNI generation. Other intracellular bacteria block phagosomal ROIs and RNIs either by neutralizing them or by synthesizing the enzymes superoxide dismutase and catalase that break down ROIs, RNIs and hydrogen peroxide.

ii) Avoiding Antibodies

Some intracellular pathogens avoid the humoral response by moving directly from one host cell to another, giving antibodies no chance to bind. For example, in mice, *L. monocytogenes* can induce the actin-based formation of a pseudopod that invaginates into a neighboring non-phagocytic cell (Plate 13-2). The neighboring cell engulfs the bacterium-containing pseudopod and confines it in a vacuole. The bacterium then uses LLO and phospholipases to break out of the vacuole and enter the cytoplasm of the new cell. Because the bacterium is never exposed in the extracellular milieu, it never becomes an antibody target.

iii) Avoiding T Cells

Some intracellular bacteria avoid stimulating T cell responses by interfering with APC function. For example, infection of DCs by *M. tuberculosis* promotes downregulation of the expression of MHC class I and II and CD1. Antigen presentation to T cells and NKT cells is thus inhibited.

E. Immunity to Viruses

I. DISEASE MECHANISMS

Viruses are stripped-down intracellular pathogens that consist of a nucleic acid genome packaged in a protein coat called a *capsid*. The viral genome may be DNA or RNA, and the capsid may or may not be covered in a membranous structure called an *envelope*. Most viruses enter a host cell by binding to a host surface receptor. Replication of the viral genome and synthesis of viral mRNAs follow, which may be carried out by host or viral enzymes, depending on the virus. However, all viruses lack protein synthesis machinery and rely on the host cell for viral protein translation and progeny virion assembly. Progeny virions released from an infected cell attack neighboring host cells and initiate new replicative cycles that lead to widespread dissemination of the virus. Progeny virions that reach the blood are free to spread systemically. Examples of diseases caused by viruses are given in Table 13-5.

Viruses cause disease both directly and indirectly. Viruses frequently kill or at least inactivate host cells, depriving the host of these cells' normal functions such that clinical symptoms appear. As well, the immune response to the viral infection frequently damages host tissues and induces inflammation, causing immunopathic disease. Clinicians classify diseases caused by viruses as either *acute* or *chronic*. When a host is initially infected with a virus, the host experiences *acute disease* in that the illness may be mild or severe (depending on the degree of pathogenicity or *virulence* of the virus) but is only short-term in duration. An effective immune response removes the virus completely from the body. However, sometimes viruses are not completely eliminated during the acute infection and remain in the body to establish *persistent* infections. The on-going low levels of viral replication associated with these persistent infections cause long-term or recurrent illnesses that are considered *chronic diseases*.

Many persistent viral infections result in *latency* rather than chronic disease. During latency, an effective cell-mediated immune response ensures that no new virus particles are assembled. The spread of the virus to fresh host cells is halted

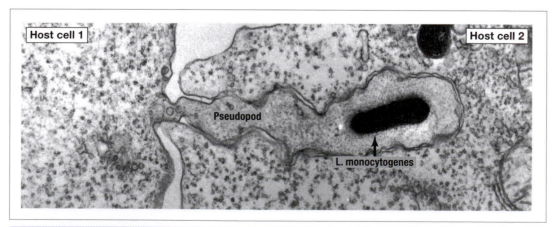

Plate 13-2 **Pseudopod Invasion**
[Reproduced by permission of Daniel Portnoy and Lewis Tilney, University of California, Berkeley.]

Table 13-5 Examples of Viruses and the Diseases They Cause

Pathogen	Disease
Adenovirus	Acute respiratory infections
Cytomegalovirus (CMV)	Pneumonitis, hepatitis
Epstein-Barr virus (EBV)	Infectious mononucleosis, Burkitt's lymphoma
Hepatitis virus (A, B, C)	Hepatitis, cirrhosis, liver cancer
Herpes simplex (HSV)	Cold sores
Human immunodeficiency virus (HIV)	Acquired immunodeficiency syndrome (AIDS)
Human papilloma virus (HPV)	Skin warts, genital warts, cervical cancer
Human T cell leukemia virus 1 (HTLV-1)	T cell leukemias and lymphomas
Influenza virus	The "flu"
Kaposi's sarcoma herpes virus (KSHV)	Kaposi's sarcoma
Measles virus (MV)	Measles
Poliovirus	Poliomyelitis, post-polio fatigue
Rabies virus	Rabies
Rhinovirus	Common cold
SARS (severe acute respiratory syndrome) virus	Severe acute respiratory syndrome
Varicella zoster virus (VZV)	Chicken pox, shingles
Variola virus	Smallpox
West Nile virus (WNV)	Flu-like illness, fatigue, encephalitis

and the host experiences no disease symptoms. The virus then persists in the body in an inactive state and does not attempt to replicate. However, if the host's cell-mediated response weakens due to aging or immunosuppression, the latent virus reactivates, replicates and again causes acute disease. For example, the reactivation of latent chicken pox virus precipitates the painful skin condition known as shingles. When latent *oncogenic* (cancer-causing) viruses reactivate, malignancy can eventually result, as in the occurrence of cervical cancer associated with persistent human papillomavirus (HPV) infection. Human immunodeficiency virus (HIV), the virus causing AIDS, enters latency in individual CD4⁺ T cells for an extended period. Eventually, HIV reactivates on a grand scale and destroys the immune system, leaving the host vulnerable to lethal infections (see Ch. 15).

II. IMMUNE EFFECTOR MECHANISMS

i) Inteferons and the Antiviral State

Production of the multifunctional cytokines IFNα, IFNβ and IFNγ is one of the earliest innate responses induced by viral infections. IFNα and IFNβ are considered type I interferons, whereas IFNγ is a type II interferon. The IFNs are important mediators in immune responses to a wide range of pathogens, a fact reflected in the overlapping functions of all three IFNs (summarized in Fig. 13-5). IFNα and IFNβ are secreted primarily by host cells infected with a virus, whereas IFNγ is initially produced by activated macrophages and NK cells and later on by activated Th1 cells. Any one of these IFNs can initiate a series of metabolic and enzymatic events in an uninfected host cell that results in it adopting an *antiviral state* (Fig. 13-6, #1). A host cell in the antiviral state can take enzymatic action to prevent an infecting virus from starting to replicate. Both the transcription and translation of viral mRNAs and proteins are inhibited.

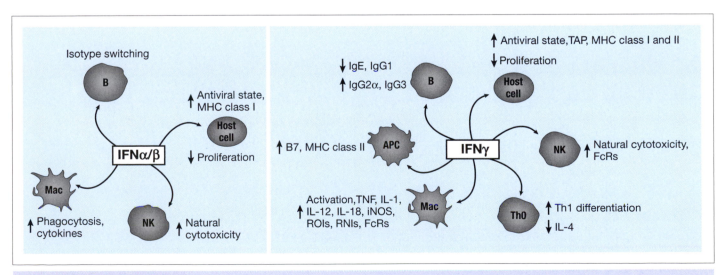

Fig. 13-5 **Major Functions of Interferons**

ii) NK Cells

Although CTLs are the prime mediators of the cell-mediated immunity needed to eliminate viruses (see later), there is often a 4–6 day delay before these cells can expand to sufficient numbers to complete the task. Where a virus causes downregulation of MHC class I on the host cell surface, direct cytolysis of infected cells by NK cells (via natural cytotoxicity) and NK production of inflammatory cytokines can supply early defense (Fig. 13-6, #2). Indeed, individuals whose NK cells are not fully functional show increased susceptibility to virus infections, especially herpesviruses. Natural cytotoxicity and inflammatory cytokine production are stimulated by all three IFNs. NK cells are also important mediators of antiviral ADCC. The upregulation of FcR expression on both NK cells and macrophages is stimulated by IFNγ.

iii) Macrophages

Macrophages activated in the course of a virus infection produce copious amounts of pro-inflammatory cytokines (Fig. 13-6, #3). The presence of IFNγ in the milieu greatly enhances this function and also allows the macrophage to express the iNOS enzyme that generates NO. This NO facilitates macrophage production of ROIs and RNIs that will aid in killing phagocytosed viruses. Macrophages can also eliminate viruses via ADCC.

iv) CD4⁺ T Cells

Whole virions or their components may be taken into a DC by clathrin-mediated endocytosis or phagocytosis. The TLRs are vital in this context, as several of them recognize viral nucleic acids (refer to Ch. 3). TLR-stimulated DCs readily process viral proteins via the exogenous pathway and display viral peptides on MHC class II to activate CD4⁺ Th cells (Fig. 13-6, #4). Th cells are important for defense against most viruses because these cells both license DCs and supply IL-2 for naïve CD8⁺ Tc cell activation. Th cells also provide the CD40L-mediated costimulation and cytokines required for B cells to mount antibody responses to viral Td antigens.

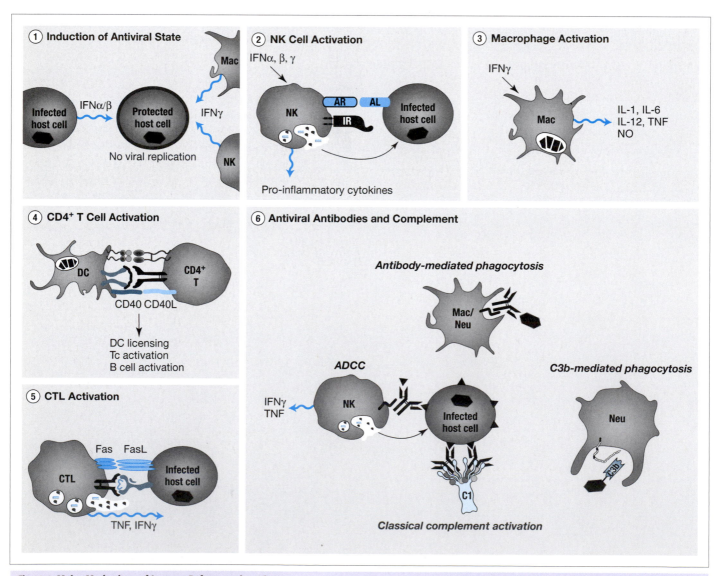

Fig. 13-6 **Major Mechanisms of Immune Defense against Viruses**

v) CD8⁺ T Cells

CTLs are crucial for immune defense against most viruses. Because these pathogens replicate intracellularly, viral antigens are displayed on MHC class I on infected cell surfaces and mark these cells as CTL targets. The effector CTLs generated from Tc cells activated in the draining lymph node return to the site of infection and kill the virus-infected cells via granule-mediated cytotoxicity, Fas-mediated apoptosis, or TNF and/or IFNγ secretion (Fig. 13-6, #5).

vi) Humoral Defense

Because a virus is an intracellular pathogen, it is often out of reach of antibodies during the primary adaptive response. Nevertheless, naïve B cells may recognize viral components displayed on the surface of an infected host cell or may encounter progeny virions as they are released from an infected cell. With the appropriate T cell help, these B cells are activated and generate plasma cells and memory B cells that are usually vital for complete resolution of the infection. Late in the primary response, neutralizing antibodies are released into the circulation that block further spread of the virus. As well, in a subsequent attack, the virus will have a harder time infecting the host because the circulating neutralizing antibodies rapidly bind to the virus and bar its access to host cell receptors. Antiviral antibodies may also initiate classical complement activation. The formation of the MAC on the surface of an enveloped virus or an infected host cell kills it, and the complement components that are produced during the cascade may opsonize extracellular virions and promote their uptake by phagocytosis (Fig. 13-6, #6). The antiviral antibodies themselves may also serve as opsonins. Finally, antibodies that have recognized viral antigens on infected host cell surfaces may engage FcRs on phagocytes and leukocytes (particularly NK cells) and provoke ADCC.

It should be noted that some viruses are combated (at least in part) by B cell responses that do not require T cell help. Viruses such as vesicular stomatitis virus (VSV) have highly repetitive structures on their surfaces that induce a Ti response. Ti responses are typically faster than Td responses because a Ti response involves only a B cell and does not require B–T cooperation. An antiviral Ti response can function early in an infection to minimize the spread of the virus until antibodies against viral Td antigens can be synthesized.

vii) Complement

As well as the classical complement activation that is part of the humoral response, surface components of virions can directly activate the lectin and alternative complement pathways. Opsonization of viruses by C3b (or C3d) promotes phagocytosis by neutrophils and macrophages (refer to Fig. 13-6, #6).

III. EVASION STRATEGIES

Viruses with small genomes count on rapid replication and dissemination to new host cells to establish an infection before the immune system can respond. Viruses with larger genomes need more time to replicate and are transmitted more slowly. Accordingly, these latter pathogens have developed ways of interfering with various components of the host immune response that allow them sufficient time to establish an infection. Once infection has occurred, many viruses hide from the immune system. Others confront the immune response head on by interfering with host cellular pathways. Evasion strategies used by viruses are summarized in Table 13-6.

i) Latency

When a virus adopts a latent state, it persists in the host cell in a defective form that renders it non-infectious for a period of time. In most cases, latency involves the inactivation of viral

Table 13-6 Evasion of the Immune System by Viruses

Immune System Element Thwarted	Viral Mechanism
Detection	Become latent
Antibodies	Alter viral epitopes via antigenic drift or shift Express viral FcR that blocks ADCC or neutralization Block B cell intracellular signaling
CD8⁺ T cells	Infect cells with very low MHC class I expression Interfere with MHC class I-mediated antigen presentation Force pMHC internalization
CD4⁺ T cells	Avoid infection of DCs Interfere with MHC class II-mediated antigen presentation Force pMHC internalization
NK cells	Express viral homologues of MHC class I Increase host synthesis of HLA-E or classical MHC class I
DCs	Block DC development or maturation Block DC upregulation of costimulatory molecules Upregulate DC expression of FasL
Complement	Block convertase formation Express viral homologues of host RCA proteins Increase expression of host RCA proteins Bud through host membrane and acquire host RCA proteins
Antiviral state	Block secretion of IFNγ Interfere with metabolic/enzymatic events that establish the antiviral state
Apoptosis	Block various steps of extrinsic or intrinsic pathways Express homologues of death receptors and regulatory molecules
Cytokines/ chemokines	Express competitive inhibitors of cytokines and chemokines Block cytokine/chemokine transcription Downregulate host cytokine/chemokine receptor expression

gene transcription needed for productive infection and the subsequent expression of new viral transcripts required for latency. Reversal from latency back to productive infection requires some type of reactivation of the productive infection genes that can only occur when the host's immune system has weakened.

Different viruses achieve latency in different ways. HIV integrates a cDNA copy of its RNA genome into the DNA of its host cell in such a way that there is limited transcription of viral genes (see Ch. 15). The DNA genomes of varicella zoster virus (VZV) and herpes simplex virus (HSV) do not integrate into the host DNA but instead form a complex with host nucleosomal proteins that blocks transcription of productive infection genes. A similar latency mechanism operates in cases of Epstein-Barr virus (EBV) and Kaposi's sarcoma herpesvirus (KSHV) infection. However, the latency of these herpesviruses is associated with the development of tumors in the host: B cell lymphomas and nasopharyngeal carcinomas in the case of EBV, and the AIDS-related Kaposi's sarcoma in the case of KSHV.

ii) Antigenic Variation

A common way for a virus to hide from the host immune system is to change its antigenic "stripe" over successive generations, expressing antigenically new forms of viral proteins that may not be recognized by an individual's existing memory lymphocytes or antibodies. This mechanism is most effective in long-lived hosts (like humans) that can sustain multiple re-infections, and is particularly important if the virus lacks the ability to become latent. The rapid modification of viral antigens through random mutations is known as *antigenic drift*. For example, like all RNA viruses, influenza virus cannot proofread its RNA genome during replication and thus sustains a high rate of mutation. The hemagglutinin (H) and neuraminidase (N) proteins present on the surface of the influenza virion are thus subtly different from viral generation to generation. These minor virus variants often replicate preferentially in the host, as they are not neutralized by antibodies raised against earlier strains. New influenza strains created by antigenic drift are responsible for localized influenza outbreaks. HIV is another virus that undergoes very rapid antigenic drift, even within a single infected individual. In this case, the mutations arise due to the highly error-prone reverse transcriptase involved in the replication of its genome (see Ch. 15).

Perhaps unique to the influenza virus is its ability to undergo *antigenic shift*. The influenza virus genome exists as eight separate single-stranded RNA segments, each of which encodes a single viral protein. With such a genetic structure, two different influenza strains that simultaneously infect a single host cell can undergo a reassortment (sometimes inaccurately called "recombination") of their genomic segments (Fig. 13-7). Virus particles containing new combinations of parental RNA suddenly arise, dramatically changing the spectrum of protein epitopes presented to the immune system. This antigenically novel flu virus is safe from antibodies and CTLs raised during previous exposure or vaccination. Antigenic shifts were responsible for four widespread influenza outbreaks in the twentieth century.

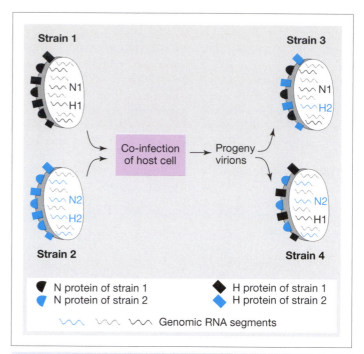

Fig. 13-7 Principle of Antigenic Shift

iii) Interference with Antigen Presentation

Antigen processing pathways offer many opportunities for a virus to sabotage immune responses and a given virus can interfere at more than one step.

iiia) MHC class I-mediated antigen presentation. Different viruses avoid activating CD8$^+$ T cells in different ways. Adenovirus blocks MHC class I synthesis in infected cells. Cytomegalovirus (CMV) and VSV infect cells that normally have very low MHC class I expression. CMV also expresses a protein that induces deglycosylation and degradation of newly synthesized MHC class I chains. A different CMV protein associates with mature peptide–MHC class I structures that do make it to the cell surface and blocks recognition by CD8$^+$ T cells. EBV produces viral proteins that resist proteolysis, meaning that peptides capable of fitting into the MHC binding groove are not easily generated. EBV also downregulates the expression of the TAP antigen transporter and so reduces peptide loading. Herpesviruses express small proteins that interfere with peptide binding to TAP on the cytosolic side of the ER. Other viruses express proteins that allow the peptide to bind to TAP but then trap the complex on the luminal side of the ER. HIV expresses an intracellular multifunctional protein called Nef (see Ch. 15) that is able to bind simultaneously to host clathrin proteins and the cytoplasmic tails of MHC class I molecules. This Nef-mediated physical connection between MHC class I and clathrin forces the internalization and lysosomal degradation of the MHC class I molecule.

iiib) MHC class II-mediated antigen presentation. Viruses have also evolved myriad ways to avoid activating CD4$^+$ T cells. Rabies virus preferentially infects neurons but is very slow to lyse these cells, meaning that viral antigens are not

easily collected by APCs until well after the virus has entered the body. The adaptive response to rabies is thus delayed. CMV and adenovirus both synthesize proteins that inhibit the intracellular signaling pathways required for MHC class II expression. Other viruses express proteins that bind to MHC class II molecules and target them for proteosomal degradation. Still other viruses interfere with MHC class II presentation after the MHC molecule has entered the endocytic system. For example, a CMV protein competes with invariant chain for binding to MHC class II, whereas HPV proteins and the HIV Nef protein disrupt the acidification of the endosomal compartments necessary for peptide generation. As it can for MHC class I, HIV Nef can induce the internalization and lysosomal degradation of cell surface MHC class II molecules.

iv) Fooling NK Cells

A virus that causes its host cell to downregulate MHC class I draws the attention of NK cells. CMV therefore expresses a viral homologue of MHC class I that engages NK inhibitory receptors and fools the NK cell into thinking it has detected normal MHC class I. The NK cell is not activated and the infected host cell is not lysed. CMV also upregulates expression of the non-classical MHC class I molecule HLA-E, which can bind to NK inhibitory receptors. In contrast, the fast-replicating West Nile virus (WNV) upregulates the expression of classical host MHC class I molecules, striving to neutralize NK cells and complete its reproduction before CTLs are generated.

v) Interference with DCs

Several viruses interfere with DC functions and thus derail T cell responses. Human T cell leukemia virus 1 (HTLV-1) infects DC precursors and prevents their differentiation into immature DCs, blocking the initiation of T cell responses. HSV-1 and vaccinia infect immature DCs and block DC maturation, whereas other poxviruses induce the apoptotic death of DCs. Measles virus upregulates the expression of FasL on an infected DC, forcing it to kill any Fas-bearing T cells it encounters. Measles virus can also cause DCs to form large aggregates called *syncytia* in which the virus replicates freely and DC maturation is stymied. When CMV infects a DC, the DC becomes tolerogenic so that it anergizes, rather than activates, any naïve T cell it encounters.

vi) Interference with Antibody Functions

Some viruses are able to interfere directly with production or effector functions of antiviral antibodies. Measles virus expresses a protein that has a negative regulatory effect on B cell activation. HSV-1 causes an infected host cell to express a viral version of FcγR that binds to IgG molecules complexed to viral antigen. The Fc portion of the antibody is rendered inaccessible so that neither ADCC nor classical complement activation can be triggered.

vii) Avoiding Complement

Viruses use many of the same mechanisms as other pathogens to avoid complement-mediated destruction. Some poxviruses and herpesviruses secrete proteins that block formation of the alternative C3 convertase. Many viruses increase a host cell's expression of the RCA proteins that regulate complement activation, preventing the infected cell from undergoing MAC-mediated lysis. Other viruses express viral homologues of RCA proteins that block MAC-mediated destruction of the virion. Still other viruses bud through the host cell membrane and acquire its RCA proteins. The RCA proteins DAF and MIRL are acquired by HIV and vaccinia in this way.

viii) Counteracting the Antiviral State

Several viruses have developed intricate mechanisms that disrupt the antiviral state. EBV expresses a soluble receptor for a growth factor essential for macrophage secretion of IFNs. In the absence of the growth factor, insufficient IFNs are produced to trigger and maintain the antiviral state. When HSV infects a cell that has already established the antiviral state, the virus expresses proteins that reverse the associated translational block, allowing viral protein synthesis to resume. Vaccinia and hepatitis C virus also synthesize proteins that disrupt the metabolic and enzymatic events needed to maintain the antiviral state. Adenovirus expresses proteins that interfere with the activity of the host's transcription factors. KSHV produces proteins that are homologous to host transcription factors but do not permit transcription of the genes required to establish the antiviral state.

ix) Manipulation of Host Cell Apoptosis

Host cell apoptosis prior to completion of replication spells viral doom. Host cell apoptosis is most commonly induced by CTL degranulation, Fas/FasL interaction, or the binding of TNF to TNFR. In addition, an infected cell will sometimes be triggered to undergo "altruistic" apoptosis (death for the good of the host) by a mechanism such as *ER stress*. ER stress results when the ER machinery of a host cell is overheated by having to pump out large quantities of viral proteins. Complex viruses with large genomes have developed ways of blocking various steps of these death-inducing pathways. Adenovirus synthesizes a multiprotein complex that induces the internalization of Fas and TNFR, removing these death receptors from the cell surface and forestalling apoptosis induced by an encounter with FasL or TNF. Several poxviruses express homologues of TNFR that act as decoy receptors for TNF and related cytokines. Adenoviruses, herpesviruses and poxviruses express multiple proteins that inhibit the enzymatic cascade necessary for apoptosis. Many viruses can increase intracellular levels of host cell survival proteins that normally prevent premature apoptosis. Alternatively, a virus may express a homologue of these survival proteins that interferes with apoptosis.

x) Interference with Host Cytokines

Early in viral infections, host cells are induced to produce copious quantities of cytokines and chemokines that support antiviral responses. Viruses therefore seek to inhibit the production or action of these molecules, or their receptors. Some poxviruses alter the local cytokine milieu and make it less favorable to the cellular cooperation that underpins an immune response. Both KSHV and adenovirus express proteins that inhibit IFN-inducible gene transcription, whereas certain poxviruses express a protein that blocks IL-1 production. Herpesviruses downregulate cytokine receptor expression, and CMV disrupts the transcription of chemokine genes. Vaccinia virus secretes interferon receptor homologues that intercept IFNα and IFNγ molecules. Poxviruses synthesize a chemokine homologue that binds to chemokine receptors on host cells but blocks the chemotaxis of lymphocytes, macrophages and neutrophils.

Inhibition of IL-12 production is a major goal of many viruses since this cytokine is crucial for Th1 differentiation and thus the antiviral cell-mediated immune response. EBV synthesizes a homologue of IL-12 that may competitively inhibit the activity of host IL-12. EBV also produces a homologue of IL-10 that suppresses IL-12 production by macrophages and IFNγ production by lymphocytes. The binding of measles virus to certain host cell receptors can also block IL-12 synthesis.

F. Immunity to Parasites

I. DISEASE MECHANISMS

Parasites include both unicellular protozoans and multicellular helminth worms, and are among the biggest killers in the pathogen pantheon. These scourges claim millions of lives every year, particularly in developing countries. Some protozoans replicate extracellularly whereas others replicate intracellularly. Helminth worms reproduce inside a host's body but outside its cells, or outside of the host entirely in a location (like a water source) where access to a host is easy. Growth and maturation of the worm then occur within the host, often causing severe and long-term damage to tissues and organs.

Many parasites have multistage life cycles, and each stage of a parasite may be able to infect a different host species. Parasites also frequently use vectors to infect their ultimate hosts. For example, humans contract malaria through the bite of an *Anopheles* mosquito infected with the protozoan parasite *Plasmodium falciparum*. Even within a given infected individual, some parasite stages may be intracellular whereas others are extracellular. All these factors can create a considerable problem from a public health point of view, since a parasite that continually changes form and/or makes use of an invertebrate or animal vector is much harder to control than a pathogen that infects humans only. In general, both cell-mediated and humoral immunity must be mobilized to conquer parasites. Examples of diseases caused by protozoans and helminth worms are given in Tables 13-7 and 13-8, respectively.

II. IMMUNE EFFECTOR MECHANISMS

Different parasites evoke different types of immune responses, depending on the size and cellularity of the invader and its life cycle. In general, protozoan parasites tend to induce Th1 responses. In contrast, helminth worm infections are usually handled by Th2 responses.

Table 13-7 Examples of Parasitic Protozoans and the Diseases They Cause

Pathogen	Disease
Entamoeba histolytica	Enteric disease
Leishmania donovanii	Leishmaniasis in viscera
Leishmania major	Leishmaniasis in face, ears, skin
Plasmodium falciparum	Malaria
Toxoplasma gondii	Toxoplasmosis
Trypanosoma brucei	African sleeping sickness
Trypanosoma cruzi	Chagas' disease

Table 13-8 Examples of Parasitic Helminth Worms and the Diseases They Cause

Pathogen	Disease
Ascaris	Ascariasis
Echinococcus	Alveolar echinococcosis
Onchocerca	African river blindness
Schistosoma	Schistosomiasis
Trichinella	Trichinosis
Wuchereria	Elephantiasis

i) Defense against Protozoans

ia) Humoral defense. All of the effector mechanisms ascribed to antibodies for defense against extracellular bacteria (refer to Fig. 13-2) apply to defense against small extracellular protozoans. Anti-parasite antibodies mediate neutralization, opsonized phagocytosis, and/or classical complement activation. Larger extracellular protozoans can be dispatched by ADCC mediated by neutrophils and macrophages.

ib) Th1 responses, IFNγ and macrophage hyperactivation. The Th1 response is critical for anti-protozoan defense because Th1 effectors are key sources of the IFNγ needed to drive macrophage hyperactivation. Like many intracellular bacteria, many protozoan parasites (e.g., *Leishmania major*) infect or are taken up by macrophages but are not destroyed within ordinary phagosomes. These organisms are resistant to or fail to induce the ordinary respiratory burst in activated macrophages. Only in hyperactivated macrophages are sufficient levels of ROIs and RNIs present to efficiently kill such parasites. In addition, TNF secreted by hyperactivated macrophages plays an important role in the control of extracellular protozoans. If hyperactivated macrophages cannot clear the infection, a granuloma is formed (refer to Fig. 13-4).

IFNγ has several other anti-protozoan effects. This cytokine: (1) is directly toxic to various forms of many protozoans; (2)

stimulates IL-12 production by DCs and macrophages, which in turn triggers additional IFNγ production by NK and NKT cells; (3) induces iNOS expression in infected macrophages, resulting in the production of intracellular NO that eliminates either the parasite itself or the entire infected cell; (4) upregulates the expression of enzymes important for phagosome maturation; and (5) upregulates Fas expression on the infected macrophage surface, rendering the macrophage susceptible to Fas-mediated apoptosis when it contacts a FasL-expressing T cell. Because Th2 cytokines such as TGFβ, IL-4, IL-10 and IL-13 inhibit IFNγ production and suppress iNOS, individuals that preferentially mount Th2 responses instead of Th1 responses are highly susceptible to diseases caused by protozoan parasites.

ic) CTLs and γδ T cells.

If a protozoan parasite escapes from a macrophage phagosome into the cytosol of a host cell, parasite antigens may enter the endogenous antigen processing system such that antigenic peptides are presented on MHC class I. The infected host cells then become targets for CTLs. However, perforin/granzyme-mediated cytolysis is not very effective against acute protozoan infections and it is CTL secretion of IFNγ that is this cell type's greatest contribution to the anti-protozoan response. Similarly, IFNγ secretion by activated γδ T cells can bolster the body's defenses during the early stages of protozoan infections. Perforin/granzyme-mediated cytotoxicity becomes important for controlling chronic stages of protozoan infections.

ii) Defense against Helminth Worms

While Th1 responses are needed to combat protozoan parasites, Th2 responses are vital for defense against large, multicellular helminth worms. For example, humans naturally resistant to *Schistosoma mansonii* express high levels of Th2 cytokines, whereas individuals susceptible to this helminth worm exhibit increased concentrations of Th1 cytokines. The anti-helminth Th2 response involves IgE, mast cells and eosinophils, a combination that does not contribute significantly to defense against any other type of pathogen. Activated CD4+ T cells are also critical for anti-helminth defense because these cells differentiate into effectors supplying the Th2 cytokines and CD40L contacts required for isotype switching to IgE by B cells (Fig. 13-8, #1). The anti-parasite IgE antibodies synthesized by the B cells enter the circulation and "arm" mast cells by binding to cell surface FcεRI. When a worm antigen engages the cell-bound IgE, the degranulation of the mast cells is triggered in close proximity to the parasite (#2). Histamine released by the mast cells causes the contraction of host intestinal and bronchial smooth muscles such that the parasite is shaken loose from its grip on the mucosae and expelled from the body. Histamine and other proteins synthesized by mast cells are also directly toxic to some helminth parasites. Circulating IgE directed against worm surface molecules may bind directly to the pathogen, attracting the attention of eosinophils expressing FcεRI molecules. The binding of the worm-bound IgE to eosinophil FcεRs triggers eosinophil degranulation and the release of substances that work directly and indirectly to kill the worm (#3). Some molecules degrade the skin of the worm, allowing other leukocytes to penetrate into its underlying tissues. These cells may also degranulate and release additional toxic proteins and peptides that kill the worm.

The Th2 cytokines IL-4, IL-5 and IL-13 are vital for defense against helminth worms. IL-4 is the main cytokine driving isotype switching in B cells to IgE. IL-5 strongly promotes the proliferation, differentiation and activation of eosinophils, and supports the differentiation of plasma cells that have undergone isotype switching to IgA production (Fig. 13-8, #4). Secretory IgA coats the mucosae and fends off further parasite attachment. IL-4 and IL-13 suppress macrophage production of IL-12, inhibiting IFNγ production and hence the development of a Th1 response (which would be largely ineffective). IL-13 is also required for the bronchial and the gastrointestinal expulsion responses.

III. EVASION STRATEGIES

A parasite that has a multistage life cycle has a variety of opportunities to thwart the immune response (Table 13-9).

i) Avoiding Antibodies

Different parasites employ different strategies to avoid antibodies. Protozoans with multiple life cycle stages can take advantage of the escape offered by antigenic variation. Just as

Table 13-9 **Evasion of the Immune System by Parasites**	
Immune System Element Thwarted	**Parasite Mechanism**
Antibodies	Have a multistage life cycle that furnishes antigenic variation Hide in macrophages Modify parasite surface proteins to cause antigenic variation Acquire host surface proteins that block antibody binding Shed parasite membranes bearing immune complexes Secrete substances that digest antibodies
Phagocytosis	Block fusion of phagosome to lysosome Escape from phagosome into cytoplasm Block respiratory burst Lyse resting phagocytes
Complement	Degrade attached complement components or cleave Fc portions of membrane-bound antibodies Force complement component exhaustion Express homologues of RCA proteins
T cells	Inhibit Th1 response by promoting IL-10 production and decreasing IL-12 and IFNγ production Secrete proteins inducing hyporesponsiveness or tolerance of T cells Interfere with DC maturation and macrophage activation

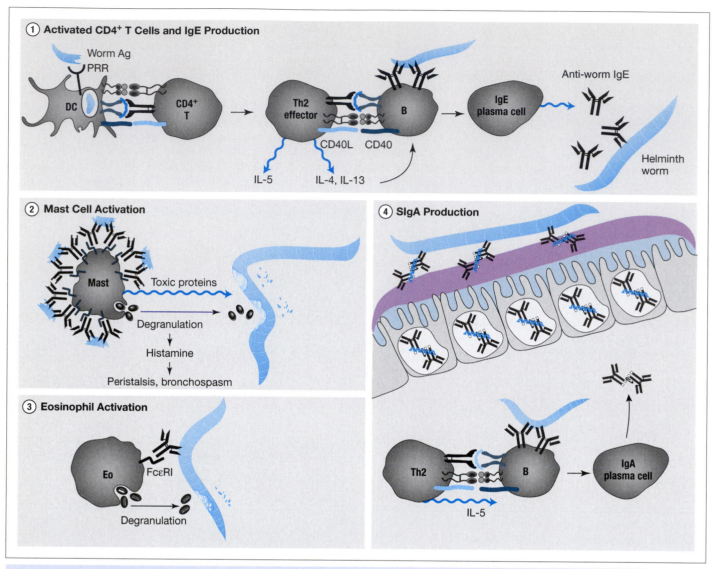

Fig. 13-8 **Major Mechanisms of Immune Defense against Helminth Worm Parasites**

the host mounts a humoral response to epitopes associated with one stage of the parasite, the organism may take on a totally different form and present a whole new panel of epitopes to the host's immune system. A lag in defense ensues while antibodies are produced to counter the new set of antigens. Other protozoans take a more direct approach. *L. major* hides from antibodies by sequestering itself within host macrophages. *Trypanosoma brucei* spontaneously modifies its expression of its *variable surface glycoprotein* (VSG), the molecule that is normally the main target of humoral responses to this parasite. There are hundreds of VSG genes but each trypanosome expresses only one VSG gene at a time. However, the trypanosome regularly shuts down expression of the first VSG gene and activates another, resulting in an altered glycoprotein coat that may not be recognized by antibodies raised against the first VSG protein. Some schistosome helminths disguise themselves by acquiring a coating of host glycolipids and glycoproteins. The dense "forest" created by

these host molecules blocks antibodies from binding to parasite surface antigens. Other helminths repel antibody attack by shedding parts of their external membranes, ejecting the immune complex of the parasite antigen and host antibody. Still other helminths produce substances that digest antibodies.

ii) Avoiding Phagolysosomal Destruction

Helminth worms are in no danger of being phagocytosed, but many protozoans have developed means of avoiding such destruction. For example, some intestinal protozoans lyse resting granulocytes and macrophages and thus minimize their chances of being phagocytosed in the first place. *Toxoplasma gondii* blocks the fusion of macrophage phagosomes to lysosomes. *Trypanosoma cruzi* enzymatically lyses the phagosomal membrane prior to lysosomal fusion and escapes to the cytoplasm of the host cell. *L. major* often remains in the phagosome but interferes with the respiratory burst.

iii) Avoiding Complement

Both protozoans and helminths can take steps to avoid complement. Certain members of both groups can proteolytically remove complement-activating molecules that have attached to their surfaces, or cleave the Fc portions of parasite-bound antibodies. For example, *L. major* can induce the release of the entire complement terminal complex from its surface. Other parasites secrete molecules that force continuous fluid phase complement activation, thereby exhausting complement components. Still other parasites express a molecule that functionally mimics the mammalian RCA protein DAF.

iv) Interference with T Cells

Members of both the protozoan and helminth groups have evolved ways of manipulating the host T cell response to favor parasite survival. For example, *P. falciparum* can promote Th cell secretion of IL-10 rather than IFNγ, resulting in downregulation of MHC class II expression and inhibition of NO production. This pathogen also expresses molecules that cause the RBCs it infects to indirectly interfere with macrophage activation and DC maturation. *L. major* expresses molecules that can bind to CR3 and FcγRs on macrophages and reduce IL-12 production by these cells. The Th1 response that would kill the protozoan is thus inhibited. Nematode hookworms secrete several proteins that induce hyporesponsiveness or even tolerance in host T cells. This state of immunosuppression allows great masses of worms to accumulate in the infected host. Other filarial worms induce the APCs with which they come in contact to decrease their surface expression of MHC class I and II and downregulate other genes involved in antigen presentation. These APCs cannot then participate effectively in T cell activation.

G. Immunity to Fungi

I. DISEASE MECHANISMS

Fungi are either unicellular and grow as discrete cells, or are multicellular and grow in a mass (*mycelium*) of filamentous processes (*hyphae*). *Dimorphic* fungi adopt a unicellular form at one stage in their life cycle and a multicellular form at another stage. All fungal cells have a cell wall like bacteria but also a cell membrane like mammalian cells. Although many fungi live most of their lives in the soil, some live commensally on the topologically external surfaces of the human body. Most fungal species are not harmful to healthy humans but immunocompromised individuals may suffer from acute fungal infections that can sometimes become persistent. When a fungus succeeds in invading the body, it usually heads for the vascular system of the target tissue. Invasion of blood vessels by a growing fungus can choke off the blood supply to the host's organs. Some fungi are *dermatophytes*, filamentous fungi that infect only the skin, hair and nails. Diseases caused by fungal infections are called *mycoses*, and examples of several are given in Table 13-10.

Table 13-10 Examples of Fungi and the Diseases They Cause

Pathogen	Disease
Aspergillus species	Respiratory infections
Blastomyces dermatitidis	Blastomycosis
Candida species	Yeast infections, vaginitis, cystitis
Cryptococcus neoformans	Meningitis, lung infection
Histoplasma capsulatum	Histoplasmosis
Pneumocystis carinii	Pneumonia and lung damage
Dermatophytes	Skin, nail and hair infections

II. IMMUNE EFFECTOR MECHANISMS

Fungal infections are controlled primarily by induced innate immunity. In the tissues, neutrophils and macrophages both carry out vigorous phagocytosis and produce powerful antifungal defensins that induce osmotic imbalance in the fungal cells (Fig. 13-9, #1). Activated neutrophils and macrophages also secrete copious quantities of IL-1, IL-12 and TNF that can be directly toxic to fungal cells. γδ T cells appear to play a poorly defined but significant role in antifungal defense at the mucosae (#2), based on the fact that mice engineered to lack γδ T cells show increased susceptibility to yeast infections. Activated NK cells stimulated by the presence of IL-12 contribute to fungal cell killing via cytokine secretion (rather than by natural cytotoxicity) (#3). TNF has a direct toxic effect on fungal cells, whereas IFNγ contributes to macrophage hyperactivation that can eventually lead to granuloma formation. Later in an infection, DCs acquire fungal antigens and activate naïve T cells to generate Th1 effectors. These T cells secrete the copious quantities of IFNγ needed to complete macrophage hyperactivation (#4). Th2 responses are comparatively rare during fungal infections and not very effective. Those patients that respond to fungi with Th2 responses instead of Th1 responses show poor resistance to these pathogens.

Antibodies are thought to contribute in only a limited way to defense against fungi that manage to invade the body. Antibody-mediated opsonization may promote phagocytosis (Fig. 13-9, #5) and thus contribute to the clearance and presentation of fungal antigens. Fungal cells are also subject to phagocytosis when opsonized by complement products. Although fungal cells can activate the complement cascade, their cell walls render them generally resistant to complement-mediated lysis.

III. EVASION STRATEGIES

Many fungi adopt different forms at different stages in their life cycles, affording them multiple opportunities to evade immune defense (Table 13-11). The cell walls and membranes

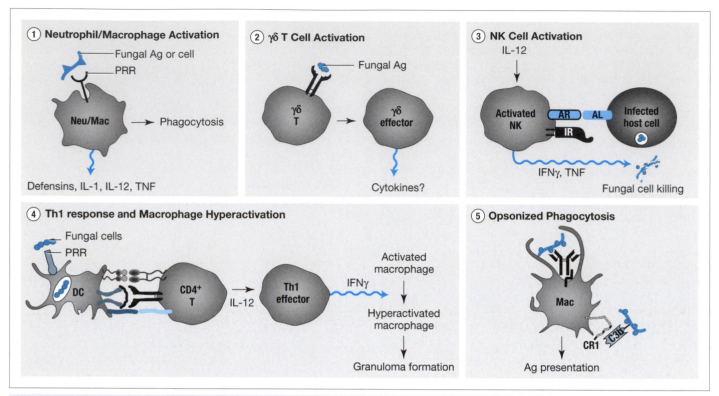

Fig. 13-9 Major Mechanisms of Immune Defense against Fungi

<table>
<tr><td colspan="2">Table 13-11 **Evasion of the Immune System by Fungi**</td></tr>
<tr><td>**Immune System Element Thwarted**</td><td>**Fungal Mechanism**</td></tr>
<tr><td>PRR recognition</td><td>Have no LPS or peptidoglycan in cell wall</td></tr>
<tr><td>Antibodies</td><td>Have a multistage life cycle
Secrete molecules blocking B cell differentiation, proliferation</td></tr>
<tr><td>Complement</td><td>Have a cell wall that blocks access to the cell membrane</td></tr>
<tr><td>Phagocytosis</td><td>Have a capsule that blocks phagocytosis</td></tr>
<tr><td>Macrophages</td><td>Secrete molecules suppressing cytokine production and B7 expression</td></tr>
<tr><td>T cells</td><td>Secrete molecules inducing ineffective Th2 response rather than Th1 response
Secrete molecules blocking T cell differentiation, proliferation
Activate regulatory T cells</td></tr>
</table>

of some fungi lack PAMPs and other structures that might trigger PRR-mediated recognition or complement-mediated lysis. Some fungi are encapsulated such that they resist antibody recognition and phagocytosis. Many fungi produce toxins and other molecules that have immunosuppressive effects. Some of these molecules promote immune deviation to an ineffective Th2 response at the expense of a Th1 response. Other molecules may inhibit the transcription of genes needed for the differentiation of activated T and B cells. Still other fungal molecules suppress lymphocyte proliferation or macrophage cytokine production. A polysaccharide in the capsule of *Cryptococcus neoformans* blocks IL-12 production by monocytes/macrophages, downregulates B7 expression by macrophages, and activates regulatory T cells.

H. Prions

Prions are the pathogens that cause spongiform encephalopathies (SEs), which are neurodegenerative diseases characterized by lesions that render the brain "spongelike." The main human SE, which is invariably fatal, is called *variant Creutzfeldt-Jakob disease* (vCJD). Animal SEs include *scrapie* in sheep and *bovine spongiform encephalopathy* (BSE or "mad cow disease") in cattle. These disorders are associated with the ingestion of infected tissues from an animal suffering from an SE. For example, a cow that consumes cattle feed made from the remains of a contaminated sheep may contract BSE, whereas a human that enjoys a hamburger made from the meat of an infected cow may eventually succumb to vCJD.

Prions are essentially transmissible proteins devoid of nucleic acid. Structurally, a prion is a conformational isomer of a normal mammalian surface glycoprotein. In the original studies of scrapie in sheep, this normal glycoprotein was denoted PrP^c

(prion protein, cellular) and the altered protein was denoted PrP^sc (prion protein, scrapie). PrP^res (prion protein, resistant to proteases) is now used to denote the altered protein in any species. When PrP^res is introduced into a healthy animal, it acts as a template for the refolding of existing host PrP^c molecules into additional copies of PrP^res. The disease-causing prion thus effectively "replicates" itself in a mass conversion of the host's PrP^c molecules to the PrP^res conformation. The misfolded PrP^res protein has profoundly altered properties compared to PrP^c. When the PrP^res protein enters neurons in the brain, it induces degeneration of this organ that is manifested as the clinical signs of SE. Intriguingly, no other part of the body appears to be affected by the presence of PrP^res.

The description just given is of the infectious form of prion disease, in which there is no mutation of the PrP^c gene of the host and no change in the amino acid sequence of the affected PrP^c proteins: the disorder is purely one of protein misfolding. However, rare cases of prion disease do arise spontaneously due to a mutation of an individual's PrP^c gene that results in production of a PrP^res protein. As long as the tissues bearing the PrP^res protein are not later ingested by another animal, there is no transmission of the disease. In rare cases, however, the mutation may occur in a germ cell such that the disease is inherited by the affected animal's offspring.

Prion infection destroys the brain without inducing either a humoral or cell-mediated adaptive response. The host's T cells are usually tolerant to the infectious PrP^res protein, as it is merely a naturally occurring self protein with modified secondary structure. By extension, in the absence of the activation of prion-specific T cells, no Td humoral response can be mounted. Furthermore, although the "foreign" conformation of PrP^res might be recognized by the BCR of a B cell, the antigen itself cannot act as a Ti immunogen because it has neither the large size nor multivalency needed to activate B cells.

This brings us to the end of our description of the mechanisms of natural immune defense against pathogens. To see some of these mechanisms "live" and in action, the reader may wish to watch the animated Immunomovie hosted on the companion Elsevier website (please see the website information in the front of this book). The Immunomovie presents an overview of molecular and cellular aspects of the basic immune responses triggered in response to a virus, an extracellular bacterium, and a helminth worm parasite. In the next chapter of this book, we discuss the "manufactured" immunity to pathogens created by vaccination.

CHAPTER 13 TAKE-HOME MESSAGE

- There are six major types of pathogens: extracellular bacteria, intracellular bacteria, viruses, parasites, fungi and prions.

- Innate immunity mediated by neutrophils, NK cells, NKT cells, γδ T cells, complement and microbicidal molecules either foils the establishment of infection or slows it down until adaptive immune mechanisms can target the pathogen more effectively.

- The adaptive elements that will be most effective depend on the nature of the pathogen: extracellular versus intracellular, small versus large, fast- versus slow-replicating.

- Most extracellular entities can be coated in antibody and cleared by antibody- and complement-mediated mechanisms. Parasitic worms are targeted by IgA and IgE antibodies that prevent the worm from anchoring in the host. IgE can trigger the degranulation of mast cells and eosinophils and the release of mediators that work to expel the worm and degrade its tissues.

- Intracellular bacteria and parasites and replicating viruses must be eliminated by cell-mediated immunity. CTLs, NK cells, NKT cells and γδ T cells secrete cytotoxic cytokines and/or carry out target cell cytolysis.

- In general, Th1 responses support cell-mediated immunity against internal threats, whereas Th2 responses are needed for humoral responses against external threats.

- Many pathogens have evolved complex strategies to evade the immune response: avoiding recognition; antigenic variation; avoiding or inactivating phagocytosis; shedding or inactivating complement components; acquiring host RCA proteins; cleaving host FcRs; inducing host cell apoptosis; and manipulating the host's immune response or cell cycle.

Section A

1) Characterize the six major classes of pathogens.

2) How is disease distinct from infection?

3) What is immunopathic disease?

Section B

1) Can you define these terms? opportunistic, invasive, keratin.

2) Outline two ways each by which the skin and mucosae defend the body's external surfaces.

3) Name four types of leukocytes that mediate subepithelial innate defense and give examples of their effector mechanisms.

Section C

1) Distinguish between exotoxins, endotoxins and antitoxins.

2) Name two diseases caused by exotoxins.

3) What is endotoxic shock and why is it considered immunopathic?

4) Describe three ways in which antibodies help protect the body against extracellular bacteria.

5) Outline two mechanisms each by which extracellular bacteria can evade antibodies; phagocytosis; complement.

Section D

1) Can you define these terms? vector, granuloma.

2) How are neutrophils, NK cells and CD4$^+$ T cells helpful in combating intracellular bacteria?

3) Why is macrophage hyperactivation effective against intracellular bacteria?

4) Briefly outline the sequence of cellular and molecular events that leads from an intracellular bacterium successfully gaining a foothold in the body to CTL-mediated elimination of cells infected with that bacterium.

5) How can antibodies be useful for the control of intracellular bacteria?

6) Name the types of leukocytes crucial for granuloma formation and outline the role each plays.

7) Outline five mechanisms by which intracellular bacteria can evade phagosomal death.

8) Outline one mechanism each by which intracellular bacteria can avoid antibodies; T cells.

Section E

1) Can you define these terms? acute disease, chronic disease, persistent infection, latency, oncogenic, iNOS, syncytia.

2) What is the antiviral state and how is it induced?

3) Give three functions that are common to all three IFNs and two unique to IFNγ.

4) How do NK cells combat viruses and why are these cells crucial for early defense?

5) Describe three ways in which CD4$^+$ T cells contribute to immune defense against viruses.

6) Describe four ways in which antibodies contribute to immune defense against viruses.

7) Describe three ways CTLs kill virus-infected cells.

8) Describe two mechanisms of latency and outline how viral latency can be reversed.

9) Distinguish between antigenic drift and antigenic shift.

10) Describe three ways each by which viruses can resist attack by CTLs and CD4$^+$ T cells.

11) Outline three ways each in which viruses can counteract the antiviral state; inhibit the induction of host cell apoptosis; interfere with host cell cytokines.

Section F

1) What do protozoan and helminth pathogens have in common?

2) How does a multistage life cycle present a challenge for the immune system?

3) How do antibodies combat protozoans?

4) Give four reasons why the Th1 response is crucial for anti-protozoan defense.

5) What CTL-mediated mechanism is most effective against protozoans and when?

6) What four types of leukocytes are involved in the Th2 response against helminth worms and why are these cells important?

7) Outline three ways in which the contents of eosinophil granules combat helminths.

8) What three cytokines are important for anti-helminth defense?

9) Outline two ways each by which protozoans can avoid antibodies; phagosomes; complement.

Section G

1) Can you define these terms? mycelium, hyphae, dimorphic, dermatophyte, mycoses.

2) Describe four mechanisms of innate defense that combat fungi.

3) Outline three ways in which the structure or products of fungi promote evasion of immune responses.

Section H

1) Can you define these terms? SE, vCJD, BSE, PrP^{res}, PrP^c.

2) What is a prion and how does it cause disease?

3) Give two reasons why prions are poorly immunogenic.

14

Vaccination

The aim of military training is not just to prepare men for battle, but to make them long for it.
Louis Simpson

Vaccination is a clinical application of immunization designed to artificially help the body to defend itself. A *vaccine* against infection is a modified form of a natural immunogen, either a pathogen or toxin. A vaccine does not cause disease when administered but induces the healthy host (the *vaccinee*) to mount a primary response against epitopes of the modified immunogen and to generate large numbers of memory B and T cells. In an unvaccinated individual (Fig. 14-1, left panel), naïve B and T cells capable of combating an infecting pathogen are present in relatively low numbers when the pathogen is first encountered. A primary immune response is all that can be mounted so that, in many cases, the individual becomes sick until antibodies and/or effector T cells can act to clear the attacker. In a vaccinated individual (Fig. 14-1, right panel), a collection of circulating antibodies and an expanded army of pathogen-specific memory B and T cells has already been generated prior to a first exposure to the natural pathogen. When the natural pathogen attacks, the circulating antibodies provide a degree of immediate protection from the invader. In addition, the memory B and T cells are quickly activated and a secondary response is mounted that rapidly clears the infection before it can cause serious illness. This type of vaccination is called *prophylactic vaccination* because it is intended to prevent disease.

Today's best known vaccination success story is the global campaign of the World Health Organization (WHO) to eradicate smallpox. In 1967, the WHO began its coordination of 200,000 health workers who took 10 years to vaccinate the world's population in its remotest corners. Beween 1976 and 1979, only one case of smallpox was recorded, leading to the declaration in 1980 that smallpox had been officially eradicated (Plate 14-1). A similar global immunization program against poliovirus is currently pushing this pathogen toward extinction. Control of several other serious diseases, including diphtheria, tetanus, pertussis, measles, mumps and rubella, has also been achieved through vaccination, at least in the developed world. Although mass vaccination programs do have a monetary cost, the expenses associated with a population becoming infected are much higher. According to the WHO,

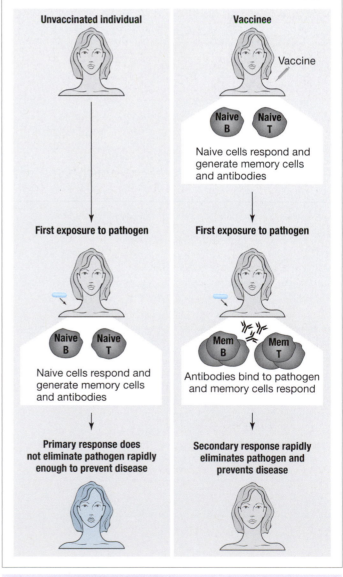

Fig. 14-1 **The Principle of Vaccination**

Plate 14-1 **WHO Declaration of Smallpox Eradication in 1980**
[Reproduced by permission of the Magazine of the World Health Organization.]

Table 14-1 Global Annual Leading Causes of Death)*

Rank	Cause	Percentage of Total	Number of Deaths
1	Cardiovascular disease	29.3	16,585,000
2	Cancers	12.6	7,115,000
3	**Lower respiratory infections**	**6.8**	**3,871,000**
4	**HIV/AIDS**	**5.1**	**2,866,000**
5	Pulmonary disease	4.7	2,672,000
6	Perinatal conditions	4.4	2,504,000
7	**Diarrheal diseases**	**3.5**	**2,001,000**
8	**Tuberculosis**	**2.9**	**1,644,000**
9	**Childhood diseases**[†]	**2.3**	**1,318,000**
10	Road traffic accidents	2.1	1,194,000
11	**Malaria**	**2.0**	**1,124,000**
12	Diabetes	1.6	895,000

*WHO 2002; entries in bold are infectious diseases. [†]The WHO includes diphtheria, pertussis, tetanus, poliomyelitis and measles in the childhood diseases category.

for every U.S. $1 million spent on global childhood vaccines, U.S. $4 million in long-term health care savings are realized. Despite these successes, there are still major challenges in vaccination, especially in the developing world. Communicable diseases represent 6 of the top 12 causes of mortality worldwide, and account for about 20% of all the world's annual deaths (Table 14-1). In some cases (e.g., measles), logistical, social or economic factors prevent delivery of vaccines that are currently available and would be effective if administered. In other cases (e.g., malaria, AIDS and tuberculosis), the elusive nature of the pathogen involved has hindered the development of an effective vaccine.

A. Vaccine Design

In designing a vaccine, several factors that can affect its success must be taken into consideration. These factors are discussed next and summarized in Table 14-2.

I. EFFICACY AND SAFETY

A successful vaccine has a high level of *efficacy*; that is, it is effective in protecting the vaccinee from disease. Accordingly, the immune response that the vaccine induces must be appro-

Table 14-2 Characteristics of a Successful Vaccine

Characteristic	Description
Vaccine efficacy	Stimulates vigorous and appropriate Td response leading to pathogen elimination by antibody or CTLs as well as development of memory T and B cells Coverage of vaccinated population is typically 80–95%
Vaccine safety	No risk of causing disease Side effects are not worse than natural disease symptoms
Pathogen	Causes acute rather than chronic disease Induces immunity upon natural exposure Undergoes little antigenic variation Does not attack cells of the immune system Does not have an environmental or animal reservoir

priate for the elimination of the pathogen of interest. For example, an extracellular pathogen is best countered by antibodies, so that the vaccine must be capable of activating B cells. Furthermore, the antibodies produced must be of an isotype known to be effective against that particular pathogen. For example, SIgA best protects the mucosae but IgG is needed to immobilize pathogens in the blood. Similarly, if a pathogen is intracellular, the vaccine should activate CD8+ T cells.

The efficacy of a vaccine is often expressed as its *coverage:* the percentage of individuals vaccinated who do not experience disease after exposure to the pathogen. No vaccine has yet proved 100% effective due to the inherent genetic variation among humans, but most standard vaccines given in childhood are effective in 80–95% of a given population; that is, the coverage of these vaccines is 80–95%. It is here that the concept of *herd immunity* comes into play. If a pathogen cannot gain a foothold in a population because most members respond to the vaccine, even individuals who do not respond to the vaccine will be protected because the chance of any individual "in the herd" being exposed to the pathogen is much reduced.

Epitopes that induce an effective anti-pathogen response are known as *protective epitopes* and are derived from *protective antigens*. It is these protective antigens that are usually chosen to formulate a vaccine. Candidate vaccine epitopes are often first identified by the analysis of gene and protein sequences of an antigen. An antiserum is then raised against the candidate epitope in an animal and shown to neutralize the pathogen *in vitro*. Only if this result is positive is *in vivo* testing carried out to definitively prove that an epitope is protective and potentially useful as a vaccine.

In addition to being effective, a vaccine needs to be safe; that is, it should have very few detrimental side effects. Such side effects range from redness and tenderness at the injection site to high fever, seizures, pneumonia, encephalitis and even death. Balancing the severity of vaccine side effects against the incidence and severity of the disease gives different results in different parts of the world. Where disease incidence is low, as in the developed world, a vaccine that results in relatively severe side effects is not used. In contrast, where disease incidence is much higher, as in the developing world, the risk of harm from the disease may outweigh the risk of harm from vaccine side effects.

Before any candidate vaccine can be administered to humans, both its safety and efficacy must be determined in a series of *preclinical* and *clinical trials*. Preclinical trials test vaccines in cell cultures or animal models, whereas clinical trials test vaccines in human volunteers (Table 14-3). Vaccine testing is not always straightforward. For certain pathogens, a good animal model (in which immune responses and side effects are analogous to those in humans) may not be available, impeding the early stages of vaccine development. Another problem arises in situations where it would be unethical to deliberately administer a highly pathogenic infectious agent (e.g., HIV) as a test challenge to human volunteers that have been given a candidate vaccine. In these cases, a clinical trial is organized in which the candidate vaccine is given to individuals already living in an area where the pathogen is endemic.

II. PATHOGEN CHARACTERISTICS

Several characteristics of the pathogen itself can have a major influence on the success of a vaccine. Good pathogen candidates for vaccination programs are those that cause acute rather than chronic disease because these pathogens tend to invoke a vigorous immune response that leaves survivors

Table 14-3 Preclinical and Clinical Trials

Clinical Trial Stage	Protocol
Preclinical	
Trial I	Test vaccine for toxicity and immunogenicity in cell cultures and non-primate animals
Trial II	Test vaccine for toxicity and immunogenicity in non-human primates
Clinical	
Phase I trial	Test vaccine for immunogenicity, dose–response range, optimal route of administration and side effects in a small number of healthy human volunteers
Phase II trial	Test vaccine for optimal route of administration and dose as well as efficacy against challenge with the pathogen in a small number of human volunteers; alternatively, vaccinate a small group of individuals likely to be exposed to the pathogen
Phase III trial	Test vaccine for safety and efficacy against challenge with the pathogen in a large group of healthy human volunteers; alternatively, vaccinate a large group of individuals likely to be exposed to the pathogen
Phase IV trial	Monitor vaccinees in the general population for rare detrimental side effects and to determine how well the vaccine reduces disease overall

with very long-lasting or even permanent immunity. A vaccine derived from such a pathogen is likely to induce a similar level of protective immunity. In contrast, the natural immune response to a pathogen that causes chronic disease is typically inadequate to clear the pathogen. It is therefore less likely that a vaccine derived from such a pathogen will be successful in stimulating a protective immune response in a vaccinee.

Pathogens that do not exhibit a high degree of antigenic variation (e.g., measles virus) are also favored as vaccine candidates because memory B and T cells and their antibodies and effector cells will continue to recognize the pathogen in successive exposures. Conversely, pathogens that evade the immune response by undergoing extensive antigenic variation (e.g., influenza virus, HIV) are poor vaccine targets. Similarly, a pathogen with many different life cycle stages and forms (e.g., the malarial parasite *Plasmodium falciparum*) can confound vaccine design because protective antigens that remain invariant from life stage to life stage are rare. HIV presents an additional and unique challenge in vaccine design as this virus decimates an individual's CD4$^+$ T cell population, the very cells needed to respond to any vaccine.

The ultimate goal of a vaccination program is to completely eradicate a pathogen from a population so that, as in the case of smallpox, vaccination against the pathogen will no longer be required. The best candidates in this regard are pathogens that infect only humans, meaning that they cannot escape into an animal or an environmental *reservoir*. A reservoir in this sense is a species or environmental niche outside of the human population in which a pathogen can survive. If a reservoir does exist, then once an entire human population is vaccinated and the pathogen has run out of human hosts to infect, the pathogen can retire to its reservoir. If the vaccination program is terminated at this point, there will soon be a new generation of individuals born who are susceptible to the pathogen. When the pathogen eventually emerges from its reservoir in search of fresh human hosts, it readily attacks the unvaccinated individuals and regains its foothold in the human population. Eradication efforts are thus thwarted.

B. Types of Vaccines

Table 14-4 summarizes the many different ways to construct a vaccine as well as the major advantages and disadvantages of each type. Note that the types of vaccines that work best for one class of pathogen may not work at all for another.

I. LIVE, ATTENUATED VACCINES

A *live, attenuated vaccine* consists of live, whole bacterial cells or viruses that are treated in such a way that their pathogenicity is reduced but their immunogenicity is retained. Because a viable, replicating pathogen is used, large quantities of the immunogen are produced in the vaccinee and both the innate and adaptive arms of the immune system are readily triggered. Moreover, most attenuated pathogens supply both B and T epitopes such that both humoral and cell-mediated adaptive responses are mounted. However, although live, attenuated vaccines are generally very effective, a single dose is not usually enough to induce the long-lasting immunity experienced by individuals who were infected with the natural pathogen. A *booster* (repeat dose of vaccine) must usually be administered to induce faster, stronger secondary immune responses and bolster protection. A disadvantage of live, attenuated vaccines compared to other vaccine types is that, in order to maintain efficacy, a "cold chain" of equipment and procedures is required to ensure that the live vaccine is kept chilled from the time of production to the moment of vaccination.

Sometimes nature creates a live, attenuated vaccine for us. The first example of a live, attenuated vaccine was Jenner's cowpox virus (refer to Ch. 1). In this almost unique case, an

Table 14-4 Types of Vaccines and Their Pros and Cons

Type	Description	Major Pros	Major Cons
Live, attenuated	Whole pathogen treated to decrease pathogenicity but maintain immunogenicity; may still replicate	Low number of doses usually very effective Minimal need for adjuvant* Supplies B and T epitopes	Cold chain required Chance of reversion of attenuating mutation
Killed or inactivated	Whole pathogen killed or inactivated to block replication but maintain immunogenicity	No possibility of reversion No cold chain required Supplies T and B epitopes	Cannot replicate so requires boosters and adjuvant Does not induce robust Tc responses
Toxoid	Chemically inactivated toxin of pathogen	No need to use whole organism	Only effective if disease caused solely by toxin
Subunit	Pathogen protein or polysaccharide purified from natural sources or synthesized using recombinant DNA methods	Avoids use of whole organism Can be manipulated to increase immunogenicity	Can be costly to produce May not be as immunogenic as natural pathogen component Does not induce robust Tc responses
Peptide	Pathogen peptide purified from natural sources or synthesized using recombinant DNA methods	Avoids use of whole organism Composition is known Very stable	Epitope size and number restricted May require coupling to a carrier protein
Recombinant DNA vector	Virus-based vector containing recombinant DNA of pathogen antigen. Vaccinee is infected with the viral vector and the pathogen DNA is transcribed and translated within the vaccinee's cells like a viral protein.	Avoids use of natural pathogen Replicates like a pathogen to produce large amounts of immunogen Supplies T and B epitopes Minimal need for boosters and adjuvant	Possible side effects due to vector components Anti-vector antibodies raised during priming may necessitate boosting with a different vector
Naked DNA	Small plasmid containing recombinant pathogen DNA. Plasmid is injected into a vaccinee and the pathogen DNA is taken up by the vaccinee's cells and transcribed and translated.	Easy and inexpensive to manipulate Induces B, Th and Tc responses Plasmid sequences may act as adjuvant	Integration of plasmid into host cell genome may induce tumorigenesis

*a substance that enhances local inflammation and immune responses to vaccine antigens.

intact, mildly pathogenic virus (cowpox) induced an immune response that protected against a related but much more deadly virus (smallpox). Usually, however, less virulent versions of the same pathogen must be created in the laboratory. Attenuating mutations of a pathogen were originally induced by culturing the pathogen for extended periods under less than optimal conditions (i.e., conditions very different from the physiological conditions in its human host). This approach selects for organisms that have undergone mutations such that they now grow well in the suboptimal setting but can no longer cause disease in the host (although they may still be able to replicate). A virus can be attenuated by introducing it into a species in which it does not replicate well (i.e., infection of an animal with a human virus), or by forcing the virus to replicate repeatedly in cells maintained in laboratory culture vessels, a protocol called *passaging*.

The most modern methods of attenuation use recombinant DNA technology to directly mutate or delete genes encoding proteins known to contribute to pathogen virulence. Complete deletion, rather than just point mutation, of a virulence gene is preferred because it decreases the danger that an attenuated pathogen may recover its virulence through reversion of the attenuating mutation. Such reversions can lead to the vaccinee coming down with the very disease the vaccine was designed to prevent.

II. KILLED VACCINES

In a *killed vaccine*, a whole bacterium, parasite or virus is killed or inactivated by treatment with gamma irradiation or a chemical agent such as formaldehyde. Used correctly, these procedures preserve the structure of the protective epitopes but remove the pathogen's ability to replicate or recover virulence. Killed vaccines are generally more stable than live vaccines and less sensitive to cold chain disruptions. However, because the treated pathogens cannot replicate, larger amounts of these vaccines must be administered in the primary dose, raising costs. In addition, because killed vaccines are dead and therefore trigger less intense "danger signals" than live vaccines, the response to a given dose is weaker and frequent boosters are required. A greater problem is that killed vaccines offer only limited protection against intracellular pathogens. Because the vaccine organism is dead, it cannot actively penetrate host cells. The processing of the organism's antigens by the endogenous antigen processing pathway and presentation on MHC class I are therefore limited. As a result, Tc cell activation and CTL generation by this route do not occur. A DC that has phagocytosed the killed vaccine may cross-present peptides derived from it but the levels of pMHC class I generated activate only limited numbers of Tc cells. Overall, killed vaccines tend to induce predominantly systemic humoral responses featuring neutralizing antibodies, a type of immune response that is not very effective against intracellular pathogens.

III. TOXOIDS

As described in Chapter 13, the disease caused by a pathogen may be entirely due to its production of exotoxins. Thus, vaccinating a host against the exotoxin protects the host from disease (if not infection). *Toxoids* are exotoxin molecules that have been chemically altered (usually by formalin treatment) such that they lose their toxicity but not their immunogenicity. Neutralizing antibodies generated in response to toxoid administration bind to the exotoxin and render it harmless. For example, vaccination or boosting with the tetanus toxoid protects against this disease for at least 5 years. Diphtheria and pertussis are also prevented by toxoid administration.

IV. SUBUNIT VACCINES

A *subunit vaccine* contains a protein or polysaccharide that has been purified from a pathogen and contains at least one protective epitope. A major advantage of such vaccines is that a whole organism is never used, avoiding any risk of reversion and the possibility of side effects due to irrelevant pathogen components.

i) Protein Subunit Vaccines

Some protein subunit vaccines are prepared using conventional protein purification techniques. However, these techniques are labor-intensive and not very efficient, making these vaccines very costly. Recombinant DNA technology has greatly simplified the synthesis of vaccines for pathogens that are difficult or impossible to grow *in vitro*, and/or whose components are a challenge to purify in sufficient amounts from *in vivo* infections. To construct a protein subunit vaccine using recombinant DNA technology, the DNA encoding the vaccine antigen is introduced into the genome of a microorganism such as yeast or *Escherichia coli*. These microbes can easily be cultured in huge volumes in the laboratory and the pathogen protein is synthesized in correspondingly high amounts along with other microbial proteins. A polypeptide of interest can then be isolated from among the newly synthesized proteins. The manipulation of the pathogen's DNA prior to its introduction into the microbe also allows for deliberate mutation that confers greater ease of purification or increased immunogenicity. The use of the recombinant DNA method of protein preparation also reduces the chance that potentially harmful pathogen components will be copurified with the protective antigen.

Protein subunit vaccines do have some disadvantages. Sometimes the production of a vaccine protein in a recombinant organism alters the conformation of the protein such that its stability is affected or protective conformational epitopes are no longer present. As a consequence, certain neutralizing antibodies may not be induced in the vaccinee. As well, like killed vaccines, subunit vaccines are not alive and so cannot penetrate host cells. The vaccine epitopes are thus inefficiently presented on MHC class I and can activate Tc cells only by cross-presentation, resulting in a relatively weak CTL response.

ii) Polysaccharide Subunit Vaccines

Subunit vaccines can also be made from pathogen polysaccharides. Polysaccharides are known to be important for the pathogenicity of many encapsulated bacteria. In addition, capsule polysaccharides are abundant on bacterial surfaces and thus are easy to purify. Importantly, polysaccharides do not have adverse side effects when injected into animals or humans. However, polysaccharides are not proteins and thus do not provide T cell epitopes. Instead, polysaccharides tend to provoke Ti responses that do not generate memory lymphocytes. Indeed, many polysaccharides do not induce immune responses of any type in children under 2 years of age. Vaccine efficacy in these cases has been greatly increased by chemically joining a bacterial capsule polysaccharide to a carrier protein capable of supplying a T cell epitope and thus provoking a Td response. The combination of carrier protein plus polysaccharide is called a *conjugate vaccine*. The diphtheria and tetanus toxoids and modified versions of these proteins are frequently used as carriers in conjugate vaccines against various pathogens.

V. PEPTIDE VACCINES

Some isolated peptides function as protective T and B epitopes and so can serve as vaccines. With a *peptide vaccine*, the precise molecular composition of the vaccine is known and there is no possibility of reversion to a pathogenic phenotype. In addition, due to the relatively small size of the vaccine agent, it is less likely that larger entities such as infectious agents or genomic material will copurify with the vaccine and contaminate it. Both natural and synthetic peptides have been explored as vaccine candidates, and peptides have been produced both by conventional purification methods and by recombinant DNA technology. Where a natural peptide is not immunogenic, it can be modified in the laboratory to become so. Once a protective peptide has been identified and purified, it is mixed with an adjuvant (see later) that encourages its uptake by a vaccinee's APCs, including DCs. The peptide is then presented and cross-presented on MHC class II and class I to naïve Th and Tc cells, respectively.

Peptide vaccines have their own set of disadvantages. Because peptides are by definition short, the epitopes contained in such vaccines tend to be small, linear and non-conformational. However, most B cell epitopes on a whole pathogen are conformational, so that the antibodies produced in response to a peptide vaccine may offer only limited protection during a natural infection. To verify the immunogenicity of a peptide representing a B cell epitope, scientists often examine the kinetics of the binding of the candidate peptide to antibodies from the serum of a patient with the disease of interest. The verification of peptides representing T cell epitopes is more complicated because the peptide and the MHC molecule that presents it are seen as a unit by the TCR. The peptide must therefore contain not only the epitope recognized by the TCR but also amino acids allowing it to bind to MHC. Different MHC alleles may bind to and present a given peptide with varying degrees of success, causing a variation in the level of immunity induced in vaccinees.

VI. DNA VACCINES

DNA vaccination involves the introduction into a vaccinee of pathogen-derived DNA sequences that direct the synthesis of immunogenic pathogen proteins *in vivo*. *DNA vaccines* include *recombinant vector vaccines* and *naked DNA vaccines*. Although several recombinant vector vaccines have made it to the early clinical trial stage, naked DNA vaccines are still highly experimental.

i) Recombinant Vector Vaccines

A recombinant vector vaccine uses an unrelated attenuated virus or bacterium as a vector to introduce DNA from the pathogen of interest into the vaccinee. These vectors can penetrate human cells and often replicate within them but do not cause disease in the host. The vaccine is "recombinant" because genes encoding the pathogen antigen of interest and a selectable marker (used for purification) are incorporated into the vector using recombinant DNA technology. After injection into a vaccinee, the recombinant vector infects host cells just like an unmodified virus or bacterium and the vaccine gene(s) is transcribed and translated like a viral or bacterial component. If the host cell infected is a DC, the vaccine protein enters the endogenous antigen presentation pathway. Vaccine peptides are thus displayed on MHC class I and Tc cells are activated. Alternatively, the vaccine protein may be released from the synthesizing host cell and activate B cells directly, or may be taken up and processed by DCs via the exogenous antigen presentation pathway and activate Th cells. Cross-presentation by DCs may also activate Tc cells.

The organism most often used as a recombinant vector is the vaccinia virus. Vaccinia is a large, complex poxvirus, with many genes that are not essential for host cell invasion and replication. Consequently, these genes can be replaced in the vaccinia genome with foreign DNA encoding protective antigens from a pathogen of interest. Recombinant vectors based on poliovirus are also being explored because the natural route of infection of this virus is through the digestive tract. Thus, vectors based on poliovirus should be able to be administered orally and so induce both systemic and mucosal responses to an immunogen of interest. However, the genome of poliovirus can accept only a small amount of foreign DNA, and at least one poliovirus strain is known to be genetically unstable. Human adenovirus is also an attractive vector candidate because this virus is easily cultured *in vitro* and large amounts of foreign DNA can be inserted into its genome without affecting viral replication.

A major problem with recombinant vector vaccines is that the vaccinee may mount an immune response not only against the inserted pathogen antigen (as desired) but also against irrelevant vector components. If anti-vector antibodies are produced in the vaccinee following the first administration (priming), and the recombinant vector vaccine is administered a second time (boosting), the anti-vector antibodies may bind to the vaccine and prevent it from accessing host cells. As a result, the booster dose of pathogen antigen is never synthesized and the host is not properly immunized. This difficulty can be avoided if the priming is done with a recombinant vector vaccine but the boosting is done with a protein subunit vaccine (which does not contain the vector). Alternatively, the boosting can be done with the same pathogen antigen incorporated into a different vector.

ii) Naked DNA Vaccines

To prepare a naked DNA vaccine, the gene encoding the pathogen antigen of interest is cloned into a plasmid that can replicate to a high copy number in *E. coli* but not at all in human cells. The bacteria are grown to large quantities and lysed, and the vaccine plasmid is purified and injected into the vaccinee. The DNA is said to be "naked" because it is not contained within a virion or bacterium. Once the plasmid is taken up by a host cell, the cell commences production of the pathogen antigen. Protective Tc responses as well as Th and B cell responses are then mounted. Concerns with these vaccines center around avoiding integration of the plasmid into the host cell genome, which might mutate a host gene. Such integration has been linked to tumorigenesis in experimental animals.

C. Adjuvants and Delivery Vehicles

Live, attenuated vaccines and many recombinant vectors often retain the capacity of a pathogen to penetrate tissues, replicate in the host and induce inflammation. For these reasons, these types of vaccines are highly immunogenic. Killed vaccines, purified subunit and peptide vaccines, and naked DNA vaccines lack these properties, meaning that they need help to induce effective immune responses. In addition, protein and DNA molecules introduced into a vaccinee may encounter extracellular proteases and nucleases that degrade the vaccine before it can be taken up by APCs. To avoid these impediments to efficacy, vaccines are often administered with substances called *adjuvants* and *delivery vehicles*. Adjuvants enhance local inflammation and thus the ability of a vaccine antigen to induce an immune response. In addition, adjuvants mediate a "depot effect" that allows antigen to continuously leak into the body and boost the response. Delivery vehicles protect vaccine antigens from degradation. In some cases, a single entity can act as both an adjuvant and a delivery vehicle.

I. ADJUVANTS

Although there are several adjuvants available for use in experimental animals, the only adjuvant currently licensed for routine use in humans is *alum*. Alum is a gel composed of aluminum hydroxide or aluminum phosphate. The inflammation induced by alum administration tends to promote antibody rather than cell-mediated responses but it is still not understood exactly how this agent works. The use of alum is not wholly without drawbacks, as it is normally injected deep into muscle. This approach ensures that vaccinees (children, in particular) are properly vaccinated but increases vaccinee discomfort. Scientists are still striving to develop a human adjuvant that is effective but painless to administer.

II. DELIVERY VEHICLES

Significant increases in vaccine-induced responses can be achieved by administering the vaccine in a non-toxic delivery vehicle. The vehicle protects the vaccine molecules from protease or nuclease degradation (increasing its persistence in the tissues) and may also act as an adjuvant (inducing inflammation). Some vehicles facilitate the display of multiple molecules of the vaccine antigen on the vehicle surface, creating a multivalent form of the antigen that increases its immunogenicity. These properties have made delivery vehicles invaluable adjuncts for experimental vaccination with subunit and DNA vaccines. To date, however, the use of delivery vehicles for human vaccination has been limited.

Several different types of delivery vehicles have been devised. *Liposomes* are prepared by mixing the vaccine antigen of interest with a suspension of phospholipids under conditions that favor the formation of a spherical membranous structure. The vaccine antigen is trapped in the aqueous center of the hydrophobic liposome. Liposomes are readily phagocytosed by DCs and macrophages, meaning that the antigen is soon processed and used to initiate T cell activation.

Virosomes are "artificial viruses" that can be used to deliver vaccine antigens directly into a host cell. A virosome is basically a liposome that displays the envelope glycoproteins of a virus on its surface. Pathogen antigens of interest are either captured within the lumen of the virosome or are chemically cross-linked to its surface. Because of its viral surface proteins, a virosome can bind to and "infect" host cells and deliver the antigen directly into the MHC class I antigen processing pathway. Alternatively, the virosome may be phagocytosed by an APC.

ISCOMs or "immunostimulating complexes" are hollow balls made of cholesterol, phospholipid and detergent. Relatively bulky protein immunogens can easily be inserted into the interior of the ball. To the immune system, the ISCOM resembles a multivalent antigen with a shape that invites phagocytosis by APCs. In addition, the detergent in the ISCOM is a powerful adjuvant.

D. Prophylactic Vaccines

At the time of writing, there are over 25 human infectious diseases for which safe and effective vaccines are available. In most developed countries, many of these vaccines are given as part of a schedule of standard childhood immunizations against endemic pathogens. Other vaccines are used only in special circumstances.

I. STANDARD IMMUNIZATIONS

Most prophylactic vaccines are designed to be administered during infancy, soon enough to prevent the onset of childhood disease but late enough to avoid the establishment of immune tolerance. Booster doses of these vaccines are usually given two to three times, with a gap of several weeks or months in between. To circumvent the need for multiple injections and repeated clinic visits, single vaccines have been created that contain antigens from several different pathogens: these are called *combination vaccines*. An adaptation of the Recom-

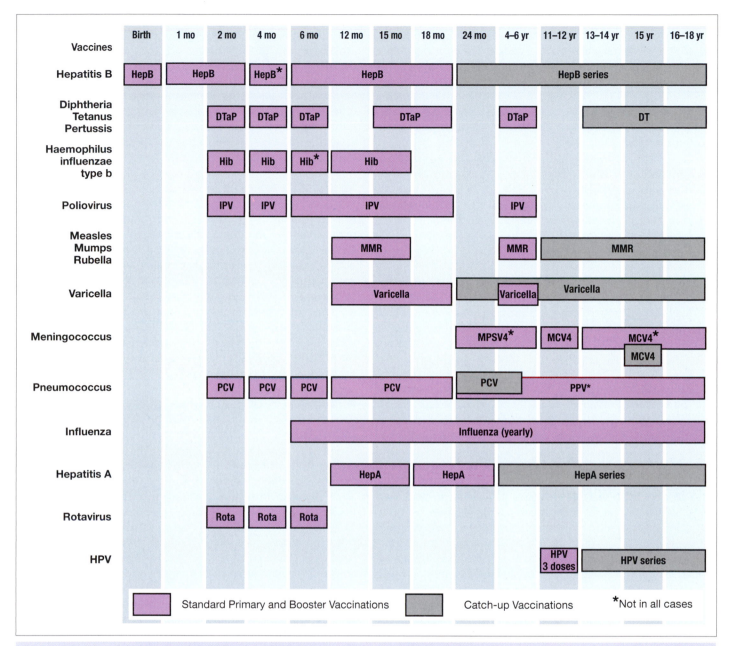

Fig. 14-2 Adaptation of the Recommended Childhood and Adolescent Immunization Schedule in the United States
[Adapted from the Recommended Immunization Schedules for Persons Aged 0–6 Years and 7–18 Years, 2007; most current information is at http://www.cdc.gov/vaccines/recs/schedules]

mended Childhood and Adolescent Immunization Schedule for the United States is shown in Figure 14-2. (Recommended vaccine schedules can vary considerably by country.) If vaccinations are missed in infancy, an individual can receive "catch-up" vaccinations later in life. There is some debate among immunologists (at least in the Western world) as to when childhood vaccination should start. Some feel that, rather than starting at birth, vaccination should be delayed until 6 months of age, when the infant's immune system has matured somewhat and the protection afforded by the maternal antibodies in the infant's circulation starts to fade.

Some vaccines, particularly live, attenuated viruses, induce very strong memory responses even after the priming dose.

Thus, boosting may only be necessary at intervals of several years (if at all) in order to sustain life-long immunity. Alternatively, the vaccinee may naturally encounter the pathogen itself often enough to keep triggering memory responses without the need for more than a single dose of vaccine. When a booster is required, it does not have to involve the same vaccine formulation as the priming dose. For example, a live, attenuated vaccine might be followed by a killed or protein subunit vaccine.

We next discuss the most common standard prophylactic vaccines and the diseases they aim to prevent. If left unchecked, many of these disorders would hospitalize, permanently disable, disfigure or kill a significant number of individuals. Thousands more would be transiently incapacitated.

i) Diphtheria, Tetanus and Pertussis

ia) Diphtheria. Diphtheria is a devastating childhood disease caused by the exotoxin of the bacterium *Corynebacterium diphtheriae*. The exotoxin inhibits protein synthesis in cells of the heart and nervous system and also induces an inflammatory response in the throat that can obstruct breathing. Disease can be prevented by adequate levels of circulating antibodies to the exotoxin. The current vaccine is a formalin-inactivated exotoxin toxoid.

ib) Tetanus. Tetanus is caused by the exotoxin produced by the bacterium *Clostridium tetani*. The exotoxin attacks neurons and causes painful muscle spasms. Because the disease is caused solely by the exotoxin, neutralizing antibodies raised in response to vaccination with an exotoxin toxoid are protective.

ic) Pertussis. Whooping cough is a highly contagious disease caused by the bacterium *Bordetella pertussis*. *B. pertussis* gravitates to the mucosae of the bronchi and produces two exotoxins that inhibit the clearance of mucus and promote the attachment of the bacteria to the respiratory tract. As a result, severe coughing is triggered that quickly debilitates young children. Almost half of *B. pertussis* infections occur in infants, and a significant proportion of these cases require hospitalization. Potentially fatal pneumonia can occur, as well as seizures and encephalopathy leading to permanent brain damage or death. The current vaccine is made from a mixture of *B. pertussis* proteins and induces the production of neutralizing antibacterial and antitoxin antibodies.

id) The combination DTaP vaccine. Children are simultaneously vaccinated against diphtheria, tetanus and pertussis through administration of the combination DTaP vaccine, consisting of *B. pertussis* proteins (P) combined with the diphtheria (D) and tetanus (T) toxoids. (The "a" in DTaP stands for "acellular." The previous DTP vaccine, which contained whole *B. pertussis* cells, proved to be harmful in rare cases. The use of an acellular preparation of *B. pertussis* proteins eliminates this problem.) Completion of a full series of DTaP doses while a child is young is important: 83% of vaccinees receiving three doses are completely protected from these diseases, but only 36% are resistant to pertussis after one dose. In addition, the risk of adverse effects of the DTaP vaccine is greater when an individual receives his or her first dose as an adolescent or an adult. Protection from diphtheria and tetanus decreases after 10–20 years so that a booster containing the diphtheria and tetanus toxoids (DT) is routinely recommended for adults every 10 years. More frequent vaccination may be appropriate for sewage workers and metal scrap yard workers. Adults and adolescents contracting *B. pertussis* infection experience far fewer ill effects than do children, removing the need to give a 10-year booster against this pathogen.

ii) *Haemophilus influenzae* Type b

Haemophilus influenzae is an encapsulated extracellular bacterium that has nothing to do with the "flu" or the influenza virus. Different strains of *H. influenzae* bear different types of capsules. *H. influenzae* with the type "b" capsule (Hib) are particularly pathogenic to humans and cause thousands of infant deaths worldwide. Hib bacteria preferentially colonize the mouth and throat but can cause fatal meningitis if an individual's immune system is sufficiently depressed. Neutralizing antibodies provide very good defense against this pathogen. Several different vaccines have been developed in which a Hib capsule polysaccharide is conjugated to a carrier protein to form a Td antigen with high immunogenicity in infants. Adults who undergo a splenectomy should also seek out the Hib vaccine.

iii) Hepatitis A Virus

The hepatitis A virus (HAV) is a non-enveloped single-stranded RNA virus that is contracted from the consumption of contaminated food and water. HepA can cause significant liver damage. Vaccination is highly effective in protecting against HAV infection and induces the production of antiviral antibodies that appear to persist for decades. Inactivated, whole virus vaccines are available in many countries, with live, attenuated vaccines being used to a more limited extent. Vaccination is particularly important for those who either live in or plan to travel to places (such as the tropics) where HAV is endemic. Adults with chronic liver disease and hemophiliacs should also ensure that they have been vaccinated against HAV.

iv) Hepatitis B Virus

The hepatitis B virus (HBV) is an enveloped DNA virus that causes severe liver damage and hepatocarcinoma. The virus is usually contracted via sexual activity or contaminated needles but can be passed from infected mother to child. HBV infection becomes chronic in about 90% of infected infants, and up to 25% of these individuals will die of HBV-associated liver disease as adults. The virus is highly infectious, meaning that chronically infected individuals can easily pass on the virus through their blood or body fluids. The current HBV vaccine is of the protein subunit type and consists of a HBV surface antigen produced using recombinant DNA techniques. Protective levels of neutralizing antibodies are produced in 95% of infants receiving the vaccine. Adults who are companions to HBV-infected individuals, hemodialysis patients, health workers, sex workers, or scientists handling human or primate tissues or blood should take special care to ensure that they are vaccinated against HBV.

v) Human Papillomavirus

In 2007, a vaccine to prevent human papillomavirus (HPV) infection was added to the list of recommended prophylactic vaccines. HPV is a non-enveloped DNA virus that causes genital warts and cervical cancer. The vaccine, currently intended for adolescent and adult females, is of the recombinant subunit type and contains purified HPV capsid proteins. The vaccine protects against the four strains of HPV that cause 90% of genital warts and 70% of cervical cancers. The use of this vaccine in males is under investigation.

vi) Influenza Virus

The influenza virus is an enveloped RNA virus that attacks human respiratory epithelial cells and causes severe acute respiratory symptoms. Influenza virus infection is particularly problematic for very young and very old individuals whose immune systems are functioning at less than full strength. Persons suffering from diabetes mellitus, alcoholism, cardiovascular disease, or from chronic respiratory disorders such as cystic fibrosis are also at increased risk. There are three types of influenza viruses, A, B and C, that can be distinguished by the antigenic characteristics of their structural proteins. Influenza type A, which can infect both humans and birds, is the most pathogenic to humans.

In a natural influenza virus infection, neutralizing antibodies directed against important viral proteins play key roles in immune defense. However, because the influenza virus undergoes antigenic drift and shift (refer to Ch. 13), it has been impossible so far to produce a vaccine that confers lifelong protection against all strains of influenza. Consequently, annual vaccination programs are undertaken in developed countries just prior to "flu season". How do the health authorities know which variant of the virus to target for vaccine production? In February of any given year, the WHO and the U.S. Centers for Disease Control study which flu viruses are emerging around the world and choose a combination of candidate strains for the production of the vaccine to be used during the next winter's flu season. For example, for the 2006–2007 flu season, it was recommended that flu vaccines contain the New Caledonia strain of influenza A virus, the Wisconsin strain of influenza A virus, and the Malaysian strain of influenza B virus. In the past, if the strains chosen turned out not to be responsible for the majority of flu cases in that year, the vaccine was not effective. Fortunately, the accuracy of "strain-picking" is now high enough that annual flu vaccination programs are generally quite helpful in reducing disease incidence and severity.

Several forms of influenza vaccine exist. Vaccines based on formalin-inactivated whole virus induce a robust activation of the immune system that can result in fever and aches. These effects are due to the immune response itself but often cause a patient to complain that "the vaccine gave me the flu". Many subunit vaccines are safe to use in infants and the elderly but provide only limited protection against infection with a live virus because only low levels of short-lived antibodies are produced. A recently developed virosome vaccine bearing components of the influenza virus appears to have superior immunogenicity to the subunit vaccine and causes fewer side effects than the whole virus vaccine.

Concerns have been raised worldwide about a possible avian influenza or "bird flu" pandemic in the near future. Over the last few years, some strains of influenza A emerging in bird populations in Asia have turned out to be highly virulent and lethal to their avian hosts. At the moment, these strains do not infect humans very well and when they do, they are transmitted from human to human in only a very limited way. However, if a bird (or human) becomes simultaneously co-infected with one of these virulent strains and a less virulent strain that efficiently infects humans, the reassortment of gene segments that occurs during antigenic shift may produce a progeny virus with both high lethality and ready transmission from human to human. Because this strain would represent a radical shift from influenza A viruses previously experienced by humans as

a species, the immunity of most of the world's population would be present only at the primary response level and a devastating death toll would likely result. Scientists are working feverishly to devise a vaccine to prevent a bird flu pandemic but currently can only guess at which epitopes will make appropriate vaccine targets.

vii) Measles, Mumps and Rubella

viia) Measles. The measles virus is a highly contagious, enveloped RNA virus that initially attacks the upper respiratory tract but then spreads via the lymphatics and blood to most other tissues. Disease symptoms include fever, a characteristic rash of red spots, and temporary immune system suppression. Potentially fatal opportunistic infections may thus gain a foothold. Those individuals who survive a natural exposure to the measles virus usually acquire lifelong protection. There are several different current vaccines for measles but all are based on live, attenuated viruses. MCV (measles-containing vaccine) contains live, attenuated measles virus alone. Live, attenuated measles virus also forms part of the combination MMR (measles, mumps, rubella) vaccine, discussed below.

viib) Mumps. The mumps virus is an enveloped RNA virus that initially causes a respiratory infection but then travels in the blood to infect and cause swelling of the salivary glands. Severe complications include sudden and permanent deafness in one or both ears, orchitis and meningoencephalitis. Childhood vaccination provides very effective protection against mumps.

viic) Rubella. Rubella (also known as German measles) is caused by the rubella virus, an enveloped RNA virus. In young children, rubella is a relatively mild disease characterized by a rash, low fever, malaise and mild conjunctivitis. Lifelong immunity results from childhood infection. The danger in rubella lies in its teratogenic effects on the developing fetus of a woman who never had the disease as a child. The placenta may become infected, allowing the virus to enter the fetal circulation and infect fetal organs. If infection occurs within the first trimester, the fetus may develop cataracts, heart disease, deafness, and sometimes hepatitis and mental retardation.

viid) MMR vaccine. The combination MMR vaccine confers long-lasting protection against measles, mumps and rubella. The current MMR vaccine is composed of live, attenuated versions of the measles, mumps, and rubella viruses. Both antibody and cell-mediated immune responses are mounted and long-lasting memory is induced. The effectiveness of the MMR vaccine has meant that measles is now a relative rarity in the developed world. This success has caused many North American parents to adopt a rather cavalier attitude toward measles, although the disease still kills hundreds of thousands of children in developing countries each year. Even in North America, local measles outbreaks in the early 1990s caused the deaths of several unvaccinated high school students.

viii) Meningococcus

The encapsulated bacterium *Neisseria meningitidis* (commonly called meningococcus) causes bacterial meningitis characterized by severe inflammation of the membranes of the brain. The disease most often strikes children and adolescents, and individuals with primary immunodeficiencies of complement components are particularly susceptible. Meningococcus is readily transmitted through respiratory secretions and can be fatal. Survivors may suffer from neurological defects leading to mental impairment, deafness or seizures. Most meningococcal disease is caused by three types of meningococcus defined by their capsular polysaccharides: groups A, B and C. Group A infections dominate in Africa, whereas group C infections are most prevalent in the Americas, Australasia and some parts of Europe. Group B outbreaks have occurred most recently in Europe, Latin America and New Zealand. Both the capsular polysaccharide and the endotoxin produced by these bacteria contribute to their virulence, and humoral immunity is the key to successful host defense.

Current vaccines are based on bacterial capsular polysaccharides that are formulated either alone (MPSV4 vaccine) or conjugated to a protein carrier (MCV4 vaccine). Either vaccine may be given to children greater than 24 months of age. Boosters are required every 3–5 years for the MPSV4 vaccine but a single dose of the conjugate vaccine generates long-lasting protective memory. Mass pre-emptive vaccination against group C meningococcus in young children and adolescents has been instituted in several Western countries. Unfortunately, vaccines based on one meninogococcal group are not effective against the other two groups.

ix) Pneumococcus

Streptococcus pneumoniae (commonly called pneumococcus) is an encapsulated bacterium that initially colonizes the upper respiratory tract. The capsule forestalls engulfment by pulmonary phagocytes such that the bacteria multiply in great numbers in the lung and then spread throughout the body. Thousands of cases of pneumonia, ear infections and meningitis are attributed to this bacterium every year. Two types of vaccines for this pathogen have been developed. The polysaccharide vaccine (PPV) contains the polysaccharides of the 23 strains of *S. pneumoniae* that represent 80–90% of disease-causing strains. However, the Ti response induced by this vaccine is ineffective in very young children. Instead, infants are given a conjugate vaccine (PCV) that contains modified diphtheria toxoid linked to the capsule polysaccharides of the 7 most common strains of *S. pneumoniae* attacking infants. When administered intramuscularly, this vaccine provokes a Td response that leads to effective protection of about 90% of vaccinated infants. Unfortunately, the conjugate vaccine induces immunity to only 50% of the strains that cause disease in older children and adults. As a result, some children particularly susceptible to *S. pneumoniae* infections may need a booster of the polysaccharide vaccine a little later in life. Individuals that have undergone splenectomy or are suffering from cardiovascular disease, diabetes or alcoholism should also ensure they are vaccinated against *S. pneumoniae*.

x) Polio

The poliovirus is a very infectious, non-enveloped RNA virus. This virus is acquired orally and replicates first in the intestinal tract but then travels in the blood to the spinal cord and central nervous system (CNS). If the virus succeeds in destroying the motor nerves, permanent muscle paralysis called *poliomyelitis* results. Poliovirus occurs in three types, called 1, 2 and 3, which are distinguished by their viral coat proteins; all three types cause the same symptoms. In a natural infection, both antibodies and cell-mediated responses are needed to eliminate the virus.

Two polio vaccines are currently available: the Salk vaccine, more commonly known now as the inactivated poliovirus (IPV) vaccine, and the live, attenuated Sabin vaccine, or oral poliovirus (OPV) vaccine. IPV is administered by intramuscular injection, requires repeated boosters, and does not induce high levels of mucosal immunity in the gastrointestinal tract. OPV is administered orally, does not require boosters, and induces significant gastrointestinal mucosal immunity. However, there is also a small chance (one case per 2–3 million doses) that one of the attenuated viruses in the OPV vaccine may revert and cause poliomyelitis in a vaccinee, a syndrome called *vaccine-associated paralytic polio* (VAPP). Because the wild poliovirus has been eliminated in North America, the risk of VAPP is greater than the risk of getting the disease, such that the OPV vaccine is no longer recommended for use in the United States. Instead, four childhood doses of IPV are given. In Europe, the IPV vaccine is given first to induce systemic protection followed by boosters of the OPV vaccine to protect the mucosae and bolster systemic immunity. Administration of the OPV vaccine alone is still recommended for vaccination programs in the few regions of the world where the wild poliovirus remains endemic and where sterile injections and repeated clinic visits are an issue. Although these programs have achieved much success, wild poliovirus can still be found in parts of southern Asia and Africa.

xi) Rotavirus

Rotavirus is a non-enveloped RNA virus that has a characteristic wheel-like shape when viewed by electron microscopy. This virus is the most common cause of severe diarrhea in children and kills over 600,000 youngsters annually around the world. Rotavirus is transmitted primarily by contact with a contaminated surface or by ingestion of contaminated food or water. An older generation of rotavirus vaccines was withdrawn due to severe adverse effects on the intestine but a newly developed vaccine, which is a live attenuated virus administered orally, has proven safe. As of 2007, it is recommended in the United States that infants be vaccinated at ages 2, 4 and 6 months with the new rotavirus vaccine.

xii) Varicella (Chicken Pox)

Varicella (chicken pox) in children is caused by the varicella zoster virus. The disease is characterized by fever and an outbreak of itchy red spots all over the body. In its acute phase, the virus is highly contagious such that almost all young children in endemic areas contract chicken pox. Although most children experience relatively mild symptoms, some cases can be complicated by high fever, pneumonia or encephalitis. Individuals first infected as adults are more likely to experience serious illness. In pregnant women, varicella infection can lead to damage of the fetal CNS or even fetal death.

Once the characteristic chicken pox rash has subsided, the affected individual will not usually experience it again. However, the natural immune response to varicella zoster only suppresses viral replication and does not completely eliminate the virus from the body. Thus, latent virus from the original infection can later become reactivated and cause the painful skin disease *shingles*. This reactivation is usually associated with extreme stress or immunosuppression.

The current varicella vaccine, which contains live, attenuated virus, is highly immunogenic and one dose is sufficient to induce antibody production in 97% of school-aged vaccinees. These antibodies persist in the circulation over several years but it is unclear whether this persistence is due to the vaccine or to regular re-exposure to varicella zoster in the community. In 2007, a booster dose of varicella vaccine at age 4–6 years was added to the recommended immunization schedule for children in the United States.

II. VACCINES FOR SPECIAL SITUATIONS

Some vaccines are recommended only for individuals in situations likely to result in significant exposure to the natural pathogen.

i) Anthrax

Anthrax is caused by the extracellular bacterium *Bacillus anthracis*, which occurs naturally in the soils of farms and woodlands. It commonly infects livestock and range animals but rarely humans. When *B. anthracis* does infect a human, the resulting anthrax disease takes one of three forms: inhalation (the most lethal), cutaneous or gastrointestinal. Early symptoms range from respiratory distress to skin lesions to fever and severe diarrhea. If not rapidly treated with antibiotics, all three types can lead to systemic bacterial infection and death.

Even where concerns about bioterrorism exist, vaccination against anthrax is recommended only for: members of the military who might encounter this organism in a battlefield context; researchers culturing this organism; those working with potentially infected animals in a region of high anthrax incidence; and those handling animal hides or wools imported from countries with lax standards for spore transfer prevention. The disease caused by *B. anthracis* is due in part to the toxins it produces. The current licensed vaccine for anthrax therefore consists of a cell-free preparation of one of the *B. anthracis* toxins. However, the schedule of administration of this vaccine is quite arduous, leading to problems with compliance among vaccinees.

ii) Cholera

Cholera is a disease of debilitating diarrhea caused by the extracellular bacterium *Vibrio cholerae*. Transmission is by consumption of water contaminated with fecal waste from infected humans, so that in regions where sanitation is good, disease incidence is minimal. When *V. cholerae* is ingested, the bacteria multiply rapidly in the gut and produce an exotoxin that induces relentless loss of water and ions from gut epithelial cells. Life-threatening dehydration can occur within hours of infection, particularly in very young children. Antibodies to the exotoxin mitigate but do not eliminate the disease. Rather, mucosal SIgA that blocks the attachment of the bacteria to the gut epithelial cells is the main defender. Cholera is therefore an obvious candidate for a mucosally administered vaccine, but the only vaccine currently available in many countries is a killed whole cell preparation administered by injection. Unfortunately, this vaccine has serious side effects, is effective in only 50% of a given population, and protects for only 6 months. Thus, this vaccine is really suitable only for foreign travelers to countries in which cholera is endemic. New types of cholera vaccines that can be given orally are starting to be used in a few developing countries. These vaccines have fewer side effects than the original vaccine and confer longer lasting protection.

iii) Plague

Plague is caused by the gram-negative bacterium *Yersinia pestis*. *Y. pestis* naturally infects rodents (especially rats) and their fleas and is transferred to humans by flea bites. At the site of the bite, the skin becomes blistered and blackened (hence, "Black Death"). Within a week of the bite, the bacteria access the draining lymph node via the lymphatics and cause high fever and a large, painful swelling of the node known as a "bubo" (hence, *bubonic plague*). From the lymph node, the bacteria then spread to the blood and organs with disastrous speed. Once the bacteria access the blood, the plague is said to be *septicemic* in form. If the bacteria reach the lungs, pneumonia occurs and *pneumonic plague* is said to be present. Unfortunately, pneumonic plague is also readily spread by the inhalation of respiratory droplets expelled by the coughing or sneezing of an infected person. In the absence of antibiotic treatment, all three forms of plague have a high fatality rate. Plague is endemic in many countries in Africa, Asia, Latin America and South America, but the risk of infection is low as long as rat-infested areas are avoided.

The current plague vaccine is based on formalin-killed whole bacterial cells. Side effects of fever, headache and pain are common and increase in severity with repeated doses. Thus, vaccination is recommended only for health care workers in endemic areas, disaster relief workers, and laboratory personnel working directly with *Y. pestis*. Control of local rodents and elimination of their habitats is usually the most cost-effective way of preventing plague outbreaks.

iv) Rabies

Rabies is caused by a slow-replicating, enveloped RNA virus. Upon entering the body, the rabies virus replicates first in skeletal muscle and connective tissue. It then spreads along the peripheral nerves to the spinal cord and CNS, where it causes progressive encephalitis. The disease is inevitably fatal in humans if left untreated. Modern rabies vaccines are based on killed viruses grown in human cell cultures or in chick embryos, and are safe and highly efficacious. Nevertheless, rabies is mainly kept under control by the vaccination of wildlife, domestic pets and stock animals. Only individuals who might expect to encounter rabies in their line of work (such as wildlife workers or veterinarians) receive the vaccine prophylactically. Because the virus replicates only slowly, unvaccinated individuals bitten by a rabid animal can obtain a post-exposure shot of rabies vaccine to trigger an effective response against the virus. The bitten individual should also receive a preparation of anti-rabies antibodies (see "Passive Immunization" later in this chapter).

v) Tuberculosis

Tuberculosis (TB) is a lung disease caused by *Mycobacterium tuberculosis*, a slow-replicating intracellular bacterium that does not produce bacterial toxins. Upon initial infection, *M. tuberculosis* causes only mild inflammation in individuals with a robust immune response. However, this pathogen is extraordinarily hard to remove, so that the immune response that is provoked is prolonged. Activated T cells start to hyperactivate macrophages, which in turn secrete cytokines that damage the lungs. Granulomas that form to wall off the bacteria appear as "tubercles" (lumps) on the lungs. Eventually, the granulomas break down and release a large proportion of the bacteria. A healthy immune system can eliminate most of the invaders at this point such that the tubercular lesions calcify and become visible in X-rays as scarring on the lung (Plate 14-2). Surviving bacteria in the lesions may become dormant for as long as 20 years. Indeed, 90% of individuals infected with *M. tuberculosis* remain clinically healthy. However, if the immune system is later compromised such that the bacteria can resume replication, a lesion may rupture and release millions of bacteria first into the lungs and then throughout the body. The patient is then considered to have "active TB". Active TB can also occur in individuals when they are first infected if the immune system is very weak. Patients with active TB are highly contagious and suffer from weight loss, loss of vigor, and coughing (often with blood). If not treated aggressively, these individuals often die of their disease.

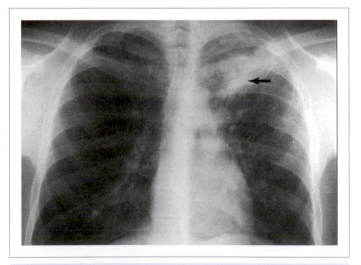

Plate 14-2 Lung Scarring in Tuberculosis
[Reproduced by permission of Ian Kitai, The Hospital for Sick Children, Toronto.]

It has been very difficult to develop an effective vaccine against TB because natural immunity to *M. tuberculosis* is still not completely understood. Normally, a Th1-dependent cell-mediated response combats this pathogen but, as described earlier, the response is not always totally protective. The only vaccine currently available to fight TB is the *bacillus Calmette-Guerin* (BCG) vaccine. BCG is a live, attenuated form of *Mycobacterium bovis*, a bacterial species that shares many antigens with *M. tuberculosis*. The original BCG strain underwent a spontaneous deletion that removed genes conferring virulence without significantly compromising immunogenicity. However, numerous mutations have since occurred in the BCG genome that have resulted in some loss of efficacy in certain human populations. Indeed, protection levels for individuals vaccinated with BCG range from zero to 80%, for unknown reasons.

In developing countries with a high prevalence of infectious TB, the WHO recommends a single dose of BCG vaccine for newborns followed by a booster at age 10–15 years. Routine vaccination for TB is not the norm in North America because the general population is at a low risk of infection; instead, TB patients are treated with drugs. BCG vaccination is recommended for North American infants and children who are continually exposed to family members with untreated or ineffectively treated TB. Scientists are currently trying to develop a more effective vaccine using *M. tuberculosis* itself but face many challenges, some of which are discussed in Box 14-1.

vi) Typhoid Fever

Typhoid fever is caused by *Salmonella typhi*, a highly invasive intracellular bacterium whose only natural host is humans. An acute infection is characterized by high fever, abdominal discomfort, malaise, and headache that can last for several weeks. Many (but not all) patients get a rash of salmon-colored spots (hence, *Salmonella typhi*). Life-threatening complications include intestinal perforation and hemorrhage. Ingested bacteria initially enter either enterocytes or M cells in the small intestine but eventually spread to macrophages throughout the body. In a small number of patients, bacteria reach the gallbladder and establish a chronic infection. These people become asymptomatic carriers who shed the bacteria in infectious form. Typhoid fever is now a rarity where good water treatment prevails but the disease still kills hundreds of thousands in developing countries. Travelers to endemic countries should be vaccinated against typhoid fever.

The original typhoid vaccine contains inactivated whole *S. typhi* cells. This vaccine is reasonably efficacious and confers protection for 5 years but has unpleasant side effects. Two more modern vaccines have been developed based on either live, attenuated bacteria or purified bacterial capsule. Although both are less toxic than the original, they are less efficacious and offer protection for only 2–3 years. Development of additional vaccines against typhoid is essential because several strains of *S. typhi* have recently become resistant to almost all antibiotics.

vii) Variola (Smallpox)

Variola virus, which causes smallpox, is no longer found in nature in any part of the world but is still feared as a bioterrorism agent. There is no effective drug for smallpox and the

Tuberculosis kills over 1.6 million people per year worldwide. The lack of a highly efficacious vaccine against *M. tuberculosis* and the rise of multi-drug-resistant strains of this bacterium prompted the WHO to declare TB to be a global health emergency in 1993. Because *M. tuberculosis* can lurk intracellularly for decades, a vaccine inducing a comprehensive immune response with long-lived memory is required. These characteristics are best evoked by a live, attenuated vaccine. Scientists are examining the genome of *M. tuberculosis* to identify genes that can be mutated to decrease the virulence or persistence of the organism without affecting its immunogenicity. However, for unknown reasons, the less virulent a mutated *M. tuberculosis* strain is, the less protective it is. Researchers are also searching for *M. tuberculosis* antigens for use in subunit or DNA vaccines but lack a reliable *in vitro* system to test for the protective efficacy of such antigens. Because of the long latency of *M. tuberculosis* in humans, clinical trials of immense length must be conducted to judge the efficacy of a vaccine. As well, no one antigen seems to protect against all stages of *M. tuberculosis* infection. Subunit vaccines based on bacterial products have thus far offered no more protection than the BCG vaccine.

TB vaccine development has also been hampered by the lack of a sufficiently accurate animal model to use for testing. Mice are not highly susceptible to *M. tuberculosis*. Although *M. tuberculosis* infection in primates resembles that in humans, primate trials are prohibitively expensive. Another challenge is the current dearth of optimal adjuvant options. Experiments with subunit vaccines in delivery vehicles have been attempted but none so far has offered protection superior to that induced by the venerable BCG vaccine.

length of protection afforded by the existing smallpox vaccine (based on live, attenuated vaccinia virus) is uncertain. Scientists worry that much of the world's population has not been immunized for more than 20 years, and that younger generations have never been immunized. Moreover, the existing vaccine is unsafe for immunocompromised individuals who, due to HIV, are now present in great numbers around the world. Even among the general population immunized in the 1960s, hundreds of complications and several deaths were recorded as a result of vaccination, representing a safety record that would be unacceptable by today's standards. Many therefore believe that a new, safer smallpox vaccine should be developed. In the meantime, most countries have opted to replenish their supplies of the existing smallpox vaccine with the goal of reserving inoculation for situations of known variola exposure. Because the virus replicates slowly, exposed individuals can be vaccinated up to 4 days later and still develop an effective adaptive response.

viii) Yellow Fever

Yellow fever is caused by a small enveloped RNA virus that is mosquito-borne and can infect both monkeys and humans. The disease is prevalent in tropical climates where mosquitoes are endemic year-round. Although some patients are asymptomatic, others experience headache, fever, vomiting and nosebleeds. In severe cases, patients have fever accompanied by hepatic, circulatory and renal failure as well as severe jaundice (hence, "yellow" fever). Many severe infections are fatal.

The yellow fever vaccine is a live, attenuated virus usually given in one subcutaneous dose. Vaccination programs have been highly successful and protection generally lasts for 10 years. Individuals living in temperate climates receive this vaccine only if traveling to an endemic area.

E. The "Dark Side" of Vaccines

Vaccines prevent much misery and millions of premature deaths worldwide. Nevertheless, like all powerful medicines,

Table 14-5 Examples of Serious Adverse Events Associated with Vaccines

Vaccine	Serious Adverse Event
Vaccines produced in chick embryos	Contaminating chick proteins in vaccine may trigger symptoms in vaccinees with allergies to eggs. Contaminating pathogens could initiate disease or tumorigenesis
Vaccines produced using preservatives or antibiotics	Contaminating molecules can trigger symptoms in allergic individuals
Live, attenuated virus vaccines	May infect fetus of vaccinated pregnant woman and have teratogenic effects
Measles vaccine	Can induce immunosuppression and thus vulnerability to opportunistic organisms
Rubella vaccine	May induce mild arthritis
Sabin OPV vaccine	May induce VAPP due to viral reversion

vaccines carry a risk of side effects. In general, because of extensive animal and cell culture testing, such risks are low and the side effects (sometimes called "adverse effects") are mild and limited to redness or pain at an injection site, sneezing or nasal congestion after intranasal administration, fatigue, or headache. In a very few instances, the adverse effects of a vaccine are more serious and have systemic consequences (see Table 14-5). Most developed countries carry out a post-licensing surveillance of adverse events that can trigger the withdrawal of a vaccine should it prove harmful in even rare circumstances. Many of the current worries about vaccine side effects will be mitigated as vaccine production and purification schemes improve, and as DNA vaccines become more cost-effective to produce.

In the early 1990s, concerns were raised about a possible association between the HepB vaccine and multiple sclerosis (MS; see Ch. 19). However, infection with wild HBV is not a risk factor for MS, and a well-controlled clinical study carried out in 2001 showed no increased risk of MS in women who had received the HepB vaccine. Also in the 1990s, a link between the MMR vaccine and the increasing rate of autism in developed countries was suggested. However, extensive examinations and retrospective assessments have confirmed that MMR vaccination does not increase the risk of autism. Despite these and other scientific studies, much misinformation about vaccines persists in the popular press. These misconceptions have led some parents to forego vaccinating their children, with unintended and often harsh consequences (see Box 14-2).

certain pathogen or toxin; or an individual is immunodeficient or immunocompromised. Passive immunization to prevent Rh disease (destruction of fetal RBCs) is outlined in Box 14-3.

The antibodies used for passive immunization are usually derived from the pooled serum of human donors who have been selected for naturally high titers of the desired antibodies. Sometimes volunteers are repeatedly immunized with a pathogen antigen of interest to generate high levels of specific IgG antibodies that are recovered as the polyclonal antibody preparation. Monoclonal antibodies tailored for use in humans are also starting to be used for passive immunization.

F. Passive Immunization

Prophylactic vaccination is also called *active immunization*, since the individual is administered a pathogen antigen and his/her body is responsible for activating the lymphocytes and making the antibodies necessary to provide defense against future assaults. In *passive immunization*, preformed specific antibodies from an exogenous source are injected into an individual that has been exposed to the pathogen. For example, an individual who has suffered a dog bite may have been exposed to rabies virus. The injured person receives a preparation of anti-rabies virus antibodies that heads off viral replication and thus the disease. Unlike vaccination, passive immunization provides immediate protection because the 7–10 day lag necessary to mount an adaptive response has been eliminated. However, because the individual's own immune system has not been stimulated, no memory is generated and protection lasts only for days to months (rather than years).

Natural passive immunization occurs when maternal anti-pathogen antibodies pass through the umbilical circulation to the developing fetus, and later to the newborn in colostrum and breast milk. Medically administered passive immunization is very useful in situations where: no vaccine exists for a pathogen; the vaccine is not 100% efficacious; an unvaccinated individual has been or expects to be imminently exposed to a

G. Future Directions

I. PROPHYLACTIC VACCINES

While much ongoing basic and clinical research is devoted to improving existing prophylactic vaccines, other efforts are focused on developing vaccines for high-profile diseases like HIV/AIDS, malaria and TB for which there are currently no effective vaccines. Some scientists believe that one reason it has been difficult to produce good vaccines for these diseases is that the causative natural pathogens do not cause robust, acute infections. A relatively low-key primary immune response ensues in the host during a natural infection with these organisms. As a result, it is difficult to formulate a pathogen-derived vaccine that will induce the production of a large enough army of memory cells to mount an effective secondary response. In addition, in the case of AIDS, HIV kills billions of CD4+ T cells such that the natural anti-HIV immune response is too weak to prevent the virus from establishing a permanent foothold in the body (see Ch. 15). The design of an AIDS vaccine that can induce an effective response is thus particularly challenging. With respect to malaria vaccine development, significant additional problems arise due to the complex life cycle of the causative parasite. These challenges are outlined in Box 14-4. Other pathogens that are priority targets for vaccine development include: group A streptococci causing flesh-eating disease and other serious ailments; bacteria and herpesviruses

Box 14-3 Passive Immunization to Prevent Rh Disease

The prevention of Rh disease is an example of passive immunization that does not involve a pathogen. Rh is a protein expressed on the RBCs of many, but not all, humans. In "Rh disease", the immune systems of mothers whose RBCs do not express Rh (Rh⁻) destroy the RBCs of a fetus that does express Rh (Rh⁺), if the fetus involved is the second (or later) Rh⁺ fetus. During a first pregnancy in which the Rh⁻ mother is carrying an Rh⁺ baby (see Figure), very few fetal cells get into the mother's circulation prior to birth and only naïve maternal anti-Rh lymphocytes are present. Any anti-Rh immune response mounted against the first Rh⁺ fetus is weak and not particularly harmful. However, the traumatic process of birth propels many more fetal cells into the maternal circulation. These fetal cells provoke a vigorous response by the maternal immune system that eliminates the fetal RBCs and generates anti-Rh memory cells. If a subsequent pregnancy also involves an Rh⁺ fetus, the few fetal cells that make it into the maternal circulation prior to birth will activate memory B cells. Maternal plasma cells are generated that produce large amounts of anti-Rh antibodies capable of crossing the placenta and destroying RBCs in the fetus. Rh disease is prevented by passively immunizing an Rh⁻ woman with an anti-Rh antibody preparation (eg., Rhogam) early during her first pregnancy and again shortly after the birth of her first child. The exogenous anti-Rh antibodies bind to the Rh antigen on any fetal RBCs that accessed the maternal circulation during birth, clearing them before they can interact with naïve anti-Rh B cells in the mother. Virtually no memory anti-Rh B cells are generated, so that if the fetus in the next pregnancy is Rh⁺, the risk of fetal damage is reduced. As insurance, Rh⁻ women are given anti-Rh antibodies throughout and after subsequent pregnancies.

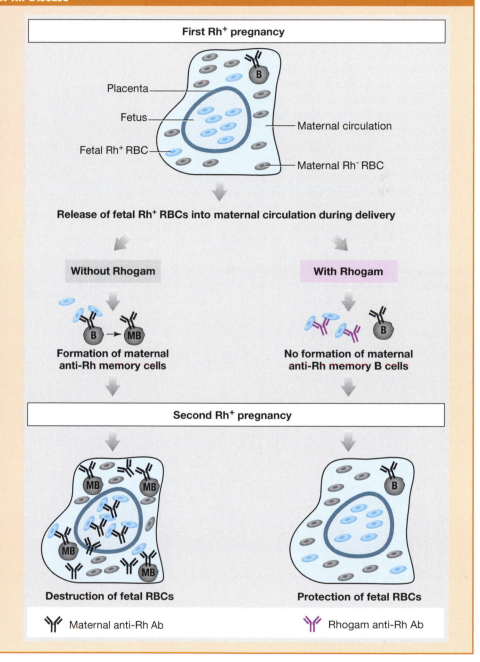

Box 14-4 The Life Cycle of *Plasmodium falciparum* and Its Effect on Malaria Vaccine Development

Malaria kills one child in Africa every 30 seconds and costs the African economy about U.S. $2 billion each year. The malarial parasite *Plasmodium falciparum* has a complex, multi-host, multistage life cycle (see Figure) that has so far stymied vaccine development. Researchers have designed prototype vaccines targeting various plasmodium stages but none has yet proved truly efficacious. One type of vaccine is designed to prevent sporozoites from reaching the liver or producing merozoites. In theory, no RBCs are lysed, precluding the onset of clinical symptoms. A second type of vaccine is intended to target the merozoites in the RBCs. The objective is to kill the infected cells before the parasite can multiply or generate gametes. Symptoms may still arise but they are mild. The third vaccine type is meant to target the formation of new parasite gametes in an infected individual and to block transmission to the next host. This vaccine does not do the vaccinee much good but could reduce the spread of the disease in the community. Early trials of these vaccine prototypes have made it clear that only a complex vaccine incorporating elements of all three will succeed in generating effective protection against malaria. Strategies in which priming with a plasmid bearing DNA for various plasmodium antigens is followed by boosting with a viral vector bearing DNA for the same plasmodium antigens that are currently being tested. The idea is to induce both humoral and cell-mediated responses against infected hepatocytes, infected RBCs and free sporozoites in the blood, as well as humoral responses that block merozoite entry into RBCs. Vaccines that incorporate mosquito stage antigens are also under development.

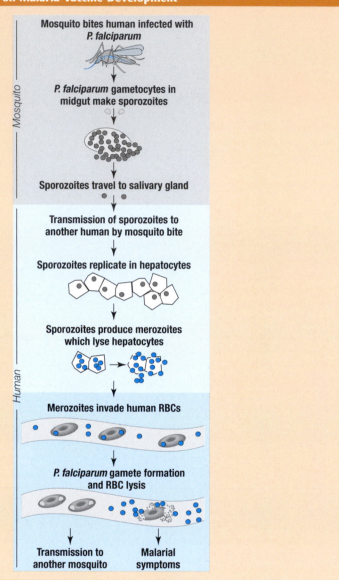

causing sexually transmitted diseases; enteroviruses causing childhood diarrhea; and hepatitis C virus causing liver damage and hepatocarcinoma.

II. THERAPEUTIC VACCINES

Therapeutic vaccines are agents designed to cure or mitigate established disease, rather than prevent it. A therapeutic vaccine that is truly efficacious does not exist as yet, but the principles of vaccination are being explored with the goal of stimulating the immune response to eliminate tumors, cure chronic infections, suppress autoimmune disease, or relieve allergy. Examples of these potential therapeutic vaccines are given in Table 14-6.

We have come to the end of our discussion of vaccines, entities that are designed to reduce the incidence and severity of many devastating infectious diseases. In Chapter 15, we describe a pathogen that destroys the immune system and has thus far thwarted vaccine designers at every turn–HIV.

Table 14-6 **Examples of Potential Therapeutic Vaccines**

Situation	The Problem	Vaccine Agent	Expected Response
Tumor therapy	Inadequate natural antitumor responses by a patient's NK cells and CTLs allow tumors to grow.	A patient's tumor cells (that express unique antigens) are stimulated *in vitro* with cytokines to promote pMHC class I display; the tumor cells are inactivated and injected back into the patient.	The unique tumor antigens upregulated on the treated tumor cells should induce effective antitumor Th, Tc and B cell responses that eliminate the tumor.
Chronic infection	Inadequate natural humoral and cell-mediated responses to a pathogen during an acute infection do not eliminate the pathogen, allowing chronic infection.	A viral antigen containing a hidden epitope expressed only in the chronic phase of an infection.	Activation of naïve lymphocytes recognizing the hidden epitope. These cells were not activated in the acute phase of the infection but may now provide the final push that eliminates the pathogen.
Indirect effects of infections	A disease is caused by an inadequate natural immune response to an underlying pathogen infection; e.g., ulcer formation and stomach cancers are caused by an underlying *H. pylori* infection.	Components of the underlying pathogen (e.g., *H. pylori*)	The immune response eliminating *H. pylori* has the effect of curing the ulcer and preventing the development of stomach cancer.
Autoimmune disease	Mechanisms of peripheral tolerance fail such that autoreactive lymphocytes are activated and destroy host tissues.	Self antigen delivered orally or delivered with a cytokine favoring immune deviation and/or activation of regulatory T cells.	Oral tolerance to the self antigen decreases the autoimmune response. Cytokine-induced immune deviation to a Th2 response or suppressive effects of regulatory T cells reduce tissue damage.
Allergy	Allergic symptoms are largely due to the production of anti-allergen* IgE antibodies that activate mast cells (see Ch. 18).	Allergen is delivered with a cytokine that favors IgG and suppresses IgE antibody production.	Anti-allergen antibodies of the IgE isotype are not produced and mast cells are not activated.

*An allergen is an antigen that causes allergic symptoms.

CHAPTER 14 TAKE-HOME MESSAGE

- A prophylactic vaccine is a modified form of a natural pathogen or its components that is given to an individual prior to pathogen exposure to establish expanded clones of long-lived, pathogen-specific memory lymphocytes. When the individual subsequently encounters the natural pathogen for the first time, a secondary rather than primary response is mounted such that he/she is far less likely to become seriously ill or die.

- Types of vaccines include live, attenuated; killed or inactivated; toxoid; subunit; peptide; and DNA vaccines.

- An efficacious vaccine induces a protective response against the pathogen in most members of the vaccinated population. A safe vaccine has side effects that pose little or no risk compared to the risk of harm from the disease itself. Vaccine efficacy and safety are formally tested in preclinical and clinical trials.

- Adjuvants are used to enhance a vaccine's immunogenicity, whereas delivery vehicles protect a vaccine from degradation.

- Prophylactic vaccines may be given to children as part of a standard immunization program or to individuals in special circumstances requiring protection from a particular pathogen.

- Passive immunization with preformed pathogen-specific antibodies can be given to prevent disease prior to or after exposure to an infectious agent.

- It is particularly difficult to design prophylactic vaccines for pathogens that become entrenched in the body, have complex life cycles, or destroy elements of the immune system.

Introduction

1) Can you define these terms? vaccination, vaccine, vaccinee, prophylactic

2) Explain why vaccination prevents disease.

Section A

1) Can you define these terms? coverage, herd immunity, protective epitope.

2) How is vaccine efficacy affected by whether the vaccine induces a cell-mediated or humoral response?

3) Why is *in vivo* testing an essential part of vaccine evaluation?

4) Give four examples of vaccine side effects.

5) How does geography affect the decision about whether or not to use a vaccine?

6) Describe the series of preclinical and clinical trials usually used to test a vaccine.

7) Give three characteristics of a pathogen that would make it a good candidate for a vaccine.

8) What is a reservoir and how can it affect vaccination programs?

Section B

1) Can you define these terms? virulence, booster, cold chain, toxoid, conjugate vaccine, vector.

2) Give two advantages and two disadvantages of using a live, attenuated vaccine.

3) Describe two ways by which a pathogen can be attenuated.

4) Give two advantages and two disadvantages of using a killed vaccine.

5) Under what circumstances is a toxoid vaccine effective?

6) Give two advantages and two disadvantages of subunit vaccines.

7) Distinguish between the two current types of DNA vaccines.

8) Distinguish between subunit vaccines and recombinant vector vaccines.

9) A clinician finds that the booster dose of a particular recombinant vector vaccine is not as effective as expected. What might be the problem and how would you solve it?

10) Outline two ways in which vaccine design and production have benefited from recombinant DNA technology.

Section C

1) Can you define these terms? adjuvant, delivery vehicle.

2) Why are adjuvants often necessary for immunizations?

3) Describe two types of delivery vehicles and how they increase vaccination success.

Section D

1) Can you define these terms? combination vaccine, IPV, OPV, VAPP, Hib, DTaP, shingles.

2) Name two diseases combated by toxoid vaccines.

3) Why must the influenza virus vaccine be administered yearly?

4) Describe the components of the MMR vaccine.

5) Why is OPV used in Africa but not in Western countries?

6) Why are some vaccines only used in special situations? Give three examples.

7) What is the advantage of using an oral vaccine to combat a pathogen like *V. cholerae*?

8) Give two reasons why it has been so hard to develop an effective vaccine for TB.

9) Outline the current dilemma concerning smallpox vaccination.

Section E

1) Give three examples of possible severe side effects of vaccines.

2) Why do some parents resist vaccinating their children?

Section F

1) Distinguish between active and passive immunization.

2) Describe how prevention of Rh disease is an example of passive immunization.

Section G

1) Why has it been so difficult to produce vaccines against HIV and malaria?

2) Give two examples of potential therapeutic vaccines.

15

HIV and Acquired Immunodeficiency Syndrome

He that will not apply new remedies must expect new evils;
for time is the greatest innovator.
Francis Bacon

Primary immunodeficiencies are inborn, whereas *acquired* or *secondary* immunodeficiencies are due to external interference in the immune system. Infections with certain viruses, severe metabolic disturbances and trauma can cause secondary immunodeficiencies, as can immunosuppressive drugs taken to treat cancer, transplant rejection, hypersensitivities or autoimmune disease. Patients with such secondary immunodeficiencies are left vulnerable to opportunistic bacterial and viral infections that can be lethal. However, the global health impact of these cases pales in comparison to that of *acquired immunodeficiency syndrome* (AIDS) due to infection with *human immunodeficiency virus* (HIV). HIV first became known to the medical community in the early 1980s when rare tumors and cases of an unusual pneumonia appeared in young, ostensibly healthy people. AIDS is now one of the top five leading causes of death of young adults around the world. The numbers associated with the AIDS epidemic are truly staggering (Fig. 15-1).

	Total number of persons living with HIV/AIDS in 2006	New HIV infections during 2006	Deaths due to AIDS in 2006
Sub-Saharan Africa	24.7 million	2.8 million	2.1 million
South & South East Asia	7.8 million	860,000	590,000
Latin America	1.7 million	140,000	65,000
Eastern Europe & Central Asia	1.7 million	270,000	84,000
North America	1.4 million	43,000	18,000
East Asia	750,000	100,000	43,000
Western & Central Europe	610,000	21,000	6,500
North Africa & the Middle East	460,000	68,000	36,000
Caribbean	250,000	27,000	19,000
Oceania	81,000	7,100	4,000
Total	**39.5 million**	**4.3 million**	**2.9 million**

Fig. 15-1 **The Global HIV/AIDS Epidemic in 2006**
[With information from www.avert.org/.]

It has been estimated that over 23 million people around the world have died of the disease, and that close to another 40 million are living with the infection. Over 4 million new HIV infections were reported globally in 2006, equivalent to almost 12,000 new infections each day. Most of these occur in developing countries, more than half affect persons younger than 25 years of age, and more than half occur in women. Almost 3 million people a year die of the disease, including over 300,000 children. The situation is worst in sub-Saharan Africa where AIDS accounts for about 20% of all young adult deaths. While a decline in infection rates has been observed in some Western countries, infection rates in Eastern Europe and Central Asia have risen by more than 50% since 2004. Much has been learned about the biology of HIV over the past 20 years but this complex and wily virus still frustrates the best efforts of researchers to produce an effective vaccine.

A. What Is HIV?

There are two known human immunodeficiency viruses: HIV-1 and HIV-2. HIV-1 can be found anywhere in the world but is most prevalent in the Western Hemisphere. HIV-2 occurs almost exclusively in Western Africa. Both HIV-1 and HIV-2 are retroviruses belonging to the lentivirus class and are *cytopathic*, meaning that they usually kill the cells they infect. A typical lentivirus persists in its host despite humoral and cell-mediated immune responses directed against it, and clinical disease is detected only after a long latency period. Although both HIV-1 and HIV-2 also infect macrophages and DCs, it is the destruction of CD4+ T cells by these viruses that causes AIDS. Because it has been more extensively studied, the remainder of this chapter focuses on the biology of HIV-1.

I. OVERVIEW OF THE HIV-1 LIFE CYCLE

HIV is transmitted primarily via the transfer of body fluids. An HIV virion most often accesses the body by breaching the mucosae during sexual contact or via intravenous drug injection involving contaminated needles. HIV can also be transmitted from mother to child before or during birth or by breast-feeding. An illustration of the HIV-1 life cycle following infection of a CD4+ T cell is shown in Figure 15-2. Once past

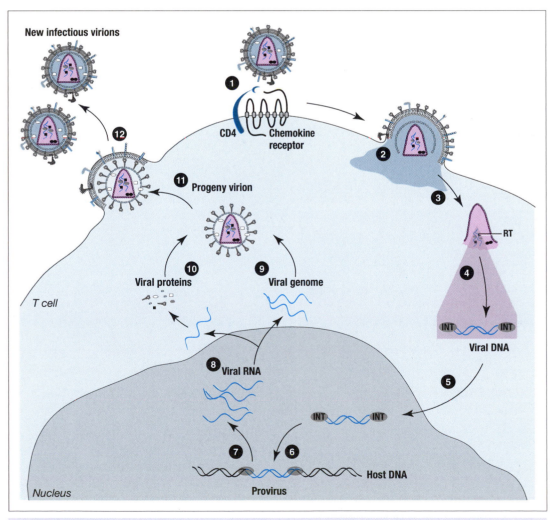

Fig. 15-2 **HIV Life Cycle**

the mucosal barrier, the entry of the virus into host cells is facilitated by the simultaneous binding of viral surface proteins to the CD4 coreceptor and a chemokine receptor (#1). The virion is then internalized (#2) and its outer protective layers stripped away by host cell enzymes (#3). Once the viral genome and its associated proteins are exposed in the cytoplasm, a viral enzyme called *reverse transcriptase* (RT) commences reverse transcription of the viral RNA genome into a DNA copy (#4). This viral DNA becomes associated with a viral *integrase* (INT) enzyme and both are transported from the cytoplasm (#5) into the host cell nucleus where the integrase mediates insertion of the viral DNA into the host cell genome (#6). This integrated form of the viral DNA is called proviral DNA or the *provirus*. The provirus may remain untranscribed in the host cell genome for a considerable time without disturbing the infected cell. However, when transcription of the provirus is finally triggered (#7), new copies of the viral RNA are produced (#8). These viral RNAs move into the cytoplasm where some serve as viral genomes (#9) and others are translated to generate the viral proteins (#10) necessary to package the viral genome and assemble the progeny virion (#11). The progeny virion buds through the host cell membrane to acquire its envelope (#12 and Plate 15-1, blue particles). The new infectious virions then spread through the body in search of fresh host cells to infect and the CD4⁺ T cell that has supported the generation of these virions dies. Importantly, other <u>uninfected</u> CD4⁺ T cells are also killed by HIV via mechanisms that remain unclear. The resulting systemic loss of CD4⁺ T cells leaves the victim vulnerable to fatal opportunistic infections and tumorigenesis.

II. HIV-1 STRUCTURE

The HIV-1 virion has a multilayered structure consisting of the *envelope*, the *matrix* and the *capsid* (Fig. 15-3). The HIV proteins and the genes from which they are derived are summarized in Table 15-1.

i) Envelope

The HIV envelope is a phospholipid bilayer acquired when the progeny virion first budded out through the membrane of an infected host cell. Embedded in the envelope are various viral and host cell proteins. Most prominent is the viral envelope "spike" made up of the gp41 and gp120 glycoproteins. These proteins are produced by the action of a host protease on a viral polyprotein precursor called Env (see later) that leaves the membrane-bound gp41 protein non-covalently associated with the non-membrane-bound, globular gp120 protein. Three gp41–gp120 units then come together to form each spike. It is the carbohydrate-rich gp120 protein that binds to host cell receptors. The HIV envelope also contains low amounts of host cell surface proteins such as the adhesion molecules

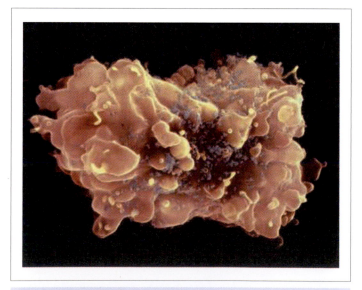

Plate 15-1 **HIV Budding from an Infected T cell**
[Reproduced by permission of Boehringer Ingelheim Pharma KG, photo Lennart Nilsson, Albert Bonniers Förlag AB.]

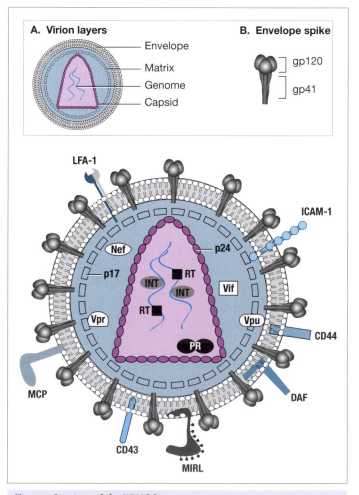

Fig. 15-3 **Structure of the HIV Virion**

Table 15-1 HIV Genes and Major Proteins

Gene	Protein	Function
	Structural/Enzymatic	
gag	p17 matrix protein	Acts as scaffolding for HIV envelope
	p24 capsid protein	Protects RNA genome
pol	Reverse transcriptase (RT)	Transcribes viral RNA into viral DNA
	Integrase (INT)	Inserts viral DNA into host DNA to form provirus
	Protease (PR)	Cleaves *gag*- and *pol*-encoded polyproteins into smaller functional proteins
env	gp41 transmembrane envelope protein	Part of envelope spike that promotes fusion with host cell membrane
	gp120 non-transmembrane envelope protein	Part of envelope spike that binds to host cell receptors to initiate viral entry
	Regulatory	
tat	Tat	Sustains host cell transcription of proviral DNA
rev	Rev	Promotes viral mRNA transport to host cytoplasm and translation
	Accessory	
vif	Vif	Promotes viral cDNA synthesis by inhibiting a host antiviral protein
vpr	Vpr	Facilitates transport of viral DNA into host cell nucleus for integration as provirus; induces cell cycle arrest; required for expression of non-integrated viral DNA.
vpu	Vpu	May promote progeny virion budding
nef	Nef	Acts in several ways to promote host cell survival and activation that support viral DNA synthesis

ICAM-1, CD44, CD43 and LFA-1, and the RCA proteins DAF, MIRL and MCP.

ii) Matrix

The matrix is a spherical layer that supports the envelope and surrounds the capsid. The matrix is composed primarily of the viral structural protein p17, which is anchored to the viral envelope and creates the scaffolding around which the envelope is wrapped. Within the matrix are the accessory proteins Vif, Vpr, Vpu and Nef.

iii) Capsid

The cone-shaped capsid protects the two RNA molecules making up the HIV genome. The capsid is composed of about 1200 molecules of the structural protein p24. Within the capsid, each RNA molecule is associated with several molecules of the viral RT and INT enzymes required for provirus generation and integration into the host genome. A viral *protease* (PR) necessary for cleaving two viral polyprotein precursors called Gag and Pol (see later) is also present. PR generates p17 and p24 from the Gag polyprotein, and RT and INT from the Pol polyprotein.

iv) Genome

The HIV genome is a single-stranded RNA (ssRNA) molecule present in two copies. During viral replication, the ssRNA molecules undergo reverse transcription to give rise to double-stranded viral DNA molecules that contain nine genes and will act as the HIV provirus once integrated into the host genome (Fig. 15-4). Like all retroviruses, the HIV genome contains the three principal genes *gag*, *pol* and *env*. These genes encode the three HIV polyprotein precursors Gag, Pol and Env from which the major structural and functional HIV proteins are derived. Like all lentiviruses, the HIV genome contains the *tat* and *rev* genes. These genes are made up of two exons each and encode the replication regulatory factors Tat and Rev, respectively. Finally, the HIV genome contains four small genes

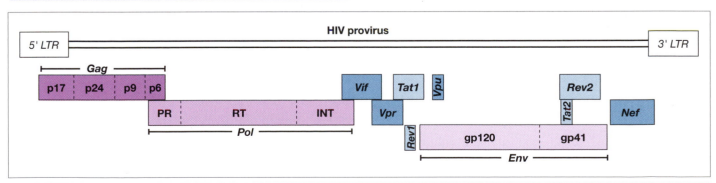

Fig. 15-4 **Genes Encoding the Proteins of HIV**
[Adapted from Girard M. and Excler J.L. (1999). Human immunodeficiency virus. In Plotkin S.A. and Orenstein W.A., Eds., *Vaccines*, 3rd ed. W.B. Saunders Co., Philadelphia.]

encoding the accessory proteins Vif, Vpr, Vpu and Nef (refer to Table 15-1 for their functions). Once the provirus is integrated in a host cell genome, it is flanked on its 5′ and 3′ ends by *long terminal repeat* (LTR) sequences. The LTRs promote transcription of the proviral genes by host cell RNA polymerase in the nucleus, as well as the translation of the viral mRNAs by ribosomes in the host cell cytoplasm.

An interesting feature of the HIV genome is that several of its genes physically overlap. Thus, depending on the reading frame used, two different mRNAs are derived from the same stretch of viral DNA and are later translated into two different viral proteins. For example, the *gag* gene overlaps the *pol* gene, and the *vif* gene overlaps the *vpr* gene.

B. HIV Infection and AIDS

I. MOLECULAR EVENTS

i) Viral Tropism

The *tropism* of a virus refers to the range of host cells the virus can infect. Although the predominant target of HIV is the CD4$^+$ T cell population, the virus can attack a wide range of other cell types that also express CD4, including macrophages, LCs and dermal DCs. However, as described earlier, the entry of HIV into cells also requires that the virus bind to a chemokine receptor. The major chemokine receptors that help CD4 to facilitate HIV entry are CCR5 and CXCR4. CCR5 is expressed by CD4$^+$ T cells, DCs and macrophages, whereas CXCR4 is expressed by CD4$^+$ T cells and DCs but not by macrophages. Unlike HIV-infected T cells, HIV-infected macrophages and DCs are quite resistant to the cytopathic effects of HIV and are not killed in great numbers during the course of the disease. Infected macrophages and DCs thus represent a significant reservoir within the host where HIV can reproduce out of reach of the immune system.

Strains of HIV that bind to CXCR4 only are known as "X4 viruses" and predominantly infect T cells (and low numbers of DCs). Historically, these X4 strains were known as "T cell- or T-tropic viruses". Strains of HIV that bind to CCR5 only are known as "R5 viruses" and infect both macrophages and T cells (and low numbers of DCs). Historically, R5 strains were known as "macrophage- or M-tropic strains" to distinguish them from the T-tropic strains. HIV strains that can bind to both CCR5 and CXCR4 are called R5X4 strains. In most patients, the HIV virions generated during the early stages of infection are R5 in character. As the infection progresses and mutations accumulate in the viral genome, these R5 strains give rise to R5X4 or X4 strains. X4 viruses usually replicate more rapidly than R5 viruses, accelerating the development of AIDS in the later stages of the infection.

ii) Viral Entry

When the HIV envelope protein gp120 binds to CD4, the conformation of gp120 is altered such that a region capable of binding to chemokine receptors is exposed. The interaction of the CD4–gp120 complex with a chemokine receptor further adjusts the conformation of the entire envelope spike such that

gp41 is brought into contact with the host cell membrane. Gp41 inserts into the host cell membrane and promotes its fusion to the viral envelope, allowing the virus to efficiently transfer into the cell's interior. HIV can also gain access to T cells by making use of a DC that has captured the virus and internalized it into a specialized intracellular vacuole. The HIV-loaded DC migrates to the local lymph node and makes contact with a CD4$^+$ T cell in the node. An intercellular interface called an *infectious synapse* is formed that facilitates the rapid transfer (by an unknown mechanism) of the virus from the DC into the T cell.

iii) Provirus Formation and Activation of Viral Replication

After fusion of the HIV envelope with the host membrane, the viral core is everted into the host cell cytoplasm. The capsid protecting the viral genome is removed by host cell proteases, leaving the viral *preintegration complex*. Within this complex, HIV RT associated with the viral RNA rapidly synthesizes the viral DNA, and matrix proteins direct the transport of this DNA, viral INT and other associated viral proteins into the host cell nucleus. Once in the nucleus, INT creates the provirus by inserting a single copy of the viral DNA into the host cell DNA; the virus is then said to be in its *latent form*. Infected T cells in this *preactivation stage* do not transcribe the provirus to any meaningful extent and very few progeny virus particles are formed. However, if the T cell is stimulated by either TCR engagement or cytokine binding, intracellular signaling is triggered that initiates new transcription within the host cell. Along with various host genes, the HIV regulatory genes start to be expressed. Expression of the HIV genes encoding structural and accessory proteins as well as enzymes then commences, followed by progeny virion production.

II. CLINICAL EVENTS

HIV is most often introduced into the body by sexual contact, and the rectal and vaginal mucosae are particularly vulnerable to HIV attack. In the earliest, asymptomatic stages of a primary HIV infection, HIV attaches to DCs and infects macrophages and CD4$^+$ T cells resident in the rectal or vaginal lamina propria. The virus-bearing DCs and infected T cells carry the virus to additional resting naïve CD4$^+$ T cells in the local lymph node, and viral replication proceeds at an exponential rate. Within 2–6 weeks of exposure, the individual may experience acute fever and an illness similar to infectious mononucleosis (Fig. 15-5A). The number of virions present in the circulation (the *viral load*) is high and significant levels of p24 can be detected in the blood (Fig. 15-5B). Infected lymphocytes increasingly convey virions to additional lymphoid tissues. It is at this point that the individual, who may not suspect HIV infection at all, is most contagious. Anti-HIV antibodies made by the infected individual are present at very low levels and are detectable only by the most sensitive methods (Fig. 15-5C). At about 4–8 weeks post-infection, the ratio of CD4$^+$ T cells to CD8$^+$ T cells is reduced drastically from its normal value of 2 : 1 due to the mass killing of CD4$^+$ T cells and the peaking

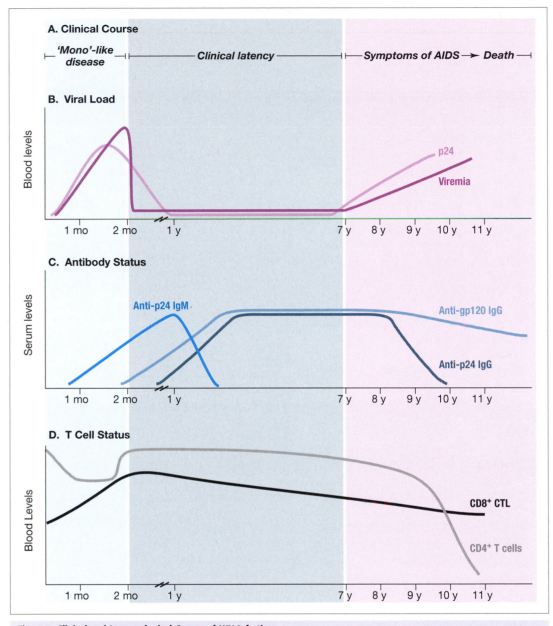

Fig. 15-5 **Clinical and Immunological Course of HIV Infection**

of the anti-HIV CTL response (Fig. 15-5D). The infection is partially contained (but not totally cleared from the body), the clinical symptoms abate, and viremia subsides.

Between 3 and 8 weeks post-infection, circulating anti-HIV neutralizing antibodies start to be detectable in the blood of infected individuals (refer to Fig. 15-5C). However, because the virus has undergone tremendous and very rapid antigenic variation (see later), many of these antibodies will fail to recognize the current array of viral epitopes. Macrophages and resting T cells are infected on a massive scale and proviruses are integrated into the genomes of these cells. During this period, whenever infected CD4⁺ T cells and macrophages are stimulated, the provirus is transcribed and translated and progeny virions are produced and released systemically. As the

virus spreads to more and more fresh T cells in the lymphoid organs, it continues to replicate and release progeny virions. Nevertheless, HIV-specific CTLs and antibodies control the infection, viremia does not develop, and the patient is not overtly ill. Meanwhile, in the lymph node, progeny virions that have attracted the attention of complement components bind to the CRs expressed on the surfaces of FDCs in the GCs and remain "trapped" on the exterior surfaces of these cells. Macrophages and T cells that come into contact with these FDCs are then soon infected. During this phase, which may last only a few months or for many years, billions of not only infected but also uninfected CD4⁺ T cells are destroyed. The mechanism by which the uninfected CD4⁺ T cells are lost is not fully understood, but scientists speculate that some form of induced programmed cell death may be involved. Despite the drastic assault on CD4+ T cells, their numbers hold relatively steady during this time due to the tremendous power of the bone marrow to produce new hematopoietic cells. Thus, although the virus is active during this phase, the patient experiences a period of "clinical latency".

As long as an HIV patient's "T4 count" remains above 200 CD4+ T cells/mm³ blood (about 25% of normal), the infected individual continues to show no clinical signs. Eventually, however, there is a collapse in the CD4+ T cell population followed by a drop in anti-HIV antibody titers and a decline in numbers of anti-HIV CTLs. Viral p24 reappears in the blood and viremia increases sharply as up to 10^{10} virions/day are produced (refer to Fig. 15-5B). As the disease approaches its late stages, lymph node architecture degenerates and the virus that was trapped on FDC surfaces is freed. The viral load continues to mount as the T4 count falls to below 200 CD4+ T cells/mm³ blood. The loss of CD4+ T cells, FDCs and the lymph node microenvironment drastically compromises the patient's ability to mount adaptive immune responses. Clinical latency ends and symptoms of AIDS appear when the host's immune system has been so disrupted that he/she can no longer fend off opportunistic pathogens. When a patient's T4 count drops to <50 CD4+ T cells/mm³, death is imminent in the absence of anti-retroviral drug therapy.

III. HIV CLASSIFICATION

i) HIV-Infected Persons

HIV-infected persons are classified using a scheme that has both an immunological status component and a clinical presentation component. A patient's classification dictates the course of treatment for his/her disease. As shown in Table 15-2, there are three categories (1, 2 and 3) that define the immunological status of HIV-positive individuals based on their CD4+ T cell count. There are also three levels of clinical presentation status based on the opportunistic diseases and

cancers diagnosed. Clinical category A patients are HIV-infected but asymptomatic, or may show mild disease. Clinical category B patients have opportunistic infections that are slightly less devastating than those in category C. A signature clinical feature of C3 HIV patients is *Kaposi's sarcoma*, an unusual tumor of connective tissues (Plate 15-2). Kaposi's sarcoma is caused not by HIV but by *Kaposi sarcoma herpesvirus* (KSHV), a virus that gains a foothold only after HIV destroys the bulk of the CD4+ T cell population. Two other frequent features of category C patients are *cachexia* or *wasting syndrome* and pneumonia caused by infection with the fungal pathogen *Pneumocystis carinii*.

ii) HIV Strains

HIV infections are characterized by extreme antigenic variation. Although a single HIV clone initiates a new infection, distinct HIV isolates with a range of genomic RNA sequences can soon be found within an individual patient. Several broad groups or *clades* of antigenically distinct types of HIV-1 have been defined based on nucleotide differences in the *env* and

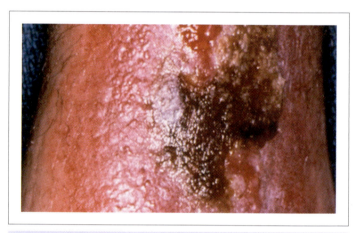

Plate 15-2 **Kaposi's Sarcoma**
[Reproduced by permission of the CDC/Steve Kraus.]

Table 15-2 **Classification of HIV-Infected Persons**

	Category A† **Asymptomatic or mild disease**	**Category B** **Moderate disease**	**Category C** **Severe disease**
Category 1*: >500 CD4+ T cells/mm³ blood	A1	B1	C1
Category 2: 200–499 CD4+ T cells/mm³ blood	A2	B2	C2
Category 3: <200 CD4+ T cells/mm³ blood	A3	B3	C3
Examples of Associated Diseases	Fever, fatigue, mono-like symptoms, swollen lymph nodes	Shingles, pelvic inflammatory disease, thrush, candidiasis	Kaposi's sarcoma, wasting syndrome, *Pneumocystis carinii* pneumonia

*Immunological status (T4 count)
†Clinical status

gag genes. (A clade is a group of related organisms descended from a common ancestor.) At the protein level, the most variable HIV molecule is gp120. At the clinical level, some minor differences in transmissibility and disease progression have been noted among HIV-1 clades, but all cause premature death in the absence of anti-retroviral drug treatment.

Three factors contribute to the generation of HIV antigenic variability. Firstly, the RT that makes the viral DNA is error-prone, meaning that it introduces mutations as it copies the RNA genome. Secondly, the very high production rate of the virus provides plenty of replication cycles and thus numerous chances for mutations to be introduced. Thirdly, the ability of viruses of different clades to recombine if they co-infect the same individual can suddenly generate a new viral genome with sequence characteristics of both parental viruses. New HIV strains that evade neutralizing antibodies and antigen-specific CTLs by altering their antigenicity are called *genetic escape mutants*.

C. Immune Responses during HIV Infection

Although HIV is not a highly infectious pathogen, it has been phenomenally successful. Part of this success stems from the fact that HIV acts only slowly on its hosts. Thus, the virus has had ample time to spread widely throughout the world's population. More importantly, HIV employs sophisticated mechanisms that permit it to both evade immune responses and destroy immune system cells. The effects of HIV on host immune responses are discussed next and summarized in Table 15-3.

I. Th RESPONSES

At least in the initial stages of an HIV infection, HIV-specific Th cells play a key role in keeping viral loads under control. The cytokine help supplied by these cells supports the differentiation of HIV-specific CTLs that kill infected macrophages, DCs and T cells and thus reduce viremia. However, in most HIV-infected individuals, the virus soon mutates to evade these CTLs and is free to infect large numbers of Th cells. The presence of the virus interferes with the expression of survival genes in Th cells and promotes their apoptosis. Those few Th cells that survive appear to have undergone anergization. As a result, only weak (if any) anti-HIV Th responses are mounted later in the infection, compromising CTL responses. Persistent viremia is then observed that usually cannot be controlled without anti-retroviral drugs. However, certain HIV-infected individuals called *long-term non-progressors* appear to be able to control their viremia without drugs. Many of these individuals retain large populations of HIV-specific Th cells capable of vigorous Th1 responses. The mechanism underlying the HIV resistance of long-term non-progressors is under investigation.

Table 15-3 Effects of HIV on Host Immune Responses

Immune Response Element	Effect of HIV Infection
Th response	Destruction of infected and uninfected CD4+ T cells
	Interference with cell survival pathways
	Anergization of remaining CD4+ T cells
CTL response	Inhibition of naïve CD8+ T cell responses due to lack of CD4+ T cell help
	Destruction of CD8+ T cell clones due to induction of CD4 expression
	Failure to develop memory CTL response due to genetic escape mutants
	Apoptosis of CD8+ T cells caused by induction of Fas on CD8+ T cells and FasL on interacting CD4+ T cells
	Compromised recognition of target cells due to downregulation of MHC class I induced by Nef
Antibody response	Poor quality of neutralizing antibody response.
	Antibodies produced may not neutralize substrains of HIV arising by antigenic variation
	Polyclonal B cell activation induced by non-specific gp120 binding leads to immune complex formation and clonal exhaustion
	Compromised naïve B cell response due to lack of CD4+ T cell help; lack of high affinity antibody production
Cytokines	Abnormal cytokine production, especially increased TNF
NK cells	Potential inhibition of natural cytotoxicity by free Tat
Complement	Capture of host RCA proteins in viral envelope while budding
	Recruitment of Factor H to virion surface
	Downregulation of complement receptors

II. CTL RESPONSES

As mentioned earlier, HIV-1 replication and viremia are initially held at bay by vigorous CTL responses. These responses are usually directed against epitopes in gp120 or in highly conserved regions of the Pol, Nef and Gag proteins. Some individuals who remain HIV-negative despite repeated exposure have been found to retain HIV-specific CTL activity. HIV-specific CTLs destroy HIV-infected cells mainly by perforin/granzyme-mediated cytolysis rather than by cytotoxic cytokine secretion. However, in most cases, the anti-HIV CTL response eventually fails because of the catastrophic loss of Th cells. The continuously mutating virus generates genetic escape mutants that present new epitopes, requiring the activation of more and more naïve CD8+ Tc cells. In the absence of sufficient Th help, these cells are not adequately activated. In addition, in the presence of HIV, the expression of CD4 is frequently induced on the surfaces of naïve CD8+ Tc cells, marking them for HIV-mediated destruction. The presence of the virus also upregulates Fas expression on CTLs, making them vulnerable to apoptosis induced by FasL-expressing CD4+ T cells. Finally, the HIV Nef protein helps to dampen the anti-HIV CTL response because Nef downregulates the expression of surface MHC class I. The resulting reduction in pMHC on the surface of an infected cell inhibits its recognition by CD8+ CTLs.

III. ANTIBODY RESPONSES

As was illustrated in Figure 15-5, humoral responses against HIV are mounted but they are not very effective. Although detectable levels of antibodies to the antigens of most other pathogens appear within the first 2 weeks after infection, the lag between HIV infection and the appearance in the blood of measurable levels of anti-HIV IgM antibodies is at least 3 weeks and often longer. Moreover, many of the antibodies produced are not able to neutralize the virus. This failure is due at least in part to the extreme antigenic variability of HIV, since neutralizing antibodies generated early in the course of the infection often do not recognize epitopes of the mutated strain present later in the infection. Indeed, as most HIV infections progress, the viral strains that emerge appear to have been selected for complete resistance to antibody neutralization.

Although HIV does not replicate in B cells, the virus has a severe impact on B cell function. Gp120 binds to a site in the V region of the Ig heavy chain (outside the antigen-binding cleft) and non-specifically activates multiple clones of B cells. As a result of this extraneous B cell expansion, only about 20% of the antibodies produced in response to HIV infection are actually directed against HIV. The non-specific antibodies spur the formation of circulating immune complexes and promote autoimmunity in many HIV-infected persons. More importantly, included among the B cell clones activated non-specifically by gp120 are those directed against many bacterial and fungal pathogens. The eventual clonal exhaustion of these B cells robs the patient's immune system of a key weapon needed to defend against these pathogens later in the infection. In addition to the loss of these B cell clones, the decimation of the CD4+ Th cell population impairs the activation of any remaining clones of naïve anti-HIV B cells.

IV. CYTOKINES

HIV infection results in an abnormal profile of cytokine secretion that contributes to many AIDS symptoms. In the early stages of the disease, HIV-infected persons have elevated blood levels of TNF, IL-1, IL-2, IL-6 and IFNα. TNF in particular is a powerful activator of the HIV provirus and induces the wasting syndrome associated with AIDS. IL-1 has been linked to AIDS-associated fever and dementia.

V. NK CELLS

HIV may also have detrimental effects on NK cells. There is some evidence that extracellular Tat protein released by HIV-infected cells may interfere with the natural cytotoxicity exerted by an NK cell once it has bound to a target cell. Thus, even though HIV downregulates MHC class I expression on infected T cells (creating a deficit that should act as a call to arms for NK cells), the warriors are disabled and the infected T cells escape death.

VI. COMPLEMENT

HIV blocks complement-mediated defense in three ways. Firstly, as new progeny viruses bud through the host cell membrane, they incorporate host RCA proteins into their envelopes and can thus forestall the deposition of the MAC. Secondly, the HIV envelope has multiple binding sites that recruit the soluble RCA protein Factor H to the virion surface. Factor H then inhibits alternative complement activation. Thirdly, HIV infection downregulates expression of CRs on host cells, so that when a complement component binds to an infected cell, there is a decreased likelihood that opsonized phagocytosis will occur.

D. Host Factors Influencing the Course of HIV Infection

In this section, we discuss transmission of HIV, as well as several host factors that influence whether an individual exposed to HIV will actually be infected or not, and whether that person progresses slowly or rapidly to AIDS.

I. TRANSMISSION OF HIV

It is not easy to get infected with HIV. The virus has to be transferred into a body fluid by sexual contact, breast-feeding, the mixing of blood during transfusions or trauma (including during birth), or the sharing of contaminated needles. Merely touching, kissing or being sneezed on by an HIV-infected person is not enough. HIV infection is easily prevented if elementary precautions are taken, such as the use of condoms, gloves, screened or cloned blood products, and clean needles and syringes. Since the tragedy in the 1980s in which HIV contamination of donated blood supplies killed thousands of hemophiliacs, almost all countries have adopted rigorous programs for screening blood donors. The current AIDS epidemic is therefore not due so much to a highly contagious virus (which HIV is not) as it is to faulty human behavior and cultural intransigence. Particularly risky behaviors are having multiple sexual partners, engaging in anal intercourse, having unprotected sex, having sex with prostitutes, abusing intravenous drugs, and getting a tattoo via a needle. If exposed to HIV via unprotected sex, uncircumcised males are at higher risk of infection, as are individuals who already have another sexually transmitted disease. Health care workers also run a certain risk of infection: those in the developed world who sustain a needle stick injury have a 0.3% chance of acquiring HIV.

Globally, about 600,000 infants per year (almost 2000 a day) are infected with HIV through mother–child transmission. Transfer of the virus occurs most commonly at delivery but is also possible during the second half of pregnancy or in the course of breast-feeding. Treatment of the mother with anti-retroviral drugs during pregnancy, delivery and after the birth can substantially reduce viremia in the mother and thus the risk of transmission to the child. The infant can also be given anti-retroviral drugs for the first 6 weeks of life. As well, HIV-infected mothers are encouraged to feed their babies formula so that breast-feeding and its associated risk of viral transmission can be avoided. Unfortunately, many HIV-infected mothers cannot access these types of preventive measures. As a result, the WHO estimates that, in 2005 alone, mother–child HIV transmission caused the deaths of 570,000 children under 15 years of age.

II. RESISTANCE TO HIV INFECTION

Two groups of people appear to be able to resist HIV infection even when directly exposed to the virus. The first group includes individuals carrying mutations in host genes required for viral entry. For example, a mutated allele of CCR5 called CCR5Δ32 has a 32 base pair deletion in the receptor nucleotide sequence that renders it non-functional. The vast majority of individuals homozygous for CCR5Δ32 do not appear to be susceptible to HIV-1 infection and indeed suffer no health deficits at all. Furthermore, many CCR5/CCR5Δ32 heterozygotes who become infected with HIV experience at least a 2 year delay in the onset of AIDS. Individuals in the other HIV-resistant group have normal CCR5 (and other chemokine receptor) genes but express particular HLA proteins. For example, there is a subset of closely related MHC class I proteins called the HLA-A2/6802 family that has been linked to HIV resistance. Expression of members of the HLA-A2/6802 family has often been observed in uninfected newborns of HIV+ mothers, uninfected sexual partners of infected persons, and in a well-studied group of repeatedly exposed but uninfected prostitutes in Kenya. It is not yet clear why HLA-A2/6802 alleles are protective. Among MHC class II molecules, proteins containing certain DRB1 β chains have been linked to HIV resistance. Researchers are investigating the possibility that these HLA subsets present highly conserved HIV epitopes that are strongly immunogenic and induce ongoing, effective Th and CTL responses.

III. CLINICAL COURSE VARIABILITY

Once an individual has become infected with HIV, additional host-related factors may have a positive effect on symptom-

free survival and clinical latency. Long-term non-progressors resist clinical disease for more than 10 years, whereas so-called "rapid progressors" succumb within 2 years of infection. What determines who will be a long-term non-progressor and who will be rapid progressor is not precisely known. However, long-term non-progressors exhibit relatively stable CD4+ T cell counts and ongoing polyclonal anti-HIV CTL responses directed against multiple HIV antigens. In contrast, rapid progressors show more fluctuation in their CD4+ T cell counts and the expansion of relatively few CTL clones. Individuals expressing HLA molecules belonging to the A2, B27, B51 and B57 families also have a more favorable clinical outcome, as do individuals fully heterozygous at all HLA class I loci. These people have an advantage in antigen presentation over their more homozygous counterparts, and thus may be able to mount broader CTL responses against HIV.

E. Animal Models of AIDS

Although *in vitro* studies have revealed important information about the HIV virion and its life cycle, studies done solely in culture dishes cannot reproduce many important aspects of *in vivo* infection and disease progression. A major challenge in studying AIDS has been to develop an adequate animal model in which to examine immune responses, test anti-retroviral drugs, and assess vaccination strategies. Clinical researchers have compiled a "wish list" of features that a system ideally should have to be a useful animal model of AIDS (Fig. 15-6).

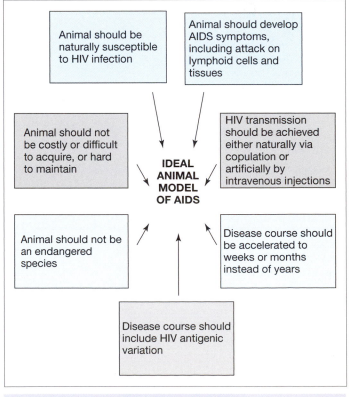

Fig. 15-6 **Features of the Ideal Animal Model of AIDS**

Several different types of animal systems have been established to study HIV infection, including primate and mouse models. Each model has its advantages and drawbacks, and none is ideal.

I. PRIMATE MODELS OF AIDS

Primate models for the study of AIDS have been developed in several species. Chimpanzees are the only animal that can be naturally infected with HIV-1. However, unlike HIV-infected humans, HIV-infected chimpanzees do not develop AIDS. As well, these animals are costly and difficult to obtain, and are an endangered species. Models based on monkeys are attractive because, unlike the apes, these animals are readily available, are not endangered, are comparatively affordable, and breed well in captivity. However, monkeys are not susceptible to HIV-1 infection. Several species of monkeys are susceptible to infection with the monkey lentivirus SIV_{mac} (simian immunodeficiency virus that infects <u>mac</u>aque monkeys) but only macaques infected with SIV_{mac} actually get sick. In addition, lentivirus infection of monkey lymphoid tissues follows a course of infection that differs significantly from that of HIV infection of human lymphoid tissues.

II. MOUSE MODELS OF AIDS

Mouse models of AIDS are frequently used to perform first stage screening of AIDS vaccine candidates for toxicity or to establish dose–response curves. Mice mature rapidly, are readily available, and are not an endangered species, such that mouse trials are significantly shorter and less expensive than similar evaluations in monkeys or full-scale clinical trials in humans. In most mouse AIDS models, human lymphoid tissue is transplanted into mouse mutants that are genetically immunodeficient (so that the mouse will not mount an immune response against the human tissue). The transplanted animals are then infected with HIV and the infection of the human lymphoid tissue is monitored. However, the course of HIV infection in this hybrid system diverges significantly from that in a whole human.

F. HIV Vaccines

I. OVERVIEW

The impact and incurability of AIDS have made the development of an HIV vaccine a global priority. In the 1990s, because the virus had been identified so rapidly, there was much optimism that an effective vaccine could be readily produced using modern biotechnology. Unfortunately, HIV's antigenic variation and multiple strategies for evading and destroying the human immune system have greatly hampered vaccine development. It remains unclear which epitopes of the virus should be included in a vaccine, what type of response (humoral or cell-mediated or both) will be most effective, where (mucosally or systemically) the response will have to be induced, and how

strong the response will have to be. Nevertheless, the small number of people who have been continually exposed to HIV but never infected, and the long-term non-progressors who have the virus but experience a long delay in developing AIDS, represent hope for the eventual production of an HIV vaccine that might induce a similar level of protective immunity to HIV in the general population.

In 2003, the Global HIV/AIDS Vaccine Enterprise was established among AIDS researchers to coordinate efforts and thereby accelerate the development of a prophylactic vaccine for HIV/AIDS. The Enterprise involves private sector and government scientists, public health officials, advocacy groups and funding agencies. Its mandate is to promote the sharing of strategies, resources and results. The Enterprise is also actively seeking ways to overcome regulatory obstacles and increase numbers of trained personnel in developing countries.

II. BARRIERS TO HIV VACCINE DEVELOPMENT

Obviously, the destruction of the very cells responsible for responding to a vaccine antigen constitutes a huge hurdle to HIV vaccine development. However, there are also other barriers, as summarized in Figure 15-7. In animal models, vaccines based on live, attenuated viruses are generally the most effective means of protecting against natural infection. However, a problem with applying this approach to a human HIV vaccine is that attenuation may not offer a sufficient guarantee of vaccine safety. In monkeys, some attenuated SIV strains have shown a remarkable talent for reversion, repairing engineered deletions in genes by use of their error-prone RT enzymes. In addition, attenuated SIV has been shown to cause an AIDS-like disease in newborn monkeys.

Although subunit vaccines would be relatively safe, the extreme variability of HIV presents enormous challenges to the

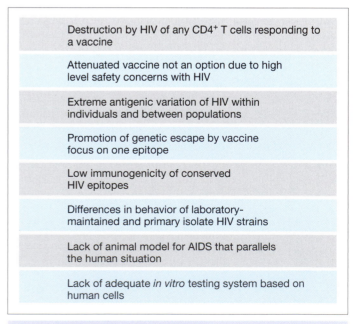

Destruction by HIV of any CD4$^+$ T cells responding to a vaccine

Attenuated vaccine not an option due to high level safety concerns with HIV

Extreme antigenic variation of HIV within individuals and between populations

Promotion of genetic escape by vaccine focus on one epitope

Low immunogenicity of conserved HIV epitopes

Differences in behavior of laboratory-maintained and primary isolate HIV strains

Lack of animal model for AIDS that parallels the human situation

Lack of adequate *in vitro* testing system based on human cells

Fig. 15-7 **Barriers to HIV Vaccine Development**

use of this approach. An HIV immunogen has yet to be identified that induces antibodies or CTLs capable of recognizing a broad range of primary viral isolates from a multitude of patients. Neutralizing antibodies recognizing one clade usually do not recognize another. In addition, focusing on one epitope tends to foster the emergence of genetic escape mutants. Moreover, it seems that the most conserved epitopes in HIV tend to be the least immunogenic, and those that are immunogenic are subject to extensive antigenic drift. These factors do not bode well for producing a vaccine that will induce protective immunity in most individuals within a population, let alone between populations.

Another technical challenge in HIV vaccine development is related to the animal systems used to grow the virus and test candidate vaccines. Laboratory-maintained strains of HIV do not behave exactly like primary HIV isolates obtained from patients, making it difficult to extrapolate results. Similarly, although monkey and mouse models have been very helpful in investigating certain aspects of HIV biology, infections in these models differ significantly from natural HIV infections of humans. Again, the promise of a vaccine tested in these models might not extend to the human situation.

G. Treatment of HIV Infection with Anti-Retroviral Drugs

Until an effective HIV vaccine is developed, the best weapon the world has against AIDS is anti-retroviral drug therapy. These drugs are responsible for the dramatic reductions in AIDS progression and AIDS-related deaths seen in the developed world since the late 1990s. There are at present four classes of licensed anti-retroviral drugs that act on proteins essential for viral spread: protease (PR) inhibitors, nucleoside RT inhibitors, non-nucleoside RT inhibitors, and fusion inhibitors (Table 15-4). Suppression of viral replication by any one of these types of inhibitors technically reduces the number of HIV replicative cycles and should decrease the patient's viral load. However, due to HIV's propensity for genetic escape, the use of only one antiviral drug at a time soon selects for a strain of HIV resistant to that drug, re-escalating the attack on the immune system. Clinicians therefore often treat HIV patients with *highly active anti-retroviral therapy* (HAART), a regimen that features combinations ("cocktails") of three or more anti-viral drugs from two different drug classes. For example, an RT inhibitor may be combined with a PR inhibitor and a fusion inhibitor. As well as firmly shutting down viral replication for an extended period, this multidrug approach greatly decreases the chance of the virus generating a substrain that is simultaneously resistant to all drugs used in the cocktail. The patient's immune system then has some breathing room so that at least some restoration of HIV-specific T cell responses, as well as responses against other pathogens, can occur. The HAART regimen has been highly successful in extending the lives of HIV patients, but only if the patient has been assiduously compliant with the prescribed regimen.

I. CLASSES OF ANTI-RETROVIRAL DRUGS

i) Protease Inhibitors

Protease inhibitors are small molecules that work by competitively binding to the active site of the HIV PR. Without PR function, the production of progeny viruses is stymied because

Table 15-4 Common Anti-Retroviral Drugs

Inhibitor Type	Examples	Mechanism	Potential Side Effects
Protease inhibitors	Indinavir, saquinavir, ritonavir, atazanavir, tipranavir	Small molecules bind competitively to active site of HIV PR	Diarrhea, hyperlipidemia, rash, dry skin, nausea, vomiting; multiple drug interactions with other medications
Nucleoside RT inhibitors	Azidothymidine (AZT; zidovudine), stavudine, lamivudine (3TC), abacavir, tenofovir	Nucleoside analogues competitively inhibit viral DNA synthesis by RT	Dysregulated lipid metabolism, lactic acidosis, bone marrow suppression, peripheral neuropathy, myopathy, mouth ulcers, nausea, vomiting
Non-nucleoside RT inhibitors	Nevirapine, efavirenz, etravirine	Molecules of diverse structure induce conformational changes to RT that inactivate it	Rash, hepatotoxicity, multiple drug interactions with other medications
Fusion inhibitors	Enfuvirtide	Peptide based on gp41 blocks virus envelope fusion to host cell membrane	Dizziness, severe allergic reaction, increased susceptibility to bacterial pneumonia
Integrase inhibitors	Experimental agents only	Agent blocks function of HIV INT and thus integration of HIV cDNA into host genome	Under investigation
Chemokine receptor inhibitors	Experimental agents only	Agent blocks access of HIV to CCR5 or CXCR4	Unacceptably toxic thus far; remain under investigation

the Gag and Pol polyproteins of the virus are not cleaved. Although PR inhibitors can be highly effective, therapy with older drugs in this class (e.g., indinavir, saquinavir) entails the taking of numerous pills that have unpleasant side effects, decreasing patient compliance. Newer generations of PR inhibitors (e.g., ritonavir, tipranavir) require fewer pills, have fewer side effects, and show greater efficacy against the virus.

ii) Nucleoside RT Inhibitors

Nucleoside RT inhibitors work on the principles of competitive inhibition and premature DNA chain termination during viral replication. The inhibitors are analogues of the usual deoxynucleosides that the RT joins together to synthesize DNA. Occupation of the active site of RT by the inhibitor blocks the elongation of the growing viral DNA chain. In addition, the presence of large numbers of molecules of the inhibitor makes it hard for the RT to find its natural substrate. Unfortunately, patients taking RT inhibitors also experience unpleasant side effects.

iii) Non-Nucleoside RT Inhibitors

Unlike nucleoside inhibitors, the non-nucleoside inhibitors act on the RT enzyme at locations distant from the active site. Although these inhibitors are diverse in structure, they all induce major conformational changes to the RT molecule that disrupt its enzymatic activity. These drugs are better tolerated by patients than either protease inhibitors or nucleoside RT inhibitors.

iv) Fusion Inhibitors

This relatively new class of anti-retroviral drug works by blocking viral entry into host cells. Enfuvirtide, the only fusion inhibitor licensed to date, is a synthetic peptide that blocks the conformational change of the HIV envelope spike that allows the virus to fuse with the host cell membrane. Thus, in the presence of enfuvirtide, HIV cannot enter the cell. A disadvantage of enfuvirtide is that it must be delivered by subcutaneous injection. Enfuvirtide is currently recommended for patients whose viral isolates show resistance to all other classes of inhibitors.

v) Integrase Inhibitors

A very new class of anti-retroviral drugs includes molecules that inhibit the function of HIV integrase, thereby blocking the integration of the HIV cDNA into the host genome. Most agents in this class are currently experimental and available only in the clinical trial setting.

vi) Chemokine Receptor Inhibitors

Several experimental agents have been produced to block the access of HIV to CCR5 or CXCR4 and thus prevent the virus from entering cells. However, major problems with toxicity and efficacy have arisen and all but one of these agents has failed to make it past phase I clinical trials.

II. LIMITATIONS OF ANTI-RETROVIRAL DRUGS

Despite the successes achieved with HAART, clinicians have come to the realization that it will be extremely difficult to completely eliminate HIV from a patient's body. The anti-retroviral drugs currently available are effective only on actively replicating HIV and do not eliminate latent virus lurking in the genomes of resting cells. HIV can maintain latency for years both inside resting T cells, resting macrophages and in other inaccessible body reservoirs such as the CNS. Moreover, HAART apparently does not stop <u>all</u> viral replication, and a very low level apparently continues that cannot be detected by current clinical assays. The highly mutable nature of HIV means that the strain lingering in the patient is constantly evolving and will likely develop resistance to existing HAART drugs. New approaches are needed that can completely suppress viral replication and attack viruses in reservoirs without harming normal cells. It is hoped that the integrase inhibitors mentioned above as well as novel agents that can degrade viral RNA will turn out to be truly effective in eliminating or incapacitating the virus wherever it hides.

Another potential limitation of existing anti-HIV treatments is that we do not yet know the side effects of long-term use of these drug cocktails. Patients may end up having to choose between physical ailments due to the virus itself and those due to HAART. Some relief might be gained if drugs that specifically block access to CCR5 or CXCR4 are successfully developed. The apparent good health of individuals who naturally lack expression of CCR5 and the resistance of these persons to HIV infection make this protein a highly promising target.

We have reached the end of our discussion of HIV and AIDS. In Chapter 16, we describe how a healthy immune system strives to fight cancer development, and how tumors evade both immune surveillance and elimination by immune system effector cells.

- HIV is a cytopathic RNA retrovirus causing AIDS in human hosts. HIV is most commonly transmitted during sexual contact or intravenous drug use involving contaminated needles.

- The interaction of the HIV envelope protein gp120 with CD4 and host chemokine receptors allows the virus to enter macrophages, DCs and CD4$^+$ T cells.

- Viral RT transcribes the viral RNA into a viral DNA that is integrated into the host genome to become the provirus. When the infected cell is stimulated, viral transcription and translation begin and copious amounts of newly synthesized virions bud from the host cell prior to its death.

- In a newly infected individual, HIV attacks DCs, macrophages and CD4$^+$ T cells. Later on, the virus mutates its gp120 protein and preferentially infects CD4$^+$ T cells. HIV also kills uninfected CD4$^+$ T cells via an unknown mechanism.

- B cells and CD8$^+$ T cells respond to HIV infection but the virus-specific antibodies and CTLs produced are ultimately insufficient to contain the virus.

- Clinically, an HIV-infected individual experiences a mononucleosis-like illness followed by an asymptomatic period of clinical latency that may last for several years.

- As the immune response fails, viral replication accelerates and the resulting massive loss of CD4$^+$ T cells leads to profound immunodeficiency and death due to opportunistic infection or malignancy.

- Many barriers continue to thwart the development of a successful HIV vaccine, including HIV antigenic variation, T cell depletion, and a lack of animal models.

- Treatment with anti-retroviral drugs extends the life of an infected individual but does not eliminate the virus.

Introduction–Section A.I

1) Can you define these terms? cytopathic, lentivirus

2) Distinguish between primary and secondary immunodeficiencies.

3) Is the number of people that die of AIDS per year worldwide closer to 3000, 300,000 or 3,000,000?

4) Give three methods by which HIV is transmitted between individuals.

5) What is a provirus and how does it function in the generation of progeny virions?

6) Which cell type is the primary target of HIV?

Section A.II

1) Sketch the structural layers of the HIV virion and label each.

2) Describe the structure of the HIV envelope spike.

3) Name four host-derived proteins that appear in the HIV envelope.

4) Describe the structure of the HIV genome and give the names and functions of its nine genes.

5) Describe the three polyprotein precursors of HIV and the structural or functional proteins derived from each.

6) What are the LTRs and what do they do?

Section B.I

1) Can you define these terms? T-tropic, M-tropic, infectious synapse

2) What molecules are required for HIV entry into a host cell?

3) Why are macrophages a reservoir for HIV?

4) Give two differences between X4 and R5 HIV strains.

5) How does gp41 facilitate entry of HIV into a cell?

6) Describe the contents of the HIV preintegration complex and how it is formed.

7) What is the preactivation stage and how is it terminated?

Section B.II

1) Can you define these terms? viral load, T4 count

2) Give a clinical and molecular description of the situation in an HIV-infected person at about 2 weeks post-exposure. How has this picture changed by about 8 weeks post-exposure?

3) Give a clinical and molecular description of the clinical latency in an HIV-infected person.

4) Give a clinical and molecular description of the situation in an HIV-infected person as latency ends.

Section B.III

1) Can you define these terms? clade, genetic escape mutant

2) Define the three major clinical categories of HIV-infected persons.

3) Outline three factors that contribute to the antigenic variability of HIV.

Section C

1) Why has HIV been such a successful pathogen?

2) Give two effects of HIV on Th cells that contribute to the eventual failure of the Th response.

3) How does the effectiveness of the CTL response change over the course of HIV infection?

4) Is the humoral response effective against HIV? Explain.

5) What effects does HIV have on B cell functions?

6) Which cytokine is a powerful activator of the HIV provirus?

7) How does HIV interfere with the NK response?

8) Give two ways by which HIV blocks complement-mediated defense.

Section D

1) Give three ways by which HIV infection can be prevented.

2) Which two groups of people are resistant to HIV infection and why?

3) Distinguish between long-term non-progressors and rapid progressors.

Sections E–F

1) Compare the usefulness of chimpanzees and monkeys as AIDS models.

2) What is the purpose of the Global HIV/AIDS Vaccine Enterprise?

3) Give three barriers to HIV vaccine development.

Section G

1) Name four classes of anti-retroviral drugs and describe how they work.

2) What is HAART and why is it used?

3) Give two limitations of the anti-retroviral drugs currently in use.

16

Tumor Immunology

WHAT'S IN THIS CHAPTER?

Do what you can, with what you have, where you are.
Theodore Roosevelt

Over 8 million new cases of cancer occur annually throughout the world. At the start of 2007, the American Cancer Society predicted that 1,444,920 Americans would be newly diagnosed with some kind of malignancy and that 559,650 Americans would die of their cancers during the year. As well as mortality, cancer imposes a huge economic burden. The direct costs associated with the treatment of cancer patients in the United States (including drugs, hospital care and home care) exceed U.S. $50 billion annually, even before the lost productivity of stricken workers is taken into account. Although research has revealed much about the origin and nature of cancers, the role of immune responses in dealing with them remains unclear. Many immunologists believe that the immune system tries to protect the host against cancer by acting as a "tumor surveillance" mechanism. However, although cancers are clearly a threat to the host, the immune system does not eliminate all tumors promptly and some cancers appear to induce tolerance rather than an immune response.

A. Tumor Biology

I. TUMORS AND CANCERS

Throughout the life of an organism, body cells divide, differentiate and die in a carefully controlled manner. Tissue-specific stem cells in various adult tissues, like the HSCs of the hematopoietic system, give rise to progressively more differentiated cell types. A balance is struck between proliferation, which allows the tissue to develop and maintain its required size and structure, and evermore restricted differentiation, which allows the tissue to function. When the cells of a tissue undergo unusual division that serves no useful function for the host, an abnormal tissue mass called a *neoplasm* ("new growth") or *tumor* ("swelling") may be created. Neoplastic cells may not actually divide any faster than their normal counterparts.

However, because a higher than normal proportion of these cells is dividing and the rate of programmed cell death is decreased, an abnormal accumulation of cells occurs that eventually may constitute a tumor.

i) Classification of Tumors

All tumors can be classified as either *benign* or *malignant*, and a malignant tumor is a *cancer*. A benign tumor is relatively slow growing and contains cells that are well differentiated and well organized (Plate 16-1A) so that the tumor is very much like the normal tissue from which it originated. Like a healthy tissue, the benign tumor is surrounded by a stromal cell framework and is nourished by blood vessels. Factors secreted by the stromal cells perpetuate the proliferation of the neoplastic cells. However, a benign tumor is securely encapsulated and the altered cells cannot break away from the cell mass and enter the blood. Benign tumors do not normally cause death, but if they do, they do so by indirect means. For example, the physical pressure exerted by a benign tumor due to its abnormal size or location may compress or damage an adjacent organ.

In contrast to benign tumors, malignant tumors are lethal to the host unless they are completely removed or killed. *High grade* tumors contain cells that are poorly differentiated and grow aggressively; these tumors are associated with poor prognoses if left untreated. Conversely, a *low grade* cancer does not grow aggressively and is generally made up of cells that are fairly well differentiated. Histologically, the cell mass of a malignant tumor is usually disorganized (Plate 16-1B) and rarely encapsulated. Moreover, these tumors may become *invasive*, moving into nearby healthy organs and disrupting their function. Malignant tumors are also frequently *metastatic*, meaning that cells from the original tumor mass (the *primary tumor*) can break away and spread via the blood to nearby or distant secondary sites. Once established and growing in their new locations, these secondary tumors are called *metastases*.

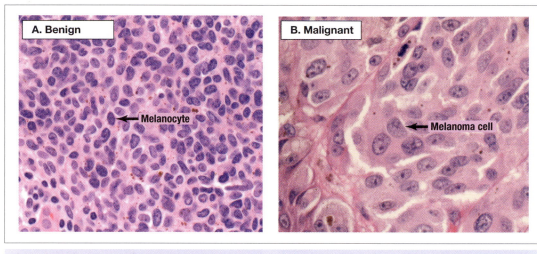

Plate 16-1 **Benign versus Malignant Skin Tumors**
[Reproduced by permission of Danny Ghazarian, Princess Margaret Hospital, University Health Network, Toronto.]

ii) Morbidity and Mortality

Although one might assume that a direct effect of malignant growth would be failure of the affected tissue, this is rarely a direct cause of death. The vital organs are very resilient and a host can often survive with only a fraction of an organ in functional condition. Instead, cancer patients usually succumb to cachexia and infection. Cachexia is "terminal wasting" and is marked by weight loss and muscle atrophy. This wasting is not completely understood but appears to be due partly to a redirection of nutrients to the growing tumor, and partly to tissue breakdown instigated by biologically active substances released from the tumor. The altered metabolism of a terminal cancer patient also renders him/her immunodeficient, so that the immediate cause of death is frequently pneumonia or septicemia caused by opportunistic microbes.

II. CARCINOGENESIS

The transformation of a cell from normal to malignant occurs in a multistep process called *carcinogenesis*. During carcinogenesis, mutations accumulate in the genes controlling cell proliferation and programmed cell death, resulting in deregulation of cellular growth. In normal cells, multiple mechanisms exist to repair mutations as rapidly and accurately as possible. If the genetic abnormalities cannot be repaired, the cell undergoes "cell cycle arrest" (is prevented from proliferating) or is induced to die. Cancer results only when these mechanisms fail, or environmental factors increase the rate of mutation such that the repair mechanisms cannot keep up. Deleterious mutations can then accumulate, leading to inappropriate gene expression that drives abnormal cell division.

At the clinical level, human cancers are described as *sporadic* or *familial*. Most cancers are sporadic in that the tumorigenic mutations occur in a somatic cell of a tissue and do not arise from alterations in the individual's germ cells. In rare cases, the germ cells of an individual exhibit an accumulation of tumorigenic mutations that causes the host to be genetically predisposed toward a particular malignancy. If cancers then arise in the affected individual's descendants due to this inherited genotype, the malignancies are said to be familial.

i) Carcinogens

Carcinogens are substances or agents, such as radiation and particular chemicals, that induce genetic mutation or deregulation, thereby promoting carcinogenesis. Table 16-1 gives examples of carcinogens and the tumors with which they are associated. Chronic inflammation, with all its cytokine mediators, is also thought to increase the chance that a cell will sustain a carcinogenic mutation. At least for some cell types, inflammation appears to induce an increased rate of mitosis and may trigger changes to cellular survival, differentiation or apoptosis. As well, ROIs and RNIs produced by macrophages activated by pro-inflammatory cytokines may be inadvertently taken up by a cell, leading to DNA damage.

Certain pathogens are thought to be potentially carcinogenic. About 15% of human cancers appear to be linked to infection with a particular pathogen (Table 16-2). For example, infections with the hepatitis B (HBV) or hepatitis C (HCV) viruses are frequently seen in liver cancer patients. However, there is no solid evidence that these pathogens are solely responsible for the genetic deregulation seen in these malignancies; that is, not every individual infected with one of these pathogens develops a cancer. Some scientists believe that, in at least some of these cases, it may be the persistent inflammation associated with chronic infection by these pathogens that drives the associated carcinogenesis. In the case of HBV and HCV, the killing of virus-infected liver cells by antiviral CTLs promotes rapid cell division aimed at liver regeneration. With this increased cell division comes an increased chance of carcinogenic mutations. As well, HBV expresses a protein called HBx that prevents the normal death of liver cells, promoting neoplasia. In the case of infection by a DNA tumor virus, the

Table 16-1 Examples of Chemical and Radioactive Carcinogens

Carcinogen	Tumor Association
Chemical Carcinogens	
Alcohol	Cancers of the liver, esophagus, larynx
Aluminum	Lung cancer
Asbestos	Cancers of the GI tract, peritoneum, lung
Benzene	Leukemia
Cadmium compounds	Lung cancer
Diethylstilbestrol	Cancers of the cervix, vagina, breast, testis
Nickel compounds	Cancers of the lung, nasal sinus
Silica crystals	Lung cancer
Soot	Lung cancer
Tobacco	Cancers of the lung, esophagus, larynx, pancreas, liver
Radioactive Carcinogens	
Plutonium-239 and decay products	Bone, liver, lung cancers
Radium-224 and decay products	Bone cancer
Radon-222 and decay products	Lung cancer
Thorium-232 and decay products	Leukemia, lung cancer
Iodine-131 and decay products	Breast and thyroid cancers, leukemia

Table 16-2 Examples of Pathogens Associated with Carcinogenesis

Pathogen	Tumor Association
Epstein-Barr virus	Burkitt's lymphoma, Hodgkin's lymphoma
Helicobacter pylori	Stomach cancer
Hepatitis B virus	Liver cancer
Hepatitis C virus	Liver cancer
Human papillomavirus	Cervical, penile cancer
Kaposi's sarcoma herpesvirus	Kaposi's sarcoma

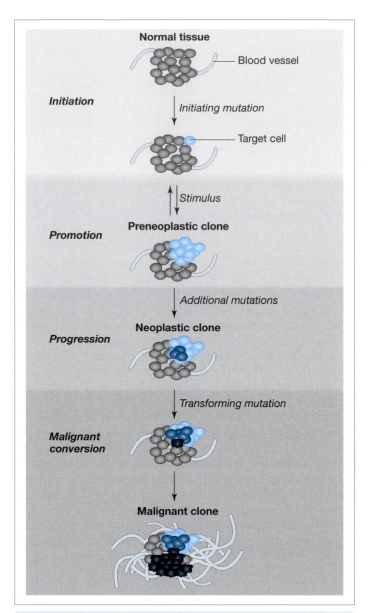

Fig. 16-1 **Four Steps of Carcinogenesis**

ii) Four Steps to Carcinogenesis

Four steps of carcinogenesis are generally necessary to establish a primary tumor: *initiation, promotion, progression* and *malignant conversion* (Fig. 16-1). In the initiation step, the DNA in the nucleus of a cell experiences a mutation. In most cases, the mutation either has no effect on the cell or is repaired such that there are no alterations to the structure or function of any cellular proteins. In some cases, however, the mutation introduces an error that is not repaired and alters the sequence of a protein involved in growth regulation, differentiation or apoptosis. In a subset of these cases, the cell in which the mutation occurred may eventually gain a growth advantage. It is this cell that is considered the *target cell* or *cell of origin* of the tumor.

Promotion involves the exposure of the target cell to a stimulus that allows the selective proliferation of this cell. An

link to cancer formation is quite clear. For example, several strains of HPV have been proven to cause cervical cancers. Certain proteins of these viruses can directly interfere with the regulatory proteins that control DNA repair mechanisms and the cell cycle.

expanding *preneoplastic* clone of genetically altered cells then develops. However, unlike initiation, promotion is completely reversible, so that if the promoting stimulus is removed, the clone will undergo *regression* (resolution of the tumor) to a cell number in keeping with that present in neighboring healthy tissue.

Progression involves additional genetic events that occur in one of the cells of the preneoplastic clone. Genetic instabilities introduced into regulatory genes or DNA repair genes during the initiation and promotion stages may predispose the genome of the cell to further alterations affecting the control of cellular growth, differentiation or apoptosis. These secondary mutations may arise spontaneously or may be caused by environmental influences impinging on the cell. At this point, the preneoplastic clone becomes a *neoplastic* clone with a significant growth advantage over normal cells. In some cases, this growth advantage depends on the presence of a hormone or pathogen, in that the tumor regresses when the hormone or pathogen is removed. In other cases, the growth advantage is due solely to the accumulated mutations and is irreversible. These clones are on the threshold of malignancy.

Malignant conversion, the final stage of carcinogenesis, is reached when one of the neoplastic cells in a state of progression acquires the transforming mutation that finally pushes the cell to adopt an invasive growth pattern. There may be structural changes affecting cell morphology and architecture, as well as biochemical changes to metabolic pathways, that enhance the tumor cell's resistance to normal death signals. The highly mutated and malignant cell gives rise to cells that grow in a totally deregulated manner and form the primary tumor. To grow beyond a minimal size, the primary tumor must then induce the formation of new blood vessels capable of bringing nutrients into the interior of the tumor mass. This process of *angiogenesis* is driven by molecules secreted by either the primary tumor or the surrounding stromal cells.

iii) Metastasis

Within a primary tumor, the malignant cells continue to acquire new mutations, some of which may confer the ability to metastasize. Metastasis requires that the transformed cells gain the ability to detach from the primary tumor mass and move on their own through the extracellular matrix (ECM) of the tissue. The metastatic tumor cells then push through the endothelial cell layer lining a local lymphatic channel or blood vessel and access the blood, circulating around the body until they adhere to a blood vessel in another part of the body. The tumor cells extravasate through the blood vessel wall and into the ECM of the new tissue. If the metastatic cells can survive in the new environment, a secondary tumor may start to grow in this location.

The migration of metastatic tumor cells is not random. Different types of cancer cells establish secondary tumors in different tissues, depending on the chemokines produced by a given tissue and the chemokine receptors expressed by the malignant cells. For example, cells in some primary breast cancers have been found to express abnormally high levels of the chemokine receptors CXCR4 and CCR7 (Fig. 16-2). These cells either force their way through the endothelium of a local blood vessel to enter the circulation, or access the lymphatic vessels draining the breast tissue. The cells are drawn to leave the blood or lymphatic circulation wherever the cells of a tissue express high levels of the chemokines SDF-1 and/or SLC, which are ligands for CXCR4 and CCR7, respectively. SDF-1 is produced by resting cells in the lungs, liver, bone marrow and lymph nodes, whereas SLC is synthesized solely by lymph node cells. Accordingly, it is in these tissues that breast cancer metastases are most commonly found.

iv) Cancer Stem Cells

Not all cells in a tumor have an equal capacity to initiate new tumors. In fact, in animal models, only a small fraction of the total cancerous cell population is capable of establishing a new tumor if transplanted into a suitable recipient. These findings have led to the *cancer stem cell hypothesis*, in which a small number of malignant cells (that arose from a single mutated cell) are responsible for sustaining the bulk of the tumor. In this scenario, the vast majority of cells in the tumor are derived from cancer stem cells and proliferate rapidly but cannot establish a new tumor on their own. As well as this ability to establish a new tumor, cancer stem cells also have the HSC-like properties of self-renewal, a slow rate of cell division, and the capacity to give rise to more differentiated "progenitor" cells.

How are cancer stem cells generated? In some cases, the cancer stem cell may result from the malignant transformation of a normal tissue-specific stem cell, the cell type responsible for maintaining a particular tissue or organ throughout the life of an organism. The long life span of a tissue-specific stem cell allows it ample time to accumulate the many mutations required for malignant conversion. These growth-deregulating changes are passed on to daughter progenitor cells that then undergo uncontrolled cell division, resulting in cancer expansion. In other cases, a cancer stem cell may arise because an oncogene has been activated in a progenitor cell that has already differentiated. This progenitor then appears to "de-differentiate" and acquire stem cell-like properties, particularly the ability to self-renew and establish new tumors.

Because of their slow cell division, cancer stem cells may escape destruction by current anti-cancer drugs, which only kill rapidly proliferating cells. Thus, cancer stem cells may survive to re-establish the tumor once the treatment has stopped, accounting for the recurrence of many cancers. In human patients, cancer stem cells have been identified in tumors of the brain, skin, colon, breast and hematopoietic cells. Recent work has shown that at least some of these cancer stem cells appear to have unique surface marker profiles, holding out hope that these cells can be isolated and studied more fully, and that new therapeutic strategies can be devised to kill them.

III. TUMORIGENIC GENETIC ALTERATIONS

Mutations to three major classes of genes involved in controlling cell growth and death are at the root of most cancers. These classes are *DNA repair genes*, *oncogenes* and *tumor suppressor genes* (Table 16-3).

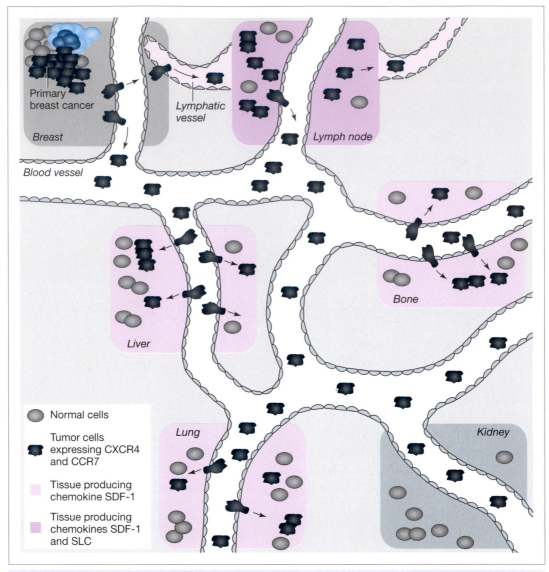

Fig. 16-2 **The Role of Chemokines in Metastasis**

Legend:
- Normal cells
- Tumor cells expressing CXCR4 and CCR7
- Tissue producing chemokine SDF-1
- Tissue producing chemokines SDF-1 and SLC

Labels in figure: Primary breast cancer, Breast, Blood vessel, Lymphatic vessel, Lymph node, Bone, Liver, Lung, Kidney

i) DNA Repair Genes

The protein products of DNA repair genes detect and fix mutations to other genes. If a DNA repair gene does not function properly, the chance is increased that genes involved in the control of proliferation, differentiation or apoptosis will acquire deregulating mutations that lead to cancer. Thus, mutations to DNA repair genes contribute indirectly to malignant transformation. For example, the skin cancers in patients with *xeroderma pigmentosum* (XP) are due to defects in the XPC gene that normally encodes a protein involved in repairing DNA damage caused by exposure to UV radiation. Another example is seen in patients suffering from *ataxia–telangiectasia* (AT). AT patients have a mutation in the DNA repair gene ATM (ataxia–telangiectasia mutated) and are predisposed to the development of childhood leukemias and lymphomas. (Interestingly, both XP and AT are considered to be primary immunodeficiencies because the mutations involved also affect immune system function.) Examples of other genes important

in DNA repair are MSH2, BRCA1 and BRCA2. MSH2 is frequently mutated in colorectal cancers, whereas mutations of BRCA1 and BRCA2 are associated with familial breast and ovarian cancers.

ii) Oncogenes

The term *oncogene* ("cancer gene") refers to a normal cellular gene that is altered or deregulated in a way that <u>directly</u> contributes to malignant transformation. Once an oncogene is identified, its normal cellular counterpart is referred to as a *proto-oncogene*. Most proto-oncogenes are positive regulators of cell growth, such as growth factors and their receptors, intracellular signaling molecules, and transcription factors. A proto-oncogene gains oncogene status when a mutation causes it to become constitutively activated. The increased activity of the oncogene drives perpetual cell division. Important oncogenes include Ras, Myc and Her-2. Ras encodes an intracellular signal transducer and is mutated in many colorectal,

Table 16-3 Genes Altered in Various Tumors

Gene Name	Tumor Association
DNA Repair Genes	
XPC	Melanoma in xeroderma pigmentosum
MSH2	Colorectal cancer
BRCA1	Breast, ovarian cancer
BRCA2	Breast, ovarian cancer
XRCC1	Breast cancer
XRCC3	Melanoma
ATM	Childhood leukemia, lymphoma
NBS1	Leukemia, breast cancer, multiple myeloma
Oncogenes	
Ras	Colorectal, pancreatic, breast, skin cancer
Myc	Breast, colon, lung cancer
Her-2	Breast cancer
Bcr-Abl	Leukemia
Jun	Bone, skin cancer
Raf	Liver, lung cancer
PKC	Skin cancer
Met	Kidney cell cancer
Tumor Suppressor Genes	
p53	Multiple advanced cancers
PTEN	Multiple advanced cancers
Rb	Retinoblastoma
APC	Colorectal cancer
VHL	Renal cell cancer
WT1	Renal cell cancer

pancreatic, lung and skin cancers. Myc encodes a transcription factor and is frequently amplified (increased number of gene copies) in cells of breast, colon and lung tumors. Her-2 encodes a member of a family of epidermal growth factor (EGF) receptor chains with tyrosine kinase activity. When Her-2 combines with another EGFR family member to form a heterodimeric receptor, Her-2 can be activated by the binding of EGF (or a related ligand) to its partner. Excessive expression of Her-2 due to amplification of the Her-2 gene is primarily associated with sporadic breast cancers.

iii) Tumor Suppressor Genes

Unlike oncogenes, tumor suppressor genes (TSGs) usually encode negative regulators of cell growth or survival so that the normal function of a TSG protein product is to discourage tumor development. TSGs become implicated in malignant transformation when they undergo "loss-of-function" mutations such that the restraining influence of the TSG product is lost. Because diploid cells have two copies of any TSG, mutation of one of them is usually not sufficient to cause malignant conversion. However, if the remaining normal copy of the TSG is later inactivated, the individual becomes predisposed to tumor formation. Examples of important TSGs are p53, PTEN and Rb. The p53 gene encodes a transcription factor that functions as a master regulator of cell cycle progression and cell death. Normal p53 function causes cells that have suffered irreparable DNA damage to die, and induces the cell cycle arrest of cells with DNA damage that can be repaired. When p53 is mutated, DNA-damaged cells can proliferate uncontrollably. Mutations of p53 have been found in close to 50% of all human cancers. PTEN is a TSG whose normal function is to negatively regulate signal transduction in several key pathways driving cell survival, proliferation and metabolism. In the absence of PTEN, cells may escape a scheduled death and go on to proliferate instead. PTEN is the second most commonly mutated gene in human cancers. The normal function of the Rb protein is to block the division of cells with genetic abnormalities. Rb was discovered through studies of the development of retinoblastoma (retinal cancer) in young children.

B. Tumor Antigens

As mentioned earlier, some scientists believe that one of the functions of the immune system is to carry out "immunosurveillance" and routinely seek out and destroy tumor cells. Such vigilance is thought to account for cases of "spontaneous regression" in which a tumor disappears apparently on its own. Even several decades ago, spontaneous regression was attributed to successful immune responses against unknown antigens expressed by tumor cells. Modern studies of mice deficient for various components of the immune system have also supported the concept of immunosurveillance, as these mutant animals show increased frequencies of spontaneous and carcinogen-induced cancers. However, other investigators question the validity of immunosurveillance and point to the fact that the more common malignancies, such as colon, breast and lung tumors, do not occur at increased frequency in immunodeficient or immunosuppressed humans. In addition, the fact that most cancers develop in immunocompetent individuals clearly shows that if immunosurveillance exists, it fails on a regular basis.

Despite these differing views, it has generally been assumed that genetic mutations associated with carcinogenesis can lead to the expression by a tumor cell of macromolecules that might be immunogenic. When these *tumor antigens* have been identified, they have turned out to be macromolecules that are abnormal in structure, concentration or location, or macromolecules expressed at an unusual time during the life of an animal. Although most tumor antigens are proteins, unusual carbohydrates that may serve as tumor antigens can arise when cells experience carcinogenic mutations in genes that control various aspects of carbohydrate synthesis and modification. Regardless of their protein or carbohydrate nature, tumor antigens generally fall into two classes: *tumor-associated antigens* (TAAs) and *tumor-specific antigens* (TSAs). However, the reader is cautioned that, despite much research effort, only a handful of TAA and TSAs have actually been isolated to date.

I. TUMOR-ASSOCIATED ANTIGENS (TAAs)

A TAA of a tumor cell is a normal protein or carbohydrate expressed in a way that is abnormal relative to its status in the healthy, fully differentiated cells in the surrounding tissue of origin. In other words, a TAA is almost always a case of the "right molecule expressed at the wrong concentration, place and/or time".

i) "Wrong Concentration"

Genetic modifications that deregulate the expression of a normal gene can result in a tumor cell that expresses a normal protein at much higher levels (up to 100-fold) than those present on a healthy cell (Fig. 16-3A). A specific example is the protein product of the oncogene Her-2. As noted earlier, the Her-2 protein forms heterodimers with other EGFR chains to form a receptor that mediates stimulation of the cell by EGF and related growth factors. Her-2 is normally expressed at low levels by various types of epithelial cells. However, due to

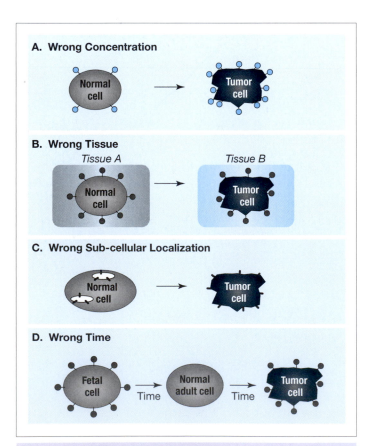

Fig. 16-3 Types of Tumor-Associated Antigens

genetic amplification that results in multiple copies of the Her-2 gene being present in the genome, the Her-2 protein is greatly overexpressed by cells of some breast cancers. The same phenomenon occurs in a low proportion of prostate, colon, lung and pancreatic cancers. In all these tumors, circulating growth factors lead to excessive activation of the Her-2 molecules present on the tumor cells, resulting in uncontrolled proliferation. In other cases of "wrong concentration", a protein that is usually expressed only on a rare cell type (and thus ouurs in the body at low levels) is present at an abnormally high concentration because the cell has become malignant and proliferated extensively.

ii) "Wrong Place"

Some TAAs arise when a protein whose expression is normally restricted to cells of tissue A is expressed inappropriately by tumor cells of tissue B; the protein appears in the "wrong place" (Fig. 16-3B). The *cancer-testis antigen* family of proteins furnishes some good examples of "wrong place" TAAs. The cancer-testis proteins are expressed solely in spermatogonia and spermatocytes in healthy individuals but are found on other cell types when these cells become cancerous. The best-studied examples of this class of TAAs are the MAGE proteins that were first discovered in melanomas. TAAs that are expressed in the "wrong place" also include those that are expressed in the appropriate tissue type but show abnormal localization (Fig. 16-3C). For example, a melanocyte-specific protein called TRP-1 (tyrosine-related protein-1) constitutes a TAA when it appears in the plasma membrane of melanoma cells, instead of in the membranes of the intracellular vesicles with which it is normally associated.

iii) "Wrong Time"

Some TAAs are *embryonic antigens*, proteins whose expression is normally restricted to fetal cells but which are expressed in adult cells that have undergone malignant transformation (Fig. 16-3D). The gene encoding the embryonic antigen is silent in normal adult tissues but is reactivated in the adult tumor cells. An example of an embryonic TAA is *carcinoembryonic antigen* (CEA), which is normally expressed only in the liver, intestines and pancreas of the human fetus but is highly associated with colon, breast and ovarian cancers in the adult. Similarly, *alpha-fetoprotein* (AFP) is normally expressed only in fetal liver and yolk sac cells but is strongly linked to liver and testicular cancers in the adult.

II. TUMOR-SPECIFIC ANTIGENS (TSAs)

TSAs are new macromolecules that are unique to the tumor and are not produced by any type of normal cell. Because of their non-self nature, TSAs should constitute *bona fide* immunogens capable of provoking an immune response. However, for reasons that are elaborated later, very few of them do. The cancer cells expressing these TSAs therefore continue to grow unchallenged.

Macromolecules constituting TSAs may be localized on the tumor cell surface or in its interior (Fig. 16-4). The new molecular structures may be derived from a simple amino acid sub-

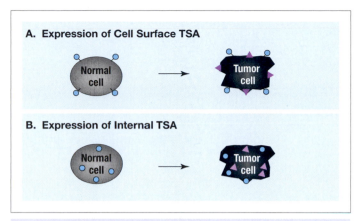

Fig. 16-4 **Types of Tumor-Specific Antigens**

stitution that alters the linear sequence and/or conformation of a protein, or from a more drastic internal rearrangement or deletion of parts of a gene. Unique proteins may also be produced when a chromosomal translocation fuses DNA sequences from two different genes, creating a new gene encoding a novel fusion protein composed of parts of two different proteins. TSAs may also be derived from DNA sequences that are not mutated but that are not transcribed in the normal way. For example, new proteins can arise from the use of a cryptic transcription or translation start site, an alternative reading frame, transcription of a gene in the opposite direction, or transcription/translation of a pseudogene sequence. Other TSAs are abnormal carbohydrates, produced when the gene encoding an enzyme involved in carbohydrate biosynthesis has undergone a transforming mutation. Certain mucins and gangliosides, which are altered in some types of melanomas and brain, breast, ovary and lung cancers, are considered to be carbohydrate TSAs. Finally, the viral proteins expressed by tumor cells infected with an oncogenic virus are TSAs.

Many TSAs are the mutated protein products of genes controlling cell growth or apoptosis. For example, mutations of the TSG p53 result in a nonfunctional protein that cannot induce cell cycle arrest or apoptosis in response to DNA damage. Another example occurs in certain head and neck cancer patients whose tumor cells show a mutation in the pro-apoptotic gene caspase-8. The mutation results in the addition of numerous amino acids to the caspase-8 protein, an alteration that interferes with its ability to induce cell death. In some melanoma patients, the tumor cells have a mutation in the cell cycle control gene Cdk4 that results in uncontrolled cell proliferation. Finally, mutated forms of EGFR are expressed by several different types of cancer cells. Abnormal signaling driven by the binding of EGF to the altered receptor drives tumor expansion.

C. Immune Responses to Tumor Cells

Despite the ongoing difficulties in identifying TAAs and TSAs, various components of both innate and adaptive immunity apparently do respond to tumor cells. These responses have been implicated both in immunosurveillance guarding against the growth of incipient tumors and in the elimination of established cancers. When regression of a tumor is induced by an immune response, the process is called *tumor rejection*.

I. ACUTE INFLAMMATION

As mentioned earlier, chronic inflammation is thought to promote tumorigenesis. In contrast, acute inflammation can have anti-tumorigenic effects. Occasionally the mere presence of premalignant or malignant cells can initiate an inflammatory response. One theory is that premalignant or malignant cells sometimes express surface molecules that can bind to the PRRs of innate leukocytes (Fig. 16-5). Another hypothesis is that molecules released from the debris of necrotic tumor cells might constitute "danger signals" that activate innate leukocytes. A good example of such molecules might be HSPs. As introduced in Chapter 7, the expression of HSPs is induced or upregulated in response to environmental stresses, including the presence of a tumor. Whatever mechanism is involved, it appears that activated neutrophils, eosinophils and macrophages play key roles in tumor rejection because these cells release the major pro-inflammatory cytokines IFNγ, IL-12 and TNF. TNF was originally identified by its ability to directly induce the hemorrhagic necrosis of a tumor. IFNγ and IL-12 drive Th1 and NK cell responses and support CTL generation, and also have powerful anti-angiogenic effects that prevent the tumor from building up the blood vessels needed for growth. In addition, IFNγ induces the formation of a collagen capsule around a tumor mass, a natural barrier against metastasis. As well as these cytokines, activated granulocytes and macro-

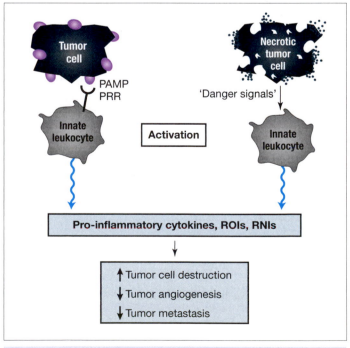

Fig. 16-5 **Role of Acute Inflammation in Combating Tumorigenesis**

phages release ROIs and RNIs that can kill cancer cells and damage tumor vasculature.

II. γδ T CELLS

Mice engineered to lack γδ T cells develop skin tumors with increased frequency and rapidity when the animals are exposed to carcinogens. This observation suggests that γδ T cells resident in the epithelial layers of the body may guard against malignancies. Whether γδ T cells perform such a function in humans and to what extent are under debate, since most common human adult cancers (such as tumors of the lung, breast, colon and prostate) arise from transformation of an epithelial cell despite the presence of γδ T cells. Nevertheless, many researchers believe that at least some γδ T cells mount anti-tumor responses following the binding of tumor antigens to their γδ TCRs. These responses may be enhanced by a type of costimulatory signaling triggered by the binding of MICA/B to NKG2D (Fig. 16-6, #1). MICA and MICB are stress molecules frequently present on human tumor cell surfaces, and many γδ T cells express the NK activatory receptor NKG2D. Once fully activated by tumor-derived ligands, γδ T cells are thought to generate anti-tumor Th-like and CTL-like effector cells.

Some scientists think that γδ T cell responses to tumors can also be activated by HSPs. As well as serving as "danger signals" for the innate response, HSPs can act as chaperone proteins with the capacity to bind to host components both intracellularly and extracellularly. HSPs may therefore associate with the TAAs or TSAs of living or necrotic tumor cells and convey them to γδ T cells. In humans, immune responses against breast cancer cells have been associated with the formation of intracellular complexes between an HSP and activated mutated forms of p53. However, the involvement of HSPs in anti-tumor γδ T cell responses *in vivo* awaits definitive confirmation.

III. NKT CELLS

Some TAA/TSAs (which may or may not be bound to HSPs) may be captured by immature DCs (IDCs) (Fig. 16-6, #2). In the correct cytokine milieu, these IDCs commence maturation in the local site and become mature DCs (MDCs). It is possible that these cells may present non-peptide antigens from the

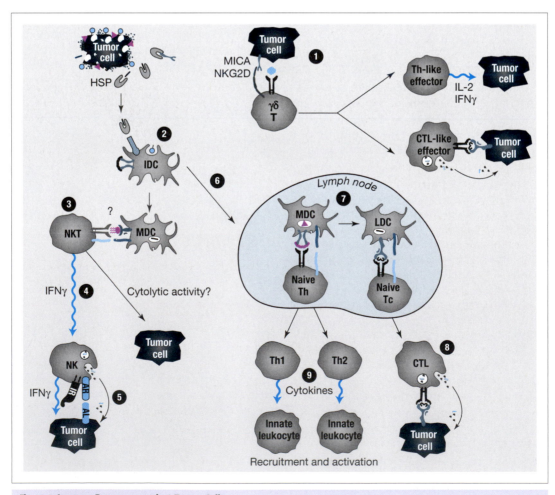

Fig. 16-6 **Immune Responses against Tumor Cells**

TAA or TSA on CD1d to NKT cells (Fig. 16-6, #3). Although it is not yet certain that NKT cells actually recognize such antigens, mutant mice engineered to lack NKT cells show increased susceptibility to chemical carcinogens. This observation suggests that activated NKT cells do provide early and important defense against tumor establishment. As described in Chapter 11, activated NKT cells rapidly express a collection of Th0 cytokines, including copious amounts of IFNγ that stimulates the anti-tumor activities of NK cells (Fig. 16-6, #4). In addition, activated NKT cells can carry out perforin-dependent cytolysis of tumor cells *in vitro*, although it has been difficult to demonstrate this capacity *in vivo*.

IV. NK CELLS

NK cells are thought to be key effectors mediating anti-cancer immunity. As described in Chapter 11, NK cells are activated when the signals transduced through their activatory receptors overpower the signals generated by their inhibitory receptors. Human tumor cells often lose expression of the MHC class I molecules that would normally act as inhibitory ligands but gain the capacity to express high levels of MICA, which acts as an activatory ligand (AL). Thus, activatory receptors (AR) such as NKG2D are engaged but NK inhibitory receptors (IR) are not, so that the NK cell delivers a cytotoxic "hit" to the tumor cell (Fig. 16-6, #5). In addition, the early burst of IFNγ production by activated NKT cells, followed by prolonged IFNγ production by NK cells, has an important anti-metastatic effect in mice. In humans, however, there have been conflicting reports on whether there is a link between decreased NK cell cytotoxicity and cancer incidence.

V. αβ T CELLS

Some IDCs that acquire TAA/TSAs mature as they migrate to the local lymph node (Fig. 16-6, #6). Within the node, the MDCs may activate naïve anti-tumor αβ Th cells, becoming licensed DCs (LDCs) that may activate naïve anti-tumor αβ Tc cells via cross-presentation (Fig. 16-6, #7). Studies in both humans and mice have provided convincing evidence that both CD4⁺ Th and CD8⁺ Tc cells contribute to tumor rejection. Researchers have identified many cloned human T cell lines with anti-tumor activity, and the transfer of syngeneic anti-tumor T cells to a tumor-bearing mouse can reduce the size of the tumor.

Many scientists believe that the primary function of activated anti-tumor CD4⁺ Th cells is to supply help to naïve anti-tumor CD8⁺ Tc cells and thus promote the generation of anti-tumor CTLs (Fig. 16-6, #8). In cancer patients, CTLs can often be found within the tumor tissue itself, leading to the designation of these cells as *tumor-infiltrating lymphocytes* (TILs) (Plate 16-2). However, it remains unclear whether TILs actually have the power to control a human cancer. In addition to supporting the anti-tumor CTL response, Th1 and Th2 effectors release cytokines that recruit and further stimulate innate leukocytes such as neutrophils, eosinophils, macrophages and NK cells (Fig. 16-6, #9).

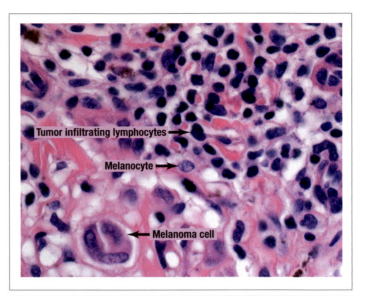

Plate 16-2 **Tumor-Infiltrating Lymphocytes in Melanoma**
[Reproduced by permission of Danny Ghazarian, Princess Margaret Hospital, University Health Network, Toronto.]

The initial assault on a tumor by αβ T cells is largely in response to epitopes of TSA/TAAs derived from the surface or interior of the tumor cells themselves. However, the resulting attack on the malignant mass frequently exposes new tumor antigens as the cells and surrounding tissues are degraded. Some of these antigens, which may have previously been hidden from the immune system, are taken up by DCs and presented as new pMHC epitopes to draw additional clones of naïve T cells to the fight. In general, the broadening of an immune response due to the increased accessibility of new antigens is called *epitope spreading* (see Ch. 19). Epitope spreading is thought to be important for sustaining the immune response against a tumor long enough for rejection to be achieved.

VI. B CELLS

It remains unclear whether B lymphocytes actively participate in immunosurveillance and tumor rejection. Anti-tumor antibodies are not often detected in cancer patients, and those that are present do not seem to be effective. However, anti-tumor antibodies can be produced experimentally and used for anti-tumor immunotherapy, as described later in this chapter.

D. Hurdles to Effective Anti-Tumor Immunity

Why does the immune system fail to prevent all tumorigenesis? For tumors expressing TAAs, central and peripheral T cell tolerance will have already been established and the malignant cells will be seen as self. This tolerance will have to be "broken"

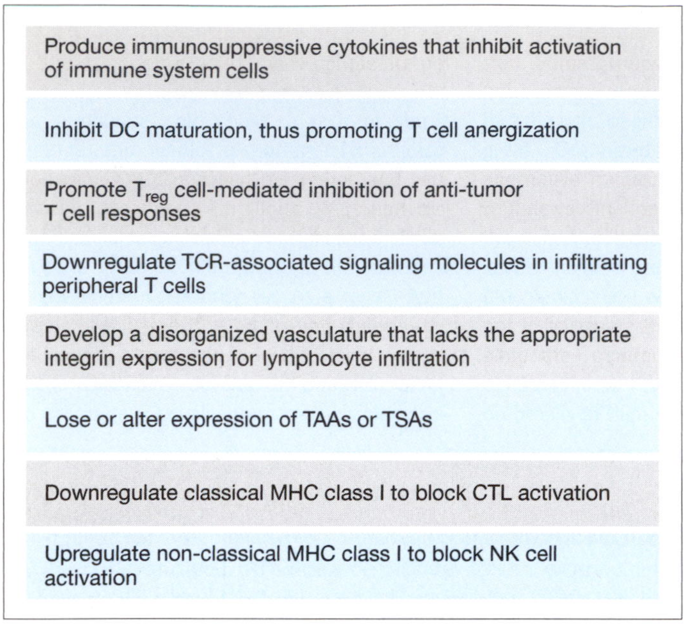

Produce immunosuppressive cytokines that inhibit activation of immune system cells
Inhibit DC maturation, thus promoting T cell anergization
Promote T$_{reg}$ cell-mediated inhibition of anti-tumor T cell responses
Downregulate TCR-associated signaling molecules in infiltrating peripheral T cells
Develop a disorganized vasculature that lacks the appropriate integrin expression for lymphocyte infiltration
Lose or alter expression of TAAs or TSAs
Downregulate classical MHC class I to block CTL activation
Upregulate non-classical MHC class I to block NK cell activation

Fig. 16-7 Major Mechanisms by Which Tumors Thwart the Immune System

in order for these TAAs to become visible to the immune system, with the added complication that self tissues expressing that TAA may come under attack. In the case of tumor cells expressing TSAs, the new macromolecules expressed may indeed constitute antigenic structures that become targets for adaptive immunity. However, other mutations acquired by tumor cells, or products that the tumor secretes, may allow it to escape immunosurveillance, suppress immune responses, and/or thwart the effector actions of immune system cells. In such cases, the tumor appears to induce tolerance to itself.

Two lines of evidence suggest that tumors can actively evade or block immune responses. Firstly, most cancer patients are highly prone to opportunistic infections. However, within weeks of removal of the tumor, the patient's immune system is often restored to full function. Secondly, animals that bear tumors respond poorly to antigens that induce vigorous immune responses in healthy animals. Mechanisms by which tumor cells suppress or evade the immune system are discussed next and summarized in Figure 16-7.

I. SECRETION OF IMMUNOSUPPRESSIVE CYTOKINES

Tumor cells often secrete immunosuppressive cytokines such as IL-10 and TGFβ that downregulate immune responses. In the presence of these cytokines, any TAAs or TSAs acquired by an immature DC are taken up in a microenvironment that does not support DC maturation. T cells that encounter these DCs are thus more likely to be anergized than activated. Indeed, patients with colorectal cancer or lymphoma often have a worse prognosis if they also have elevated plasma IL-10 levels. *In vitro*, IL-10 treatment decreases MHC class I expression on melanoma cell lines, inhibits TAP function, suppresses DC activity, and blocks CTL-mediated lysis of tumor cells. Similarly, elevated plasma TGFβ is associated with a broad

range of human cancers. TGFβ, which was first named for its ability to promote the malignant transformation of fibroblasts, stimulates angiogenesis and cellular invasion of the ECM. *In vitro*, TGFβ blocks the functions of CTLs and NK cells and suppresses antibody synthesis.

II. PROMOTION OF REGULATORY αβ T CELL RESPONSES

The weak response of T cells to TSAs may be due in some cases to suppression by the regulatory CD4⁺CD25⁺ T$_{reg}$ cell subset. As described in Chapter 10, T$_{reg}$ cells are able to suppress the effector functions of activated conventional T cells. When T$_{reg}$ cells are depleted from mice transplanted with tumors, tumor rejection is enhanced. *In vitro*, T$_{reg}$ cells nonspecifically inhibit the anti-tumor responses of NK cells, Th cells and CTLs. It is not clear how these T$_{reg}$ cells are activated or whether this inhibition is mediated by cytokines or intercellular contacts.

III. INHIBITION OF T CELL SIGNALING

Some tumor cells can downregulate the expression of TCR-associated signaling molecules in peripheral T cells that have infiltrated the cancer. An anti-tumor T cell in this population therefore cannot be activated to attack its target. It is unknown how the tumor accomplishes this feat, but the establishment of an inhibitory intercellular contact is suspected.

IV. AVOIDANCE OF RECOGNITION

In addition to inactivating the immune system cells they encounter, tumors are often able to evade immune scrutiny in the first place. At the anatomical level, the capillaries that wind in and out of a tumor mass and support malignant cell growth are disorganized and frequently lack expression of the integrins needed to facilitate T cell extravasation into the tumor site. At the molecular level, the tumor may simply never express a TAA or a TSA that can be recognized by immune system cells, or may lose the expression of its TAA or TSA due to selection pressure by the immune system. In addition, the downregulation of MHC class I by many tumor cells, which should invite elimination by NK cells, can be offset in some cases by tumor cell expression of non-classical HLA-E, HLA-F or HLA-G molecules. These MHC class Ib proteins can bind to NK inhibitory receptors and prevent an NK cell-mediated assault on the tumor. Significantly, HLA-G expression on tumor cells is upregulated by IL-10.

E. Cancer Therapy

I. CONVENTIONAL THERAPIES

The first step in cancer therapy is almost always the surgical removal of the complete primary tumor, if possible. This approach is clearly not an option for very diffuse cancers (such as leukemias), in which the tumor cells circulate throughout the body, nor for many metastatic cancers, in which tumors may be present in many small and hidden sites. Neither is surgery advisable when the tumor is located such that the efforts to remove it would irreparably damage a vital body structure, as in many types of brain cancers. In these cases, clinicians turn to chemotherapy and/or radiation therapy. Both these techniques predominantly affect rapidly dividing cells while sparing the resting or slowly dividing cells that comprise the majority of normal cells in the body. Chemotherapy and/or radiation therapy are also often used in conjunction with surgery in an effort to kill any tumor cells that were missed and prevent re-establishment of the tumor.

i) Chemotherapy

Chemotherapy is the use of pharmaceutical drugs to kill tumor cells in a cancer patient. The underlying principle is that a chemotherapeutic agent primarily affects only those cells that are growing faster than most normal cells, or cells that have a metabolic imbalance. Chemotherapeutic drugs work by directly or indirectly damaging the replicating DNA of the dividing tumor cell, inhibiting DNA synthesis, preventing cell division, or blocking the access of the tumor cells to a necessary growth factor. Most chemotherapy drugs are alkylating agents, antimetabolites, glucocorticoids or plant alkaloids. Because combinations of two or more chemotherapeutic drugs often work synergistically, patients are usually treated with "cocktails" containing anywhere from two to six agents, depending on the type and stage of the tumor. The use of more than one agent also greatly reduces the chance of the tumor becoming resistant to drug treatment, since the malignant cells would have to acquire resistance-conferring mutations to multiple drugs at once.

Unfortunately, by its very nature, conventional chemotherapy is not very specific and fast-growing normal cells such as those in bone marrow are also damaged. Immunity to pathogens therefore plunges. Cells of the gastrointestinal tract are also highly proliferative, so that chemotherapy almost inevitably causes nausea and vomiting. In addition, because the liver and kidneys tend to accumulate these powerful drugs and their metabolites, hepatic and nephrotic toxicity are common side effects. To address these issues, scientists have recently developed several molecular agents that more specifically target cancer cells. In some cases, these agents have demonstrated higher efficacy and lower toxicity than conventional chemotherapeutics. Two examples of such drugs are *imatinib mesylate* (Gleevec[R]), a small molecule that inhibits the activities of abnormal kinases in cancer cells, and *bortezomib* (Velcade[R]), a proteasome inhibitor that interferes with tumor cell functions and growth and also induces apoptosis. These drugs are discussed in more detail in Chapter 20.

ii) Radiation Therapy

The principle of radiation therapy is that photons and subnuclear particles emitted by radioactive substances generate ROIs that cause severe damage to the DNA of tumor cells with which they interact. The tumor cells then die when they attempt to initiate DNA replication. Radiation oncologists can use sophisticated radiation machines to externally deliver "hits" of damaging energy to tumors that are hidden deep within the body. Alternatively, internal delivery (*brachytherapy*) can be achieved by implanting a metal "seed" containing the radioactive material either directly in the tumor or very close to it. In this latter method, high doses of radiation are delivered for short periods of time, decreasing detrimental side effects on patients. Most radiation side effects are due to the impact of radiation on fast-growing normal tissues, such as the gastrointestinal mucosae, the skin and the bone marrow. Nausea, vomiting, hair loss and bone marrow suppression (failure to generate new hematopoietic cells) are all commonly observed side effects in radiation-treated cancer patients.

II. IMMUNOTHERAPY

Although chemotherapy and radiation therapy have improved the survival of many cancer patients, there are several problems with these approaches. The high levels of drugs and radiation that are applied in an effort to achieve sufficient killing power at the tumor site may encourage the rise of drug-resistant and radiation-resistant populations of tumor cells. In addition, systemic or local toxicity can be significant because there is often no way to confine the drug or radiation treatment precisely to the tumor site. This problem is compounded if the tumor has metastasized. A more recent concern is that the biology of the cancer stem cells thought to drive tumorigenesis appears to render them resistant to current modes of chemotherapy and radiation therapy. Clinical researchers have thus turned to the immune system and devised *immunotherapies* that can theoretically kill cancer cells with high specificity. The idea is to either create an immune response against the tumor, or augment an existing immune effector mechanism that is attempting to eliminate the tumor. Immunotherapy is already being used to complement conventional cancer treatments and may even replace some of them in the future.

Antibody-based immunotherapy generally involves the laboratory production of a tumor-specific monoclonal antibody (mAb) that is then administered to a cancer patient. (A monoclonal antibody is an antibody derived from a single B cell clone; see Appendix F.) T cell-based immunotherapy involves either *adoptive T cell therapy* or the administration of a *cancer vaccine*. In adoptive T cell therapy, samples of the patient's T cells are isolated, activated *in vitro*, and transferred back to the patient to provide anti-tumor protection. In contrast, a cancer vaccine is injected directly into the patient in an attempt to generate an anti-tumor T cell response *in vivo*. The obvious antigens to target for anti-tumor immune responses of any type are the TAAs and/or TSAs expressed by cancerous cells, and researchers were originally optimistic that they could induce clinically effective antibody and T cell responses specific for such antigens. However, it has proved extremely challenging to identify TSAs, and although TAAs might serve as alternative targets, TAA-directed antibodies and lymphocytes have the potential to attack healthy tissues. As alternatives to TAA- and TSA-based therapies, researchers are testing strategies based on cytokine secretion. Also under examination are agents that disrupt tumor angiogenesis or block the supply or action of a required tumor growth factor. A great hope for the future is that unique antigens may be identified on cancer stem cells that can be used to generate antibodies or T cells capable of destroying them.

i) Monoclonal Antibody-Based Therapies

Several novel cancer therapy techniques have taken advantage of the specificity and ease of manipulation of mAbs. MAbs that are used in the clinic are usually produced by immunizing mice with the human antigen of interest and then "humanizing" the immunoglobulins produced using genetic engineering techniques; see Appendix F. In theory, a therapeutic mAb that binds to a TAA/TSA on the surface of a tumor cell may in turn be bound by FcRs of a macrophage, neutrophil, eosinophil or NK cell. The tumor cell would then be destroyed by ADCC or complement-mediated lysis. Alternatively, mAbs may be used to sequester growth factors needed by the tumor to continue its expansion, or to interfere with angiogenic factors and thus reduce the blood supply to the malignant mass. Other mAb therapies are based on *immunoconjugates* in which a tumor-specific mAb is covalently linked to a cytotoxic molecule (such as a toxin) that can kill a cell on contact. Hypothetically, use of a TSA-directed mAb conjugated to such a molecule would ensure that normal cells are spared while the toxin is targeted to cancer cells in both the primary tumor and in metastases.

ia) Unconjugated mAb administration.
Unconjugated mAbs can be used on their own to combat human cancer *in vivo* in those rare cases where a TSA or TAA can be identified on the tumor cell and a mAb can be raised against the TSA/TAA in mice (Fig. 16-8A and Table 16-4). For example, a successful anti-tumor mAb treatment has been devised for the 25% of breast cancer patients whose tumors overexpress Her-2. The drug Herceptin is a humanized mAb that binds to the Her-2 protein and both promotes NK cell-mediated ADCC and interferes with Her-2-mediated signals driving tumor cell proliferation. Another useful mAb called rituximab is directed against the surface marker protein CD20. CD20 is expressed on certain immature and mature B cells but not plasma cells. Anti-CD20 mAb treatment is now widely used in conjunction with chemotherapy to control B cell lymphomas in which the cancerous lymphocytes express significant levels of CD20 (see Ch. 20). Because CD20 is not expressed on plasma cells, these cells are spared and the patient is able to maintain adequate levels of circulating Igs to fight infection. MAbs have also been developed that attack the tumor vasculature, targeting molecules that are expressed at higher levels on tumor blood vessels than on normal blood vessels.

Another strategy that makes use of an unconjugated mAb is called *CTLA-4 blockade*. The regulatory molecule CTLA-4 was introduced in Chapter 9. CTLA-4 is structurally related to the T cell costimulatory molecule CD28 and also binds the B7 ligands but has a negative effect on T cell activation. Toward the end of an immune response, expression of CTLA-4 is upregulated on activated effector T cells such that this molecule competes with CD28 for binding to B7. As a result, signal 2 declines, T cell activation is slowly abrogated, and the immune response subsides. Researchers speculate that one

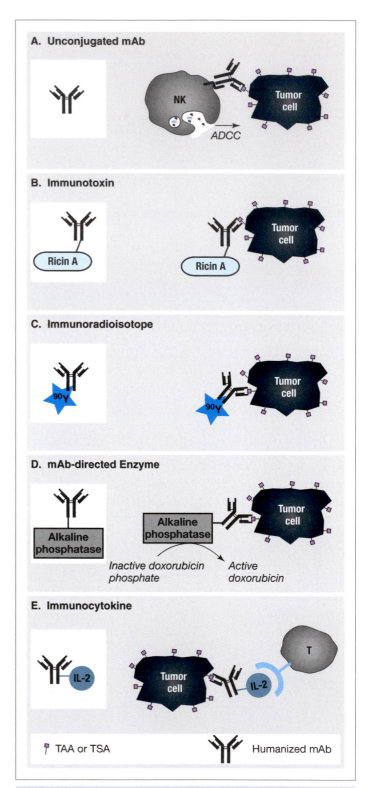

Fig. 16-8 Examples of Immunotherapy with Monoclonal Antibodies

reason certain cancers are allowed to progress is that the T cell responses against them are eventually brought to a natural halt by CTLA-4. Thus, mAbs that bind to CTLA-4 and prevent it from binding to B7 and damping down T cell activation may help to sustain existing anti-tumor T cell responses. An impor-

Table 16-4 Examples of Unconjugated Monoclonal Antibodies Used for Immunotherapy

Trade Name	Generic Name	Specificity	Mode of Action	Cancers Treated
Rituxin[R] or MabThera[R]	Rituximab	Anti-CD20	Facilitates killing of CD20-bearing B cells	B cell lymphoma
Herceptin[R]	Trastuzumab	Anti-Her-2	Blocks proliferative signaling; also promotes NK cell killing	Breast cancer
Avastin[R]	Bevacizumab	Anti-VEGF*	Prevents tumor angiogenesis	Lung cancer, metastatic colorectal cancer
Erbitux[R]	Cetuximab	Anti-EGFR[†]	Facilitates killing of tumor cells bearing EGFR	Metastatic colorectal cancer, advanced head and neck cancers

*Vascular endothelial growth factor.
[†]Epidermal growth factor receptor.

tant caveat with this approach is that the CTLA-4-mediated suppression that normally controls autoreactive T cells in the periphery will also be lost, potentially resulting in autoimmune disease. An anti-CTLA-4 mAb has been tested in clinical trials as an adjunct treatment for advanced prostate cancers and melanomas. However, it seems that it will be tricky to get the right balance between achieving tumor regression and avoiding autoimmunity: several patients that received repeated administration of the anti-CTLA-4 mAb showed some tumor regression but also signs of autoimmune disease.

ib) Immunoconjugates. As noted earlier, immunoconjugates are chimeric proteins in which mAbs are linked either chemically or via genetic engineering to lethal effector molecules such as toxins, drugs, radioisotopes, enzymes and cytokines. The idea is that the specificity of the mAb brings the effector molecule directly to the tumor site, sparing normal tissues. Sometimes the immunoconjugate is internalized by the tumor cell and the toxic portion kills the cell by disrupting intracellular survival pathways. In other cases, the effector molecule is delivered within the tumor cell mass or its vasculature but remains external to the individual tumor cell. Some examples of immunoconjugates are given in Table 16-5.

As attractive as the immunoconjugate concept is, it faces two major challenges. Firstly, there are very few known TSAs to raise a mAb against, making it difficult to construct truly tumor-specific agents. Secondly, the preparatory research in animal models frequently overestimates the efficacy and underestimates the toxicity of a drug or toxin in humans. Despite these difficulties, researchers are pursuing several therapeutic approaches based on immunoconjugates.

When the molecule linked to the mAb is a toxin, the immunoconjugate is known as an *immunotoxin* (Fig. 16-8B). For example, mAbs have been joined to bacterial toxins such as pseudomonas exotoxin (PE) and diphtheria toxin (DT) as well as to the "A" subunit of the plant toxin ricin. Once internalized by the tumor cell, the toxin inhibits protein or nucleic acid synthesis or damages the genomic DNA. Many immunotoxins are extremely potent, requiring only one molecule of toxin per tumor cell to kill. This efficacy means that very few immunotoxin molecules have to reach the tumor site to have a therapeutic effect. However, it has been difficult to control the immunogenicity and half-life of immunotoxins, and to achieve adequate penetration of solid tumors. Clinical trials of immunotoxins in human cancer patients are ongoing.

When a radioisotope such as iodine-131 or yttrium-90 is linked to the mAb, the resulting immunoconjugates are called *immunoradioisotope*s (Fig. 16-8C), and the use of these agents to treat cancer patients is called *radioimmunotherapy* (RIT). For example, a yttrium-90-labeled mAb has been constructed that binds to the CEA protein often expressed by colon, pancreatic, breast, ovarian and thyroid cancers. The targeting of this immunoconjugate to these tumor cells has been encouragingly precise, and effective clinical responses have been observed in cancer patients. One difficulty with RIT is that the entire body (and particularly the bone marrow) is exposed to the radioisotope while the mAb is circulating to find its target. Researchers are developing administration strategies that more quickly concentrate the immunoradioisotope at the tumor site so that it does minimal damage to normal tissues. Another drawback to RIT is that radioisotope released from the mAb

Table 16-5 Examples of Conjugated Monoclonal Antibodies Used for Immunotherapy

Trade Name	Conjugate	Specificity	Mode of Action	Cancers Treated
Zevalin[R]	Y-90	Anti-CD20	Facilitates killing of CD20-bearing B cells	B cell lymphoma
Bexxar[R]	I-131	Anti-CD20	Facilitates killing of CD20-bearing B cells	B cell lymphoma
Mylotarg[R]	Calicheamicin (cytotoxic antibiotic)	Anti-CD33*	Facilitates killing of CD33-bearing hematopoietic cells	Acute myeloid leukemia

*CD33 is an adhesion protein found on leukemic cells and immature myeloid cells.

during catabolism can accumulate in the liver or kidney, damaging these organs.

Another type of immunoconjugate is created when a mAb is linked to an enzyme and is used for ADEPT: *antibody-directed enzyme/pro-drug therapy* (Fig. 16-8D). In this situation, an anti-tumor mAb is conjugated to an enzyme capable of converting an inert pro-drug into an active cytotoxic drug. For example, alkaline phosphatase conjugated to a mAb is capable of converting inactive doxorubicin phosphate to the active anti-cancer drug doxorubicin. The immunoconjugate is given to the patient in advance of the pro-drug, giving the mAb–enzyme complex time to specifically localize in the tumor site and to be cleared from other areas of the body. The pro-drug is then administered so that when the toxic metabolite is generated, it is (theoretically) confined to the tumor mass where the mAb–enzyme is bound. In practice, it is sometimes difficult to clear the mAb–enzyme complex from all non-cancerous tissues. Another challenge is that, since the immunoconjugate in this case contains an enzyme (which is a relatively large protein), the immunoconjugate may itself be immunogenic. An immune response may be induced that inactivates the enzyme before it can catalyze pro-drug conversion.

Some immunoconjugates are *immunocytokines* (ICKs), chimeric proteins in which an anti-TSA/TAA mAb is linked to a cytokine that either has anti-tumor properties or can stimulate immune system cells in the tumor site (Fig. 16-8E). (ICKs are not internalized like immunotoxins.) Here, the plan is to prolong the half-life of the cytokine and bring it directly to the malignant mass. For example, one well-studied ICK contains IL-2 fused to a mAb directed against EpCAM, an epithelial adhesion molecule upregulated on colon cancer cells. Use of this immunoconjugate in both humans and mice has shown promise for the control of colon cancer cell metastasis. ICKs containing IL-2 have also worked well in some T cell leukemia patients.

ii) Cancer Vaccines

iia) Pathogen-based cancer vaccines.
When a cancer is associated with a particular pathogen infection, it may be possible to devise a prophylactic vaccine that can block the infection and thus prevent subsequent tumorigenesis. At the moment, the most common pathogen-based cancer vaccines are those directed against the oncogenic viruses. Although it is still debated whether these viruses are themselves carcinogenic, it is generally agreed that infected cells expressing oncoviral antigens often later become malignant. Thus, these antigens are logical targets for vaccine development. For example, vaccines containing proteins of the HBV virus reduce not only the incidence of hepatitis but also liver cancer. Similarly, a vaccine against HPV has been proven to decrease the incidence of cervical cancer.

iib) TSA/TAA-based cancer vaccines.
Two non-viral TSAs that have been considered as cancer vaccine candidates are the proteins encoded by the oncogene Ras and the TSG p53. In theory, distinctive peptides from the mutated versions of these proteins should be presented on the surface of tumor cells and should act as targets for CTLs. *In vitro* assays have demonstrated the presence in mice of CTLs capable of responding to purified mutated Ras peptides (when appropriately presented). Unfortunately, however, these anti-Ras CTLs did not respond to intact tumor cells. In addition, although immune responses against mutated p53 peptides occur in experimental animals, they have not been detected in human cancer patients. What about TAAs? Although a vaccine directed against a TAA might in theory provoke an attack against normal cells, in reality, many TAAs are expressed in non-vital organs and the limited immune destruction of healthy cells in these locations does not threaten the survival of the host. For example, small numbers of melanoma patients immunized with melanoma TAAs have shown some regression of their tumors without significant damage to healthy tissues. In addition, some colon cancer patients have shown clinical improvement after vaccination with CEA, an embryonic TAA that is not normally expressed on adult tissues. However, neither of these vaccines has proved to be efficacious in the few large Phase III clinical trials that have been carried out.

A key difficulty with TSA/TAA cancer vaccines is that many TAAs and TSAs simply are not very immunogenic. Moreover, the DCs naturally present at a tumor site do not seem to be able to activate naïve anti-tumor T cells, most likely because of a lack of adequate "danger signals" and the presence of immunosuppressive cytokines secreted by the tumor cells. Anergization (rather than activation) of responding TSA/TAA-specific lymphocytes may thus be induced. Indeed, in a few cases, administration of a TSA/TAA vaccine has actually enhanced tumor cell growth.

One way to improve the effectiveness of cancer vaccines may be to deliver them using DCs that have been forced to mature *in vitro*. For example, IDCs isolated from a mouse can be treated *in vitro* with the growth factor GM-CSF to trigger maturation (Fig. 16-9A). The MDCs are then mixed with tumor antigens that may be purified TAAs or TSAs, or contained in a whole tumor cell lysate. In a process called *electroporation*, the culture is subjected to a mild electrical current that opens tiny pores in cell membranes. The tumor antigens enter the MDCs such that these cells are said to be "pulsed" or "loaded." The loaded MDCs are then administered as the vaccine to a tumor-bearing, syngeneic mouse in the hopes of sparking an anti-cancer response. Because electroporation introduces the tumor antigens directly into the MDC cytoplasm, peptides derived from these antigens are presented on MHC class I and theoretically provoke a CD8$^+$ T cell response. If some of the tumor antigens enter the DC via clathrin-mediated endocytosis, the relevant peptides may be presented on MHC class II such that CD4$^+$ T cells are also activated. A variation on this approach is in early clinical trials in humans. The patient's own tumor cells and IDCs are isolated, the tumor cells are lysed, and the lysate is used to load the patient's IDCs. The IDCs are stimulated to mature *in vitro* and then returned to the patient with the goal of provoking an enhanced T cell response. Unfortunately, the actual rejection of primary tumors in patients treated in this way has been minimal.

Another way to ensure that the tumor antigens making up a cancer vaccine are ultimately presented by mature rather

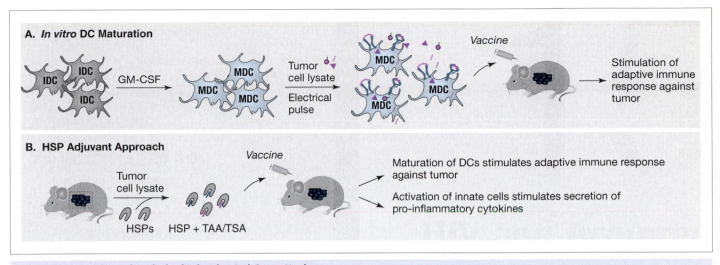

Fig. 16-9 Two Experimental Methods of Enhancing Anti-Cancer Vaccines

than immature DCs may be to administer the vaccine using HSPs as an adjuvant (Fig. 16-9B). Because HSPs are naturally released by stressed cells, they may serve as "danger signals" that promote DC maturation *in vivo* and induce leukocytes to secrete pro-inflammatory cytokines. In this microenvironment, TSA/TAA-specific lymphocytes that encounter the mature DCs presenting TSA peptides should be activated rather than anergized. An adaptive response should then be mounted against the tumor. This experimental approach has met with considerable success in mouse models and has now moved into the early clinical trial stage in humans.

iii) Adoptive T Cell Cancer Therapy

In contrast to cancer vaccines, which are a form of active immunization, adoptive T cell cancer therapy is a form of passive immunization in which anti-tumor T cells are transfused into a cancer patient in the hopes that these cells will attack and destroy the tumor and prevent its recurrence. The T cell population that is transfused may be polyclonal and contain T cells specific for a variety of (unknown) tumor antigens, or may be an enriched population selected for a particular T cell surface marker phenotype, or may be a single clone selected by co-culture with APCs loaded with a known tumor antigen. The process is usually initiated by harvesting T cells from a cancer patient's peripheral blood or draining lymph node. The T cells acquired often include Th cells and T_{reg} cells as well as CD8$^+$ T cells. Alternatively, TILs present in the tumor tissue itself may be isolated. The harvested T cells are greatly expanded in number and activated by culture with cytokines and APCs. The activation of these T cells may be enhanced by exposure to stimulatory antibodies that bind to CD28 or CD3. The expansion/activation process can take several weeks, and in the meantime, the cancer patient is treated with chemotherapeutic drugs that deplete his/her body of lymphocytes to accommodate the incoming population of T cells. The activated T cells are then transfused into the patient, whereupon they home to the tumor site. The transfused patient is then monitored for signs of tumor regression.

There has been some clinical success with adoptive T cell therapy in a small number of cases. For example, in one study of melanoma patients, T cells that had infiltrated each tumor were isolated, expanded in the laboratory, and transferred back to the original patient. Most of these patients subsequently showed accumulations of melanoma-reactive T cells in their peripheral blood and enjoyed some degree of tumor regression. Early clinical trials are also under way to explore adoptive T cell cancer therapy as a treatment for Her-2$^+$ breast cancer. However, adoptive T cell therapy is currently extremely expensive, labor-intensive, lengthy, and difficult to scale up to the Phase III clinical trial level.

iv) Cytokine-Based Therapies
iva) Pure cytokine administration. Although it might be expected that cytokines should have anti-tumor effects, it has been surprisingly difficult to exploit them. For example, IFNγ promotes CTL responses, activates macrophages, and has anti-angiogenic effects. Accordingly, IFNγR-deficient mice show increased susceptibility to chemically induced tumor development. However, although high IFNγ levels have been found in spontaneously regressing melanomas, they have also been detected in progressing renal, breast and ovarian cancers. In the clinic, administration of IFNγ has not improved the lives of metastatic melanoma patients. Similarly, administration of IL-10 has induced regression of melanoma and breast cancer metastases in some animal models but has had no effect in others. Although IL-2 promotes the expansion of tumor-specific T cells *in vitro* and counteracts the effects of immuno-suppressive cytokines, intravenous injections of IL-2 have induced clinical tumor regression in only 20% of treated renal carcinoma and melanoma patients. Moreover, the dose of IL-2 that can be safely administered to humans is quite limited because of its toxicity. IL-12 has considerable anti-tumor activity due to its ability to promote NO production by macrophages, stimulate NK and NKT cells, and drive IFNγ secretion. However, injection of moderate doses of IL-12 into human cancer patients has yielded only mixed results, and large doses

of IL-12 are toxic. IL-4 blocks the expansion of tumor cell lines *in vitro* but has not improved cancer patient survival. Similarly, the growth factor GM-CSF is a powerful stimulator of DC maturation and function *in vitro*, and enhances ADCC-mediated tumor cell killing by granulocytes. However, GM-CSF injected into a cancer patient is cleared very rapidly from the circulation, limiting GM-CSF's ability to access tumor sites *in vivo*.

The reader may wonder why, with its name "tumor necrosis factor", TNF is not the most obvious choice for direct therapeutic administration of a cytokine to cancer patients. Unfortunately, the systemic dose of TNF necessary to have an effect on a cancer patient's tumor also causes liver toxicity, organ failure, and hypotension. Better results have been obtained using localized injection of TNF in conjunction with IFNγ and chemotherapy.

ivb) Engineered cytokine secretion. It is speculated that pure cytokines have rarely been effective in controlling tumor growth because the dose of cytokine has not been delivered directly to the tumor in a sustained manner. The ICK approach described earlier is one attempt to remedy this problem because the antibody that is conjugated to the cytokine will guide it to the tumor. Another method may be to use a transfection approach in which tumor cells isolated from an individual are genetically engineered to express a cytokine gene. When these modified tumor cells are returned to the same individual, the cells home back to the primary tumor and commence secreting the cytokine. For example, mice injected with tumor cells engineered to express GM-CSF show increased DC activity and develop strong anti-tumor Th1 and Th2 responses. The production of NO and anti-angiogenic factors by activated macrophages is also enhanced. In human melanoma patients, trials have been carried out in which the patient's tumor cells were isolated, engineered *in vitro* to express GM-CSF, and returned to the patient. The modified cells homed back to the primary tumor and created a local milieu of GM-CSF that induced the infiltration of the tumor by activated lymphocytes, neutrophils and eosinophils. These leukocytes then worked together to destroy the tumor vasculature. In addition, vigorous antibody responses against melanoma surface antigens and CTL responses against intracellular epitopes were induced. Clinical trials are ongoing to refine and expand the utility of this currently labor-intensive therapy.

We have reached the end of our discussion of whether and how the immune system attempts to fight cancer. We move now from a field in which researchers try to induce immune responses to sustain life to one in which they try to suppress them to sustain life: human tissue transplantation.

CHAPTER 16 TAKE-HOME MESSAGE

- **Tumors are cellular masses that form in tissues due to abnormal cell survival and cell division coupled with resistance to programmed cell death.**

- **A benign tumor contains well differentiated cells, grows slowly and is not usually life-threatening.**

- **A malignant tumor or "cancer" is composed of poorly differentiated cells, grows in an uncontrolled and invasive way, and is lethal if not treated.**

- **Metastasis is the spreading of a cancer to secondary locations that may be distant in the body.**

- **A cancer arises when a cell accumulates multiple mutations in DNA repair genes, oncogenes and/or tumor suppressor genes, resulting in deregulated growth. The frequency of these mutations is increased by carcinogens such as chemicals, irradiation and certain pathogens.**

- **TAAs are normal cellular proteins expressed at abnormal concentration, tissue location or developmental stage, whereas TSAs are macromolecules unique to the tumor and not expressed by any normal cell.** [handwritten: *Tumor associated antigen*]

- **Immunosurveillance by the immune system may detect TAAs and TSAs and implement defense against tumors via the actions of NK, NKT and T cells, inflammatory cells, and cytokines such as TNF, IFNγ and IL-12.**

- **Tumors may not be recognized by the immune system if the tumor cells express only TAAs and no TSAs.**

- **Tumor cells may actively evade or suppress immune responses by: losing expression of a TSA to which an immune response has been initiated; secreting immunosuppressive cytokines; promoting regulatory T cell-mediated suppression of anti-tumor responses; inhibiting T cell signaling; expressing non-classical MHC class I molecules that shut down NK cells; having a vasculature that discourages T cell extravasation.**

- **Immunotherapies are being explored that can specifically kill tumor cells or induce an augmented anti-tumor immune response. These strategies include the use of mAbs, cytokines, immunoconjugates, cancer vaccines and adoptive T cell therapy.**

Section A.I

1) Can you define these terms? neoplasm, metastasis, metastases

2) What alterations to rates of cell survival, division and apoptosis might lead to the growth of a tumor?

3) Distinguish between the terms "tumor" and "cancer".

4) Give three differences between benign and malignant tumors.

5) How do cancers usually cause death?

Section A.II–III

1) Can you define these terms? cell cycle arrest, carcinogen, cell of origin, angiogenesis, proto-oncogene

2) Distinguish between sporadic and familial cancers.

3) Name three classes of carcinogens and give an example of each.

4) How is chronic inflammation thought to be carcinogenic?

5) Describe the four steps of carcinogenesis.

6) Why is metastasis not random?

7) Briefly outline the cancer stem cell hypothesis.

8) Name the three classes of genes commonly mutated in cancers and describe how each leads to tumorigenesis. Give an example of each class of gene.

Section B Introduction

1) Can you define these terms? spontaneous regression, tumor antigen

2) Give two reasons why some researchers believe in the effectiveness of immunosurveillance.

3) Give two reasons why other researchers doubt the effectiveness of immunosurveillance.

4) What are the two major classes of tumor antigens?

Section B.I

1) Complete the following: "A TAA is almost always a case of the right molecule expressed at the wrong ?, ? or ?."

2) What is a cancer-testis antigen? Give an example.

3) What is an embryonic antigen? Give an example.

Section B.II

1) How is a TSA different from a TAA?

2) Describe four ways a TSA can arise.

Section C

1) What is tumor rejection?

2) How might acute inflammation help to combat tumorigenesis?

3) Describe the potential contributions of $\gamma\delta$ T, NKT, NK and $\alpha\beta$ T cells cells to anti-cancer responses.

4) What is epitope spreading and how might it help during an anti-cancer response?

5) What contribution does the humoral response make to anti-cancer responses?

Section D

1) Outline two lines of *in vivo* evidence that suggest tumors actively evade or block immune responses.

2) Describe three ways in which tumor cells might suppress or evade the immune system.

Section E.I

1) Describe the three conventional modes of cancer treatment.

2) Why do chemotherapy and radiation therapy have such devastating side effects?

3) Chemotherapy and radiation therapy do not appear to be effective against cancer stem cells. Why not?

Section E.II

1) Why might immunotherapy be an improvement over current conventional approaches for treating cancers?

2) Give two reasons why it has been difficult to induce clinically effective anti-cancer responses with immunotherapy.

3) Describe the two major ways in which monoclonal antibodies are used for immunotherapy.

4) What is a CTLA-4 blockade?

5) Give four examples of immunoconjugates and describe how each functions.

6) Describe the two basic types of cancer vaccines. Are they effective? If not, why not?

7) Outline two ways that might be used to improve ineffective cancer vaccines.

8) What is adoptive T cell cancer therapy and how does it work?

9) Describe two types of cytokine-based immunotherapies.

Bonus Question (Appendix F)

1) Describe how a humanized mAb is produced and why.

17

Transplantation

WHAT'S IN THIS CHAPTER?

War is the only game in which it doesn't pay to have the home-court advantage.
Dick Motta

The term "transplantation" typically brings lifesaving surgeries to mind: kidney transplants, heart and lung transplants, and skin grafts for burn victims. These procedures are called *solid organ transplants*. However, the *bone marrow transplants* (BMTs) and *hematopoietic cell transplants* (HCTs) used to treat malignancies like leukemias are also tissue transplants, as are blood transfusions. In all these cases, the goal is to supplement or replace a patient's non-functional tissue or cells with healthy tissue or cells from a donor. Indeed, tissue transplantation is now the preferred treatment for many serious disorders of the heart, lung, kidney, liver and hematopoietic system. However, unless the donor and recipient of a transplant are genetically identical, the donated tissue is viewed by the immune system of the recipient as non-self. Without drug-induced suppression of the recipient's immune system, the donated tissue is destroyed by recipient leukocytes in an immune response called *graft rejection* (Plate 17-1).

Researchers use specific terms to describe the genetic relationship between an organ donor and a prospective recipient. Individuals who are *syngeneic* are identical at all genetic loci, such that transplantation may be undertaken without fear of graft rejection. Individuals who are *allogeneic* have different alleles at various genetic loci so that transplants between such persons will be rejected in the absence of immunosuppression. In the simplest of transplant scenarios, the donor and recipient are the same patient, as in the case of a burn victim whose own healthy skin is used to replace a damaged section of skin elsewhere on his/her body. This type of syngeneic, self–self tissue transfer is called an *autologous graft*. Syngeneic grafts that involve the transfer of tissue between two genetically identical individuals (such as between human identical twins or two mice of the same inbred strain) are called *isografts*. Conversely, tissue transfers between genetically different members of the same species (such as two humans who are not identical twins) are called *allografts*. (In immunological circles, the relationship of individuals' MHC genotypes are often of overriding importance, so that the terms "syngeneic" and "allogeneic" may sometimes be used to denote genetic identity or non-identity at the MHC only.) *Xenografts* are tissue transfers between members of two different species, as in the case of a pig organ transplanted into a human.

A. The Molecular Basis of Graft Rejection

The strong, rapid responses that are principally responsible for allograft rejection are due to the recognition of allelic differences in antigens encoded by the *major histocompatibility loci* (MHC). Thus, the term *allorecognition,* as it is most commonly used, refers to recipient immune responses mounted against donor MHC. Certain genes outside the MHC also show a low degree of allelic variation and may also be involved in graft rejection responses. These latter genes came to be called *minor histocompatibility loci* based on their influence in solid organ transplant situations, where allelic differences at these loci result in slower, milder anti-graft responses relative to those triggered by MHC differences. In the case of BMTs, differences at the minor histocompatibility loci can cause rapid graft rejection.

I. IMMUNE RECOGNITION OF ALLOGENEIC MHC MOLECULES

How does an allogeneic MHC molecule (allo-MHC) on a graft provoke lymphocyte responses? For B cells, the mechanism of allo-MHC recognition is straightforward. The allo-MHC molecules of the donor often have a slightly different conformation than recipient MHC molecules, so that the recipient has B cells with BCRs that recognize allo-MHC as they would any incoming non-self protein. The recipient's B cells are activated and produce *alloantibodies* directed against the allo-MHC mole-

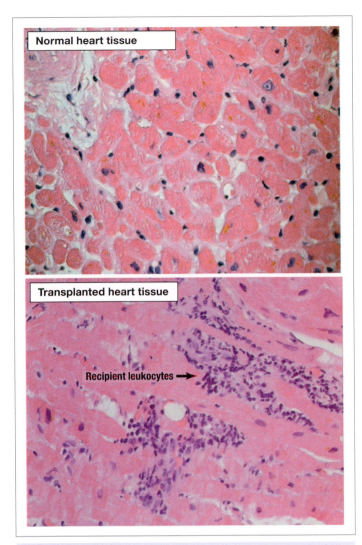

Normal heart tissue

Transplanted heart tissue

Recipient leukocytes →

Plate 17-1 **Graft Rejection**
[Reproduced by permission of Jagdish Butany, University Health Network/Toronto Medical Laboratories, Toronto General Hospital.]

cules displayed on the graft surface. For T cells, the situation is more complex because TCRs recognize epitopes composed of peptide bound to MHC. There are two mechanisms that account for T cell responses to allo-MHC: *direct allorecognition* and *indirect allorecognition*.

i) Direct Allorecognition

Direct allorecognition occurs when a recipient's T cells recognize pMHCs involving not <u>self</u>-MHC, as would occur during a pathogen attack, but <u>allo</u>-MHC molecules expressed by cells of the graft. What is the basis for this apparently contradictory recognition of non-self MHC? T cells are selected in the thymus through the mechanisms of central tolerance such that the lymphocytes released to the periphery recognize complexes of [self-MHC + X], where X is a non-self peptide. Any T cells recognizing [self-MHC + Y], where Y is a self peptide, are either eliminated in the thymus or silenced in the periphery by the mechanisms of peripheral tolerance. In a transplant situation, the cells of an allograft are covered with allo-MHC mol-

ecules loaded with a wide variety of peptides. Most of these peptides are considered "self" with respect to both the donor and recipient because they are derived from proteins that are invariant (*monomorphic*) in a given species. Thus, from the recipient's point of view, the peptide–MHC complexes in the transplanted tissue can be described as [allo-MHC + Y]. However, T cells do not distinguish the individual components of such complexes; they recognize the overall shape. If the conformation of an [allo-MHC + Y] epitope looks like some combination of [self-MHC + X], a recipient T cell specific for [self-MHC + X] will be activated by a donor cell bearing [allo-MHC + Y] (Fig. 17-1). It should be noted that, due to polymorphism in the peptide-binding groove, allo-MHC and self-MHC molecules may bind and present different peptides from the same protein. Thus, a monomorphic protein in an allograft may also give rise to pMHC complexes in which both the allo-MHC and the peptide contribute to a structure that looks like [self-MHC + X]. In all these cases, the response of the recipient's T cells to the allograft is really a form of cross-reactivity that occurs when peptide presented on allo-MHC looks like non-<u>s</u>elf peptide presented on <u>s</u>elf-MHC.

How does allo-MHC in a grafted tissue such as a kidney manage to activate cross-reactive T cells, since naïve T cells can be activated only in lymph nodes? During transplantation, donor DCs expressing allo-MHC travel along with the donated organ and are introduced into the recipient. These donor DCs can migrate out of the graft into the recipient lymph node draining the transplant site. Naïve recipient Th and Tc cells expressing cross-reactive TCRs can then be activated by the donor DC within the node and generate T cell effectors as well as memory T cells. The Th effectors support the activation of anti-graft B cells and Tc cells within the lymph node, and the differentiated progeny of these cells home back to the graft and attack it.

T cell responses to allogeneic cells are much more intense than T cell responses to pathogens, and the mechanism of direct allorecognition explains why. A vast array of peptides are presented simultaneously and continuously by the allo-MHC molecules of the graft, allowing for the simultaneous and continuous activation of a wide range of naïve T cell clones. In contrast, a pathogen attack results in the presentation of a relatively small array of pathogen-derived peptides that stimulates only a limited number of naïve T cell clones. It has been estimated that, while only 1 in 10,000 T cell clones

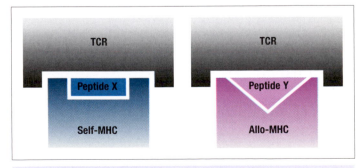

TCR

TCR

Peptide X

Peptide Y

Self-MHC

Allo-MHC

Fig. 17-1 **Direct Allorecognition at the TCR–pMHC Interface**

responds to any given pathogen, as many as 1 in 50 T cell clones can be activated by allo-MHC.

ii) Indirect Allorecognition

In indirect allorecognition, naïve recipient T cells are activated by <u>recipient</u> DCs that have acquired peptides derived from MHC proteins of the donor (Fig. 17-2). When cells of an allograft die, some of their component proteins are shed into recipient tissues and are taken up by recipient APCs (including DCs). Alternatively, recipient APCs may enter into the grafted tissue and acquire shed donor proteins. Most of these donor proteins are encoded by genes that are monomorphic within a species, so that peptides derived from these donor proteins are seen as "self". If acquired by recipient DCs, these self peptides are presented on self-MHC and do not provoke an immune response because the naïve recipient T cells recognizing this combination were deleted during the establishment of central tolerance. However, some of the shed donor proteins will be allo-MHC molecules, and peptides derived from allo-MHC are seen as non-self. If these non-self peptides are acquired by recipient DCs, they may be presented and cross-presented on self-MHC to naïve Th and Tc cells, respectively, in the local lymph node. Effector Th cells are generated that home back to the allograft and recognize complexes of allo-MHC peptide bound to self-MHC that are presented by additional recipient APCs. If the recipient and donor are syngeneic at some MHC class II loci, then donor APCs from within the grafted tissue may also present allo-MHC peptides that can be recognized by Th cells. Similarly, if the recipient and donor are syngeneic at some MHC class I loci, Tc cells stimulated by indirect allorecognition in the lymph node can generate CTLs that home back to the transplant site and attack the allograft.

II. IMMUNE RECOGNITION OF MINOR HISTOCOMPATIBILITY ANTIGENS

Within a given species, subtle allelic variation may exist in genes outside the MHC. In the case of a transplant between individuals who are allogeneic at such a locus, the proteins encoded by the donor allele and the recipient allele will differ slightly in amino acid sequence. If this difference leads to graft rejection, this protein is considered to be a *minor histocompatibility antigen* (MiHA). The corresponding gene is then classified as a *minor histocompatibility gene*. Where an organ donor and recipient are allogeneic at an MiHA locus, antigen processing of MiHA proteins in donated tissue (by either donor or recipient APCs) gives rise to *minor H peptides* that may combine with MHC molecules on recipient APCs or graft cells to form pMHC structures that look like [self-MHC + X] to recipient T cells. Thus, even when an organ donor and recipient are MHC-matched, graft rejection can occur due to differences in MiHAs.

Some MiHAs result from allelic variation in "housekeeping" genes within a population. Housekeeping genes are expressed in virtually all cell types of a species and generally encode proteins that maintain basal metabolic functions. Other MiHAs result from differences in the expression of tissue-specific genes, such as certain mitochondrial proteins and proteins that differ between males and females of a species. Examples of human and murine MiHAs are given in Table 17-1.

Immune responses to MiHAs are illustrated in Figure 17-3. In this example, a chromosome N contains a gene P that occurs in two allelic forms: P1 and P2. Let us suppose that the recipient

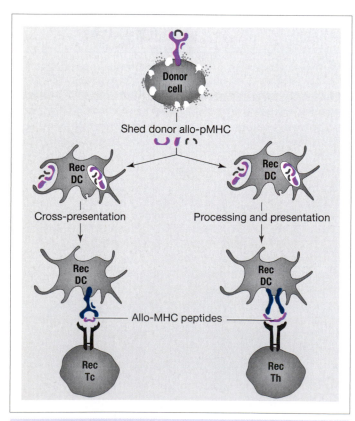

Fig. 17-2 **Indirect Allorecognition**

Table 17-1 **Examples of Human and Murine Minor Histocompatibility Antigens**		
MiHA name	**Protein/Gene**	**Localization**
Human		
HY-B7	SMCY	Y chromosome
HY-A1	DFFRY	Y chromosome
HY-DRB3	DBY	Y chromosome
HA-2	Class I myosin	Chromosome 6
HB-1	HB-1	Chromosome 5
Mouse		
COI	Cytochrome oxidase	Mitochondrial DNA
ND-1	NADH dehydrogenase	Mitochondrial DNA
β2m	β2-Microglobulin	Chromosome 2
H-YK^k	SMCY	Y chromosome

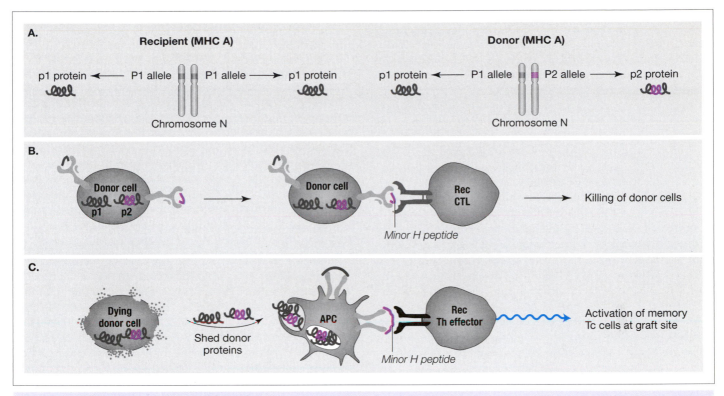

Fig. 17-3 **Recognition of Minor Histocompatibility Antigens**

and the graft donor are syngeneic at the MHC but partially allogeneic at the P locus: P1/P1 and P1/P2, respectively (Fig. 17-3A). A donor cell synthesizes both the P1 and P2 proteins endogenously so that peptides derived from them are presented on MHC class I (Fig. 17-3B). Recipient CTLs recognize the P2 peptide–MHC class I complexes as foreign and attack the donor cells. At the same time, the breakdown of dying donor cells releases P1 and P2 proteins into the extracellular milieu, where these molecules can be taken up by APCs (donor or recipient) and presented on MHC class II (Fig. 17-3C). Recipient Th effector cells recognize the P2 peptide–MHC class II complexes as foreign and release copious cytokines that support the activation of memory Tc cells at the graft site. Although Th-produced cytokines could also theoretically support B cell activation, the conformational difference between P2 and P1 is likely too small to be recognized by the BCRs of recipient B cells. Humoral responses to MiHAs are thus uncommmon.

As mentioned earlier, differences in minor histocompatibility genes are generally associated with a slower form of graft rejection. Why is the response to an MiHA weak compared to that invoked by allo-MHC? An MiHA expressed by a donor cell will give rise to a very limited number of minor H peptides with the potential to form only a few pMHC structures that recipient T cells will recognize as foreign. Indeed, the proportion of recipient T cells responding to an MiHA difference is usually no greater than that responding to a pathogen-derived antigen. In contrast, expression of allo-MHC leads to the activation of a large number of T cell clones due to TCR cross-reactivity.

B. Solid Organ Transplantation

I. IMMUNOLOGY OF SOLID ORGAN TRANSPLANT REJECTION

Transplantation of solid organs is a traumatic event from the body's point of view. Tissues are surgically resected and blood vessels are damaged. All these events initiate responses from innate leukocytes even in a syngeneic transplant situation. Adhesion molecules are upregulated in the injured endothelium of the graft, promoting the infiltration of activated recipient macrophages and other inflammatory cells that produce IL-1 and IL-6. Because these events are non-specific and do not usually result in loss of the grafted tissue, they are not considered graft rejection. However, they likely play an important role in setting the stage for graft rejection.

Graft rejection proper is caused by adaptive immune responses against alloantigens in the incoming organ. Once the organ is transplanted, both donor APCs (including immature DCs) from within the graft and recipient APCs that subsequently infiltrate the donated organ will eventually travel to the local lymph node. Because of the inflammatory milieu surrounding the graft, any DCs that have captured alloantigens are induced to mature. Donor DCs can then activate naïve T cells via direct allorecognition, while recipient DCs mediate indirect allorecognition. Both donor and recipient DCs can be involved in the presentation of peptides derived from MiHAs. In all these cases, the DCs present a pMHC that naïve recipient Th cells can recognize. (It should be noted that sometimes the alloreactive T lymphocytes stimulated during a first exposure to a graft may in fact be memory T cells. Such cells are generated in cases where a naïve T cell was stimulated <u>prior to allograft transplantation</u> by exposure to a pathogen that provided [self-MHC + X] pMHC structures that cross-react with [allo-MHC + Y].)

Once a recipient's Th cells are activated, they generate Th effectors that provide T cell help for the activation of naïve anti-graft B cells and Tc cells in the lymph node. The activated B cells generate plasma cells that produce antibodies directed against the allo-MHC proteins of the graft. These alloantibodies enter the circulation to travel to the graft and kill its cells via classical complement activation and/or ADCC. Although the alloantibodies produced may be directed against either MHC class I or MHC class II of the donor, the majority of the clinical damage will be due to the alloantibodies that recognize MHC class I because, unlike MHC class II, these proteins are widely expressed on most cell types, including the cells making up the graft. Tc cells that become activated generate CTLs that home to the graft, recognize allo-MHC present on graft cells, and kill these cells via multiple mechanisms of cytotoxicity. If the recipient and donor share identity at some MHC class I loci, CTLs specific for minor H peptides will also be able to attack graft cells directly. In addition, CTLs directed against minor H peptides complexed to self-MHC class II may attack recipient APCs that have infiltrated the graft, and in so doing, collaterally damage graft cells. Lastly, Th effectors can home to the allograft and release damaging cytokines, particularly IFNγ.

II. CLINICAL GRAFT REJECTION

Clinically, rejection of solid organ transplants is classified as *hyperacute*, *acute cellular*, *acute humoral*, or *chronic*. The clinical and histological features of these forms of graft rejection and their proposed underlying mechanisms are summarized in Table 17-2.

i) Hyperacute Graft Rejection

<u>Hyperacute rejection</u> (HAR) is the destruction of a graft almost immediately after transplantation. Solid organs are vascularized structures and HAR is usually caused by the presence in the recipient of <u>preformed</u> antibodies that recognize antigens on the endothelial cells of blood vessels within the transplanted organ (Fig. 17-4A). When these preformed antibodies bind to these antigens, they rapidly initiate complement activation via the classical pathway such that the vascular network of the graft is destroyed <u>within minutes</u> by MAC deposition. In general, no cellular infiltration is seen.

The chance of HAR is greatest when a recipient possesses preformed anti-ABO blood group antibodies. Although ABO antigens are defined as antigenic structures present on an individual's erythrocytes (see Blood Transfusions, later in this chapter), they are also expressed on vascular endothelial cells, making them a major target within donated tissue. HAR can also be triggered by preformed antibodies recognizing non-self MHC. Such antibodies are common in the circulation of prospective recipients who have been previously exposed to allo-MHC molecules, as in the case of individuals who have had a previous graft, blood transfusion or pregnancy. In the transplantation world, patients who harbor preformed antibodies likely to cause HAR are said to be "highly sensitized" or "hyperimmunized". The chance of HAR of a graft can be greatly decreased by careful screening of the recipient and potential donor for blood group and MHC mismatches (see later). In addition, the recipient may undergo a pre-transplant procedure called *plasmapheresis* in which the recipient's blood

Table 17-2 Types of Allograft Rejection

	Hyperacute (HAR)	Acute Cellular (ACR)	Acute Humoral (AHR)	Chronic (CGR)
Time for graft rejection	Within minutes	Within days or weeks	Within days or weeks	Within months
Targets on donor cell	ABO blood group Ags*; allogeneic MHC	Allogeneic MHC	Allogeneic MHC	Allogeneic MHC
Mechanism	Preformed Ab† that induce death by MAC formation	Direct allorecognition by recipient CTLs that exert cytoxicity and secrete cytokines	Indirect allorecognition by recipient CD4+ T cells that promote allo-Ab production	Indirect allorecognition; cytokine-induced ischemia
Cellular infiltration	None	Lymphocytes	Neutrophils	Lymphocytes and neutrophils
C4d staining	Yes	No	Yes	Yes
Fibrosis	No	No	Yes	Yes
Treatment	Pre-transplant plasmapheresis; no effective treatment once rejection is under way	Immunosuppressive drugs, steroids	Immunosuppressive drugs, post-transplant plasmapheresis	None effective

*Ag, antigen.
†Ab, antibody.

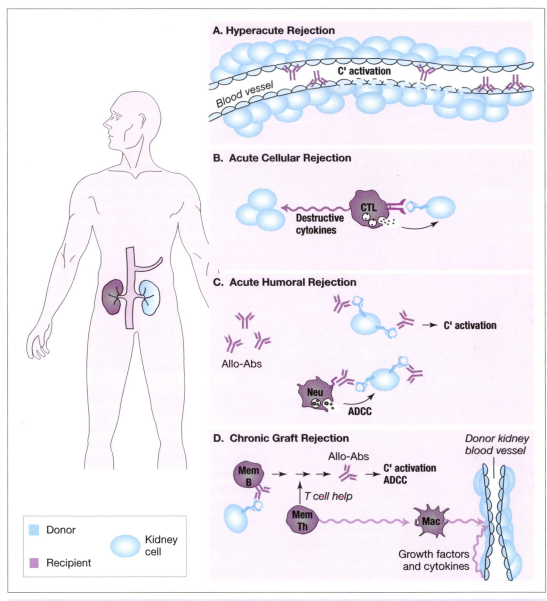

A. Hyperacute Rejection

Blood vessel   C' activation

B. Acute Cellular Rejection

CTL

Destructive
cytokines

C. Acute Humoral Rejection

C' activation

Allo-Abs

Neu

ADCC

D. Chronic Graft Rejection

Donor kidney
blood vessel

Mem
B   Allo-Abs

C' activation
ADCC

T cell help

Mem
Th

Mac

Growth factors
and cytokines

Donor

Recipient

Kidney
cell

Fig. 17-4 **Types of Clinical Graft Rejection**

is passed through a machine that removes all antibodies before returning the blood to the recipient. If the levels of the offending antibody are successfully diminished to below those required for HAR, the donated organ is less likely to be rejected, at least initially.

ii) Acute Graft Rejection

There are two types of acute graft rejection, *acute cellular* and *acute humoral*, that generally occur within a few days or weeks of transplantation. These types of rejection are distinguished by their underlying mechanisms.

iia) Acute cellular rejection (ACR). ACR is almost always the result of direct allorecognition of mismatched donor MHC by alloreactive recipient CTLs (Fig. 17-4B). The greater the number of MHC differences between the donor and recipient,

the more Tc clones are mobilized and the faster the rejection mediated by effector CTLs. The CTLs destroy the graft both by secreting destructive cytokines such as TNF and by initiating perforin- and granzyme-mediated killing. NK-mediated cytotoxicity may also take a toll, as the incoming organ may not express the self-MHC needed to bind to NK inhibitory receptors. However, the actual participation of NK cells in graft rejection has been difficult to document *in vivo*. Histologically, ACR is characterized by lymphocyte infiltration into the transplanted organ and inflammation of its blood vessels.

iib) Acute humoral rejection (AHR). AHR is usually due to humoral responses initiated via indirect allorecognition by recipient Th cells. The cytokines produced by the resulting Th effector cells support the activation of anti-graft B cells and the synthesis of anti-graft alloantibodies. These antibodies destroy graft cells via ADCC or classical complement (C') activation (Fig. 17-4C). The histological features of AHR include neutrophil (but not lymphocyte) infiltration into the graft, vascular inflammation, fibrosis, and necrosis of blood vessel walls. AHR can also be detected by staining for C4d, a degradation product of the complement component C4b. C4d tends to accumulate on capillaries in a graft undergoing AHR, whereas ACR is generally negative for C4d staining. The effects of AHR can be mitigated if the recipient undergoes post-transplantation plasmapheresis to remove the alloantibodies.

iii) Chronic Graft Rejection

Chronic graft rejection (CGR) of solid organs is defined as the loss of allograft <u>function</u> several months after transplantation. The transplanted organ may still be in place, but persistent immune system attacks on the allo-MHC expressed by its component cells have gradually caused the organ to cease functioning. Precisely what causes CGR is unknown but the most important clinical predictor of CGR is a prior, transitory episode of acute rejection.

Both alloantibodies and cell-mediated responses are involved in CGR, and the indirect pathway of allorecognition appears to be particularly important. In response to allogeneic donor peptide presented on recipient APCs, recipient memory T cells are activated and differentiate into Th effectors. These Th effectors produce cytokines that either directly damage the grafted tissue or constitute T cell help for recipient B cells producing alloantibodies (Fig. 17-4D). Over half of chronically rejected organs exhibit C4d deposition in the capillaries of the transplant. In a critical part of CGR that is characteristic but poorly understood, activated recipient T cells also induce monocytes and macrophages that have infiltrated the graft, as well as endothelial cells of the graft's blood vessels, to produce numerous growth factors and cytokines. These molecules drive both smooth muscle cell proliferation and ECM protein synthesis. The proliferating smooth muscle cells effectively shrink the lumens of the graft's blood vessels, causing *ischemia* (local loss of oxygen to a tissue). The increase in ECM proteins also results in fibrosis that further compromises graft function.

Thus, CGR is histologically characterized by fibrosis, collagen deposition, and a loss of blood flow within the graft.

III. GRAFT-VERSUS-HOST DISEASE (GvHD) IN SOLID ORGAN TRANSPLANTS

Transplantation, although a lifesaving procedure, can be a double-edged sword. Sometimes donor immune system cells traveling along with a transplanted tissue attack recipient tissues, such that "graft-versus-host disease" (GvHD) is said to have occurred. The epithelial cells of the skin, liver and intestine of the transplant recipient are the prime targets of GvHD. As a result, the recipient may start to lose weight as well as liver, lung and/or gut functions. Onset can be acute (and sometimes fatal) or chronic. Although GvHD rarely occurs in solid organ transplants, it is a huge problem for patients undergoing HCT since, by definition, the transplant is of donated immunocompetent cells that disperse throughout the body. GvHD is discussed in more detail in the HCT section of this chapter.

C. Minimizing Graft Rejection

Graft rejection can be minimized by a combination of approaches that rely on MHC matching, alloantibody analysis and immunosuppression. Researchers are also currently experimenting with treatments designed to increase a recipient's immunological tolerance of an incoming graft. Various sources of donated human organs are outlined in Box 17-1, whereas the possibility of using animal organs for human transplantation is explored in Box 17-2.

I. HLA MATCHING

The first step in a transplantation protocol is to determine the MHC haplotype of the prospective recipient. Since MHC molecules in humans are known as HLAs (human leukocyte

---

Box 17-1 Types of Organ Donation

Originally, organs for transplantation were harvested only from persons whose hearts had stopped beating (*cadavers*). Unfortunately, the organs of a cadaver remain healthy for only a few minutes or hours before the lack of oxygen kills their cells. Persons who suffer "brain death" (but retain heart and lung function) may also be organ donors. With life support machinery, a brain-dead donor's organs remain healthy while family consent to organ donation is obtained, HLA haplotypes are matched, and the donated organ is transferred to the recipient's hospital. However, only a small percentage of deaths involve brain death and the number of patients on transplant

waiting lists greatly exceeds the organ supply. Clinicians have thus increasingly reached out to "living related" donors. Healthy humans can survive with only one kidney, and the amazing regenerative powers of the liver mean that individuals can often donate sections of this organ with minimal risk to themselves. There are two advantages to living related donors: (1) family members will likely share at least some HLA alleles with the prospective recipient, reducing the chance of graft rejection; (2) the transplant can be planned in advance and proceed without delay. Because of fears of undue pressure or calls for financial compensation, offers of organs from living, <u>non</u>-related

donors are usually declined. At least for kidney transplants, the use of living related donors has meant that reliance on kidneys from cadavers has plunged: from 1988 to 2000, the percentage of kidney transplants that involved cadaveric donors dropped from 80% to 60%. With the use of living related donors and improved immunosuppressive drug regimens, about 85% of American kidney transplant patients are still enjoying life 5 years after receiving their new organs. Clinical research is ongoing to find ways to achieve the same results for other types of organ transplants.

antigens), this procedure is known as "HLA typing". HLA typing is carried out by examining the collection of HLA alleles expressed on a recipient's leukocytes. Once a donor organ becomes available, the HLA haplotype of the donor is similarly determined so that HLA mismatches with the recipient can be evaluated. Since only 1 in 400,000 unrelated persons are fully matched at the major HLA loci, almost all transplants (except those involving twins or siblings) are done under conditions of at least some HLA mismatching. Depending on the type of transplant being performed and the alternatives open to the patient (i.e., dialysis in the case of kidney failure), the doctors involved decide what degree of mismatch, if any, is acceptable before proceeding with the transplant. Heart, lung, liver and kidney transplants can now be done under conditions of significant HLA mismatching due to the control of graft rejection by modern immunosuppressive drugs.

i) HLA Typing by Complement-Dependent Cytotoxicity (CDC)

Complement-dependent cytotoxicity (CDC) was for many years the major means by which HLA typing was carried out. In this technique, blood leukocytes from the individual being typed are mixed in an array of microwells with defined collections of anti-HLA antibodies (Fig. 17-5). The anti-HLA antibodies used are obtained from individuals exposed to allogeneic cells via a previous transplant, transfusion or pregnancy. Also added to the wells are complement and a visible dye (such as trypan blue) that is excluded from viable cells. In any given well, if the cells express HLA alleles recognized by the added anti-HLA antibodies, the cells are perforated by classical complement activation such that the dye can enter the cells and turn them blue. (In the example shown in Figure 17-5, the individual being typed has the hypothetical HLA3 haplotype and not an HLA1, HLA2 or HLA4 haplotype.) HLA mismatching of a prospective donor and recipient is thus revealed by different patterns of reactivity across the same array of antibody collections. Although labor-intensive, CDC assays are quite rapid in that results can be obtained within 3 hours. Such speed is essential to the successful transplantation of organs from accident victims.

ii) HLA Typing by DNA-Based Methods

iia) Fragment-based techniques. The discovery in the 1970s of restriction enzymes that cut DNA in a sequence-specific way gave rise to another class of tissue typing techniques. When a given piece of DNA is digested with a particular restriction enzyme, it is cleaved in a sequence-dependent manner into "restriction fragments" of distinctive length. The fragments can then be separated by size via electrophoresis through an agarose gel, and visualized by hybridization to a tagged probe specific for the gene of interest. In the case of polymorphic genes, the number and location of restriction enzyme cleavage sites can vary between alleles so that the lengths of fragments generated by one allele are often slightly different from those generated by another allele. Such a difference is referred to as a *restriction fragment length polymorphism* (RFLP). The HLA compatibility of a prospective recipient and potential donors can be established by comparing the RFLP patterns of HLA genes. Although a sensitive technique, RFLP analysis is quite labor-intensive and relatively slow, such that this method has largely been replaced by more modern technologies.

iib) PCR-based techniques. The most frequently used methods of tissue typing today rely on the *polymerase chain reaction* (PCR) (see Box 17-3). PCR allows the amplification of specific DNA sequences such that large quantities of a DNA of interest can be generated from a miniscule sample. Using PCR, it has been possible to accurately determine the nucleotide sequences of most known human MHC class I and II alleles and to identify the specific variations that make them unique. As a result, clinicians can now identify the HLA alleles possessed by a given individual in a highly specific way.

PCR can be used in two ways to define the HLA haplotype of prospective transplant recipients and donors. In some cases,

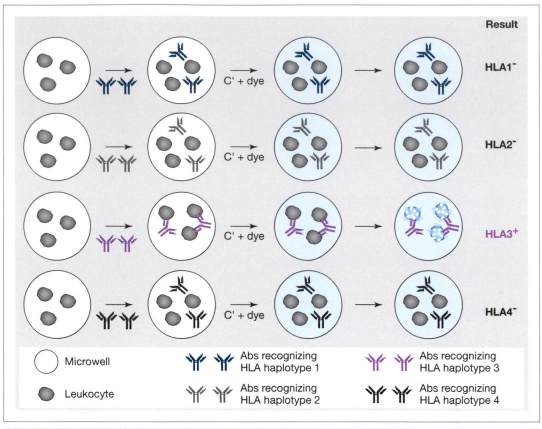

Fig. 17-5 **HLA Typing by Complement-Dependent Cytotoxicity (CDC)**

Box 17-3 The Polymerase Chain Reaction (PCR)

In the early 1980s, a technique called the *polymerase chain reaction* (PCR) was developed that allows researchers to examine the DNA of cells that are available only in very limited numbers. PCR can amplify a single DNA molecule into more than a billion essentially identical copies in just a few hours. Any piece of DNA can be amplified as long as the sequences of short stretches of DNA flanking the sequence of interest are known. The first step is to synthesize a set of *PCR primers*, which are oligonucleotides that are complementary to the short flanking sequences and provide initiation points for the synthesis of the copied DNA strands. A sample of the DNA of interest is heated to denature the double-stranded structure into two single strands, which are then combined with an excess of the PCR primers. Once the PCR primers bind to their complementary sequences on the DNA, the "primed DNA template" is ready to be copied. Incubation of the primed DNA template with the four deoxynucleotide precursors (A, C, G, T) and a heat-resistant DNA polymerase results in the extension of the primers one nucleotide at a time, all the way toward the end of each DNA template. By repeating the denaturation–extension procedure for multiple cycles, amplification of the "target" DNA, that small stretch of DNA of interest between the primers, is readily achieved. The Table shows the exponential increase in DNA target copies produced during a 32-cycle PCR: in only 3 hours, over a billion copies of the target DNA are made. This relative bounty of DNA allows the gene fragment of interest to easily be sequenced and characterized.

Cycle Number	Number of Copies of the Original DNA
3	2
10	256
15	8192
20	262,144
25	8,388,608
32	1,073,741,824

the total DNA from an individual's cells is mixed with generic primers that allow PCR amplification of all HLA alleles. The DNA containing the amplified HLA sequences is split into multiple aliquots and each is mixed with a labeled oligonucleotide probe known to be complementary to a particular HLA allele. Hybridization of a particular probe to the recipient's DNA confirms that the recipient expresses that HLA allele. Alternatively, aliquots of an individual's DNA can be separately mixed with a series of allele-specific primers under PCR reaction conditions. If PCR amplification occurs in a particular tube, it indicates that the primer was able to bind to the individual's DNA and that the individual possesses that particular HLA allele. Several technical variations of PCR-based HLA typing methods exist, and some can give results within 3 hours. Although these approaches are highly sensitive and accurate, they are technologically complex to perform and relatively costly.

II. ALLOANTIBODY ANALYSIS

Transplant survival is greatly decreased if the recipient has circulating preformed alloantibodies that could potentially mediate HAR. Thus, it is essential to determine if such antibodies are present in the recipient prior to the transplant, a process referred to as *cross-matching*. Originally, a recipient's serum was tested for the presence of anti-HLA antibodies by assessing the serum's reactivity against panels of cells, the so-called *panel reactive antibody* (PRA) test. However, because there are literally hundreds of MHC molecules and no test cell expresses just one HLA allele at a time, this type of test does not reveal the exact specificity of the anti-HLA antibodies in the recipient's serum. Today, molecular methods allow the screening of recipient serum for the presence of antibodies against individual HLA alleles. In such tests, various HLA molecules are separately synthesized *in vitro* using recombinant DNA technology. These HLA molecules are then individually coated onto synthetic beads and mixed with samples of the recipient's serum to allow the binding of antibodies that recognize a particular HLA allele. The pattern of serum reactivity with this collection of HLA-coated beads indicates HLA haplotypes that should be avoided in a donor because the recipient possesses alloantibodies recognizing these MHC molecules; that is, the recipient exhibits a "positive cross-match". When an appropriate donor is identified, more specific testing of the recipient's serum is carried out against a sample of the donor's cells to make certain that there is no positive cross-match between them. If a transplant must be made between a recipient and donor exhibiting a positive cross-match, the recipient undergoes pre-transplant plasmapheresis to remove the preformed alloantibodies.

III. IMMUNOSUPPRESSION

Graft rejection can be combated by immunosuppressive drugs designed to derail the alloreactive response to graft antigens. These drugs have made transplants between unrelated individuals possible, saving the lives of countless patients. However, the same immune system machinery that is blocked to avoid graft rejection is also normally responsible for tumor surveillance and dealing with pathogens. Thus, an immunosuppressed patient is liable to develop malignancies or come down with an opportunistic infection. In addition to these risks, the various drugs used for immunosuppression are often associated with serious medical complications. Physicians thus must attempt to maintain a balance between transplant survival and the patient's tolerance of detrimental side effects. The most commonly used immunosuppressive drugs, how they work, and their associated side effects are outlined in Table 17-3.

IV. INDUCTION OF GRAFT TOLERANCE

The ultimate objective of transplantation researchers is to induce permanent tolerance to an allogeneic graft in the absence of immunosuppression. Some scientists are seeking ways to interfere with the metabolism, proliferation and/or migration of attacking lymphocytes, whereas others are investigating approaches designed to anergize or delete only those recipient T cells that might attack the graft. Several strategies of the latter type are described here but the reader is cautioned that they are all experimental at this point.

i) Bone Marrow Manipulation

If thymocytes capable of recognizing particular allo-MHC alleles could be negatively selected in a recipient's thymus, central tolerance to a particular donor's tissue would be achieved. The first step in implementing such a strategy is to establish a state of *mixed chimerism* in a recipient's bone marrow such that both recipient and donor HSCs carry out hematopoiesis. To prevent destruction of the donor HSCs by recipient T cells, however, the patient must be treated in such a way that peripheral lymphocytes are destroyed but HSCs are preserved. *Myeloablative conditioning* is a regimen of vigorous chemotherapy and/or irradiation that accomplishes this goal but is highly toxic to most patients. *Non-myeloablative conditioning* is a less severe regimen of chemotherapy and/or irradiation that has the same result but is generally better tolerated by patients. Once the recipient has been conditioned, he/she intravenously receives a *donor cell infusion* consisting of donor bone marrow cells processed so as to enrich the number of HSCs. Despite being allogeneic, the donor HSCs are not rejected by the recipient thanks to the depletion of mature T cells achieved by the conditioning. Instead, the infused donor HSCs travel to the recipient's bone marrow and engraft in this location alongside the recipient's HSCs. Hematopoiesis in the recipient then generates progenitors of both recipient and donor origin. These progenitors seed the depleted recipient thymus, giving rise to thymic DCs that express donor MHC alongside thymic DCs expressing recipient MHC. Developing thymocytes then undergo negative selection following encounters with both donor and recipient pMHCs. Some immunologists say that the thymocytes become "educated" to recognize both donor and recipient MHC molecules as self. Thus, when lymphocytes in the reconstituted recipient encounter a transplanted solid organ from the same donor, they are tolerant to the MHC differences expressed by the donor cells and graft

Table 17-3 Immunosuppressive Drugs Commonly Used to Prevent Transplant Rejection

Drug	Nature	Mechanism	Outcome	Side Effects
Cyclosporine A (CsA)	Fungal compound	Inhibits a phosphatase essential for interleukin transcription, especially IL-2	Responses of T and B cells to allogeneic cells are dampened	Cardiovascular disease, diabetes, toxicity to kidneys and CNS, periodontal swelling, dysregulated hair growth
Tacrolimus (TAC)	Bacterial compound	Same as CsA	Same as CsA but more effective	Similar but milder than those of CsA
Mycophenolate mofetil (MMF)	Derivative of mycophenolic acid	Blocks purine synthesis and thus DNA replication in lymphocytes*	T and B cells cannot proliferate	Nausea, vomiting, diarrhea, low leukocyte and platelet counts
Sirolimus (rapamycin)	Bacterial compound	Blocks enzyme required for signaling leading to IL-2-induced proliferation; blocks pro-inflammatory cytokine and growth factor expression	T cells cannot proliferate; B cell activation and antigen uptake by DCs are also inhibited	Anemia, low leukocyte and platelet counts, dysregulated lipid metabolism
Malononitrilamide (MNA)	Low molecular weight molecule	Blocks pyrimidine synthesis and thus DNA replication in lymphocytes*; inhibits intracellular signaling required for Ab and selectin expression	T and B cells cannot proliferate; Ab production and lymphocyte extravasation are inhibited	Nausea, chills, dizziness, headache
Fludarabine	Nucleoside drug precursor	Converted to a purine analogue that inhibits DNA replication and transcription in lymphocytes	T and B cells cannot proliferate; may induce apoptosis	Fatigue, fever
CAMPATH-1H[†]	Humanized mAb recognizing CD52	Binds to T cells and promotes their death by ADCC, complement or direct induction of apoptosis	T cells are depleted	Chills, nausea, muscle spasms, fever

*Other cell types use an alternative pathway to synthesize purines and pyrimidines and so are not affected by these drugs.
[†]See also Chapter 20.

rejection is less likely to occur. Some success with this approach has been achieved in animal models and in early clinical trials involving transplant patients.

ii) Thymic Manipulation

Some researchers have tried to induce central T cell tolerance to alloantigens by introducing donor cells directly into the thymus of an animal whose mature peripheral T cells have been depleted by irradiation or anti-lymphocyte antibody treatment. The presence of donor pMHC structures on cells in the thymus should result in deletion of thymocytes with the potential to mediate allorecognition of a graft from this same donor. In addition, the presence of alloantigen in the thymus may induce the generation of T_{reg} cells recognizing alloantigenic peptides presented by recipient APCs. In theory, these allo-specific T_{reg} cells could subsequently migrate to the periphery and perhaps suppress rejection of grafts expressing the alloantigen. In practice, this strategy could be challenging to implement in a patient whose thymus has already involuted or sustained damage due to previous drug treatment.

iii) Costimulatory Blockade

As described in Chapter 10, peripheral tolerance is induced when antigen-specific T cells are anergized, and anergy is achieved when T cells are activated in the absence of costimula-

tion. Artificial interference with the delivery of costimulatory signals is known as establishing a *costimulatory blockade*. In theory, the advantage of a costimulatory blockade in a transplant situation is that, rather than targeting all T cells, the anergization should be largely focused on T cells activated by an encounter with specific antigens of the graft. Costimulatory blockades have been established experimentally in mice by injecting anti-B7, anti-CD40L or anti-ICOS mAbs that inhibit the delivery of T cell activation signal 2. Another type of costimulatory blockade exploits the properties of the T cell regulatory molecule CTLA-4. In Chapter 16, we described the experimental inhibition of the function of CTLA-4 in order to prolong T cell responses against tumor cells. In the transplantation context, a soluble fusion protein made up of CTLA-4 plus the constant region of an Ig molecule has been generated to promote CTLA-4 function and downregulate responses of recipient T cells directed against grafted tissue. Does costimulatory blockade work? It seems that no one blockade agent on its own can induce permanent tolerance to a graft, but where a combination of these reagents has been used in animal models, there has been some temporary improvement in graft survival.

iv) Modulated DCs

Another way to induce graft tolerance via T cell anergy might be to generate tolerogenic or modulated DCs (refer to Ch. 10).

The idea is to deliver donor antigens to immature DCs *in vivo* under conditions that prevent DC maturation. The DCs should then be capable of anergizing any T cells they encounter, preventing an alloreactive response. In one experimental approach, researchers biochemically linked a donor antigen of interest to a mouse mAb that recognized a DC-specific marker. Administration of this mAb to a recipient mouse allowed the mAb to convey the donor antigen to immature DCs in lymphoid tissues and facilitate antigen internalization without triggering maturation. Alternatively, since immature DCs routinely process apoptotic cells to acquire self antigens without triggering maturation, isolated apoptotic donor cells have been injected intravenously into mice to promote the acquisition of donor antigens. It may also be possible to isolate a recipient's DCs and culture them *in vitro* under conditions (e.g., with IL-10) that would modulate them. Once returned to the transplant recipient, these modulated DCs would be expected to anergize, rather than activate, any naïve allo-MHC-specific T cells they encounter. The modulated DCs might also trigger the differentiation of regulatory T cells that could suppress alloreactive T cells. This strategy is under investigation in animal models.

v) Regulatory T Cells

As described in Chapter 10, CD4+ and CD8+ regulatory T cell subsets are thought to play an active role in controlling immune responses, particularly those directed against autoantigens. However, evidence is accumulating that these cells can also suppress alloreactive responses in a transplantation situation. Transplantation researchers are experimenting with strategies designed to isolate sufficient quantities of purified regulatory T cells and manipulate their activation and functions in the hopes of increasing graft tolerance. Regulatory T cells normally occur in relatively low numbers *in vivo*, but recent work suggests that expansion and/or purification of regulatory T cells *in vitro* may allow these cells to become therapeutically useful in the near future.

vi) Gene Therapy

"Gene therapy" is the term used to describe highly experimental clinical protocols in which the purified DNA of a particular gene is introduced into the somatic cells of a patient in an effort to ameliorate disease. Transplantation researchers have tried various types of gene therapy in an effort to decrease graft rejection. Transplantation is actually an ideal venue for gene therapy because transplantation already calls for manipulation of tissue outside the body, and the introduction of a gene or gene regulator into isolated donor cells is easily accomplished. For example, in mice, a plasmid expressing an immunosuppressive cytokine can be introduced into a heart allograft to suppress the anti-graft response of the recipient. Other gene therapy approaches are aimed at modulating donor DCs within a graft so that naïve alloreactive T cells that access graft tissue are anergized rather than activated. This strategy shows promise, as the survival of some mouse allografts has been greatly improved by engineering donor DCs to express an inhibitor that blocks CD28–B7 interaction. In other cases, manipulation of a mouse allograft to express the FasL gene

has led to better transplant survival because the FasL on the donor cells induces the apoptosis of Fas-bearing alloreactive recipient T cells. Translation of these tactics to the human situation lies in the future.

D. Hematopoietic Cell Transplantation

Hematopoietic cell transplants (HCTs) are carried out when a patient has a disease involving hematopoietic cells, such as leukemia (see Ch. 20). In the past, such ailments could only be treated by the painful and risky transplantation of whole bone marrow. Although BMT is still sometimes used to restore the hematopoietic systems of cancer patients subjected to high dose chemotherapy, other conditions are usually treated by HCT, which involves the much less stressful collection of donor HSCs from peripheral blood. Prior to the HCT procedure, the number of HSCs that can be isolated from the donor's blood is increased by giving the donor repeated injections of purified growth factors known to drive HSC proliferation. The HSCs then "spill out" from the bone marrow into the blood and are harvested by *leukapheresis*. During leukapheresis, blood passes out of a patient's body into a machine that collects leukocytes but returns erythrocytes and other blood components to the patient's circulation. The harvested leukocytes (including the critical HSCs) are then frozen in preparation for the transplant. Meanwhile, the prospective recipient undergoes non-myeloablative conditioning to partially empty his/her hematopoietic compartment of leukocytes. This conditioning prolongs the survival of donor leukocytes that are subsequently infused into the recipient, allowing the donated HSCs to repopulate the recipient's immune system.

I. GRAFT REJECTION IN HCT

In the absence of any pre-transplant conditioning of the recipient, an HCT would require near-identity in donor and recipient HLA haplotypes to prevent a graft-destroying response by the recipient's lymphocytes. Under these circumstances, HCT donors would be effectively limited to the patient's HLA-matched siblings, and very few such transplants would be possible. The use of pre-transplant myeloablative conditioning of the patient removes any lymphocytes with allogeneic specificity and has allowed the expansion of the pool of potential transplant donors to include unrelated, allogeneic individuals. However, myeloablative conditioning is associated with a high level of toxicity that can only be withstood by younger patients. Because many patients requiring HCT are older people with hematopoietic cancers (see Ch. 20), clinical researchers are currently focusing on non-myeloablative conditioning regimens that are less toxic but still effective in circumventing rejection of the allogeneic HCT.

It should be noted that the ABO blood group compatibility that is important for solid organ transplants is much less relevant for HCTs due to the lack of a graft vasculature to protect. Indeed, about 30% of allogeneic HCTs are performed under conditions of ABO incompatibility.

II. GRAFT-VERSUS-HOST DISEASE (GvHD) IN HCT

As mentioned earlier, GvHD occurs when donor cells in the transplanted tissue attack recipient cells and tissues. Because the transplanted tissue in an HCT consists of competent immune system cells, GvHD is a significant issue for HCT recipients. In particular, the HLA mismatching in allogeneic HCT has the potential to lead to acute and severe GvHD that can be fatal. Once activated by recognition of an MHC or MiHA mismatch, alloreactive donor T cells undergo clonal expansion. Pro-inflammatory cytokines and chemokines are released that recruit donor and recipient macrophages and other cell types that do not discriminate between donor and recipient targets. TNF secreted by these cells is thought to be responsible for the metabolic wasting associated with many cases of GvHD. However, the bulk of tissue destruction in GvHD is caused by Fas- or perforin-mediated cytotoxicity exerted by donor CTLs. Helper Th1 responses directed against MiHA presented on MHC class II may also contribute. Unfortunately, GvHD is promoted by the myeloablative and non-myeloablative conditioning regimens used to prepare recipients for HCT. The tissue damage caused by the high doses of drugs used to empty the hematopoietic compartment promotes an inflammatory response and the production of cytokines that both damage tissues and encourage donor T cells to attack recipient tissues.

In the absence of any intervention, acute GvHD occurs in 75% of HCT recipients with a mismatch at a single MHC locus, and in 80% of recipients with three mismatches. Even among recipients completely matched at all MHC loci, about 30–50% will develop clinically significant GvHD due to MiHA mismatches. The easiest way to prevent GvHD in HCT would be to deplete the donor cell population of all mature T cells, either by use of lymphocyte-depleting antibodies or chemical inhibitors. However, the same mature T cells that carry out GvHD are also responsible for the beneficial "graft-versus-leukemia" effect seen in patients undergoing HCT for treatment of certain hematopoietic cancers (see later and Ch. 20). Thus, researchers are investigating ways to dampen but not obliterate T cell function in the graft. In mice, a delay between non-myeloablative conditioning of the recipient and donor cell infusion reduces the incidence and severity of GvHD, but this approach is not always very practical for human patients. Future strategies may include the prevention of donor T cell recognition of recipient antigens, or the anergization of donor T cells. In addition, because much of the damage of GvHD is due to cytokines, scientists are exploring ways to block the production or action of these molecules using various antibodies or antagonists. Indeed, modest reductions in GvHD in animals have been observed when TNF is neutralized.

Another approach to allogeneic HCT that minimizes GvHD is the use of umbilical cord blood instead of bone marrow as a source of HSCs. Cord blood is collected at birth, frozen and stored in a cord blood bank that can later be accessed for allogeneic HCT. Because cord blood is collected from newborns that have yet to be exposed to many foreign antigens, most T lymphocytes present are naïve cells rather than memory cells, and thus are less easily activated. However, although this T cell naiveté reduces the severity of any GvHD invoked by the transplant, it may also be responsible for the high frequency of serious post-transplant infections observed following the use of this approach.

III. GRAFT-VERSUS-LEUKEMIA (GvL) EFFECT

It has been repeatedly observed that leukemia patients who receive an allogeneic HCT suffer a relapse of their disease if, in an effort to reduce GvHD, all donor T cells are eliminated prior to the transplant. It seems that, although residual donor T cells may contribute to GvHD, they can also turn their alloreactivity against any leukemic cells arising in the recipient after the HCT. This phenomenon is called the *graft-versus-leukemia* (GvL) effect. Thus, when deciding how extensively to deplete the donor HCT preparation of T cells, the physician must weigh the benefit of GvHD prevention against the value of the GvL effect. It is thought that the GvL effect results from responses against MHC and MiHA differences between hematopoietic cells of the donor and the recipient, rather than responses against specific tumor antigens. Ideally, clinicians would like to preserve the donor T cell response to recipient hematopoietic cells (maintaining GvL) while preventing the donor T cell response to recipient epithelial cells (preventing GvHD). Strategies to achieve this balance are under investigation in preclinical and clinical trials.

E. Blood Transfusions

One of the most common tissue transplants is blood transfusion, in which a preparation of blood cells (but not serum) donated by healthy volunteers is injected into the bloodstream of injured or diseased individuals or those undergoing surgery. However, the RBCs of different individuals frequently express different antigens, so that two individuals may be of *incompatible blood types*. If the blood cells of an incompatible individual are transfused into a recipient by mistake, a *transfusion reaction* results because the recipient's circulation contains preformed antibodies specific for donor RBC antigens (see later). These antibodies immediately attack the transfused RBCs and induce their complement-mediated destruction. Clinically, the symptoms of a transfusion reaction may be as mild as a headache or wheezing, or as serious as tissue necrosis. In the most severe cases, the recipient's blood pressure falls and the blood vessels constrict, resulting in renal failure and shock. Transfusion reactions account for about 1 patient death per 500,000 units transfused in the United States. It should be noted that transfusion-related GvHD, in which donor leukocytes transferred with the transfused blood attack recipient tissues, can also occur. The incidence of transfusion-related GvHD is very rare in developed countries because, prior to transfusion, leukocytes are removed from the donated blood by either irradiation or physical depletion. In countries where leukocyte removal is not practised, 5% of cases of transfusion-related GvHD result in death.

The majority of severe transfusion reactions result when the blood of an incorrect ABO blood type is transfused into a

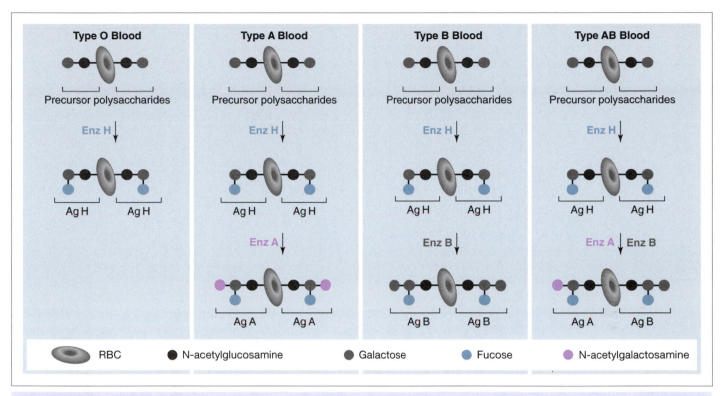

Fig. 17-6 **Structures of ABO Blood Sugars**

Table 17-4 **Blood Group Antigens and Antibodies**

Blood Type of Recipient	Genotype	Sugars Added by Transferases	Antigens on RBCs	Anti-ABO Antibodies in Circulation	Blood Type of Compatible Donors
A	AA or AO	Fucose; NAGA*	A	Anti-B	A, O
B	BB or BO	Fucose; GAL†	B	Anti-A	B, O
AB	AB	Fucose; NAGA; GAL	A and B	None	A, B, AB, O
O	OO	Fucose	H	Anti-A and anti-B	O

*N-Acetylgalactosamine.
†Galactose.

recipient. As introduced earlier, the ABO antigens are glyco-proteins expressed predominantly on the surfaces of RBCs. Every individual produces a precursor glycoprotein containing N-acetylglucosamine and galactose to which terminal sugars are added by a series of enzymes (Fig. 17-6). Every individual also expresses a fucose transferase enzyme (Enz H) encoded by the H gene. The form of the glycoprotein that results from the addition of the fucose by the H enzyme is called *antigen H*. Individuals of blood type A express an additional transferase (Enz A) that attaches an N-acetylgalactosamine residue to the H antigen. The resulting glycoprotein is called *antigen A*. In contrast, individuals of blood type B express a different transferase (Enz B) that attaches a galactose residue to the H antigen, creating *antigen B*. Individuals of blood type AB possess both A and B transferases and produce both antigens

A and B. Individuals of type O do not express either the A or B transferase and only antigen H is found on their RBCs (Table 17-4).

Because of the differences in the terminal sugars on antigens A and B, these proteins constitute Ti antigens in a person of an incompatible blood type. During the development of central tolerance, an individual of blood type A loses the B lympho-cytes recognizing antigen A (due to negative selection) but retains the B cells recognizing antigen B. Unfortunately, certain common intestinal bacteria express glycoproteins with epit-opes closely resembling those of antigens A and B. Thus, long before any transfusion, the B cells in the type A individual that recognize antigen B are activated by cross-reaction with the bacterial epitopes, and anti-B antibodies are synthesized and released into the circulation. These cross-reactive antibodies are then maintained at significant levels by the ongoing presence of the bacteria. Thus, when a transfusion of blood of type B is attempted in the individual of blood type A, the preformed cross-reactive antibodies in the recipient's circulation immediately attack the transfused RBCs. A transfusion reaction results.

The ABO blood group antigens are not the only antigens capable of provoking transfusion reactions. A prospective recipient should also be screened for the expression of the highly immunogenic Rh antigen described in Chapter 14. A severe anti-Rh Ab response may occur in a transfusion situation if a recipient is Rh$^-$ but the donated blood is Rh$^+$, and if the Rh$^-$ recipient possesses circulating anti-Rh antibodies due to prior exposure to Rh$^+$ cells during pregnancy or tissue trans-plantation. There are other minor blood group antigens that may be mismatched between donor and recipient but physicians do not routinely screen for these alloantigens because they are not highly immunogenic. Thus, these alloantigens invoke recipient immune responses that are mild and of little clinical significance.

We have come to the end of our discussion of transplantation and move next to a description of allergy and other hypersensitivities. Many of the concepts learned in Chapters 16 and 17 overlap those relevant to the control of hypersensitivities (Chapter 18) and even autoimmune disease (Chapter 19). Clinical researchers hope that the lessons learned in one discipline will have direct application to the others.

CHAPTER 17 TAKE-HOME MESSAGE

- Allogeneic MHC molecules on transplanted tissue trigger strong immune responses that may cause rapid graft rejection, whereas molecules encoded by allogeneic minor histocompatibility loci usually trigger weak anti-graft responses.

- Direct allorecognition occurs when the TCRs of recipient T cells specific for complexes of [self-MHC + non-self peptide] cross-react with complexes of [non-self MHC + self peptide] expressed by cells in a graft.

- Indirect allorecognition occurs when allogeneic MHC molecules shed from a graft are processed by recipient APCs and presented to recipient Th cells.

- Hyperacute graft rejection (very fast) is mediated by preformed alloantibodies.

- Acute graft rejection (fast) is mediated by CTLs (acute cellular rejection) or B cells (acute humoral rejection).

- Chronic graft rejection (slow) is mediated by cytokines and other molecules induced by CTL- and Ab-induced graft destruction.

- In GvHD, donor lymphocytes in the graft attack recipient tissues.

- MHC matching between a recipient and donor improves transplant and patient survival and is optimized by tissue typing techniques.

- ABO blood group matching between recipient and donor and testing of the recipient's serum for graft-specific alloantibodies minimize the chance of HAR during solid organ transplantation.

- Long-term allograft acceptance is currently achieved through drugs with general immunosuppressive effects. A key goal of transplantation researchers is to induce specific graft tolerance in the recipient.

- HCT involves the transfer of healthy HSCs from a donor to a recipient.

- The risk of GvHD is significant in allogeneic HCT but can provide the benefit of the GvL effect to leukemia patients.

- Blood transfusions are a common form of transplantation and require ABO blood group compatibility for success.

Introduction

1) One type of tissue transplantation involves solid organs. What are three other types?

2) What is graft rejection and why does it occur?

3) Distinguish between transplants that are syngeneic, allogeneic and xenogeneic.

4) Distinguish between an isograft and an allograft.

Section A.I

1) Define "allorecognition" as the term is commonly used in the transplantation context.

2) Slower, milder graft rejection is due to differences in what type of genetic locus?

3) What are the two mechanisms that account for T cell responses to allo-MHC and what are the consequences of these responses?

4) Why are T cell responses to an allogeneic transplant much stronger than T cell responses to a pathogen?

Section A.II

1) What are minor H peptides and why do they provoke a graft rejection response?

2) Give two examples of minor histocompatibility loci.

3) Why are humoral responses not usually a component of rejection due to minor histocompatibility antigen differences?

4) How does the strength of a T cell response to a minor histocompatibility antigen compare to that of a T cell response to a pathogen? Explain.

Section B

1) Give an overview of how T cells, B cells, DCs and other APCs are involved in the rejection of a solid organ transplant.

2) Name the four types of clinical graft rejection and state their rates of onset.

3) Contrast the underlying mechanisms of the four types of clinical graft rejection.

3) In transplantation terms, how does an individual become hyperimmunized? Why is hyperimmunization a barrier to successful transplantation and how can the problem be avoided?

4) What is the most important clinical predictor of CGR?

5) Define graft-versus-host disease.

Section C

1) Give two sources of organs for human transplantation.

2) Describe three potential difficulties associated with the use of animal organs for human transplants.

3) Briefly outline the principles behind three types of HLA typing.

4) What is a positive cross-match and how is it determined?

5) What are the pros and cons of the use of immunosuppressive drugs in a transplant situation?

6) Briefly describe three types of immunosuppressive drugs and how they work.

7) Why would induction of graft tolerance be beneficial for a transplant recipient?

8) What is mixed chimerism?

9) What is the difference between myeloablative and non-myeloablative conditioning?

10) What is a donor cell infusion?

11) Describe three ways in which graft tolerance might theoretically be induced.

Section D

1) What is leukapheresis and why is it helpful for HCT?

2) For what types of disorders is HCT a treatment option?

3) How does the requirement for HLA matching between donor and recipient compare for HCT versus a solid organ transplant? ABO matching?

4) Why is GvHD more important in HCT than in solid organ transplants?

5) What is the graft-versus-leukemia effect and why is it useful?

Section E

1) What is a transfusion reaction?

2) What is the mechanism underlying ABO blood group incompatibility?

3) Sketch out a table defining the ABO blood groups and indicating who may give blood to whom and why.

18

Immune Hypersensitivity

WHAT'S IN THIS CHAPTER?

Govern a great nation as you would cook a small fish. Do not overdo it.
Lao Tzu

The immune response is usually viewed as helpful because it protects against pathogen attack. In most cases, the secondary response to a pathogen is so effective that the individual does not get sick at all. In this chapter, we examine what happens when a primary response is followed by a secondary response that hurts rather than helps the individual. Such disorders are called *immune hypersensitivities* and immunologists classify them into four types based on their underlying mechanisms: *type I, IgE-mediated hypersensitivity; type II, direct antibody-mediated cytotoxic hypersensitivity; type III, immune complex-mediated hypersensitivity;* and *type IV, delayed type hypersensitivity.* Key features of these disorders are summarized in Table 18-1 and discussed in detail in the following sections.

All hypersensitivities develop in two stages: the *sensitization* stage and the *effector* stage. The sensitization stage is the primary immune response to an antigen whereas the effector stage is a secondary immune response. In this context, hypersensitivity (HS) is defined as any excessive or abnormal secondary immune response to an antigen. A first exposure to an antigen causes most individuals to mount a normal primary response that is followed by normal secondary response upon a subsequent exposure. It remains unknown why a first exposure to an antigen results in a primary response that sensitizes some individuals, who then experience a hypersensitivity reaction upon a subsequent exposure to the same antigen.

bodies directed against certain innocuous antigens in the environment. These antigens, which are commonly referred to as *allergens*, are typically soluble proteins that are components of larger particles such as pet dander or tree pollen (Table 18-2). Most people encountering such antigens produce IgM, IgG or IgA antibodies that successfully clear the antigens without causing any symptoms. However, in individuals who make IgE antibodies to these antigens, reactions are triggered that lead to side effects that can range from itching and swelling to breathing difficulties and even shock or death. The response to the allergen is generally very rapid and occurs within 30 minutes of the encounter, so that type I HS is also known as "immediate" HS. Why only some people produce IgE antibodies to allergens is a mystery.

Another term for allergy is *atopy*, so that clinicians speak of "atopic patients" and "atopic reactions". There are two types of atopic reactions, systemic and local. A *systemic* atopic response is called *anaphylaxis* and affects the entire body (see later). In a local atopic reaction, the allergic symptoms depend on the anatomical location of the affected tissue and are generally confined to that tissue. For example, a local IgE-mediated response to an allergen in the nose is manifested as *atopic rhinitis* (hay fever), while a local response in the airway and lungs that results in the inflammation of these tissues is called *atopic asthma*. In the skin, local atopic responses take the form of *atopic dermatitis* (eczema) or *atopic urticaria* (hives). It is not understood why some antigens cause localized reactions whereas others have systemic effects.

A. Type I Hypersensitivity: IgE-Mediated or Immediate

I. WHAT IS TYPE I HS?

Type I hypersensitivity is what most people think of as "allergy". Allergies occur in individuals who express IgE anti-

II. MECHANISMS UNDERLYING TYPE I HS

i) Sensitization Stage

The sensitization stage of a type I HS reaction is initiated when an allergen first penetrates a skin or mucosal barrier (Fig. 18-1, #1), is collected by an immature DC (IDC, #2), and is conveyed to the local lymph node (#3). Within the node, the now mature

Table 18-1 Types of Hypersensitivity (HS) and Their Key Characteristics

Type	Type I HS	Type II HS	Type III HS	Type IV HS
Common name(s)	IgE-mediated HS, immediate HS, allergy, atopy	Direct antibody-mediated cytotoxic HS	Immune complex-mediated HS	Delayed type HS, cell-mediated HS
Immune system mediator	Antibody (IgE)	Antibody (IgG or IgM)	Antibody (IgG or IgM)	Effector T cells, macrophages
Time to symptoms	<1–30 min	5–8 hr	4–6 hr	24–72 hr
Mechanism	Allergens cross-link IgE bound on mast cells and basophils and induce degranulation	IgG or IgM bind to cell-bound antigen; cell is destroyed by phagocytosis, complement activation or ADCC	Immune complexes trigger complement activation; phagocyte FcR engagement leads to release of lytic mediators	Effector T cells produce IFNγ and other cytokines promoting macrophage hyperactivation
Examples	Asthma, hay fever, eczema, hives, food allergies, anaphylaxis	Hemolytic anemias, Goodpasture's syndrome	Arthus reaction, aspects of rheumatoid arthritis (RA) and systemic lupus erythematosus (SLE)	Lesions of TB and leprosy, poison ivy, farmer's lung

Table 18-2 Examples of Common Allergens

Allergen Name*	Scientific Name of Source	Common Name of Source
Amb a 2	Ambrosia artemisiifolia	Ragweed
Api m 1	Apis mellifera	Bee venom
Ara h 2	Arachis hypogea	Peanut
Bet v 1	Betula verrucosa	Birch tree pollen
Can f 1	Canis familiaris	Dog dander
Der p 1	Dermatophagoides pteronyssinus	House dust mite
Pen a 1	Penaeus aztecus	Shrimp
Phl p 5	Phleum pratense	Timothy grass

*Allergens are named according to the first three letters of the genus of the organism from which the antigen is derived combined with the first letter of the species and a number indicating the order of discovery. For example, Amb a 2 is the second allergen derived from *Ambrosia artemisiifolia*.

DC (MDC) presents processed allergen to a naïve Th cell. In the presence of IL-4, the activated Th cell differentiates into Th2 effectors. These cells then supply T help to naïve allergen-specific B lymphocytes in the node that have bound intact or fragmented allergen to their BCRs. The activated B cells and Th2 effectors commence the expression of tissue-specific homing receptors, leave the node via the efferent lymphatic (#4), enter the blood via the thoracic duct, and home back to the tissue into which the allergen first entered (#5). In this site, the Th2 cells produce copious amounts of IL-4, IL-5 and IL-13 that influence isotype switching toward IgE (rather than IgG or IgA) in the activated B cells (#6). Thus, in allergic individu-

als, plasma cells (PC) producing anti-allergen IgE antibodies can be found in the area of allergen penetration (#7). Some of these antibodies bind to allergen, and some bind to high affinity FcεRI receptors expressed on the surfaces of mast cells in the immediate area. The remaining IgE antibodies filter through the tissues and eventually enter the circulation via the lymphatic system (#8). Many of these circulating IgE molecules eventually bind to FcεRI molecules on the surfaces of basophils in the blood (#9) and additional mast cells in more distant tissues (#10). These mast cells and basophils soon are coated with allergen-specific IgE and become bombs waiting to be triggered by a subsequent encounter with the allergen. The sensitized mast cells and basophils can remain "armed" in this way for an extended period.

At this point, the reader may well ask, "But if allergens are generally innocuous, what supplies the "danger signal" that induces DCs to mature and activate naïve anti-allergen T cells?" Researchers speculate that allergens in fact are associated with some kind of cellular stress or damage that provokes a primary response in everyone. However, in non-atopic individuals, this response takes the form of harmless IgG production rather than IgE. Another unsolved mystery relates to the association of atopy with immune deviation to the Th2 phenotype. It seems that an atopic person's lymph nodes contain abnormally high local concentrations of IL-4 that induce naïve T cells to differentiate into Th2 effectors upon activation. These cells then secrete the Th2 cytokines that promote isotype switching to IgE in activated B cells and profoundly influence the effector actions of other immune system cells responding to the allergen. In mouse models, type I HS reactions to allergens can be mitigated by inducing immune deviation to Th1.

ii) Effector Stage

If an allergen enters the body of an atopic person a second time during the period that the mast cells and basophils are

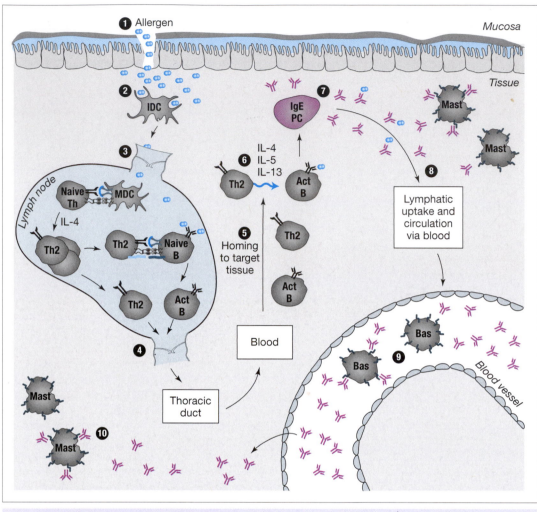

Fig. 18-1 **Sensitization Stage of Type I Hypersensitivity**

armed, the effector stage is triggered. The effector stage takes place in two phases: the *early phase reaction* and the *late phase reaction*. Because sensitized mast cells are already present in the target tissue, the mediators released by these cells in response to allergen binding are the main drivers of the early phase reaction. Sensitized basophils must be recruited from the blood by chemokines released by the allergen-activated mast cells, and so make their contribution to the late phase reaction. Other leukocytes drawn to the site of allergen accumulation, particularly eosinophils, also play their roles during the late phase reaction. Although all type I HS reactions have the same underlying biphasic mechanism, the symptoms seen vary widely because the mediators released affect different cell types in different locations. A list of mediators contributing to type I HS appears in Table 18-3.

iia) Early phase reaction. The early phase reaction of type I HS is mediated primarily by the degranulation of sensitized mast cells in the target tissue. Mast cells are abundant in the skin, in the loose connective tissue surrounding blood vessels, nerves and glandular ducts, and in the mucosae. In the lungs,

mast cells are often found surrounding the blood vessels as well as in the bronchial connective tissues and alveolar spaces. Mast cell degranulation most likely evolved to combat parasites, and the symptoms induced by the granule contents (coughing, sneezing, tearing of the eyes, scratching of the skin, cramping of the gut and diarrhea) are designed to expel these types of pathogens.

As illustrated in Figure 18-2, the early phase reaction is triggered by an encounter between sensitized mast cells (Sens Mast) and the same allergen that triggered the sensitization phase. The allergen again penetrates the mucosae of the target tissue (#1) and binds to IgE molecules that were captured by the FcεRI molecules on the surfaces of mast cells during sensitization. Upon allergen-induced aggregation of the FcεRI molecules (#2), intracellular signaling is triggered that causes the immediate degranulation of the activated cells (#3) with the release of histamine, serotonin, chemotactic factors and proteases. The fact that these fast-acting mediators are <u>preformed</u> and stored in intracellular granules accounts for the "immediate" nature of type I HS responses and the rapid onset of initial symptoms. Symptoms are then sustained for several hours by the action of newly generated mediators that require some type of synthesis for their formation, including cytokines (particularly TNF, IL-1 and IL-6) and chemokines produced by the mast cells (#4). In addition, after degranulation, mast cells start to break down such that the action of various enzymes on the plasma membrane generates *platelet-activating factor* (PAF), *leukotrienes* and *prostaglandins* (#5). Together, all these mediators produce the tissue-specific symptoms of an allergic response (#6). Furthermore, the chemokines released by the mast cells set the stage for the late phase reaction by inducing the expression of new adhesion molecules on the activated endothelium of local blood vessels (#7).

How do the mediators released during the early phase cause allergic symptoms? Histamine and PAF bind to their specific receptors on the smooth muscle cells supporting the blood vessels and induce them to relax, expanding the diameter of the blood vessel lumen (vasodilation) and increasing blood flow to the local area. Simultaneously, histamine and leukotrienes induce the contraction of the endothelial cells lining the blood vessels (increased vessel permeability), creating opportunities for cells and plasma proteins such as complement components to leak out of the circulation into the tissues. The action of histamine on sensory nerves causes the itching of eczema and the sneezing of hay fever. Histamine also induces the increased mucus secretion in the bronchioles that is characteristic of asthma. As well as binding to endothelial cells and inducing smooth muscle relaxation, PAF triggers the activation of platelets and the release by these cells of additional inflammatory mediators.

iib) Late phase reaction. About 4–6 hours after the initiation of a type I HS reaction, leukocytes expressing the appropriate

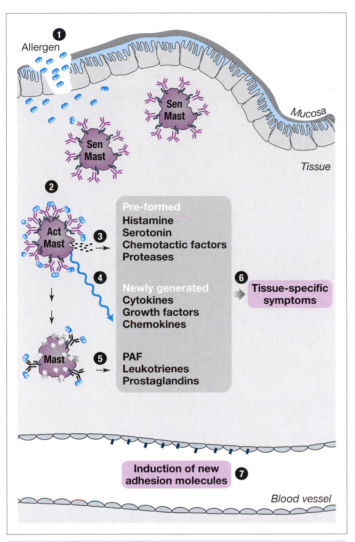

Fig. 18-2 **Effector Stage of Type I Hypersensitivity: Early Phase Reaction**

Table 18-3 Major Mediators Contributing to Type I Hypersensitivity

Mediator	Cellular Source	Promotes
Histamine	Mast cell and basophil granules	Vasodilation, vessel permeability, bronchial smooth muscle contraction, mucus production, itching, sneezing
Serotonin	Mast cell and basophil granules	Vasodilation, bronchial smooth muscle contraction
Chemotactic factors	Mast cell and basophil granules	Chemotaxis of eosinophils and neutrophils
Proteases	Mast cell and basophil granules	Mucus production, basement membrane digestion, increased blood pressure
Cytokines and growth factors	Mast cell, basophil synthesis	Mobilization and activation of immune system cells; initiation and maintenance of inflammatory response
Platelet-activating factor (PAF)	Mast cell membrane breakdown; eosinophil granules; synthesis by neutrophils, macrophages	Platelet aggregation and degranulation, pulmonary smooth muscle contraction
Leukotrienes	Mast cell membrane breakdown; eosinophil granules; synthesis by neutrophils, macrophages	Vasodilation, increased vessel permeability, bronchial smooth muscle contraction, mucus production
Prostaglandins	Mast cell membrane breakdown; synthesis by neutrophils, macrophages	Platelet aggregation, pulmonary smooth muscle contraction
Major basic protein	Eosinophil granules	Mast cell degranulation, smooth muscle contraction, death of respiratory epithelial cells
Eosinophil-derived neurotoxin	Eosinophil granules	Death of myelinated axons and neurons
Eosinophil cationic protein	Eosinophil granules	Death of respiratory epithelial cells

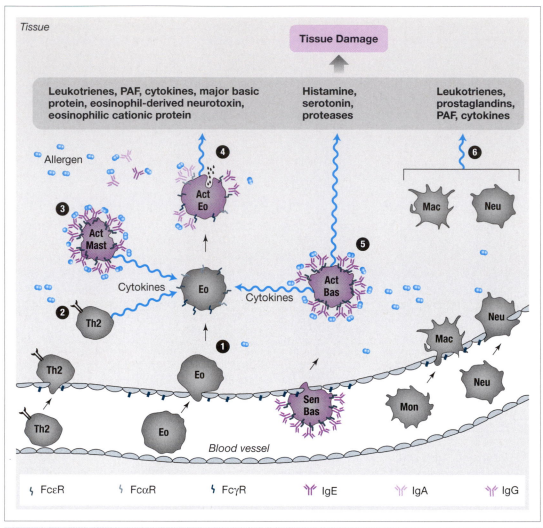

Fig. 18-3 Effector Stage of Type I Hypersensitivity: Late Phase Reaction

receptors bind to the new cell adhesion molecules expressed on the activated endothelium. These leukocytes include eosinophils, sensitized basophils, monocytes and neutrophils. The cells exit the circulation and, in response to local chemotactic factors, migrate into the allergen-polluted tissue (Fig. 18-3, #1). Eosinophils are particularly important for the late phase reaction. Th2 effectors that have been drawn to the scene (#2), as well as resident tissue mast cells that have become activated (#3), release cytokines that stimulate eosinophil chemotaxis and prepare these cells for activation. Eosinophils bear FcεRs, FcαRs and FcγRs that can interact with allergen that has already bound to IgE, IgA or IgG, respectively. Upon engagement of any of these FcRs by Ig–allergen complexes, the eosinophils degranulate and flood the surrounding tissue with leukotrienes, PAF, cytokines and eosinophil-specific mediators (#4). These molecules have potent activity against parasites and viruses but are also toxic to tissue cells and cause significant damage. Airway epithelial cells are especially sensitive to this type of assault, such that the clinical symptoms of asthma can be attributed mainly to eosinophil activation induced by a type I HS reaction.

The late phase reaction is also characterized by the activation of sensitized basophils that have been drawn from the blood into the allergen-laden tissue. Upon the binding of allergen to the IgE–FcεRI complexes on the basophil surface (Fig. 18-3, #5), the cell is triggered to produce many of the same mediators as those released by mast cells during the early phase reaction. Activated basophils also secrete large quantities of IL-4 and IL-13 that act on eosinophils.

Finally, the macrophages and neutrophils that infiltrate the allergen-laden tissue collectively release additional cytokines and other mediators that directly or indirectly cause tissue damage (Fig. 18-3, #6). These cells are important sources of PAF, leukotrienes and prostaglandins in inflamed tissues.

III. EXAMPLES OF TYPE I HS

i) Localized Atopy

Most people experience allergies as a local type I HS reaction affecting a specific target tissue such as the skin or bronchial passages. The sensitized mast cells that are triggered in these instances usually lurk among the epithelial cells lining the target tissue. It remains unclear why some atopic people have a tendency to develop a particular clinical manifestation (such as urticaria) while others tend to develop another affliction (such as asthma).

ia) Allergic rhinitis (hay fever). Allergic rhinitis, the prototypical example of a type I HS, results when airborne allergens such as ragweed pollen or mold spores are inhaled. Sensitized mast cells resident in the upper respiratory tract, the conjunctiva of the eyes, and the nasal mucosae are triggered to degranulate and release pro-inflammatory mediators in these locations. The mediators cause the characteristic symptoms of hay fever: coughing, tearing and itching of the eyes, sneezing, and the blockage of nasal passages. About 20% of individuals in the developed world suffer from allergic rhinitis.

ib) Atopic asthma. Atopic asthma is a type I HS reaction that occurs in the lower respiratory tracts of 10–20% of children and adults living in developed countries. Inhalation of antigen triggers degranulation of sensitized mast cells resident in the nasal or bronchiolar mucosae. The resulting release of pro-inflammatory mediators results in the production of copious amounts of mucus that constrict the bronchioles (sometimes severely). The patient soon complains of tightness in the chest and begins to wheeze or gasp for air. Asthma can be fatal if an acute attack totally blocks the airway. Histologically, the airway of an asthmatic patient appears chronically inflamed with mast cells, eosinophils, lymphocytes and neutrophils (Plate 18-1). In addition, the basement membrane of the airway is increased in thickness, the bronchiolar smooth muscle layer is enlarged, and increased mucus is present. Over 50 distinct inflammatory mediators are associated with asthma symptoms, and high levels of pro-inflammatory cytokines are found in the lung secretions of asthmatic patients.

ic) Atopic urticaria (hives). Atopic urticaria is a type I HS reaction in which sensitized skin mast cells degranulate and release mediators that cause swollen, reddened patches on the skin known as hives or "wheal and flare reactions" (Plate 18-2). The whitish wheal in the center of the hive is composed of leukocytes that have escaped the blood vessels due to the increased "leakiness" of these channels. The flare is the ring of redness seen surrounding the wheal due to increased blood flow into this area. The intense itching and pain of hives are caused by the stimulation of skin nerve endings by histamine. Eventually, the hives may become confluent such that the reaction covers a large area of the body. As well, urticaria is frequently accompanied by prominent swelling beneath the mucosal and cutaneous layers in the site of allergen exposure. Allergies to latex, hair chemicals, food additives, insect bites and some drugs are associated with acute urticaria.

id) Atopic dermatitis (eczema). Atopic dermatitis is a type I HS reaction in the skin that results excessive dryness and an itchy rash that is more scaly than in urticaria. Eczematous lesions may affect different parts of the body at different ages. Atopic dermatitis tends to be more chronic in nature than

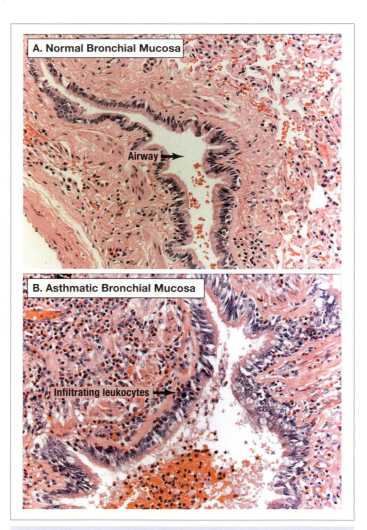

Plate 18-1 **Normal versus Asthmatic Bronchial Mucosae**
[Reproduced by permission of David Hwang, Department of Pathology, University Health Network, Toronto General Hospital.]

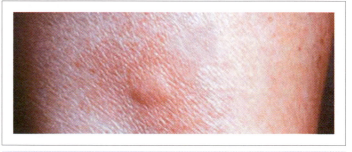

Plate 18-2 **Atopic Urticaria (Wheal and Flare Reaction)**
[Reproduced by permission of the Mayo Foundation for Medical Education and Research.]

urticaria and is often associated with respiratory allergies later in life. Individuals with atopic dermatitis also tend to be more susceptible to skin infections because the barrier function of the skin is compromised by the eczematous lesions.

ie) Food allergies. Food allergies result from IgE-mediated reactions to allergens in consumed foods. Although any food can cause an allergic reaction, 90% of food allergies have been linked to peanuts, soy, milk, eggs, wheat or fish. "Intolerances" to milk and alcohol can resemble allergies in some of their symptoms but are not IgE-mediated and are therefore not allergies. True food allergies are manifested in a variety of ways. Some food allergens cause a type of urticaria that takes the form of burning or itching of the tongue, lips and throat. Severe mucosal edema of the mouth and pharynx may occur. Other food allergens do not affect the mouth but trigger sensitized mast cells in the gut. Mediators released by these mast cells act on the gut smooth muscles, causing them to contract. Vomiting, nausea, abdominal pain, and cramping and/or diarrhea may result. Large numbers of eosinophils may infiltrate into the gastric and intestinal walls. In other cases, the mediators may increase the permeability of the gut mucosae such that a food allergen enters the circulation. Depending on where it ends up, the allergen may induce an asthmatic response or urticaria or eczema in a site distant from the mouth or gut. The only proven therapy for a food allergy is elimination of that food from the diet; however, mild food allergies are often outgrown with time.

ii) Systemic Atopy: Anaphylaxis

Anaphylaxis is a type I HS response with striking systemic consequences. Clinically, anaphylaxis is a form of extreme shock (hence, "anaphylactic shock") that can kill within minutes of exposure to the triggering antigen. Anaphylactic shock is most frequently observed in individuals sensitized to insect stings, peanuts, seafood or penicillin.

During anaphylaxis, large quantities of inflammatory mediators and vasodilators are released into the circulation by activated mast cells and basophils, causing rapid dilation of blood vessels throughout the body. Respiration immediately becomes difficult and is followed by a dramatic drop in the victim's blood pressure and extensive edema in the tissues. Patients have been known to report a "feeling of doom" at this point. The lungs may fill with fluid, the heart may beat irregularly, and control of the smooth muscles of the gut and bladder is often lost. Constriction of the bronchioles may cause lethal suffocation of the victim unless treatment with epinephrine (adrenaline) is started immediately. Sometimes the clinical course is biphasic, in that severe symptoms initially appear and then seem to resolve for 1–3 hours. Symptoms then return with a vengeance and can kill the patient if treatment is not sought again immediately.

IV. DETERMINANTS ASSOCIATED WITH TYPE I HS

i) What Makes an Antigen an Allergen?

Despite intensive study, it has not been possible to identify a single characteristic common to all allergens. They are diverse in their structures and biochemical properties, enter into the body in different ways, act at different concentrations, and interact with different molecules or cell types once within the body. It is also not known why an atopic individual might be allergic to cat dander but not dog dander, nor why allergy to birch tree pollen is common but allergy to pine tree pollen is extremely rare.

ii) What Makes an Individual Atopic?

The development of atopy clearly depends on an individual's genetic background because particular alleles of certain polymorphic genes are strongly associated with a predisposition to atopy. However, studies of atopy in identical twins have revealed that both twins are allergic in only 60% of cases. Thus, other non-genetic factors must also be involved. For example, it is believed that the environment in which an individual is raised can influence the competence of the immune system and/or how it responds to an allergen. In particular, the circumstances under which an individual first encounters an allergen can affect whether that person becomes sensitized. Furthermore, for unknown reasons, a genetically and environmentally predisposed person may still not respond to an allergen unless that allergen is encountered in the context of a triggering event. The major determinants thought to play a role in atopy are summarized in Figure 18-4 and discussed in the following sections.

iia) Genetic determinants.
The familial nature of allergy was first noticed in the mid-1800s. In a family in which both parents are atopic, 50% of the children will have allergies, compared to only 19% of children in a family with no history of atopy. Several examples of the 13 chromosomal regions and more than 20 genes that are thought to contribute to allergy in humans are shown in Table 18-4. The chromosomal region

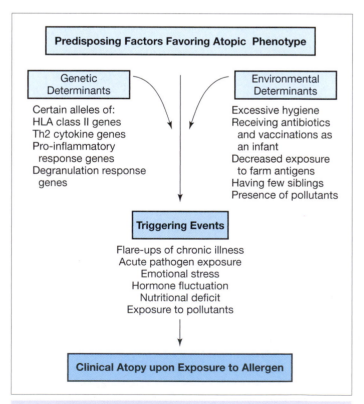

Predisposing Factors Favoring Atopic Phenotype

Genetic Determinants

Certain alleles of:
HLA class II genes
Th2 cytokine genes
Pro-inflammatory
  response genes
Degranulation response
  genes

Environmental Determinants

Excessive hygiene
Receiving antibiotics
  and vaccinations as
  an infant
Decreased exposure
  to farm antigens
Having few siblings
Presence of pollutants

Triggering Events

Flare-ups of chronic illness
Acute pathogen exposure
Emotional stress
Hormone fluctuation
Nutritional deficit
Exposure to pollutants

Clinical Atopy upon Exposure to Allergen

Fig. 18-4 **Determinants Favoring Atopy**

Table 18-4 **Examples of Human Genes Potentially Contributing to Atopy**

Chromosomal Region*	Candidate Gene(s)	Proposed Role or Function
3q27	Bcl-6	Regulation of Th2 responses
5q31–33	IL-3, IL-4, IL-5, IL-9, IL-13, TGFβ	Immune deviation to Th2; isotype switching to IgE; activation of eosinophils, basophils and mast cells
6p21–22	HLA-DR2, HLA-DR4, HLA-DR7, TAP TNF	Antigen processing and presentation Inflammation
11q13	FcεRI	Mast cell degranulation
12q14–24	IFNγ, iNOS, SCF, mast cell growth factor	Regulation of IL-4 transcription and thus isotype switching to IgE; inflammatory response; mast cell stimulation
14q11–13	TCRA, TCRD NF-κB	T cell recognition of allergen Transcription factor driving pro-inflammatory gene expression
16p	IL-4Rα	Component of IL-4 receptor

*p and q indicate the short and long arms of a chromosome, respectively. Thus, "3q27" means "chromosome 3, long arm, 2nd block, 7th band" as seen in standard karyotyping.

5q31–33 is of special interest because it contains a cluster of genes that encode cytokines promoting Th2 differentiation. Patients with certain polymorphisms in this region have more Th2 cells in the relevant tissues than non-allergic individuals, and these Th2 cells produce greater than normal amounts of IL-4. Particular polymorphisms in the IL-4Rα and IL-13 genes have been correlated with asthma.

iib) Environmental determinants. In the 1800s in Europe, the group of disorders that we now call "allergies" were known as "the rich and noblemen's disease". Even today, clinicians continue to report a steady increase in the frequency of atopy in the developed world compared to the less developed world. In the United States alone, over 5000 deaths and 500,000 hospitalizations are attributed annually to asthma attacks, and the annual cost of caring for asthma patients is estimated to be over U.S. $6 billion. In contrast, the prevalence of atopic disease remains very low in developing countries. It is not clear why an elevation of living standard appears to precipitate an increase in the incidence of atopy. One factor relevant to current times may be the emphasis in more developed cultures on energy conservation in the home. The upgrading of windows, insulation and carpeting may have decreased draftiness but has also tremendously reduced the exchange of inside and outside air. As a result, occupants of such homes may suffer increased exposure to house dust mites and other allergens. Another factor may be the exposure in more developed countries to higher levels of pollutants that promote allergen-specific IgE production and increase sensitization rates.

Atopy may also be increasing in the developed world due to an obsession with "germs". In more advanced countries, childhood infections are prevented by vaccination or rapidly resolved with antibiotics, and personal and food hygiene are fastidious. As a result, a child may not encounter sufficient pathogens in infancy to "train" his/her immature immune system to respond appropriately to antigens. This concept forms the basis of the *hygiene hypothesis*, which posits that increased hygiene leads to underexposure to pathogens and a consequent skewing of immune responses toward the Th2 phenotype. It is this imbalance in Th responses that is thought to later lead to inappropriate responses to normally innocuous antigens (atopy). Although an attractive theory, the hygiene hypothesis cannot be completely valid because individuals infected with helminth worms (who mostly live in developing countries) mount very strong Th2 responses against these parasites but show no signs of increased atopy. In addition, a corresponding decrease in developed countries of Th1/Th17-mediated autoimmune diseases has not been observed (see Ch. 19). Some immunologists now believe that early exposure to pathogens helps to establish not only Th1, Th2 and Th17 anti-pathogen responses but also regulatory responses mediated by Th3 cells. In the absence of sufficient Th3 cells, neither the Th1 or Th17 cells driving autoimmune disease, nor the Th2 cells mediating atopy, may be adequately suppressed. Such a theory could account for the observation that both atopy and autoimmune diseases are on the rise in the developed world.

There is other evidence supporting the notion that vigorous antigen exposure in early childhood prevents atopy. Children raised on farms are less likely to suffer from allergies than are children living in non-farm households in the same rural area. It seems that farm children are exposed from a young age to higher concentrations of animal antigens and bacterial endotoxin than are their non-farm peers, and that this wide-ranging exposure may be protective. In an urban parallel, children who are raised with multiple siblings or who attend day care (environments with an increased risk of infection) tend to have a lower incidence of atopy than do only children or those kept at home in their early years.

iic) Triggering events. Many allergic responses appear to be associated with a triggering event. Clinicians have noted that flare-ups of certain chronic illnesses, acute pathogen infections, emotional stress, fluctuating hormone levels, nutritional deficits, and exposure to pollutants can all be associated with the onset of type I HS reactions. The common factor here may be a weakening of the immune system that leaves the body more accessible to allergen entry and sensitization. If an infection is the trigger, then a pathogen molecule itself may be acting as an allergen in that the infected individual mounts an IgE response to the offending protein. The response successfully clears the pathogen but leaves the individual with atopic symptoms. For example, many flare-ups of atopic dermatitis are associated with the presence of certain skin fungi, and IgE antibodies directed against proteins from these organisms are found in the sufferer's circulation.

V. THERAPY OF TYPE I HS

The appropriate therapy for type I HS depends on an individual's circumstances and exposure to the allergen. Sometimes the easiest way to prevent an allergic response is to minimize contact with the allergen. Antibiotics can reduce the presence of pathogens that trigger or exacerbate flare-ups of particular allergies, while emollients can help to rehydrate dry skin and thus reduce the chance of allergen entry. However, such simple approaches are not always effective and many patients must resort to more involved therapies.

i) Antihistamines

Because a principal mediator of symptoms released during the effector phase of an allergic response is histamine, molecules that mitigate its effects are often used to treat relatively mild allergies such as hives and hay fever. Antihistamines work by binding to histamine receptors on target organs so that histamine released by degranulating mast cells and basophils cannot trigger symptoms. Antihistamines are normally taken at the onset of symptoms but if an individual knows that he/she is about to be exposed to an allergen, the prior use of antihistamines can prevent symptom development.

ii) Lipoxygenase Antagonists

Leukotrienes generated in the late phase of the allergic response mediate or promote many aspects of atopy, particularly bronchoconstriction and eosinophil infiltration. Because lipoxygenase antagonists block the generation of leukotrienes from mast cell membranes, the inhalation of these agents can provide relief from asthma symptoms and improve pulmonary function. Lipoxygenase antagonists can be given orally and usually have minimal side effects.

iii) Bronchodilators

Acute asthma attacks are often treated by inhalation of rapidly acting bronchodilators (delivered using inhalers or "puffers"). These agents are usually aerosolized drugs that block mast cell degranulation and induce smooth muscle relaxation. However, bronchodilators do not address the underlying inflammatory response.

iv) Corticosteroids

Corticosteroids have powerful anti-inflammatory effects because they indirectly inactivate a wide range of transcription factors. Both cytokine production and the expression of adhesion molecules required for the entry of inflammatory cells into target tissues are thus inhibited. Low dose corticosteroids are often used in a cream form to treat atopic dermatitis, as nose drops to treat rhinitis, and as an inhaled agent to treat asthma. Corticosteroids have very few side effects when used at a low dose to treat atopic conditions, but high dose, long term or oral use of corticosteroids can have severe consequences, including immunosuppression.

v) Cromones

Cromones are effective anti-inflammatory drugs that have fewer side effects than corticosteroids. Intranasal cromones can alleviate symptoms of allergic rhinitis and asthma, whereas oral cromones are useful for treating food allergies. The mechanism of action of these compounds is unclear but they appear to block mast cell degranulation and inhibit macrophage and eosinophil functions.

vi) Hyposensitization

Hyposensitization (or *desensitization*) is a procedure in which an allergy sufferer receives subcutaneous injections of ever-increasing amounts of the purified allergen every week or month for a period of up to 3–5 years. In many cases, the atopic individual eventually ceases to show an allergic reaction to the allergen. Hyposensitization works well for individuals with atopic rhinitis or urticaria, or who are allergic to insect venom, but is not as effective for those suffering from asthma or eczema. Hyposensitization is almost always started only after a child has passed his or her fifth birthday, and careful monitoring is necessary as induction of anaphylaxis has been observed in rare cases.

The end result of hyposensitization is the conversion of a harmful IgE anti-allergen response to a harmless IgG4 response but the underlying mechanism is still not completely understood. As illustrated in Figure 18-5, the success of the procedure may depend on the final balance of T cell subsets present and the functionality of each subset's member cells. One hypothesis is that continuous dosing with allergen may promote the differentiation of allergen-specific Th1 cells at the expense of Th2 cells. Cytokines (such as IFNγ) produced by these Th1 cells then inhibit isotype switching to IgE. Another theory is that the allergen-specific Th2 cells that secrete cytokines favoring mast cell activation and IgE production may be anergized by the repeated stimulation with allergen under conditions in which costimulation is minimal. Lastly, hyposensitization may gradually increase the number and/or activation of Th3 cells that express high levels of IL-10 and TGFβ. *In vitro*, these immunosuppressive cytokines promote isotype switching to IgG4 and dampen T cell, eosinophil and mast cell functions. In reality, all these mechanisms may make a contribution to hyposensitization.

vii) Experimental Approaches

Because many conventional allergy treatments have significant side effects, immunologists have been seeking new methods of derailing the atopic response. Although the administration of

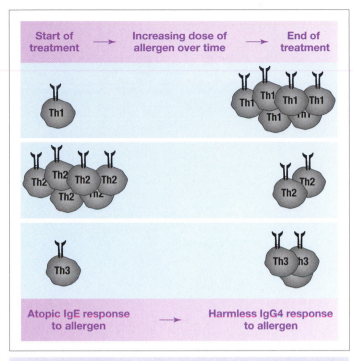

Start of treatment → **Increasing dose of allergen over time** → **End of treatment**

Atopic IgE response to allergen → **Harmless IgG4 response to allergen**

Fig. 18-5 **Hyposensitization**

Th1 cytokines might seem a logical way to counterbalancing a dominant Th2 response, these molecules are toxic in the amounts needed to see an improvement in atopic symptoms. Instead, scientists have turned to agents such as: small, inorganic molecules or mAbs that disrupt intracellular signaling triggered by engagement of FcεRI; an anti-IL-5 mAb that blocks IL-5-mediated stimulation of eosinophil differentiation and function; anti-IL-4 and anti-IL-4R mAbs that sequester IL-4 or block IL-4R so that Th2 differentiation and isotype switching to IgE are inhibited; small molecules that target the transcription factors needed for the expression of IL-4-dependent genes; an anti-TNF mAb that prevents TNF from activating transcription factors contributing to the HS response; an anti-CCR3 mAb that blocks the receipt of chemokine signals by eosinophils, Th2 cells and basophils; and an anti-B7 mAb that preferentially inhibits costimulation of allergen-specific Th2 cells. At the present time, neither the side effects of these types of treatments nor their effectiveness and specificity in patients are known.

B. Type II Hypersensitivity: Direct Antibody-Mediated Cytotoxic Hypersensitivity

I. WHAT IS TYPE II HS?

During a type II HS response, clinical damage is sustained when antibodies bind directly to antigens on the surfaces of cells and induce their lysis. These antibodies, which are sometimes referred to as "pathological" antibodies because they harm the individual, are mainly of the IgM or IgG isotype. In some cases of type II HS, the pathological antibodies attack leukocytes or RBCs, which are mobile cells. In other cases, the antibodies bind to cells that are "fixed" as part of a solid tissue. The antigen recognized may be a foreign entity that has become "stuck" in some way on the surface of a mobile or fixed cell, or may be an autoantigen. In the latter case, the pathological antibodies are *autoantibodies* that, due to a failure in tolerance mechanisms, are free in the periphery to bind to self epitopes. The type II HS reaction that ensues in this situation is then manifested as part of an autoimmune disease (see Ch. 19). Examples of type II HS include some forms of anemia, blood transfusion reactions, reactions to certain drugs, some platelet disorders, and some types of tissue transplant rejection.

II. MECHANISMS UNDERLYING TYPE II HS

The mechanisms of cell lysis and tissue damage involved in type II HS are the same as those triggered when IgG or IgM antibodies bind to pathogens. Examples of type II HS reactions against mobile or fixed cells are shown in Figure 18-6A and 18-6B, respectively. Panel A depicts a case of autoimmune hemolytic anemia (see later), in which an individual produces IgM or IgG antibodies that recognize a surface antigen on his/her own RBCs (#1). Within the blood vessel, the binding of the antibody activates the classical complement pathway and initiates the deposition of C3b on the RBC surface (#2). In some cases, the cascade proceeds to MAC formation and RBC lysis (#3). Alternatively, RBCs complexed to IgM can travel to the liver where destruction is mediated by resident Kuppfer cells via C3b-opsonized phagocytosis (#4). If the RBCs are bound by IgG antibodies, they are usually destroyed in the spleen by resident macrophages employing FcγR-mediated ADCC (#5).

The case of a type II HS response to an antigen fixed in a tissue is illustrated for a kidney in Figure 18-6 (panel B). Antibodies recognizing an antigen present on individual kidney cells bind to it and initiate classical complement activation, leading to the establishment of a chemokine gradient and C3b deposition on the kidney cells (#1). In some cases, this C3b deposition leads to direct kidney cell destruction via MAC formation (#2). In other cases, the C3b deposited on the kidney cells is bound by the CR1 receptors of neutrophils drawn to the site by the chemokine gradient. However, because the kidney cells are "fixed" in the tissue and cannot be engulfed, the neutrophils experience "frustrated phagocytosis" that causes them to release their cytotoxic contents toward the kidney cells (#3). Alternatively, the antibody-bound kidney cells can be damaged by either neutrophils or NK cells employing FcR-mediated ADCC (#4). Sometimes the chemokine gradient established by complement activation also recruits mast cells and basophils to the site. These cells can be triggered to degranulate (in an IgE-independent manner) and may contribute to the tissue damage.

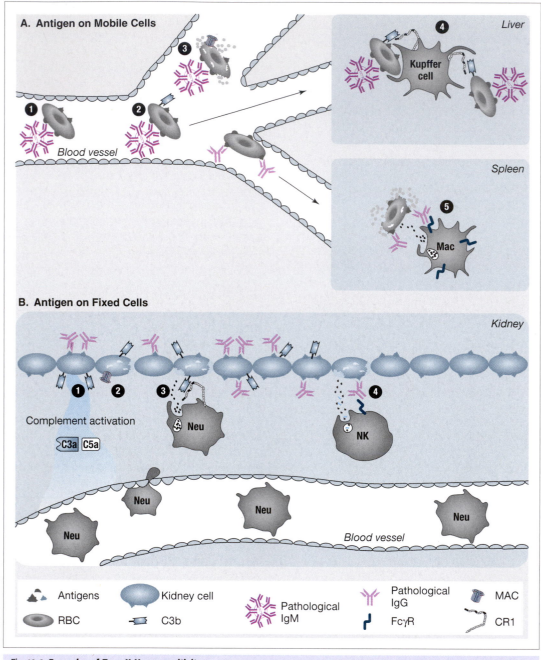

A. Antigen on Mobile Cells

Blood vessel

Liver

Kupffer cell

Spleen

Mac

B. Antigen on Fixed Cells

Kidney

Complement activation

C3a C5a

Neu

NK

Neu

Neu

Neu

Neu

Blood vessel

Antigens	Kidney cell	Pathological IgM	Pathological IgG	MAC
RBC	C3b		FcγR	CR1

Fig. 18-6 **Examples of Type II Hypersensitivity**

III. EXAMPLES OF TYPE II HS

i) Hemolytic Anemias

Hemolytic anemia is the lytic destruction of RBCs caused by a pathological antibody in a type II HS reaction. The destruction may occur within the blood vessels or within the spleen or liver, and may be due to autoantibodies or alloantibodies.

ia) Autoimmune hemolytic anemias. As mentioned before, autoimmune hemolytic anemias occur when the individual

makes antibodies directed against epitopes on his or her own RBCs (refer to Fig. 18-6A). The autoantibodies involved are classified as being either "warm" or "cold" depending on the temperature at which they show optimal reactivity when tested in the laboratory. Warm autoantibodies are most potent around 37°C and have reduced effects at lower temperatures, whereas cold autoantibodies bind effectively only below 37°C. Most autoimmune hemolytic anemias are due to warm auto-antibodies. In acute onset cases, the anemia strikes in a sudden, potentially life-threatening way. Treatment of warm autoim-mune hemolytic anemias usually involves glucocorticoids to inhibit the inflammatory response. For patients with cold auto-antibodies, the onset of winter may be enough to trigger episodes of acute hemolysis. The extremities of these patients feel cold and turn blue due to the lack of RBCs available to transport oxygen to these tissues. The primary therapy for these people is to avoid exposure to cold temperatures.

ib) Alloimmune hemolytic anemias.

Alloimmune hemolytic anemias occur when an individual has circulating antibodies directed against foreign RBC antigens and is then exposed to the corresponding allogeneic RBCs. The interaction of the preformed antibodies with the allogeneic RBCs may induce hemolysis. The resulting clinical symptoms constitute a type II HS reaction. Two of the best known examples of type II HS are transfusion reactions and Rh disease (refer to Ch. 17 and 14, respectively). Transfusion reactions occur when the incoming donor RBCs encounter naturally occurring anti-ABO antibodies in the recipient. Rh disease occurs when Rh$^+$ RBCs of a fetus are destroyed by maternal anti-Rh antibodies that were induced by a previous Rh$^+$ pregnancy.

ii) Thrombocytopenia

A patient with *thrombocytopenia* has an abnormally low number of platelets in the blood and thus exhibits impaired blood clotting. Thrombocytopenia is the most common cause of abnormal bleeding and is caused by decreased platelet production, increased platelet destruction, or abnormal distribution of platelets within the body. Type II HS reactions mediated by anti-platelet antibodies can contribute to increased platelet destruction and thus cause immune system-mediated thrombocytopenia. A prominent clinical feature of thrombocytopenias is the development of *purpura*, areas of purplish discoloration on the skin caused by leakage of blood into the skin layers. As with hemolytic anemias, type II HS thrombocytopenias may be either autoimmune or alloimmune in nature.

iia) Autoimmune thrombocytopenias.

Autoimmune thrombocytopenias are caused by an autoantibody attack on an individual's own platelets and may be acute or chronic. The disorder is mediated most often by IgG antibodies (less often by IgM). In both cases, the binding of autoantibodies directed against platelet antigens triggers platelet destruction via phagocytosis. Acute autoimmune thrombocytopenia is mainly a disease of children and young adults. Curiously, the incidence of this form of the disorder follows that of viral infections and so peaks in the winter. Symptoms are usually rapidly resolved without intervention. Chronic autoimmune thrombocytopenia affects mainly adults. This disorder causes symptoms that are similar to those of acute autoimmune thrombocytopenia but last longer than 6 months. Patients experience sporadic bleeding, with each episode lasting days or weeks. Chronic autoimmune thrombocytopenia is not associated with prior viral infections. Steroids are the main therapy when required.

iib) Alloimmune thrombocytopenias.

A good example of an alloimmune thrombocytopenia is *neonatal thrombocytopenia*, a rare disorder mediated by maternal alloantibodies specific for the platelet surface antigen PLA1. A pregnant, PLA1$^-$ woman becomes sensitized by the PLA1$^+$ fetus she is carrying and produces anti-PLA1 IgG antibodies that cross the placenta and destroy fetal platelets. The neonate then suffers from thrombocytopenia after birth and develops purpura. The principal danger of this disease lies in the potential for intracranial bleeding affecting the neonatal brain. Standard therapy for neonatal thrombocytopenia usually involves giving the newborn a platelet transfusion as well as corticosteroids to reduce inflammation.

Another alloimmune thrombocytopenia is *post-transfusion purpura*, which occurs when PLA1$^-$ individuals receive PLA1$^+$ platelets during a blood transfusion. Although some PLA1$^-$ individuals naturally harbor antibodies recognizing PLA1, post-transfusion purpura occurs most often in PLA1$^-$ patients who have been sensitized to PLA1 during prior pregnancies or transfusions. The patient's anti-PLA1 antibodies initiate platelet destruction, resulting in the sudden onset of thrombocytopenia. Luckily, this disease is usually self-limited but intracranial hemorrhage can be a concern. Therapy may involve steroids or plasmapheresis to remove the pathological antibodies.

iii) Antibody-mediated Rejection of Solid Tissue Transplants

An example of type II HS against a fixed cellular target is the hyperacute graft rejection (HAR; refer to Ch. 17). HAR occurs within minutes or hours of organ transplantation when the recipient has pre-existing alloantibodies directed against MHC molecules expressed on cells of a donated organ. These antibodies are usually present because of a previous pregnancy, organ or bone marrow transplant, or blood transfusion.

iv) Goodpasture's Syndrome

Goodpasture's syndrome is an autoimmune disease caused by autoantibodies that recognize a collagen protein found in the basement membranes of the glomeruli in the kidney and the alveoli in the lungs. The autoantibodies trigger classical complement activation that damages epithelial and endothelial cells in the target organs, causing lung hemorrhage and inflammation of the renal glomeruli. Patients present with transient kidney dysfunction, bleeding in the lungs, and blood in the sputum and urine. Although permanent lung damage is rare, the damage to the kidneys can be severe and long-lasting and may result in renal failure if left untreated. The worst cases of Goodpasture's syndrome are fatal due to lung hemorrhage and respiratory failure. In Plate 18-3, fluorescently labeled anti-IgG antibodies have been used to highlight the typical linear deposits of anti-collagen IgG along the basement membrane of the renal glomerulus in a Goodpasture's syndrome patient. Therapy for Goodpasture's syndrome usually involves a combination of plasmapheresis, corticosteroids and immunosuppressive drugs.

v) Pemphigus

Pemphigus is an autoimmune disease characterized by potentially fatal blistering of the skin and mucosae that promotes dehydration and infection. This disorder is caused by autoantibodies (usually IgG or IgA) that attack adhesion proteins

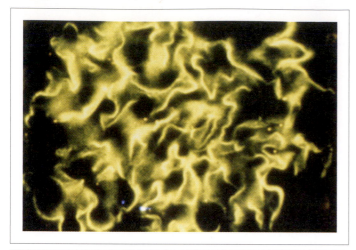

Plate 18-3 **Immunofluorescent Detection of Anti-Collagen Antibodies in the Kidney in Goodpasture's Syndrome**
[Reproduced by permission of William G. Couser, University of Washington, and Michael P. Madaio, Temple University School of Medicine.]

called *desmogleins*. Desmogleins "glue" keratinocytes together to form intact upper epidermal layers, and do the same for mucosal epithelial cells to form mucosae. Autoantibody binding to desmogleins not only induces separation of the epidermal or mucosal layers but also allows the release of a protease that causes blisters (Plate 18-4). These blisters are exceedingly painful and just touching affected skin can cause it to peel off, leaving the individual vulnerable to infections. Patients are usually middle-aged or elderly and corticosteroid administration is the standard treatment.

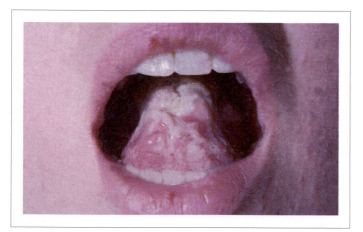

Plate 18-4 **Mouth Blisters in Pemphigus**
[Reproduced by permission of Vijay Chaddah, Grey Bruce Health Services, Owen Sound, Ontario.]

C. Type III Hypersensitivity: Immune Complex-Mediated Hypersensitivity

I. WHAT IS TYPE III HS?

During a normal humoral response against a soluble antigen, molecules of antigen and specific antibody bind together to form complexes. These complexes remain small and soluble because complement component C1q binds to the Fc regions of the participating antibodies. This binding has two effects: (1) the classical complement cascade is triggered and leads to clearance of the antibody-bound antigen by phagocytes, and (2) the deposition of C3b interferes with the growth of the antigen–antibody lattice structure. However, inefficient removal or unchecked expansion of these complexes can allow the antigen–antibody lattice to become a large and insoluble *immune complex* (IC). Because such ICs are too large to phagocytose and clear from the blood, they often become lodged in narrow channels in the body and provoke an immune response that collaterally damages surrounding cells. The clinical outcome of this response depends on the site of IC deposition but commonly involves inflammation of the renal glomeruli (glomerulonephritis), blood vessels (vasculitis) or joints (arthritis). The detrimental immune responses underlying these conditions are known as type III HS reactions.

Many type III HS reactions are clinical complications of infections with certain pathogens, such as those causing meningitis, malaria or hepatitis. A bacterium, parasite or virus may supply large amounts of an antigen that persists after the individual's immune system has dealt with the actual pathogen. Circulating antibodies may bind to these persistent antigens and form large ICs that are deposited in various tissues, causing symptoms that are distinct from those due to the pathogen itself. Some drug "allergies" may also be due to type III HS reactions. In these cases, the symptoms persist as long as the drug the individual is taking induces and binds to the pathological antibodies. Type III HS is also often found in patients expressing autoantibodies. The autoantibodies combine with soluble autoantigens, which might be proteins, glycoproteins or even DNA, that are naturally and abundantly present at all times. Finally, in rare cases, a cancer patient may make antibodies to tumor antigens shed into the blood by cancer cells. Unless the tumor resolves naturally or is forced to do so by medical intervention, exposure to such antigens is continuous and relatively long term. ICs that form between the antibodies and the tumor antigens may cause type III HS symptoms.

II. MECHANISM UNDERLYING TYPE III HS

The mechanism by which ICs cause type III HS is illustrated in Figure 18-7. Antigen–antibody pairs in the blood (#1) cross-link to form insoluble ICs that "get stuck" in a narrow body channel such as a capillary (#2). The presence of the ICs induces inflammation and activates complement, coating the ICs and nearby endothelial cells in C3b. Endothelial cells are damaged by MAC formation (#3) such that the endothelial layer becomes leaky, allowing the ICs to penetrate into the underlying tissue (#4). The presence of the ICs in the tissue triggers additional complement activation that coats tissue cells in C3b (#5), precipitating their destruction by MAC for-

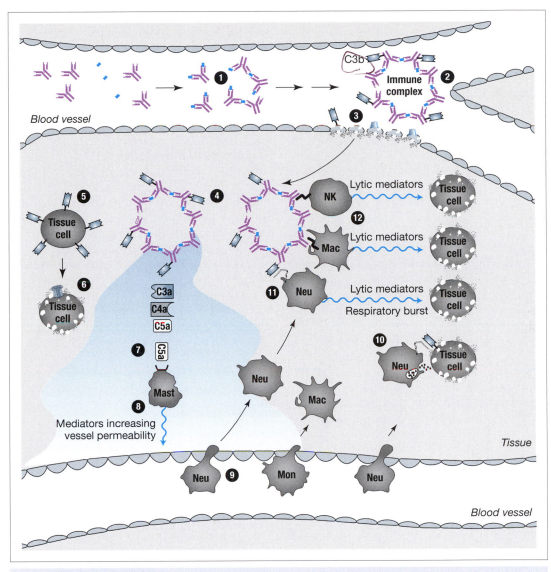

Fig. 18-7 **Type III Hypersensitivity**

mation (#6). Complement activation also generates anaphylatoxins that summon mast cells from the surrounding tissue to the site of IC deposition (#7). The mast cells are activated either by anaphylatoxins binding to complement receptors or ICs binding to FcRs. These mast cells release mediators that, along with the complement-derived anaphylatoxins, increase the permeability of the local blood vessel (#8) and facilitate the extravasation of neutrophils and other leukocytes (#9). Tissue cells coated in C3b are then destroyed by the degranulation of CR1-bearing neutrophils and macrophages that experience frustrated phagocytosis (#10). Similar damage is done to bystander cells via cytotoxic molecules released by neutrophils that bind to ICs via C3b–CR1 interactions (#11). Finally, neutrophils, macrophages and NK cells that bind to ICs via their FcRs release lytic mediators that collaterally kill tissue cells (#12).

III. EXAMPLES OF TYPE III HS

Type III HS reactions are classified by whether the inflammation they induce appears locally or systemically.

i) Localized Type III HS

In a localized type III HS reaction, the ICs involved are deposited only where antibody and antigen first encounter one another during the effector phase; the ICs do not get a chance to circulate systemically. Only local tissue damage characterized by pain, redness, and swelling is observed in what clinicians call an *Arthus reaction*. This reaction typically takes the form of localized vasculitis due to deposition of ICs in dermal blood vessels. Localized type III HS reactions in patients are comparatively rare, although researchers sometimes induce Arthus reactions in animals to study the mechanisms underlying this type of HS.

ii) Systemic Type III HS

In a systemic type III HS reaction, symptoms may appear in sites that are far removed from the original site of antigen–antibody contact, and multiple sites may be affected simultaneously. The parts of the body involved tend to be those that inadvertently trap large ICs circulating in the body. Joints, capillaries and the renal glomeruli are particularly susceptible to IC deposition, such that systemic type III HS disease often involves some combination of arthritis, vasculitis and glomerulonephritis.

Systemic type III HS is frequently associated with repeated exposure to an environmental or drug antigen. Persistent exposure to the offending antigen can cause the IC deposition and the inflammation to become chronic. Treatment of chronic type III HS caused by an environmental or drug antigen requires the identification of the offending antigen and limitation of the patient's exposure to that substance.

As well as being provoked by foreign antigens, systemic type III HS can result from the accumulation of ICs involving autoantibodies recognizing soluble self antigens. In most of these cases, the type III HS reaction is just one manifestation of a complex autoimmune disease (see Ch. 19). For example, patients with *systemic lupus erythematosus* (SLE) produce many different types of autoantibodies, including those directed against DNA, nucleoproteins, cytoplasmic antigens, leukocyte antigens, and clotting factors. Similarly, in *rheumatoid arthritis* (RA), autoantibodies to the patient's own IgG molecules can be found in the circulation. In both SLE and RA, the deposition of ICs occurs systemically so that, along with their other symptoms, patients experience joint inflammation, vasculitis and, in the case of SLE, kidney disease.

D. Type IV Hypersensitivity: Delayed-Type or Cell-Mediated Hypersensitivity

I. WHAT IS TYPE IV HS?

Type IV HS reactions do not occur until about 24–72 hours after exposure of a sensitized individual to the antigen (hence, "delayed-type" hypersensitivity or DTH). Whereas type I, II and III HS are all antibody-mediated, type IV HS results primarily from the tissue-damaging actions of effector T cells and macrophages. The delay in this type of HS is due to the time required for T cell activation and differentiation, cytokine and chemokine secretion, and for the accumulation of macrophages and other leukocytes at the site of exposure. Although type IV HS was originally defined as involving the responses of Th cells and hyperactivated macrophages, this definition has now expanded to include the tissue damage caused by CTLs and other cell types. Common examples of type IV HS include *chronic DTH reactions, contact hypersensitivity, and hypersensitivity pneumonitis,* all of which are described below. In addition, the cell-mediated responses to autoantigens in certain autoimmune diseases constitute a type IV HS reaction. Finally,

some clinicians consider certain forms of chronic graft rejection to be type IV HS reactions because an ongoing cell-mediated response causes immunopathological damage to the transplant recipient.

II. EXAMPLES OF TYPE IV HS AND THEIR MECHANISMS

i) Chronic DTH Reactions

Chronic DTH reactions are initiated by antigens derived from agents that are unusually resistant to elimination by the immune system. Such agents include persistent intracellular pathogens (e.g., those causing tuberculosis, leprosy, leishmaniasis), certain non-infectious agents (e.g., in silicosis, berylliosis), and some unknown agents (e.g., in Crohn's disease, sarcoidosis). An example of how a chronic DTH reaction to a persistent pathogen develops is shown in Figure 18-8. In this illustration, a pathogen has penetrated the skin of a sensitized individual (#1) and has infected a host cell. Pathogen antigens released by the infected cell are taken up by an APC (#2), processed, and presented to memory Th cells (#3). At the same time, macrophages activated by the pathogen produce IL-12 and IL-18 (#4) that promote the differentiation of effector Th1 cells that are called, in this context, T_{DTH} cells (#5). The cytokines secreted by activated macrophages also stimulate NK cells to secrete large amounts of IFNγ (#6), which acts on the macrophages to further upregulate their production of IL-12. The production of T_{DTH} cells is thus sustained. In the presence of persistent antigen, the T_{DTH} cells produce IFNγ and other cytokines and chemokines that recruit and activate additional macrophages in the site (#7). Hypersensitivity arises when the macrophages (and sometimes NK cells) relentlessly secrete pro-inflammatory cytokines that damage host keratinocytes (#8). Clinically, the skin over the site of antigen entrenchment becomes red and inflamed. As the response persists, the macrophages may become hyperactivated and initiate granuloma formation around entities that cannot be eliminated (#9). If a granuloma forms in an organ such as the liver or a lung, it can cause a lesion that may interfere with that organ's function, leading to serious liver disease or respiratory failure. Corticosteroid treatment is most often used to treat chronic DTH reactions.

ii) Contact Hypersensitivity

Contact hypersensitivity (CHS), sometimes called "contact dermatitis", is a secondary immune response to a small, chemically reactive molecule that has bound covalently to self proteins in the uppermost layers of the skin. Examples of CHS include the patchy rash and intense itching that follow a plunge into a patch of poison oak or poison ivy, and the local skin irritations experienced by individuals sensitive to drugs, metals, cosmetics, or industrial or natural chemicals. The alteration of self proteins by the binding of a CHS antigen present in these substances generates a "non-self" entity that can be thought of as a *neo-antigen* ("new" antigen). Some CHS neo-antigens are created when the chemically reactive molecule oxidizes self

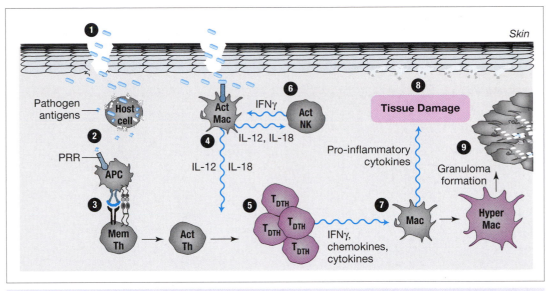

Fig. 18-8 Type IV Hypersensitivity: Chronic DTH Reaction

proteins. In other cases, a metal may form a stable metallo-protein complex that is particularly provocative to macro-phages. Sometimes a chemical will have to be metabolized in the liver before the reactive component is available for neo-antigen formation. Genetic polymorphisms in the enzymes responsible for hepatic metabolism may thus predispose an individual to developing CHS.

CHS is illustrated in Figure 18-9, in which a misguided sensitized individual has touched a poison ivy plant (#1). Poison ivy contains a CHS antigen called urushiol in its roots, stems and leaves. Upon contact with human skin, urushiol molecules penetrate the protective keratinocyte layers and bind covalently to reactive groups present on local self proteins, forming a neo-antigen (#2). The presence of the neo-antigen induces skin cells to release numerous cytokines and chemo-kines (#3) that draw leukocytes from the circulation (#4) into the affected area of the skin and activate resident macrophages to secrete IFNγ (#5). Neo-antigen shed by skin cells is taken up by LCs (#6) and peptides derived from it are cross-pre-sented to memory Tc cells that were generated during the sensitization stage. (Th cells are involved in the sensitization stage but do not appear to play a significant role in the effector stage of type IV HS.) CTL effectors are generated that damage skin cells displaying neo-antigen-derived pMHCs (#7). The CTL attack is facilitated by cytokine-mediated upregulation of adhesion molecules on keratinocyte surfaces. Activated CTLs also contribute to the high levels of IFNγ in the local milieu. IFNγ upregulates Fas expression on keratinocytes, making these cells vulnerable to apoptotic destruction by CTLs bearing FasL. IFNγ also stimulates the degranulation of mast cells and basophils (#8) and thus the release of vasodilators, chemokines and lytic mediators (such as TNF and proteases). These mole-cules collectively damage the skin and increase the access of

leukocytes to the site, facilitating the T cell attack. Mast cells appear to play a major role in CHS, since natural mouse mutants lacking mast cells show impaired CHS reactions. The primary mode of treatment of CHS is avoidance of the inciting antigen. If exposure does occur, corticosteroid cream may be applied to the affected area.

iii) Hypersensitivity Pneumonitis

Hypersensitivity pneumonitis (HP) is a type IV HS reaction in the lung caused by prolonged exposure to an inhaled antigen. A wide range of antigens (including certain chemicals and microbial components) can trigger HP, but the resulting response in the lung usually takes the same form and occurs in three distinct clinical phases: the acute phase, the subacute phase and the chronic phase. When a sensitized individual inhales the offending antigen, macrophages are activated in the lungs either because these cells ingest the antigenic particles directly or because of complement activation in the local envi-ronment. Within 48 hours, chemokines secreted by the mac-rophages promote an influx of neutrophils followed by T cells. Memory Th cells commence differentiation into Th1 effectors secreting copious cytokines. At this stage, the affected indi-vidual likely experiences influenza-like symptoms that resolve quickly if exposure to the antigen ends. If a formal diagnosis of HP is made, corticosteroids may be given to dampen the inflammatory response. However, if the acute stage of HP goes unrecognized and exposure to the antigen continues, the sub-acute phase ensues during which the macrophages become hyperactivated and initiate the formation of a granuloma around the offending antigen in the lung. Overt symptoms are rare for several days or weeks but fatigue and cough then reappear and progress steadily. If there is still no clinical inter-vention and exposure continues, the chronic phase of HP sets

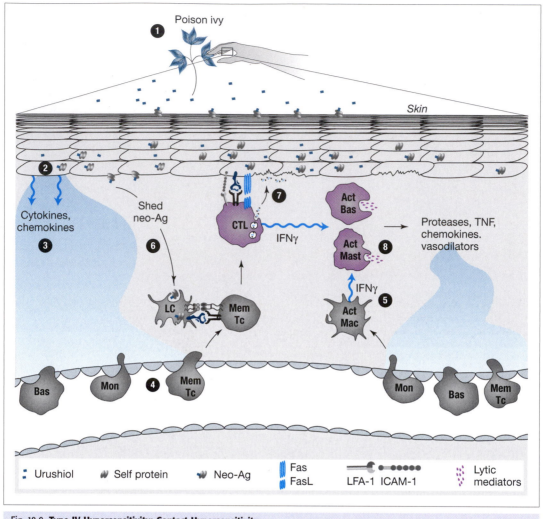

Fig. 18-9 Type IV Hypersensitivity: Contact Hypersensitivity

in and the lung tissue is damaged by the same mechanisms as operate in chronic DTH responses. Activated macrophages in the lungs also secrete large amounts of TGFβ that encourage fibrosis. At this point, the damage to the lungs may be irreversible and eventually fatal despite corticosteroid treatment.

The antigens causing HP are generally proteins derived from microbes, fungi, plants or animals. In the past, exposure to triggering antigens occurred mainly in the workplace, giving rise to afflictions such as "farmer's lung" and "cheese washer's lung". In modern times, however, most HP cases arise from non-occupational exposure to inhaled antigens. Thus, HP is now associated with disorders like "sauna-taker's disease" and "hot tub lung". However, only a small proportion of

individuals exposed to a given HP-inducing antigen actually develop HP. Those who are disease-free mount a harmless IgG response to inhaled antigens with no cellular involvement. In some instances, individuals with HP recall having had an acute respiratory infection just prior to the onset of symptoms, but the factors that control HP susceptibility remain obscure.

This concludes our description of immune hypersensitivity. We move now to a discussion of autoimmune disease, one of the great mysteries of immunology. It is still unknown why the immune system sometimes attacks an individual's own tissues and how such attacks can best be prevented.

- The development of hypersensitivity occurs in two stages: the sensitization stage and the effector stage. The sensitization stage is a primary immune response to an antigen and has no clinical consequences. The effector stage is a secondary immune response that is deleterious to the individual.

- Type I hypersensitivity (allergy or atopy) is mediated by IgE antibodies specific for antigens that are normally non-pathogenic (allergens).

- In the sensitization stage of type I HS, a first exposure to allergen triggers the production of allergen-specific IgE that binds to FcεRs of mast cells and basophils. In the early phase of the effector stage, re-exposure of the sensitized mast cells to the allergen triggers the immediate release of preformed mediators via degranulation as well as new synthesis of inflammatory molecules. Eosinophils, sensitized basophils and other leukocytes drawn to the site mediate the late phase of the effector stage via granule release and the production of additional mediators.

- The major mediators characteristic of type I HS induce vasodilation, increased blood vessel permeability, smooth muscle contraction and leukocyte recruitment.

- Local atopic responses can result in dermatitis, urticaria, asthma or rhinitis.

- Anaphylaxis is a systemic atopic response that can be life-threatening and requires immediate epinephrine administration.

- Type II hypersensitivity is mediated by direct antibody-mediated cytotoxicity. IgG or IgM antibodies bind to antigenic epitopes on fixed or mobile cells and trigger ADCC, complement activation and/or phagocytosis.

- Type III hypersensitivity occurs when soluble antigen binds to IgG or IgM to form large insoluble immune complexes. Deposition of these complexes in narrow body channels triggers inflammatory responses that damage underlying tissues.

- Type IV hypersensitivity is mediated by T cells and macrophages that infiltrate a site of antigen exposure and induce a delayed form of inflammatory tissue damage.

Introduction

1) Define "immune hypersensitivity".

2) Name the four types of immune HS.

3) All immune HS develop in what two stages?

Section A.I–II

1) Give four names used to refer to type I HS.

2) Describe the sequence of events making up the sensitization stage of type I HS.

3) What difference in cytokine production by atopic and non-atopic persons is relevant to type I HS?

4) Distinguish between the two phases of the effector stage of type I HS.

5) Give three examples each of preformed and newly generated mediators relevant to type I HS.

6) Describe three effects of mediator release that lead to allergic symptoms.

7) What contribution do eosinophils make to type I HS?

Section A.III

1) Distinguish between localized and systemic atopy.

2) What are the clinical names for the following? hives, eczema, hay fever.

3) What is a "wheal and flare reaction" and what causes it to develop?

4) Distinguish between atopic dermatitis and atopic urticaria.

5) In a person with a food allergy, which body sites may show atopic symptoms? What are these symptoms?

Section A.IV

1) What three factors may combine to determine whether a person will have an atopic response to an allergen?

2) Give three examples of genes associated with atopy and discuss why they may be relevant.

3) Outline the hygiene hypothesis.

4) Give three examples of triggering events with respect to atopic responses.

Section A.V

1) Give four examples of commonly used allergy therapies and how they work.

2) Give four examples of experimental allergy therapies and how they might work.

Section B

1) What is a pathological antibody and how does it cause type II HS? Is IgE relevant?

2) What cytolytic mechanisms are operating in type II HS responses against mobile cells? Against fixed cells?

3) Give two examples each of autoimmune and alloimmune type II HS disorders and describe the epitopes targeted by the pathological antibodies.

4) Why is hyperacute graft rejection considered to be a case of type II HS?

5) What protein is targeted in Goodpasture's syndrome? In pemphigus?

Section C

1) What is an immune complex?

2) How does the triggering of type III HS differ from that of types I and II HS?

3) Give three sources of antigen that might trigger type III HS.

4) What is an Arthus reaction?

5) Give two examples of a systemic type III HS disorder.

Section D

1) What is the main difference between type IV HS and types I–III?

2) Why are type IV HS responses comparatively delayed in onset?

3) Outline the mechanism underlying chronic DTH reactions.

4) Outline the mechanism underlying contact HS.

5) Outline the mechanism underlying hypersensitivity pneumonitis.

19

Autoimmune Diseases

We have failed to grasp the fact that mankind is becoming a single unit, and that for a unit to fight against itself is suicide.
Havelock Ellis

A. What Is an Autoimmune Disease?

In the early days of immunological studies, scientists believed that the immune system would simply be incapable of mounting anti-self immune responses because they would harm the body. In 1900, Paul Ehrlich labeled this concept as "horror autotoxicus", meaning that the immune system should have a "horror of" (and therefore avoid) being "autotoxic". However, even at that time, there were isolated reports suggesting that autoreactivity could definitely occur. Clinical and experimental evidence for this phenomenon slowly accumulated in the 1950s, and by the 1960s, it was generally accepted that several diseases, including *systemic lupus erythematosus* (SLE) and *multiple sclerosis* (MS), resulted from "autoimmunity". It has since become clear that autoimmunity itself is not unusual and is not the same as "autoimmune disease". In fact, every healthy person is autoimmune to a limited degree, as demonstrated by the low levels of anti-self antibodies that can be found in every individual. The concentrations of these autoantibodies can even increase as a consequence of inflammation or infection without the individual experiencing any clinical effects. Autoimmune disease arises only when autoimmunity causes clinical damage to a self tissue. Currently, there are over 80 different autoimmune diseases that vary widely in incidence and phenotype.

The clinical symptoms associated with autoimmune disease are considered immunological in origin because they are due to the activation and effector functions of autoreactive lymphocytes. Autoreactive Th cells respond to activation by releasing cytotoxic cytokines such as IFNγ, TNF and IL-17. Th cells also supply T help for autoantibody production by autoreactive B cells and for CTL generation by autoreactive Tc cells. Sometimes autoantibodies cause autoimmune disease by directly deregulating or disrupting the function of the *target tissue* expressing the specific self antigen under attack. For example, in *myasthenia gravis* (MG) and *Graves' disease* (GD),

the binding of autoantibodies to particular cell surface receptors causes dysfunction of the muscles or thyroid gland, respectively. In other cases, such as *Goodpasture's syndrome* (GS) and *pemphigus* (PG), autoantibody binding leads to the destruction of the targeted cells and damage to bystander tissues that is manifested as type II hypersensitivity (refer to Ch. 18). Type III hypersensitivity may also be a component of an autoimmune disease if autoantibodies bind to a self antigen and form large insoluble complexes that travel throughout the body in the circulation. If these complexes lodge in the body's narrow channels, inflammatory tissue damage can be triggered at distant sites where the self antigen is not even expressed. SLE furnishes a good example of this latter type of autoimmune disease.

In addition to the destruction caused by autoantibodies, autoreactive Tc cells receiving help from autoreactive Th cells generate CTLs that kill healthy cells expressing the self antigen. The cytokines released by these CTLs damage bystander cells and help to sustain the activation of the APCs required for autoreactive Th responses. The cell-mediated tissue destruction caused by autoreactive Th effectors and CTLs constitutes a form of type IV hypersensitivity. The autoimmune form of *diabetes mellitus* is considered by many to be an example of such a disorder.

Clinicians have traditionally categorized autoimmune diseases into two broad classes: (1) *organ-specific* autoimmunity, in which a particular anatomical site is targeted for immune destruction, and (2) *systemic* autoimmunity, in which the immune response is not restricted to a particular organ or tissue. However, this categorization is based on the clinical manifestations of the disease rather than on the expression pattern of the targeted self antigen. In some cases, an antigen may be ubiquitously expressed but the autoimmune response to it may occur only in one organ. Why the attack is limited in this way is unknown.

B. Mechanisms Underlying Autoimmune Diseases

How does an autoimmune disease arise? Negative selection (refer to Ch. 5 and 9) during lymphocyte development removes the vast majority of autoreactive lymphocyte clones during the establishment of central tolerance. Any autoreactive clones that do escape to the periphery are normally controlled by the mechanisms of peripheral tolerance (refer to Ch. 10). Thus, as summarized in Figure 19-1, four things must happen for an autoimmune disease to develop: (1) an autoreactive clone in the thymus must escape elimination by central tolerance mechanisms and be released to the periphery; (2) the escaped autoreactive clone must encounter its specific self antigen in the periphery; (3) the peripheral tolerance mechanisms designed to regulate autoreactive lymphocyte responses must fail; and (4) the response by the autoreactive clone must result in clinical damage.

Rarely, an autoimmune disease is caused by a defect in central tolerance mechanisms. For example, *autoimmune polyendocrinopathy candidiasis ectodermal dystrophy* (APECED) is a disease that results from mutations to a transcriptional regulator called *autoimmune regulator* (AIRE). In the absence of AIRE function, there is a defect in the expression of certain self antigens during negative selection in the thymus. As a result, T cells bearing TCRs specific for these self antigens cannot be eliminated and are released to the periphery where they attack multiple tissues.

Most cases of autoimmune disease occur due to inadequacies in peripheral tolerance. In the sections that follow, we describe several mechanisms that are believed to contribute to the activation of autoreactive lymphocytes in the periphery and thus set the stage for the onset of autoimmune disease. These mechanisms, which may not act in a mutually exclusive way, include: inflammation; molecular mimicry by pathogen antigens; inherent defects in regulatory T cells, APCs, B cells, conventional T cells, cytokines or complement; and epitope spreading.

I. INFLAMMATION

Many episodes of autoimmune disease appear to occur soon after infection with a pathogen. One school of thought is that the inflammatory milieu that arises in the course of a pathogen infection "breaks" peripheral tolerance. As described in Chapter 10, there are two key mechanisms that contribute to the maintenance of peripheral tolerance. Firstly, if an autoreactive T cell binds to its cognate pMHC presented by a tolerogenic DC, the autoreactive T cell is anergized due to a lack of costimulation and no longer responds to the antigen. Secondly, any autoreactive T cells that are not anergized and attempt to initiate activation are soon suppressed by regulatory T cells. Both of these measures may be undermined when pathogen invasion generates a local inflammatory environment. At the mechanistic level, it is believed that the DNA and components of infecting microbes plus any host DNA released upon pathogen-induced cell death constitute "danger signals" that can engage the TLRs of immature DCs that have accumulated in the target tissue. These DCs then commence maturation, upregulate costimulatory molecules, and migrate to the local lymph node where they display pMHCs derived from both pathogen antigens and self antigens to naïve T cells, including to autoreactive T cells. In the inflammatory environment, a naïve autoreactive T cell may be activated because the DC it encounters is mature. In addition, the large amounts of pro-inflammatory cytokines secreted by the mature DC may allow an autoreactive T cell to escape the control of regulatory T cells and to become activated. In any case, T cell effectors are generated that migrate back to the target tissue expressing the self antigen. These effectors then destroy tissue cells either directly or by facilitating autoantibody production. Moreover, because the self antigen is continuously present as part of a tissue, the immune response cannot mop up the antigen in the same the way as it would a pathogen. The unwanted autoimmune attack is perpetuated and causes an autoimmune disease.

Even in the absence of a pathogen infection, DC maturation contributing to autoimmune disease may occur if cells experience stress or physical trauma. Cells that have become necrotic due to mechanical injury, transformation or other forms of stress frequently release stress molecules with effects on DCs. For example, *in vitro*, HSPs can mimic the effects of an inflammatory milieu on DCs, inducing these cells to produce pro-inflammatory cytokines and upregulate MHC class II and B7 molecules. Again, autoreactive T cells that interact with such DCs and recognize pMHCs presented by these cells may initiate an autoimmune response that leads to clinical symptoms.

Support for the inflammation theory of autoimmune disease initiation is growing. Mature DCs have been found in the damaged tissues of autoimmune patients, although it is not clear whether these DCs arrive in the lesions as mature cells or are induced to mature once they arrive. As well, in some animal models, clinical signs of autoimmune disease do not develop if DCs are prevented from accumulating in the target tissue. Chronic inflammation may also be relevant to the observation that many autoimmune disease patients have a higher risk of developing *lymphomas*, which are solid tumors composed of cancerous lymphocytes (see Ch. 20). The relationship between autoimmunity and cancer is briefly explored in Box 19-1.

- Escape of autoreactive clone from thymus
- Encounter of autoreactive clone with self antigen
- Failure of peripheral tolerance mechanisms
- Infliction of clinical damage by autoreactive clone

Fig. 19-1 Events Required for Autoimmune Disease

For reasons that are still not understood, patients with an autoimmune disease have an increased risk of developing cancer compared to the general population. Lymphomas are the most common malignancy observed in these patients, but leukemias and tumors of the kidney, lung, breast and GI tract also occur. Some of these cancers do not appear until as long as 20 years after the initial diagnosis of the autoimmune disorder. Intriguingly, the converse is also true, in that many cancer patients develop clinical features of autoimmunity (particularly autoantibody production) many years after the initial diagnosis of the tumor.

How could autoimmunity lead to cancer? It is thought that the destructive chronic inflammation associated with many autoimmune diseases may increase the level of carcinogenic molecules (such as ROIs) in the microenvironment. These reactive substances could act on a nearby cell in an early stage of carcinogenesis and push it down the path to malignant conversion. Alternatively, the relentless proliferation of an autoreactive lymphocyte activated by abundant self antigen might provide more opportunities for oncogenic mutations to occur and accumulate.

How might a cancer lead to autoimmunity? Lymphocytes directed against tumor-related antigens can sometimes be found in the blood of cancer patients. Although some of the antigens targeted by these lymphocytes may be TSAs (refer to Ch. 16), others may be TAAs, such as cell cycle control proteins and tumor suppressors. Self tissues expressing these proteins might then come under autoimmune attack. In addition, the physical disruption associated with the growing tumor may expose previously hidden autoantigens that then attract the attention of the immune system.

The therapies used to rid cancer patients of their tumors may also induce the onset of autoimmune disease. In addition, treatments that make a patient more vulnerable to infection might allow the invasion of a pathogen associated with autoimmune disease (see later in this chapter).

II. MOLECULAR MIMICRY BY PATHOGEN ANTIGENS

In addition to triggering inflammation, pathogens may also contribute to the onset of an autoimmune disease via *molecular mimicry*. Molecular mimicry is deemed to occur when a component of a pathogen bears an epitope that resembles an epitope derived from a self antigen. Thus, some T and/or B cells in an individual may bear antigen receptors that recognize both the self epitope and the pathogen epitope. In the absence of infection by the relevant pathogen, such potentially autoreactive lymphocytes are held in check by peripheral tolerance. However, if the pathogen infects the individual, the presentation of a relatively large amount of the cross-reactive pathogen epitope in the inflammatory milieu created by the infection may break peripheral tolerance and trigger activation of the autoreactive lymphocyte. An attack on host tissues expressing the self epitope then ensues. Several examples of pathogen amino acid sequences that resemble regions of self proteins thought to be involved in autoimmune diseases are shown in Figure 19-2.

Both T cell- and B cell-mediated responses can be triggered by molecular mimicry. The first step leading to T cell molecular mimicry is the acquisition of a pathogen or its components by

Autoimmune Disease	Cross-reacting sequences	Origin of peptide
Rheumatic fever	QKMRRDLEE	Human myosin
	KGLRRDLDA	*Streptococcus* cell wall protein
Multiple sclerosis	VVHFFKNIV	Human myelin basic protein
	VYHFVKKHV	Epstein-Barr virus protein
Rat equivalent of multiple sclerosis	YGSLPQKSQRTQDEN	Rat myelin basic protein
	YGCLLPRNPRTEDQN	*Chlamydia pneumoniae* protein
Mouse equivalent of inflammatory bowel disease	NIISDA	Mouse heat shock protein
	NAASIA	Mycobacterium protein

Fig. 19-2 Examples of Peptide Sequences Potentially Involved in Molecular Mimicry
[With information from Rohm A. P. *et al.* (2003). Mimicking the way to autoimmunity: an evolving theory of sequence and structural homology. *Trends in Microbiology* **11**, 101-105; Quinn A. *et al.* (1998). Immunological relationship between the class I epitopes of streptococcal M protein and myosin. *Infection and Immunity* **66**, 4418.]

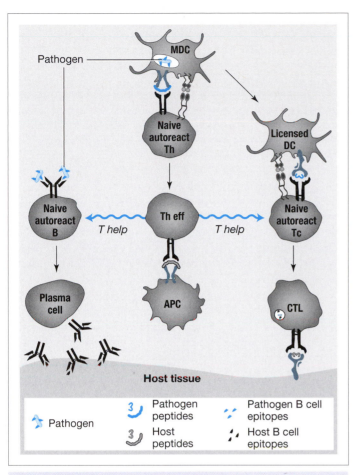

Fig. 19-3 **Cellular Model of Molecular Mimicry**

III. INHERENT DEFECTS IN IMMUNE SYSTEM COMPONENTS

i) Abnormalities in Regulatory T Cells

Because regulatory T cells play a key role in maintaining peripheral tolerance, abnormalities in the numbers or functions of these cells can lead to autoimmune disease. Particularly implicated are the T_{reg} cells that primarily use intercellular contacts to suppress the activation of effector T cells. For example, failure of T_{reg} cells to develop in an animal, or artificial depletion of T_{reg} cells that already exist, allows activated autoreactive T cells to escape normal shutdown and to cause autoimmune symptoms. In humans, some autoimmune disorders have been associated with normal numbers of T_{reg} cells of decreased suppressive activity, or with decreased numbers of T_{reg} cells of normal suppressive activity. A very rare disease called *immune dysregulation, polyendocrinopathy, enteropathy X-linked* (IPEX), which has a significant autoimmune component, is caused by mutations in the Foxp3 gene required for T_{reg} development. In most cases, IPEX patients have an absence or severe deficiency of T_{reg} cells. The same is true for *scurfy* mice, which have a natural mutation of the Foxp3 gene.

ii) Abnormalities in APCs

At least in animals, some autoimmune diseases are associated with aberrations in the generation and/or function of APCs. In one rodent autoimmunity model, the activated macrophages present showed enhanced powers of migration and pro-inflammatory secretion that were linked to increased non-specific damage to self tissues. It was hypothesized that this tissue damage exposed otherwise sequestered self antigens and resulted in the unintended activation of autoreactive lymphocytes. In another rodent autoimmunity model, the abnormal DCs present showed reduced expression of MHC class II and B7. Although this deficit meant that naïve autoreactive T cells were less likely to be activated, it also impeded the activation of the regulatory T cells necessary to block autoimmune disease development. Indeed, the onset of autoimmune symptoms in susceptible rodents is hastened, rather than suppressed, if DCs are treated such that the interaction between B7 and CD28 is blocked.

iii) Abnormalities in B Cells

As described in Chapter 5, the random process of V(D)J recombination during B cell development in the bone marrow naturally generates some B cells with BCRs recognizing self antigens. The majority of these autoreactive B cells are eliminated by negative selection in the bone marrow during the establishment of B cell central tolerance. Autoreactive B cells that escape negative selection undergo receptor editing in a second effort to generate a non-autoreactive BCR. B cells that fail this step, escape to the periphery, and have their BCRs engaged by self antigen usually undergo Fas-mediated apoptosis or are anergized. Abnormalities in the function of any of the proteins associated with these various B cell maturation checkpoints might result in the survival of autoreactive B cells that would normally have been removed. Although evidence

an immature DC and the induction of DC maturation. As illustrated in Figure 19-3, a mature DC (MDC) presents pathogen peptides on MHC class II to naïve Th cells bearing TCRs that recognize not only these pMHC epitopes but also a structurally similar pMHC derived from a self antigen expressed by healthy host cells. The activation of these autoreactive Th cells results in both the licensing of the DC and the generation of Th effector cells that supply help for the activation of naïve autoreactive Tc cells. The activated Tc cells in turn generate CTLs whose effector actions eliminate the pathogen but also damage healthy tissues routinely presenting the self pMHC. In the case of B cell-mediated molecular mimicry, a pathogen may supply B cell epitopes that activate naïve B cells (that receive help from Th effectors). The plasma cell progeny produce antibodies that eliminate the pathogen but may also recognize similar B cell epitopes present on non-infected host cells. Healthy tissues are damaged as the antibodies attack host cells displaying the cross-reactive epitope.

The molecular mimicry theory is supported by the identification of T cells and antibodies that respond to both pathogen and self antigens *in vitro*. However, it remains to be conclusively demonstrated that this type of attack is responsible for clinical disease in patients.

of such abnormalities in human patients is sparse, studies of mouse models have revealed a broad spectrum of mutations affecting B cell survival, development and activation that can give rise to an autoimmune phenotype. For example, SLE-like symptoms develop in mice bearing mutations that decrease Fas expression.

iv) Abnormalities in Conventional T Cells

Abnormalities in T cell signaling and activation can also be associated with autoimmune disease, although the underlying connections are unknown. It may be that these defects cause autoreactive T cells to adopt a "pre-primed" state such that they can be activated more easily than normal T cells.

v) Alterations to Cytokine Expression

Overexpression or deficiency of some cytokines as well as defects in cytokine signaling can lead to autoimmune disease, presumably because these abnormalities can skew the differentiation of Th effector subsets so that inflammatory responses predominate. In animal models, upregulation of IL-12 production by APCs can trigger a susceptible mouse to develop autoimmune disease. It is thought that the IL-12 increases the number of Th1 cells present, and thus the production of IFNγ and IL-2 that drive macrophage activation and consequently TNF-mediated inflammation. Similarly, an overproduction of IL-6, TGFβ and/or IL-23 can skew Th differentiation toward Th17 cells and the secretion of IL-17 and IL-6. These latter cytokines are increased in the inflammatory lesions of human MS and RA patients, and in a mouse model of RA, inhibition of IL-17 or IL-6 blocks the development of autoimmune symptoms. The implication is that overproduction of Th1 or Th17 cytokines, or of any cytokine promoting Th1 or Th17 differentiation, may facilitate the development of autoimmune disease. In addition to those cytokines, some RA patients show increased amounts of IL-15 and chemokines in inflamed tissues, and the IL-12-related cytokine IL-27 has been associated with glomerulonephritis in human SLE patients. The links between these cytokines and autoimmune disease remain under investigation.

The influence of a cytokine can change as an autoimmune disease progresses. For example, some RA patients treated with anti-TNF antibodies to clear TNF from their inflamed joints initially experience clinical relief but later show signs of exacerbated disease. It seems that the same pro-inflammatory cytokines that promote damage to self tissues also trigger regulatory mechanisms that slowly gain control and shut down autoreactive lymphocytes. This observation has important implications for the management of patients treated with therapies that target cytokines.

vi) Defects in the Complement System

Individuals with mutations of various complement components often exhibit systemic autoimmune symptoms. Without the function of the C1, C2 or C4 complement components, levels of C3b are suboptimal and clearance of immune complexes is reduced. Although complexes involving pathogens will form, these are present only transiently because the pathogen is soon cleared from the body by cells of the immune system. In contrast, self antigens by definition are present continuously. Immune complexes composed of self antigen and autoantibody may thus persist and can lead to the chronic presence of large networks of immune complexes that can become trapped in numerous body channels. These complexes then trigger systemic autoimmunity in the form of a type III hypersensitivity reaction.

IV. EPITOPE SPREADING

Epitope spreading is thought by many researchers to drive the progression of autoimmune diseases and to contribute to recurrent flare-ups. As introduced in Chapter 16, epitope spreading is the term used to describe the phenomenon in which the immune system appears to expand its response beyond the immunodominant epitopes first recognized by T and B cells. In the context of autoimmunity, the original self-reactive response may damage tissues such that new epitopes that were originally "cryptic" (hidden) become accessible. These epitopes would previously have been either totally sequestered from the immune system, or processed by DCs in amounts insufficient to activate T cells, so that lymphocytes recognizing these epitopes would not have been deleted during the establishment of central tolerance nor anergized in the periphery. Upon damage-induced exposure, molecules containing the cryptic epitopes become available for the first time for uptake by DCs. In an inflammatory milieu, autoreactive T and B lymphocytes directed against these epitopes are activated and expand the autoreactive attack to additional self tissues where these proteins are expressed. There might then be virtually no limit to the duration of the autoimmune response, because new cryptic epitopes might be continually revealed as tissue destruction proceeded.

Proven cases of epitope spreading leading to autoimmune disease progression in humans are few. However, one example may occur in PG, in which blistering in the mouth almost always precedes blistering in the skin. The autoantibodies causing the initial damage recognize a desmoglein protein called desmoglein-3 in the mouth mucosae. It is not until the attacks on desmoglein-3 expose epitopes on the related skin protein desmoglein-1 that autoantibodies against this latter protein are produced and skin blistering commences. The autoimmune response thus appears to spread from epitopes of desmoglein-3 to include epitopes of desmoglein-1.

There is evidence from mouse models supporting the epitope spreading theory. In mice prone to autoimmune diabetes, the initial damage to the insulin-producing pancreatic islets arises from an autoimmune attack on the islet enzyme *glutamic acid decarboxylase* (GAD). As the disease progresses, however, lymphocytes directed against different islet cell proteins become activated. Similarly, in mouse models of MS, autoreactive responses are mounted against a series of cryptic epitopes in a reproducible sequence that can be attributed to epitope spreading.

C. Examples of Autoimmune Diseases

An alphabetical listing (by acronym) of 20 of the most common autoimmune diseases is presented in Figure 19-4. In this figure, the "Dominant sex" has been indicated because some autoimmune diseases occur predominantly in either males (M) or females (F). "Disease pattern" indicates whether the clinical course of a particular autoimmune disease is considered acute (A) or chronic (C), and whether it involves relapses and remissions (R/R) rather than a steady progression. "Tissues affected" is self-explanatory, while "Self-Ag" indicates the self molecules against which the response is directed (where known). In the sections that follow, we provide more detailed information on ten of the better studied human autoimmune diseases.

I. SYSTEMIC LUPUS ERYTHEMATOSUS (SLE)

SLE is a systemic autoimmune disease that affects the skin, joints, kidney, lung, heart and brain. A pattern of relapse and remission is common, with an unpredictable frequency of flare-ups. SLE patients exhibit a characteristic rash that gives the disease its name: The rash is red in color (erythematous) and is concentrated on the cheeks such that some patients have a "wolfish" appearance (*lupus* is Latin for "wolf") (Plate 19-1). Multiple elements of the immune system may be disrupted in SLE patients such that these individuals are usually vulnerable to opportunistic infections.

A signature feature of SLE is the production of high levels of "antinuclear" autoantibodies, which are directed against double-stranded DNA and small nuclear proteins. Other autoantibodies in SLE patients recognize non-nuclear entities such as IgG, complement components and membrane phospholipids. All these autoantibodies form immune complexes that accumulate first in the blood and eventually in the target tissues, triggering damaging inflammation. The production of these autoantibodies has been linked to abnormalities in B cell development and activation. Increased numbers of B cells at all stages of differentiation can be found in the circulation of SLE patients, and these B cells tend to be more sensitive than normal B cells to the effects of cytokines. SLE patients also have higher levels of serum IL-10 than do normal individuals. IL-10 is known to stimulate B cell proliferation and differentiation.

II. RHEUMATOID ARTHRITIS (RA)

RA is caused by an autoimmune attack on antigens expressed in the synovial tissue and cartilage of the joints. In the early phase of RA, the patient typically experiences morning stiffness in the affected joints. As the RA becomes severe, cartilage destruction and bone erosion may deform the digits (Plate 19-2) and inflammation of the cervical spine may lead to progressive crippling. Activated macrophages and DCs extravasate from the local blood vessel into the joint and produce large quantities of pro-inflammatory cytokines, particularly TNF. Ligaments, tendons and bones may also suffer degradation due to the action of proteases secreted by the activated macrophages. The blood vessels within the inflamed joint soon take on the characteristics of HEVs capable of facilitating the extravasation of lymphocytes. Eventually, low numbers of activated CD4$^+$ Th effectors (including Th17 cells) and CD8$^+$ CTLs infiltrate RA joints and produce cytokines such as TNF and IL-17 that help to perpetuate the inflammation. RA synovial tissues also contain GCs that are "ectopic", meaning that these structures have developed in the wrong tissue. The plasma cells in these abnormal GCs produce autoantibodies directed against antigens in the synovial membrane and cartilage. A distinctive (but not exclusive) feature of RA is the presence in a patient's serum of *rheumatoid factor*. Rheumatoid factor is not a single entity but rather a collection of autoantibodies that are directed against the patient's own IgG molecules.

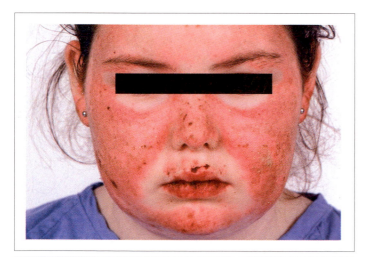

Plate 19-1 **Characteristic Rash in Systemic Lupus Erythematosus**
[Reproduced by permission of Rae Yeung, The Hospital for Sick Children, Toronto.]

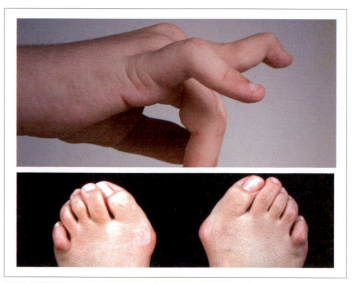

Plate 19-2 **Joint Deformation in Rheumatoid Arthritis**
[Reproduced by permission of Rae Yeung, The Hospital for Sick Children, Toronto.]

	Autoimmune disease	Dominant sex	Disease pattern	Tissue(s) affected	Self antigen (if known)
APS	Anti-phospholipid syndrome	F	C	Blood clots at multiple sites	Glycoproteins of prothrombin activator complex
ARF	Acute rheumatic fever	F ≅ M	C	Heart muscle and valves, kidney, CNS, joints	Myosin protein in heart muscle
AS	Ankylosing spondylitis	M	C	Tendons, bones, ligaments, joints	Fibrocartilage-derived Ag
CD	Crohn's disease	M = F	C R/R	Walls of colon and small intestine	?
GBS	Guillain-Barré syndrome	M = F	A	Peripheral nerves	Neuronal glycolipids and gangliosides
GD	Graves' disease	F	C R/R	Thyroid gland	TSHR and other thyroid gland proteins
GS	Goodpasture's syndrome	M	A	Kidney, lung	Collagen in basement membrane
HT	Hashimoto's thyroiditis	F	C	Thyroid gland	TSHR and other thyroid gland proteins
ITP	Immune thrombocytopenia purpura	M = F (children) F (adults)	A (children) C (adults)	Platelets	Platelet membrane glycoproteins
MG	Myasthenia gravis	F (30–50 yrs) M (70–80 yrs)	C RR	Muscles	Acetylcholine receptors

Fig. 19-4 (Part 1) Examples of Human Autoimmune Diseases

III. ACUTE RHEUMATIC FEVER (ARF)

Acute rheumatic fever primarily affects cells in the heart muscle, heart valves, kidney and CNS but also compromises the joints. ARF follows a relapsing/remitting pattern with major clinical signs of fever, a distinctive rash, carditis, arthritis and neurological effects. Some patients present with very rapid heartbeat or even acute cardiac failure, giving ARF a high mortality rate, particularly in developing countries. In many cases, the autoimmune symptoms of ARF appear 2–6 weeks after infection with certain strains of group A streptococcal bacteria, including those causing "strep throat". These bacteria express a cell wall protein called M antigen that researchers believe resembles an epitope found in the human heart protein *myosin* (refer to Fig. 19-2). Thus, ARF may be established most often by molecular mimicry. In the presence of large quantities of M antigen, autoreactive lymphocytes that are specific for human heart myosin epitopes (and that are normally held in check by peripheral tolerance mechanisms) may break tolerance and become activated by epitopes of the bacterial M antigen. Once activated, these lymphocytes then attack heart tissues as well as streptococci expressing the M antigen epitope. Indeed, antibodies recognizing the M protein are present at high levels in patients with ARF, and subsets of T cells isolated from these patients recognize pMHCs containing M protein peptides *in vitro*. Furthermore, relapses of ARF are often triggered by a subsequent infection with streptococcus.

IV. TYPE 1 DIABETES MELLITUS (T1DM)

Diabetes results when the body either does not produce enough insulin or its cells cannot respond to it normally. As a consequence, glucose cannot get into cells properly and blood sugar levels rise. Symptoms of diabetes include excessive hunger and thirst, increased urination, weight loss, fatigue and blurred vision. If blood glucose levels rise too far, the patient can fall into a life-threatening diabetic coma. When diabetes is controlled inadequately for an extended period, the elevated sugar

Autoimmune disease		Dominant sex	Disease pattern	Tissue(s) affected	Self antigen (if known)
MS	Multiple sclerosis	F	C R/R	Brain, spinal cord	Myelin basic protein, oligodendrocyte proteins
PG	Pemphigus	M=F	C	Mucosae, skin	Desmoglein proteins in skin and mucosae
PM	Polymyositis	F	C R/R	Muscles	Aminoacyl tRNA synthetases, dsDNA, small nuclear proteins
PS	Psoriasis	M = F	C R/R	Skin	?
RA	Rheumatoid arthritis	F	C	Joints, tendons, ligaments, bone	Synovial and cartilage proteins, IgG
SLE	Systemic lupus erythematosus	F	R/R	Skin, joints, kidney, lung, heart, brain	dsDNA, small nuclear proteins, IgG, complement
SS	Sjögren syndrome	F	C	Exocrine glands (lacrimal, salivary)	M3 receptor, ICA69 protein in salivary and lacrimal glands
T1DM	Type 1 diabetes mellitus	M = F	C	β-islet cells of pancreas	GAD, insulin, other β-islet cell antigens
TTP	Thrombic thrombocytopenia purpura	F	A R/R	Platelets	von Willebrand factor (clotting)
UC	Ulcerative colitis	M = F	C R/R	Inner wall of colon	?

Fig. 19-4 (Part 2) Examples of Human Autoimmune Diseases

in the blood damages the vascular endothelium in multiple tissues. This damage can result in blindness, potentially fatal heart attacks and strokes, nerve damage, kidney failure or tissue necrosis necessitating limb amputation. There are two major types of diabetes mellitus: type 1 diabetes mellitus (T1DM), in which the pancreas does not produce enough insulin, and type 2 diabetes mellitus (T2DM), in which cells become resistant to insulin. Only T1DM, which accounts for about 20% of diabetic patients, is considered an autoimmune disease.

T1DM results from an autoimmune attack on the insulin-producing β-islets of the pancreas. The pancreas is invaded by numerous leukocytes, including macrophages, B cells, and CD4$^+$ and CD8$^+$ T cells. Antibodies and CTLs directed against various β-islet cell antigens combine forces to destroy the islets, resulting in a failure in insulin production. Antibodies and T cells recognizing epitopes of the β-islet enzyme GAD or the tyrosine phosphatase IA-2 are often found in T1DM patients, as are antibodies recognizing insulin.

V. MULTIPLE SCLEROSIS (MS)

Multiple sclerosis is an autoimmune disease that primarily affects the brain and spinal cord. The autoimmune attack is directed against the myelin sheath surrounding the nerve axons, as well as against cells called *oligodendrocytes* that make the myelin. Axon function may be lost due to the impaired myelination, ultimately leading to the inhibition of nerve impulse transmission. MS patients thus suffer from widespread motor weakness and sensory impairments. The nerve fibers in the brains and spinal cords of MS patients show areas of *plaque* (*sclerosis*) where normal tissue has hardened into scar tissue (Plate 19-3). These plaques are largely responsible for the localized loss of neurological function. The disease course of MS varies widely: Some patients show very few symptoms whereas others are severely disabled within months. A relapsing/remitting pattern that persists for years is common but some patients experience chronic, progressive disability.

MS is thought to stem from the activation of autoreactive T cells that can recognize pMHCs involving peptides derived

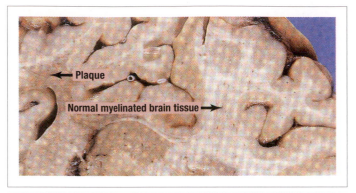

Plate 19-3 Plaque Formation in Multiple Sclerosis
[Reproduced by permission of R.A. Cooke and B.S. Stewart, *Colour Atlas of Anatomical Pathology*, Third ed. 2004 Churchill Livingston.]

from proteins of the myelin sheath. Three such proteins are *myelin basic protein* (MBP), *proteolipid protein* (PLP) and *myelin oligodendrocyte glycoprotein* (MOG). Some scientists believe that at least some cases of MS may result from molecular mimicry in which an epitope of an EBV protein closely resembles an epitope in MBP (refer to Fig. 19-2). No matter how they are activated, the autoreactive T cells soon generate effector T cells that apparently upregulate certain adhesion molecules that allow them to cross the blood--brain barrier, along with activated B cells and other inflammatory cells. Indeed, elevated numbers of Th17 cells, which are thought to make a major contribution to the inflammation, have been found in the cerebrospinal fluid of MS patients. Autoreactive CTLs responding to various peptides presented by oligodendrocytes are believed to produce cytokines and proteases that collectively damage the myelin sheath. As well, these cytokines induce microglia (macrophage-like cells resident in the brain) and infiltrating neutrophils to secrete pro-inflammatory mediators and cytokines that further contribute to myelin destruction. Nitric oxide produced by these phagocytes blocks nerve conduction pathways and contributes to the structural damage. Sites of demyelination have been found to contain elevated levels of complement products as well as autoantibodies directed against MBP, PLP or MOG.

Remission of MS disease is thought to occur when anti-inflammatory cytokines and growth factors produced by cells in the inflammatory infiltrate offset the autoimmune attack and allow the oligodendrocytes to remyelinate the damaged nerves. However, over time, the buildup of scar tissue around the nerves, coupled with the accumulating assaults on the oligodendrocytes, may prevent remyelination so that the patient is not able to fully recover from an MS episode. If the sclerosis becomes severe, the patient enters the chronic progressive stage of MS.

VI. ANKYLOSING SPONDYLITIS (AS)

Ankylosing spondylitis is a chronic inflammatory disease of bone and joints, particularly of the spine. Affected individuals first show symptoms of chronic lower back and hip pain that can persist for years. The normally elastic tissue in the tendons and ligaments of the joint is eroded and replaced first by fibrocartilage and finally by bone. When this bone replacement occurs in the spine, the lower vertebrae fuse together, causing irreversible and serious damage to the spinal column. Mobility is compromised, chest expansion may be reduced, and the patient's posture is often characteristically altered. The target tissue that undergoes autoimmune attack in AS appears to be the fibrocartilage supporting the sites where tendons and ligaments attach to the bones of the joints. Affected AS joints contain elevated numbers of plasma cells, macrophages, lymphocytes (particularly CD8+ T cells) and mast cells. The serum of AS patients often shows elevated IgA, IL-10 and acute phase proteins but, curiously, not autoantibodies.

VII. AUTOIMMUNE THYROIDITIS: GRAVES' DISEASE (GD) AND HASHIMOTO'S THYROIDITIS (HT)

There are two major types of autoimmune thyroiditis: *Graves' disease* (GD) and *Hashimoto's thyroiditis* (HT). In GD, the autoimmune attack on the thyroid gland causes it to become hyperactive and overproduce the hormone *thyroxine*. In HT, the autoreactive response against the thyroid gland results in insufficient thyroxine production.

Normal thyroxine production and the putative pathogenesis of HT and GD are illustrated in Figure 19-5. In the normal thyroid gland (panel A), thyroid follicles are composed of thyroid follicle cells surrounding a proteinaceous substance called *colloid*. When *thyroid-stimulating hormone* (TSH) produced by the pituitary gland enters the thyroid gland, the hormone binds to TSH receptors (TSHR) expressed by the thyroid follicle cells (panel A, detail). In response to TSHR signaling, *thyroglobulin* stored in the colloid is imported into the follicle cells and cleaved to produce thyroxine. Thyroxine released from the follicles enters the blood vessels and travels in the circulation to control the body's metabolic rate.

In GD patients (Fig. 19-5B), autoantibodies directed against TSHR overstimulate TSHR signaling, resulting in overproduction of thyroxine and clinical hyperthyroidism. GD patients exhibit hand tremors, insomnia, weight loss and rapid heart beat. As well, the thyroid glands of GD patients are infiltrated by T cells that respond to pMHCs derived from TSHR and other thyroid autoantigens. Sometimes autoreactive T cells of unknown specificity infiltrate into the intraocular muscles and orbital tissues around the eyes of GD patients. This infiltration induces inflammation that causes these tissues to swell, resulting in a distinctive bulging of the eyes known as *Graves' ophthalmopathy*.

In HT patients (Fig. 19-5C), autoantibodies trigger the destruction of the follicular cells or bind to TSHR (in a region different from that targeted in GD) and block TSHR signaling. Insufficient thyroxine is produced, which results in clinical hypothyroidism. HT patients complain of depression, fatigue, weight gain and dry, rough skin. In severe cases, the thyroid gland becomes greatly enlarged in size (goiter) as it attempts to make large amounts of thyroxine precursors to compensate for the thyroxine deficiency. NK cells and autoreactive CTLs

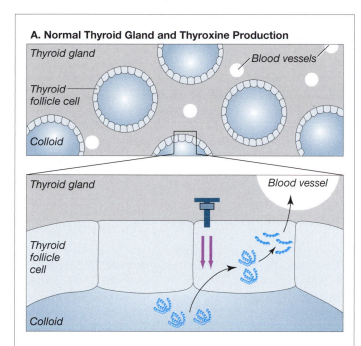

A. Normal Thyroid Gland and Thyroxine Production

Thyroid gland

Blood vessels

Thyroid follicle cell

Colloid

Thyroid gland

Blood vessel

Thyroid follicle cell

Colloid

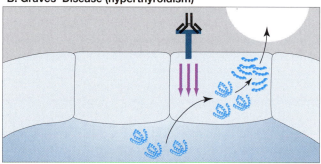

B. Graves' Disease (hyperthyroidism)

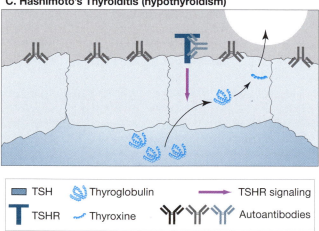

C. Hashimoto's Thyroiditis (hypothyroidism)

■ TSH	🦠 Thyroglobulin	➝ TSHR signaling
T TSHR	∿ Thyroxine	Y Autoantibodies

Fig. 19-5 Putative Pathogenesis of Autoimmune Thyroiditis

may also infiltrate HT thyroid glands, contributing to gland destruction via ADCC and perforin/granzyme-mediated cytotoxicity.

VIII. MYASTHENIA GRAVIS (MG)

Myasthenia gravis is a rare autoimmune disease that results in severe and specific muscle weakness. Onset may be early or late and clinical presentation is heterogeneous. The disease is usually chronic in character and a relapsing/remitting pattern is common. Weakness in the facial muscles may cause a patient to have difficulty talking and swallowing, whereas respiratory muscle weakness can lead to life-threatening breathing difficulties. MG is caused by autoantibodies that recognize an *acetylcholine receptor* expressed on muscle cells. Acetylcholine is a chemical messenger necessary for the transmission of nerve signals. The binding of the autoantibodies to the acetylcholine receptor prevents the binding of acetylcholine and thus interferes with the transmission of electrical impulses across the neuromuscular junction between a neuron and a muscle cell. Without transmission of this signal, the muscle is unable to contract.

IX. GUILLAIN-BARRÉ SYNDROME (GBS)

Guillain-Barré syndrome is a rare disease resulting from acute autoimmune attack on the peripheral nerves. Autoantibodies directed against gangliosides and glycolipids attack neurons in the peripheral nerves, inducing acute inflammation that demyelinates the nerve fibers and reduces electrical impulse transmission. The patient may first notice tingling in the feet or hands that rapidly (within hours) spreads up or down the body. Blurred vision, clumsiness, fainting and swallowing difficulties may occur. Within days or weeks, the muscle weakness can progress to extensive paralysis. In the worst cases, the respiratory muscles are paralyzed and the patient has to be put on a respirator. Serum TNF is usually elevated and infiltrating lymphocytes and macrophages are present in the peripheral nerves. Despite the serious clinical picture of GBS, a vast majority of patients recover without treatment. GBS may be a case of molecular mimicry because, in two-thirds of cases, onset occurs following vaccination against or recovery from a GI or respiratory infection, in particular with *Campylobacter jejuni*. Antibodies that recognize both the LPS of *C. jejuni* and human nerve gangliosides have been isolated from GBS patients.

X. INFLAMMATORY BOWEL DISEASE (IBD): CROHN'S DISEASE (CD) AND ULCERATIVE COLITIS (UC)

Inflammatory bowel disease is actually a family of autoimmune disorders affecting the large and/or small intestines. The two major types of IBD are *Crohn's disease* (CD) and *ulcerative colitis* (UC). CD affects all layers of the wall of the colon and the small intestine, whereas UC affects only the innermost layer of the wall of the colon. In CD, there may be healthy patches of bowel interspersed with diseased patches, whereas in UC, the entire colonic lining is usually affected. Both CD and UC are characterized by chronic, relapsing/remitting inflammation of the intestinal lining that causes loss of appetite, weight loss, diarrhea and fever. The scarring and swelling associated with IBD can result in obstructions that lead to abdominal cramps and vomiting. IBD is thought to arise from inappropriate inflammatory responses to unknown antigens furnished by commensal organisms normally present in the gut (and thus considered "self"). The intestinal lamina propria of IBD patients contains significant numbers of CD4$^+$ T cells secreting high levels of Th1/Th17 or Th2 cytokines in CD and UC, respectively. Although autoantibodies are sometimes detected in IBD patients, B cells do not appear to play a large part in this family of diseases.

D. Determinants of Autoimmune Diseases

Earlier in this chapter, we described the basic events that must occur before autoimmunity and autoimmune disease can develop. In this section, we discuss several variables or *determinants* that influence whether or not any of these events occurs. Firstly, the genetic make-up of an individual plays a critical role in determining his/her susceptibility to autoimmune disease. Allelic variation in genes encoding proteins important for immune system function can result in the failure of peripheral tolerance mechanisms. However, mere possession of an autoimmune-prone genotype is usually not enough for overt disease to appear. For many autoimmune diseases, the actual manifestation is linked to an external trigger such as an encounter with a particular pathogen, drug, chemical or toxin. Alternatively, a change in hormonal status may serve as an internal trigger of autoimmunity in genetically predisposed individuals.

An overview of how genetics, external triggers and hormonal influences can set the stage for an autoimmune disorder appears in Figure 19-6. In an individual with a normal genotype (left panel), most autoreactive lymphocytes are eliminated in the thymus by negative selection during the establishment of central tolerance. Those autoreactive lymphocytes that escape to the periphery are kept under control by the mechanisms of normal peripheral tolerance. An encounter with an autoantigen, even in the presence of an external trigger, almost always results in deletion or anergy of the cell. Even if an autoreactive lymphocyte manages to achieve limited activation, it is suppressed by regulatory T cells that ensure that the cell does not continue to respond to the autoantigen and damage self tissues. In an individual with a genotype predisposed to autoimmunity (Fig. 19-6, right panel), normal central tolerance is usually in place but the mechanisms of peripheral tolerance are dysregulated. Autoreactive lymphocytes that escape to the periphery and encounter autoantigen in the presence of a trigger undergo uncontrolled activation and effector cell generation. The actions of these autoreactive effectors result in the tissue damage and clinical symptoms that constitute an autoimmune disease.

The involvement of multiple determinants in the occurrence of autoimmunity makes it challenging to elucidate how these diseases develop in humans because humans are both genetically outbred and subject to a diverse array of external conditions. Sometimes a trend in a particular autoimmune disease can be observed if the population being studied is restricted by geographical location or ethnic background (see Box 19-2). However, to gain more specific insights into the mechanisms of autoimmunity, scientists study inbred animal models of autoimmune diseases in which the various determinants can be more easily controlled (see Box 19-3).

I. GENETIC PREDISPOSITION

The most convincing evidence that predisposition to autoimmune disease depends on genetics comes from studies of identical twins. When one identical twin develops autoimmune symptoms, there is a 12-60% chance that the other twin will develop the same ailment. In contrast, the chance of the same autoimmune disease developing in two fraternal twins is only about 5%. Inherited predisposition to autoimmunity is also seen in families without twins. Strikingly, in these cases, the actual form of the autoimmunity may vary in different members of the same family. It is also very common for patients suffering from one type of autoimmune disease to later develop another autoimmune disease. These observations suggest that what is often inherited is a general susceptibility to autoimmunity rather than a single defect associated with a particular autoimmune disease.

The number and identity of the genes involved in human predisposition to autoimmunity are not precisely known, and different individuals have different collections of predisposing alleles. Moreover, not every human who has a particular predisposing allele experiences autoimmune disease even when exposed to the same environmental influences, and the course of a given autoimmune disease can vary greatly from person to person. For example, the increased frequency of T1DM in close relatives of a diabetic individual is consistent with a genetic basis for this disorder; however, the disease can skip whole generations.

Due to their direct involvement in T cell responses, the most important genes conferring predisposition to autoimmunity are the HLA genes (Table 19-1). Perhaps the best illustration of HLA association with human autoimmunity can be found in AS. Over 90% of Caucasians with AS express one of a small collection of unusual alleles belonging to the HLA-B27 family. It should be noted, however, that an association between an autoimmune disease and a particular HLA molecule may not hold true for all populations. For example, an allele called HLA-DRB1*04 is tightly linked to RA development in Caucasians but not in Hispanic or African populations. Finally, to complicate matters further, some HLA proteins confer protection against the development of autoimmune disease even when another HLA molecule linked to increased autoimmune susceptibility is present. For example, HLA-DQ8 is strongly associated with the development of T1DM. However, if an HLA-DQ8-expressing individual also expresses HLA-DQ6,

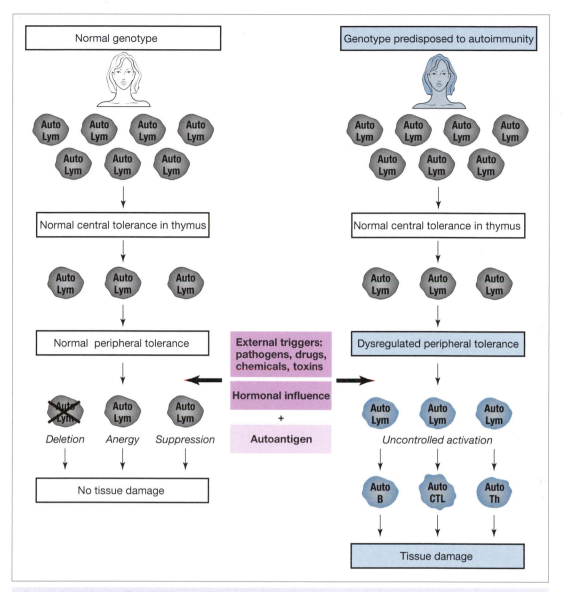

Fig. 19-6 **Determinants Influencing the Fate of Autoreactive Lymphocytes**

Table 19-1 **Examples of HLA Alleles Associated with Autoimmune Diseases**

Autoimmune Disease	HLA Allele	Autoimmune Disease	HLA Allele
Ankylosing spondylitis	HLA-B27	Myasthenia gravis	HLA-B7, B8, DR2, DR3, DR7
Crohn's disease	HLA-DRB1	Polymyositis	HLA-DR3
Goodpasture's syndrome	HLA-DR2	Rheumatoid arthritis	HLA-DR4
Graves' disease	HLA-DR3	Sjögren syndrome	HAL-DR3
Hashimoto's thyroiditis	HLA-DR3, DR5	Systemic lupus erythematosus	HLA-DR2, DR3
Multiple sclerosis	HLA-DR2	Type 1 diabetes mellitus	HLA-DR3, DR4, DQ2, DQ8

Box 19-2 **Population Trends in Human Autoimmune Diseases**

Many autoimmune diseases appear to vary in incidence by region or by ethnic group. For example, Southern European and Asian countries have a lower incidence of T1DM and MS than do Northern European countries, and SLE incidence is higher in natives of African and Caribbean countries than in Europeans and North Americans. Within the United States, African-Americans and Hispanics have a higher risk of developing SLE than do Caucasians but a lower risk of T1DM and MS. Such variation may sometimes be due to the uneven prevalence in particular ethnic groups of a gene linked to a given autoimmune disease. Similarly, differences in diet or in the occurrence of a triggering pathogen or chemical agent due to geographic factors may influence the frequency of an autoimmune disease. In other cases, however, the reasons for variation in autoimmune disease incidence by region or ethnic group are not obvious.

It is also curious that the incidence of certain prototypical autoimmune diseases, including CD, T1DM and MS, has rapidly increased in developed countries since the 1970s (see Figure). Some researchers believe that this increase may be related to the "hygiene hypothesis" described in Chapter 18. To recap, this theory posits that excessive zeal in maintaining sanitary living conditions may skew the relative frequencies of various Th cell subsets in a child's developing immune system. A deficit in Th3 cells coupled with an increase in Th1 or Th17 differentiation might promote autoimmunity. In this light, the observation that SLE incidence is much higher in Black Americans than in West Africans (who share ethnicity but not environment) makes sense. However, contradictory evidence also exists, and although some animal studies have supported the hygiene hypothesis with respect to autoimmunity, others have not. The rise in autoimmunity in the developed world thus remains a conundrum.

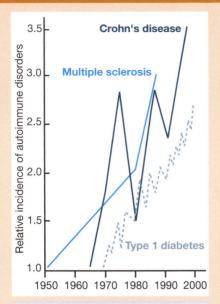

[Figure adapted Adapted from Bach J.F. (2002). The effect of infections on susceptibility to autoimmune and allergic diseases. *New England Journal of Medicine* **347**, 911-920.]

T1DM is far less likely to develop. How this protection works is not known.

In addition to HLA alleles, certain alleles of polymorphic genes involved in inflammation, complement activation or function, clearance of antigen, or T cell regulation have also been linked to autoimmunity (Table 19-2). For example, polymorphisms linked to SLE development occur in the CR1, HSP70 and TNF genes (among others). Similarly, in human RA patients, particular alleles of the IL-1, IL-1R and IFNγ genes occur at increased frequency. However, it has yet to be definitively proven that any of these polymorphisms actively contributes to the development of an autoimmune disease.

A non-HLA gene of particular relevance to several autoimmune diseases is Runx-1, which encodes a subunit of a transcription factor called CBF. CBF binds to a DNA motif (called the "Runx-1 binding site") that occurs throughout the human genome. Runx-1 binding sites that have undergone mutations have been pinpointed as autoimmune susceptibility loci. For example, mutations in a Runx-1 binding site on chromosome 2 are associated with SLE, and mutations of a Runx-1 binding site on chromosome 17 have been linked to RA.

Sometimes a region of a chromosome may have been identified as containing a susceptibility locus for an autoimmune disease but the genes remain a mystery. For example, AS is associated with specific genetic susceptibility regions that have been mapped to chromosomes 1, 2, 6, 9, 10, 16 and 19. Researchers are working to isolate the culprit genes in these regions.

II. EXTERNAL TRIGGERS

External stimuli, including chemical agents and pathogens, show tantalizing links to autoimmune disease onset or flare-ups in both humans and animal models. However, convincing

Table 19-2 **Examples of Non-HLA Genes Associated with Autoimmune Diseases**

Autoimmune Disease	Gene
Ankylosing spondylitis	Unknown genes on chromosomes 1, 2, 6, 9, 10, 16 and 19
Crohn's disease	NOD2, cation transporters
Graves' disease	CTLA-4
Guillain-Barré syndrome	FcγRII, TNF
Hashimoto's thyroiditis	CTLA-4
Rheumatoid arthritis	IL-1, IL-1R, MBL, TNFR2, ICAM-1, IFNγ, FcRs
Systemic lupus erythematosus	CR1, HSP70, FcγII, FcγIII, IL-6, IL-10, TNF, TNFR2, C1, C4
Type 1 diabetes mellitus	Insulin, CTLA-4, IL-12

Box 19-3 **Animal Models of Autoimmune Diseases**

Because the complex series of physiological changes involved in the development of any human autoimmune disease cannot be recreated *in vitro*, animal models are essential for the *in vivo* study of these ailments. Some strains of animals are susceptible to autoimmune disease induced by various treatments (Table 1), allowing researchers to investigate the cellular and molecular events leading to the initiation and maintenance of autoimmunity. Other researchers examine the genetics of autoimmunity in mutant animals, usually taking one of two approaches: (1) examine the genome of an animal spontaneously displaying autoimmunity and determine which genes have sustained mutations; or (2) mutate a gene and see if it causes autoimmune disease. With respect to the first method, several strains of animals spontaneously develop autoimmunity and display defects in known genes (Table 2). With respect to the second method, mouse models of several autoimmune diseases have been created via genetic engineering (Table 3). For example, mice have been generated in which the murine gene encoding a cytokine or complement component has been deleted. In other mutant strains, the mouse genome has been modified to transgenically express a human HLA allele associated with autoimmunity.

To date, over 40 genes have been identified whose absence or manipulation correlates with autoimmune symptoms in rodents. As expected, these genes include those encoding MHC alleles, complement components, costimulatory molecules, cytokines, cytokine receptors, FcRs, intracellular signaling proteins and pro- and anti-apoptotic molecules. However, although these animal models are very useful, their development of autoimmune symptoms likely reflects only an approximation of the genetic and cellular events that occur in humans who develop autoimmune disease.

Table 1 Examples of Animal Models in Which Autoimmune Disease Is Induced

Animal Model	Induction Protocol	Parallel Autoimmune Disease in Humans
EAE mouse	Peptide from oligodendrocyte protein plus adjuvant	Multiple sclerosis
CIA mouse	Collagen II plus adjuvant	Rheumatoid arthritis
EAT mouse	Peptides from thyroid proteins plus adjuvant	Hashimoto's thryoiditis
EAMG mouse	Peptides from an acetylcholine receptor protein plus adjuvant	Myasthenia gravis

Table 2 Examples of Animals with Natural Mutations Leading to Autoimmune Disease

Animal Model	Genes Mutated	Parallel Autoimmune Disease in Humans
NZB/W mouse	Sle1, Sle2, Sle3, Sle6	SLE
lpr mouse	Fas	SLE, Sjögren syndrome[†]
gld mouse	FasL	SLE, Sjögren syndrome
NOD mouse	*iddm* genes (> 20)	Type 1 diabetes mellitus
BB-DP rat	Ian5	Type 1 diabetes mellitus
OS chicken	Several loci	Hashimoto's thryoiditis

[†]See Figure 19-4.

Table 3 Examples of Genetically Engineered Mouse Models of Autoimmune Disease

Mouse Model	Parallel Autoimmune Disease in Humans
IL-10 gene deleted	Inflammatory bowel disease
Complement component C1q deleted	SLE
TNF transgenic (i.e., TNF is overexpressed)	Rheumatoid arthritis
IL-7 transgenic	Inflammatory bowel disease
HLA-DRB1*1501 transgenic	Multiple sclerosis
HLA-DQB1*0302 transgenic	Type 1 diabetes mellitus
HLA-B*2705 transgenic	Ankylosing spondylitis
P14 TCR transgenic[‡]	Type 1 diabetes mellitus

*Indicates a particular MHC allele. Thus, 1501 is a defined allele of the HLA-DRB1 gene. The HLA-DRB1*1501 allele is associated with predisposition to MS.

[‡]The T cells of these mice express a TCR that recognizes a viral antigen (gp33) that was engineered to be expressed only on the surface of mouse β-islet cells in the pancreas. The gp33 antigen becomes a "self" protein of the mouse pancreas in this situation and molecular events underlying the attack by the T cells on the β-islets can be studied.

proof that an encounter with an environmental stimulus actually triggers the initial onset of human autoimmunity is lacking.

i) Chemical Agents

Certain chemical and pharmaceutical agents appear to precipitate particular autoimmune symptoms. The use of hair dyes, smoking, glue sniffing, or exposure to silica dust or other toxins have been linked to episodes of RA, SLE, HT or GD. Heavy exposure to paint thinners has been blamed for some cases of MS, whereas a weak association has been noted between pesticide exposure and RA. Excess iodine intake may be a triggering factor for some cases of HT or GD, and an episode of allergic rhinitis may bring on GD in certain patients. Exposure to UV radiation has been linked to flare-ups of SLE. Other autoimmune diseases may initiate in response to drug treatment. For example, thiol-containing drugs and sulfonamide derivatives, as well as certain antibiotics and non-steroidal anti-inflammatory drugs, appear to trigger the onset of PG. Several pharmaceuticals have been associated with the initiation of an SLE-like syndrome in a small proportion of patients, and the administration of other drugs prescribed to combat high blood pressure or irregular heartbeat have been followed by the onset of SLE-like symptoms. Anti-TNF agents prescribed as anti-inflammatories have also been associated with the appearance of antinuclear antibodies. Fortunately, in these cases, the disease is often mild and of short duration, and usually resolves when the drug is withdrawn.

ii) Infections

As mentioned earlier in this chapter, infections with certain pathogens appear to provoke the initiation or intensification of some autoimmune diseases in genetically predisposed individuals. Whether this connection is attributable to molecular mimicry and/or chronic inflammation remains under debate. Severe bacterial or viral infections frequently appear to trigger an increase in autoreactive antibodies or CTLs that can lead to a flare-up of quiescent autoimmune disease or an exacerbation of existing symptoms.

With respect to viruses, a wide range of viral infections have been associated with flare-ups of SLE. The development of GBS often follows infection with herpes simplex virus, cytomegalovirus or EBV, and infections with respiratory viruses may trigger GD or acute *immunopathic thrombocytopenic purpura* (ITP; refer to Fig. 19-4). The onset of acute ITP may also be preceded by varicella infection. The vacuolar appearance of muscle cells in patients with *polymyositis* (PM; refer to Fig. 19-4) suggests that an (unknown) virus may trigger this autoimmune disease. Some researchers believe that, especially in children, T1DM is triggered by a virus (particularly rubella virus or Coxsackie B4 virus); however, this theory is controversial. Whether MS is triggered by a viral infection is also unclear. About 30% of new cases of MS appear to emerge following infection with some kind of virus, and relapses in some MS patients seem to be triggered by adenovirus or GI infections. As well, as stated earlier, molecular mimicry appears to exist between human MBP and an EBV protein. However, no one virus has been consistently associated with MS.

Infections with various bacterial species have also been linked to autoimmune diseases. The most striking example is the development of ARF following recovery from infection with a virulent member of the group A streptococci. Similarly, episodes of AS and the onset of GBS frequently follow an infection with *C. jejuni*. Infection with *Klebsiella pneumoniae* or *Escherichia coli* may also be risk factors for AS but these links remain under debate. GD has been associated with *Yersinia enterocolitica* infections, and mycoplasma infections are thought to sometimes precipitate the onset of RA or CD. A connection between *Helicobacter pylori* and ITP has been proposed because some ITP patients infected with this organism show improvement in their ITP symptoms following antibiotic treatment. In patients with *Sjögren syndrome* (SS; refer to Fig. 19-4), the exocrine glands (including the salivary glands) are destroyed by autoimmune attack. SS patients often show alterations of the normal microbial flora of the mouth, but whether these changes cause SS or are consequences of the reduced salivation in these individuals is unclear.

III. HORMONAL INFLUENCES

Many autoimmune diseases show a gender bias. For example, women account for 88% of SLE patients and 90% of SS patients, and RA affects three times as many women as men. GD and HT are also predominantly found in women. In contrast, AS and GS patients are usually males. These findings suggest that sex hormones can play a major role in inducing autoimmune disease onset in genetically predisposed individuals. Scientists hypothesize that hormone expression may somehow "reactivate" previously tolerized lymphocytes so that they can attack self tissues. For example, in a mouse model of SLE, the administration of estrogen was shown to block B cell tolerization and to result in autoimmune symptoms. In human SLE patients, estrogen metabolism is often abnormal. Moreover, flare-ups of SLE are frequently associated with changes in hormonal status, such as during pregnancy or the initiation of hormone replacement therapy. Similarly, significant numbers of HT and GD patients first develop their disease in the postpartum period, a time of major hormonal changes. Intriguingly, pregnant RA and MS patients frequently experience an improvement in their autoimmune symptoms during their last trimester.

The glucocorticoid hormones responsible for the "fight or flight" response to acute stress may also be relevant to autoimmune disease etiology. Animals that produce lower levels of these hormones show increased susceptibility to autoimmunity. Indeed, when experimentally exposed to a stress such as hypoglycemia, SLE patients produce lower levels of serum glucocorticoids than do normal individuals. Psychological stress may also be associated with symptoms of autoimmune disease. For example, marital difficulties, job stress and economic hardship have all been tentatively linked to the onset of RA or PG. Scientists speculate that, in normal individuals subjected to stress, the production of glucocorticoids may be increased to restrain autoreactive lymphocytes. In individuals predisposed to autoimmunity, this increase in glucocorticoids may not occur or is insufficient to prevent autoimmune disease.

E. Therapy of Autoimmune Diseases

Both conventional and immunotherapeutic strategies are used to treat autoimmune diseases. Conventional therapies include anti-inflammatory and immunosuppressive drugs that are helpful but usually only alleviate symptoms and do not address the underlying cause of the autoimmunity. There are also concerns about the potentially toxic side effects of long term use of these drugs. Other conventional treatments can either blunt the effects of autoantibodies or mechanically remove them from a patient's blood but, again, are not a cure. Immunotherapies use aspects or components of the immune response to ameliorate disease. These approaches generally attempt to eliminate or control the autoreactive lymphocytes or autoantibodies causing the damage. However, the development of these agents has been slow largely because results obtained in animal models have rarely translated well to the human situation. Lack of efficacy of a candidate agent or even toxicity and exacerbation of autoimmunity have been reported in some clinical trial participants. Even the results of human clinical trials must be cautiously interpreted because many human autoimmune diseases have a relapsing/remitting disease pattern. Thus, the observed resolution of an autoimmune symptom may or may not be the direct result of the candidate treatment. Such hurdles have made the development of new therapeutics for autoimmune diseases a tortuous process. Current conventional and immunotherapeutic approaches to autoimmune disease treatment as well as several experimental approaches are discussed in the following sections.

I. CONVENTIONAL THERAPIES

The mainstays of conventional autoimmune disease treatment are anti-inflammatory agents, immunosuppressive drugs and technologies that non-specifically control autoantibodies. For autoimmune diseases that have a strong association with a particular pathogen infection, antibiotics or vaccines may be beneficial. For an autoimmune disorder that has a disease-specific symptom, a treatment to alleviate that symptom may improve a patient's quality of life. For example, the excessive blood clotting in patients with *anti-phospholipid syndrome* (APS; refer to Fig. 19-4) can be treated with anticoagulant drugs. Similarly, individuals with T1DM can regulate their blood sugar with insulin injections, and MS patients can be given antispasmodic drugs to reduce muscle spasms.

i) Anti-Inflammatory Drugs

Although there is no known cure for the arthritis associated with RA or the spinal fusion of AS, various treatments aimed at relieving the inflammation driving the bone and joint symptoms can be applied. Traditionally, such patients are treated with anti-inflammatory agents such as aspirin and ibuprofen. The administration of corticosteroids that block the transcription of the major pro-inflammatory cytokines TNF and IL-1 (and presumably IL-17) can also help in many autoimmune diseases. Such agents improve joint function in RA and AS, decrease relapses of MS, reduce inflammation in SLE, ameliorate muscle weakness in MG and PM, control bleeding in GS, and soothe the blistering of PG. However, prolonged use of large doses of corticosteroids can have detrimental side effects, including decreased resistance to infection and development of osteoporosis (brittle bones).

ii) Immunosuppressive Drugs

Drugs that inhibit lymphocyte proliferation are used to treat many autoimmune diseases, including severe cases of RA, MG, PM, SS, GS and PG. Some of these drugs were described in Table 17-3 as agents used to suppress transplant rejection. However, these drugs are non-specific in their effects such that all activated lymphocytes, including non-autoreactive T and B cells needed to protect the autoimmune patient from infection or cancer, may be affected. Thus, prudent use of these immunosuppressive therapies is required to avoid increasing the risk of life-threatening infections or tumorigenesis. Very promising results have been obtained in RA patients treated with leflunomide, a malononitrilamide that improves patient symptoms without increasing opportunistic infections. Cyclosporine A can be helpful in increasing the platelet count in ITP patients, and mycophenolate mofetil has improved symptoms in some SLE cases.

iii) Non-Specific Control of Autoantibodies

Many autoimmune disease patients benefit from plasmapheresis, the mechanical procedure described in Chapter 17 in which a machine is used to remove all antibody proteins (including autoantibodies) from the blood before returning it to the patient. Patients with *thrombic thrombocytopenia purpura* (TTP; refer to Fig. 19-4) receive the greatest benefit from plasmapheresis, whereas patients with MS, GBS, MG, ITP, GS and PG show more modest clinical improvement. Another method used to control autoantibodies is an infusion of *intravenous immunoglobulin* (IV-IG). IV-IG is a preparation of antibodies of multiple specificities that have been pooled from a group of healthy donors. Exactly how IV-IG reduces autoimmune symptoms is unclear but it has proved useful for the treatment of some cases of MS, MG, GBS, ITP and PM. Scientists speculate that the mixture of exogenous antibodies may interfere with the access of the patient's autoantibodies to self tissues and/or may inhibit inflammatory responses. Very recent work suggests that IV-IG may influence the function or availability of FcRs on cell types expressing these receptors.

II. IMMUNOTHERAPY

Current immunotherapeutic strategies for the treatment of autoimmune diseases are based on relatively general interference with innate and/or adaptive responses to antigens (Table 19-3). Although such deliberate tinkering with the immune system affects responses to pathogens and tumor cells and thus leaves the patient vulnerable to potentially serious infections and/or cancer development, the hope is that responses to autoantigens will also be inhibited, providing the patient with relief from autoimmune symptoms. In future, it may be possible to treat autoimmunity more precisely by establishing tolerance to the relevant autoantigen (see Box 19-4).

Table 19-3 Examples of Immunotherapeutics Used to Treat Autoimmune Diseases

Target	Effector Molecule	Mechanism	Diseases Treated*
Cytokine Blockade			
TNF	Anti-TNF mAb or TNFR blocking agent	Inhibition of TNF signaling	RA, AS, IBD, PS
IL-1	IL-1R antagonist	Inhibition of IL-1 signaling	RA
Cytokine Administration			
DCs	IFNβ	Inhibition of autoantigen presentation and T cell migration	MS
DCs	IL-10, TGFβ	Inhibition of DC maturation	PS
Targeting of Autoreactive T Cells			
CD52	Anti-CD52 mAb (CAMPATH-1H)	T cell depletion	MS, SLE, ITP, RA
CD3	Anti-CD3 mAb	T cell anergization; iT$_{reg}$ induction	T1DM, UC, MS
CD25 (IL-2Rα chain)	Anti-CD25 mAb	Inhibition of T cell activation	PG, PS, ITP, T1DM
CD4	Anti-CD4 mAb	T cell anergization	RA, PS
CD28	Soluble CTLA-4 protein	T cell costimulatory blockade	MS, SLE, RA, PS
CD2 (LFA-2)	LFA-3–IgG fusion protein	T cell costimulatory blockade; ADCC promotion	PS
CD49d (VLA-4)	Anti-VLA-4 mAb	Disruption of T cell trafficking	IBD, MS
CD11 (LFA-1)	Anti-CD11 mAb	Disruption of T cell trafficking	PS
Targeting of Autoreactive B Cells			
CD20	Anti-CD20 mAb	B cell depletion	SLE, RA, MG, ITP
CD22	Anti-CD22 mAb	B cell depletion	SLE, RA, MG, ITP
CD40L	Anti-CD40L mAb	B cell costimulatory blockade	SLE, ITP
CD52	Anti-CD52 mAb (CAMPATH-1H)	B depletion	MS, SLE, ITP, RA

*Includes investigational treatment.

Box 19-4 Induction of Tolerance as Therapy for Autoimmune Diseases

There were originally high hopes for new autoimmune disease therapies based on "vaccinating" patients with their own autoantigens under tolerizing conditions. Whereas a conventional vaccination is intended to induce an effective immune response, autoimmune "vaccination" is designed to anergize or exhaust autoreactive lymphocytes. Theoretically, such elimination or inactivation of lymphocytes can be achieved if the patient is exposed to the offending autoantigen either in a benign context or via an unusual route of administration. However, approaches that have been successful in inducing systemic tolerance to self antigens in rodents (such as the feeding of an antigen to induce oral tolerance) are often ineffective in humans. For example, no efficacy was demonstrated in a group of MS patients given oral MBP, or in a group of RA patients given oral collagen. Tolerization has also been tried as a means of preventing autoimmune disease progression. The U.S. Diabetes Prevention Trial (2003) examined whether small, regular doses of insulin could prevent the development of T1DM in individuals who were at high risk but still healthy at the start of the trial. The hope was that the continuous administration of insulin might induce tolerance to it and thus reduce β-islet cell destruction. Unfortunately, although successful in rodents, this strategy did not prevent or delay overt diabetes in humans.

DNA or peptide vaccination has been also been explored in animals as a possible means of inducing tolerance to autoantigens. For example, in a mouse model of MS, paralysis was reversed if the affected animals were given DNA encoding an MBP peptide. Other researchers have vaccinated MS patients with low doses of purified peptides that are tolerogenic versions of the MBP peptides associated with autoimmunity. Some of these MS patients enjoyed modest clinical improvement. Vaccination with altered peptides derived from insulin and desmoglein-3 are under investigation for treatment of T1DM and PG, respectively.

Another currently experimental approach to inducing tolerance to autoantigens involves promoting the interaction of tolerogenic, self antigen-presenting DCs with autoreactive T cells. In the same vein, murine B cells have been modified so that they present autoantigen peptides on MHC class II in a tolerogenic context. It is thought that the diminution of disease symptoms seen in animal models based on these approaches may involve the induction of regulatory T cells. It may also be possible in the future to isolate a patient's regulatory T cells, culture them *in vitro* to expand their numbers, and infuse them back into the patient. The hope here is that this army of regulatory T cells might suppress the activities of autoreactive T cells. Early studies in animal models have shown that such a strategy can ameliorate or even prevent the induction of autoimmunity.

i) Cytokine Blockade

New agents have been developed that block the harmful inflammation of autoimmune diseases while causing fewer side effects than traditional anti-inflammatory or immunosuppressive drugs. Many RA and AS patients, whose disease is largely due to excessive TNF, have benefited from treatment with anti-TNF mAbs or agents that block access to TNFR. Patients experience decreased inflammation in their affected joints and improved quality of life. Anti-TNF agents are also helpful in cases of IBD and psoriasis (PS; refer to Fig. 19-4). An interesting corollary to anti-TNF therapy is that T_{reg} cells, which appear to be present in normal numbers but inactive in at least some RA patients, recover their ability to inhibit cytokine production by effector T cells.

RA patients that fail to respond to measures countering TNF are thought to have disease driven by other cytokines, most likely IL-1, IL-6, IL-15, IL-17 and/or GM-CSF. Joint symptoms in some of these patients have been partially relieved by treatment with a protein that antagonizes IL-1R, and the effectiveness of anti-IL-6 and anti-IL-15 mAbs is under investigation. In mouse models of RA and MS, treatment with a mAb that blocked either IL-17 or GM-CSF activity decreased the severity of symptoms. However, it remains to be seen whether measures targeting IL-6, IL-15, IL-17 or GM-CSF will be safe and useful for the treatment of human RA and MS patients.

ii) Cytokine Administration

Although some cytokines play a role in driving autoimmunity, others can be effective in countering it. For example, large numbers of MS patients have been successfully treated with IFNβ. The original rationale for this approach was that IFNβ would control the unknown virus that was thought to be triggering MS. However, it now appears that IFNβ reduces MS relapse rates by opposing the effects of pro-inflammatory cytokines on DCs and by downregulating the expression of MHC class II that promotes presentation of autoantigens. In addition, IFNβ is a powerful inhibitor of the matrix metalloproteinases used by activated autoreactive T cells to invade brain tissue.

Administration of the immunosuppressive cytokines IL-10 and TGFβ is beneficial for some autoimmune disease patients (particularly for those with PS), although toxicity has been an issue for others. IL-10 and TGFβ block the maturation and migration of DCs and inhibit the expression of pro-inflammatory cytokines by APCs. In patients benefiting from treatment with these cytokines, inflammation was suppressed, the interval between relapses was lengthened, and the number of relapses was reduced.

iii) Targeting of Autoreactive T Cells

Other immunotherapeutic approaches have been focused more directly on autoreactive T cells (keeping in mind that activated, non-autoreactive T cells are often affected as well). Significant success has been achieved with CAMPATH-1H, a mAb that binds to a receptor called CD52 expressed on the surfaces of most lymphocytes. Binding of CAMPATH-1H to CD52 on an autoreactive T cell then dooms it to ADCC- or complement-mediated destruction. The symptoms of autoimmune diseases as varied as RA, ITP, MS and SLE have been ameliorated following treatment with CAMPATH-1H. Another approach that has shown promising results in clinical trials is the use of a modified anti-CD3 mAb to anergize T cells in T1DM patients. Long-term positive effects on the preservation of pancreatic islet cells have been observed in these studies. There is also some evidence that anti-CD3 antibodies promote the generation of iT_{reg} cells (refer to Ch. 10). Other mAbs investigated as potential therapies include anti-CD25 mAbs, which bind to the IL-2Rα chain and inhibit the effects of IL-2, and anti-CD4 mAbs that block T cell functions. Anti-CD25 mAbs have proved useful for the treatment of some cases of PG, PS and ITP, whereas anti-CD4 mAbs have achieved limited success in trials for treatment of RA and PS.

Another strategy is to interfere with the costimulation of an autoreactive T cell. A soluble form of the CTLA-4 protein that negatively regulates CD28 signaling has been used to treat patients with PS, RA, SLE or MS; some benefit in some cases has been noted. Other researchers have devised a fusion protein composed of part of an IgG molecule and the extracellular domain of LFA-3, the CD2 receptor. This agent simultaneously binds to CD2 on T cells and FcγRs on phagocytes and NK cells, promoting the death of the T cells by ADCC. Improvement in some PS patients treated with this agent has been observed. Finally, agents targeting the adhesion molecules VLA-4 and LFA-1 have been tried as means of altering the trafficking of autoreactive T cells. An anti-LFA-1 mAb has been helpful for treatment of some cases of PS, but the use of a related agent that initially showed promise for the treatment of MS, RA and IBD has been suspended due to significant toxicity. Anti-VLA-4 mAbs have been helpful in some cases of IBD and MS but have shown unacceptable toxicity in other cases.

A caveat with targeting T cells is that, as well as affecting autoreactive, anti-pathogen and anti-tumor T cells, the activation and/or function of regulatory T cells may be impeded. Since regulatory T cells normally inhibit autoreactive CD4$^+$ and CD8$^+$ T cells, autoimmune disease symptoms may be exacerbated rather than mitigated.

iv) Targeting of Autoreactive B Cells

Autoreactive B cells not only differentiate into plasma cells producing damaging autoantibodies but also may serve as powerful APCs driving the activation of autoreactive memory T cells. Antibody-mediated depletion of autoreactive B cells has been explored for the treatment of several autoimmune diseases. The binding of mAbs directed against B cell markers induces the destruction of B cells (both normal and autoreactive) by either complement-mediated cytolysis, ADCC or the induction of apoptosis. Accordingly, the administration of mAbs recognizing the B cell markers CD20 or CD22 has been of clinical benefit to some SLE, RA, MG and ITP patients. Only mild side effects have been observed, and increased infection due to a general impairment of the humoral response does not appear to be a problem. CAMPATH-1H has also been helpful for some patients because CD52 is found on B cells as well as T cells. Treatment with anti-CD40L mAb has been investigated as a means of interfering with the CD40/CD40L engagement that is critical for antibody production. In SLE patients treated with anti-CD40L mAb, titers of circulating autoantibodies were decreased and glomerulonephritis was mitigated. Anti-CD40L mAbs have also been used to ameliorate symptoms of ITP. In some cases, however, unacceptable toxicity has occurred.

We have come to the end of our discussion of autoimmune diseases, a topic with many questions remaining. We move to another subject about which there are still several mysteries: cancers of immune system cells. Whereas Chapter 16 addressed the role of the immune system in fighting tumorigenesis, Chapter 20 discusses what happens when immune system cells themselves undergo malignant transformation.

CHAPTER 19 TAKE-HOME MESSAGE

- An autoimmune disease is a pathophysiological state in which an individual's tissues are damaged as a result of an attack by the immune system. A large number of autoimmune diseases have been identified that vary widely in their symptoms.

- Four events are required for the development of an autoimmune disease: (1) an autoreactive lymphocyte clone in the thymus must escape elimination by central tolerance mechanisms and be released to the periphery; (2) the escaped autoreactive clone must encounter its specific self antigen in the periphery; (3) the peripheral tolerance mechanisms designed to regulate autoreactive lymphocyte responses must fail; and (4) the response by the autoreactive clone must result in clinical damage.

- The breaking of peripheral tolerance may be due to: inflammation; molecular mimicry by pathogen antigens; inherent defects in immune system cells, cytokines or complement; and/or epitope spreading.

- Activated autoreactive Th, Tc and B cells produce cytokines, CTLs and autoantibodies, respectively, that cause tissue damage. This damage may be localized or systemic, depending on the nature and distribution of the self antigen.

- Many of the peripheral tolerance defects underlying autoimmune diseases involve proteins encoded by polymorphic genes. Thus, an individual's genotype, particularly the identity of his/her HLA alleles, may predispose him/her to developing an autoimmune disease.

- In someone genetically prone to autoimmunity, the balance may be tipped toward autoimmune disease by triggers such as pathogen infection, toxin exposure or altered hormone levels.

- Some autoimmune disease therapies are aimed at alleviating symptoms, whereas other approaches seek to control or eliminate autoreactive lymphocytes.

Section A

1) How do autoimmunity and autoimmune disease differ?

2) Describe the roles of Th cells, B cells and CTLs in causing the tissue damage associated with autoimmune disease.

3) Distinguish between organ-specific and systemic autoimmune diseases.

Section B

1) What four events must occur for autoimmune disease to develop?

2) Name four mechanisms that are thought to contribute to autoimmunity.

3) Describe how inflammation's effects on DCs may promote autoimmunity.

4) What type of cancer frequently arises in patients with an autoimmune disease?

5) What is the theory of molecular mimicry?

6) Give four examples of inherent immune system defects that might favor autoimmunity.

7) How is epitope spreading thought to perpetuate an autoimmune response? Give an example.

Section C

1) Can you define these terms? relapsing/remitting, rheumatoid factor, sclerosis.

2) Give two examples of proteins that are targets of autoantibodies in systemic lupus erythematosus.

3) What is the primary target tissue in rheumatoid arthritis?

4) Rheumatic fever and Guillain-Barré syndrome are often cited as examples of molecular mimicry. Why?

5) Which type of diabetes is the autoimmune form? In what percentage of patients is it seen?

6) How is the tissue damage of multiple sclerosis caused and what is the clinical outcome?

7) What is the primary target tissue in ankylosing spondylitis?

8) Compare the clinical features and mechanisms of the two types of autoimmune thyroiditis.

9) What is the protein targeted in myasthenia gravis and how does its inactivation cause symptoms?

10) Compare the clinical features and mechanisms of the two major types of inflammatory bowel disease.

Section D

1) What three major determinants govern whether an individual develops autoimmunity?

2) Why do some autoimmune diseases appear to vary by geographic location?

3) Describe three animal models of autoimmune diseases.

4) Give three examples of genes linked to autoimmune diseases.

5) Give three examples of chemical agents and the autoimmune diseases to which they may be linked.

6) Give three examples of pathogens and the autoimmune diseases to which they may be linked.

7) How might hormonal changes trigger episodes of autoimmune disease? Give two examples.

Section E

1) Give three examples of conventional approaches used to treat autoimmune diseases.

2) What are two ways in which the effects of autoantibodies can be non-specifically controlled?

3) Cytokines can either be administered or their effects blocked to treat autoimmunity. Give an example of each circumstance and its therapeutic rationale.

4) Give three examples each of how autoreactive T and B cells can be targeted to treat autoimmunity.

5) What is a major disadvantage of current immunotherapeutic strategies?

6) In the future, how might vaccination with a self antigen mitigate or prevent autoimmune disease? Give two examples of approaches that have been attempted.

20

Hematopoietic Cancers

Where there is the need for a controller, a controller of the controller is also needed.
Tadeusz Kotarbinski

A. Overview of the Biology and Treatment of Hematopoietic Cancers

I. WHAT ARE HEMATOPOIETIC CANCERS?

In Chapter 16, we discussed how the immune system attempts to deal with cancerous cells. In this chapter, the cancers that arise from the malignant transformation of immune system cells are examined. We have chosen to call these tumors "hematopoietic cancers" (HCs) to distinguish them from the non-hematopoietic cancers (NHCs) described in Chapter 16. HCs account for about 8–10% of all cancer diagnoses in the developed world and a similar percentage of cancer deaths. The tumor biology we described in Chapter 16 remains mostly relevant here despite the fact that cancers of the immune system concern hematopoietic cells that are inherently mobile and not fixed like those of body organs. However, there are unique aspects to the biology of HCs that make them a fascinating area of study in their own right. For the purposes of this book, we will focus on the three main types of HCs: *leukemias, myelomas* and *lymphomas*. The development of a leukemia is called *leukemogenesis*, while that of a myeloma is *myelomagenesis*, and that of a lymphoma is *lymphomagenesis*.

Leukemias are tumors that arise from the transformation of a hematopoietic cell in the blood or a hematopoietic precursor in bone marrow (BM). In the latter case, the cancerous progeny of the transformed cell usually make their way into the blood. Thus, leukemias most often occur as "liquid tumors" that are manifested as greatly increased numbers of myeloid, lymphoid or (more rarely) erythroid lineage cells in the blood and BM. Myelomas are tumors of fully differentiated plasma cells that are present either as solid masses or as dispersed clones in the BM, blood or tissues. Unlike normal plasma cells, which do not divide after they differentiate, myeloma cells continue to proliferate in an uncontrolled way and synthesize large amounts of Ig chains. Lymphomas develop from transformed lymphocytes that may travel through the body but most often form distinct stationary masses of lymphocytes. These "solid tumors" are generally found in the lymph nodes, thymus or spleen. The relative frequencies of HCs in North America are illustrated in Figure 20-1.

As was true for the NHCs described in Chapter 16, the specific cell that undergoes malignant transformation is known as the *target cell* or *cell of origin*. For HCs, the target cell may be either a developing hematopoietic precursor or a mature cell type. The specific terminology used by hematologists to describe these malignancies is based on the nature of the target cell involved and is outlined in Table 20-1.

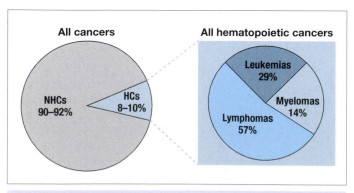

Fig. 20-1 Relative Frequencies of Hematopoietic Cancers in North America

Table 20-1 **Hematopoietic Cancer Terminology**		
HC Term	**Target Cell**	**Target Cell Description**
Myeloid or myelogenous*	Myeloid cell	Any cell of the myeloid lineage
Myeloblastic	Myeloblast	Myeloid precursor
Myelocytic	Myelocyte	Mature myeloid cell
Lymphoid	Lymphoid cell	Any cell of the lymphoid lineage
Lymphoblastic	Lymphoblast	T or B precursor cell
Lymphocytic	Lymphocyte	Mature T or B cell

*Historical term still in use for some malignancies.

II. HEMATOPOIETIC CANCER CARCINOGENESIS

Like NHCs, HCs are clonal in nature and arise when a target cell accumulates genetic alterations to DNA repair genes, oncogenes and/or TSGs. Also like NHCs, many HCs appear to be driven by cancer stem cells (refer to Ch. 16). In fact, researchers were studying a form of HC when they found the first evidence pointing to the existence of cancer stem cells. These investigators discovered that a very rare cell type in the mouse leukemia they were examining retained the HSC properties of self-renewal and differentiation. Importantly, mouse transplantation studies showed that only these leukemic stem cells were capable of transferring the leukemia to a susceptible recipient. The bulk of the leukemia cells were more differentiated leukemic blasts that were unable to self-renew or establish new malignancies in recipients. Today, researchers are actively working on defining the characteristics of cancer stem cells in a variety of HCs.

Although many HCs show relatively subtle genetic aberrations such as small chromosomal deletions or point mutations, an estimated 50% of leukemias and lymphomas are associated with major chromosomal disruptions. These disruptions frequently take the form of translocations involving the abnormal exchange of genetic material between two different chromosomes. Because the same translocation appears in many patients with the same type of leukemia or lymphoma, these translocations are called *recurring chromosomal translocations* (see Box 20-1 and Box 20-2). Not surprisingly, recurring translocations associated with HC development affect chromosomes in regions where genes regulating cellular growth, differentiation and apoptosis are located. Sometimes the exchanged chromosomal fragments come together in such a way that a new genetic entity (which may be a functional gene) is created.

In addition to translocations, exposure to environmental carcinogens, radiation or chemicals can promote HC development (just as was true for NHCs). For example, cigarette smoking and exposure to industrial solvents increase the risk of leukemogenesis. A higher incidence of leukemia has also been recorded in firefighters exposed to burning plastics, as well as in populations living close to the sites of the atomic bomb explosions in Japan in 1945 and the nuclear reactor

Box 20-1 Recurring Chromosomal Translocations

Recurring translocations are responsible for many HCs. In the example shown in the Figure, panel A depicts one chromosome each of the hypothetical chromosome pairs 1 and 2 at metaphase. These chromosomes sustain breaks in their DNA (panel B) that allow the reciprocal transfer of genetic material between chromosome 2 and chromosome 1 (panel C). The resulting translocated chromosomes (panel D) may contain existing genes that have been disrupted or repositioned, or new genes created by the fusion of sequences from both chromosomes. In the example shown in panel E, the translocation positions a hypothetical transcription factor originally on chromosome 1 downstream of a hypothetical gene expression enhancer normally situated on chromosome 2. Alternatively, the translocation may result in the fusion of these sequences. Deregulated expression of the transcription factor induced by the enhancer may help to drive malignant transformation. It should be noted that not every reciprocal translocation is itself sufficient to cause an HC, and not every person bearing one of these translocations is fated to develop an HC. Often the translocation is only an initiating event, and subsequent mutations to oncogenes, TSGs and/or DNA repair genes are needed to complete malignant conversion.

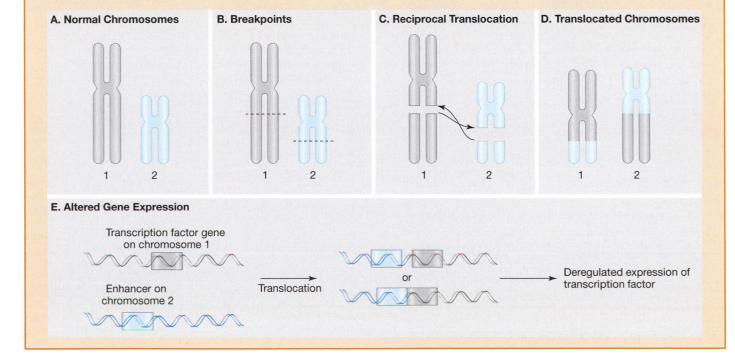

A. Normal Chromosomes

B. Breakpoints

C. Reciprocal Translocation

D. Translocated Chromosomes

E. Altered Gene Expression

Transcription factor gene on chromosome 1

Enhancer on chromosome 2

Translocation

or

Deregulated expression of transcription factor

Box 20-2 Examples of Reciprocal Translocations Leading to HCs

Examples of reciprocal translocations associated with HCs are given in the Table. In a designation such as t(12;22)(p13;q11), the "t" indicates "translocation," while the "12" and "22" in the first set of parentheses refer to the two chromosomes involved listed in ascending order. In the second set of parentheses, p13 indicates that the break in chromosome 12 took place in the first block, third band (originally defined by karyotypic staining methods) of the short ("p") arm of the chromosome. (The "p" stands for the French word "petit," meaning small.) Similarly, q11 indicates that the break in chromosome 22 took place in the first block, first band of the long ("q") arm. These breakpoints are sometimes expressed as 12p13 and 22q11.

Hematopoietic Cancer	Translocation	Genes Involved	
Acute myeloid leukemia	t(8;21)(q22;q22)	ETO (DNA binding protein)	CBF2A (transcription factor subunit)
	t(12;22)(p13;q11)	TEL (transcription repressor)	MN1 (nuclear protein)
Chronic myelogenous leukemia	t(9;22)(q34;q11)	Abl (tyrosine kinase)	Bcr (kinase/GTP exchange protein)
	t(9;12)(q34;p13)	Abl (tyrosine kinase)	TEL (transcription repressor)
B cell acute lymphoblastic leukemia	t(8;14)(q24;q32)	c-myc (transcription factor)	Igh (Ig heavy chain)
	t(12;21)(p13;q22)	TEL (transcription repressor)	CBF2A (transcription factor subunit)
T cell acute lymphoblastic leukemia	t(1;7)(p32;q35)	TAL1 (transcription regulator)	TCRB (TCRβ chain)
	t(8;14)(q24;q11)	c-myc (transcription factor)	TCRA (TCRα chain)
Chronic lymphocytic leukemia	t(14;19)(q32;q13)	Igh (Ig heavy chain)	Bcl-3 (transcription coactivator)
Myeloma	t(11;14)(q13;q32)	Cyclin D1 (cell cycle regulator)	Igh (Ig heavy chain)
	t(4;14)(p16;q32)	FGFR3 (growth factor receptor)	Igh (Ig heavy chain)
Non-Hodgkin's lymphoma	t(14;18)(q32;q21)	Igh (Ig heavy chain)	Bcl-2 (cell survival protein)
	t(3;14)(q26;q32)	Bcl-6 (transcription repressor)	Igh (Ig heavy chain)

meltdown in Chernobyl in 1986. Sadly, leukemias can develop as secondary tumors following intensive radiation or chemotherapy applied either as a treatment for a primary cancer (HC or NHC), or in preparation for a BMT or an HCT.

III. CLINICAL ASSESSMENT AND TREATMENT OF HCs

i) Characterization and Diagnosis

HCs are usually categorized by their grade and whether they cause acute or chronic disease. High grade HCs grow aggressively and cause acute disease that kills rapidly (in the absence of treatment) due to the accumulation of cancerous cells and the failure of the BM to maintain normal hematopoiesis. These types of malignancies often result when early precursor cells (*blasts*) of a particular hematopoietic lineage become transformed. Medium and low grade HCs tend to cause chronic disease that has a slower course and symptoms that may appear to come and go. The transformed cells in these malignancies are often quite well differentiated or even mature.

The 5-year survival rate of HC patients has improved greatly over the past few decades (Fig. 20-2). A great deal of this success is attributable to aggressive chemotherapy and radiation treatment regimens as well as the increased use of immunotherapies and HCTs. However, more accurate diagnoses of

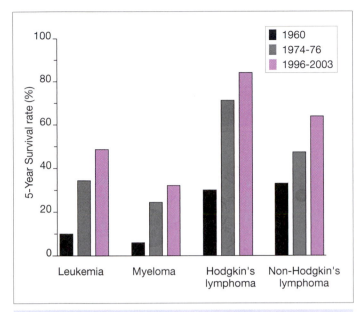

Fig. 20-2 Five-Year Survival Rates for Hematopoietic Cancer Patients
[Adapted from the Surveillance, Epidemiology and End Results (SEER) Program 1979–2000 and 1973–2003, and *National Cancer Institute: Myeloma Biology and Management*, 2nd ed. Oxford University Press, 1998.]

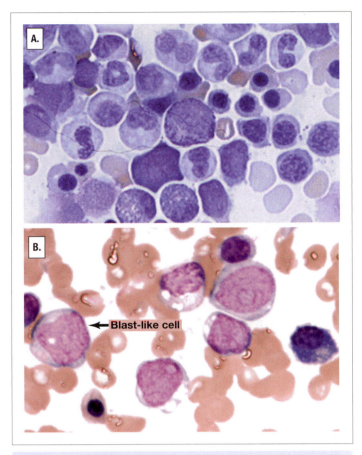

Plate 20-1 Bone Marrow Aspirates: Normal (A) and Hematopoietic Cancer (B)
[Reproduced by permission of Doug Tkachuk, Department of Pathobiology and Laboratory Medicine, University of Toronto.]

malignant conditions have also helped to improve clinical outcomes. One of the methods now commonly used to evaluate HCs is *immunophenotyping*, which involves preparing cells from the patient's BM and blood and staining these cells with tagged antibodies specific for surface marker proteins. The particular array of markers expressed on a given cancer cell allows an oncologist to more precisely identify the cell type that originally underwent transformation. Another tool of the modern HC diagnostic trade is *fluorescence in situ hybridization* (FISH), used for the examination of tumorigenic chromosomal translocations. In FISH, probes specific for a chromosomal region of interest are synthesized by incorporating an easily detected fluorochrome into a DNA fragment of complementary sequence. One or more of these fluorescently tagged probes can then be applied to whole cells or to a chromosomal spread on a microscope slide to visualize one or more whole chromosomes or a specific region of a single chromosome. Major aberrations become clearly visible, pointing the observer to a probable diagnosis. Finally, the identification of unique RNA patterns expressed by malignant cells has increasingly been used to definitively classify HC subtypes. All these techniques help the physician to arrive at a more accurate diagnosis of the HC. The treatment is then chosen that is most appropriate to an individual case and therefore offers the best chance of extending survival.

ii) Chemotherapy and Radiation

Standard chemotherapy and radiation therapy are the first modes of treatment offered to newly diagnosed HC patients. The response to these treatments is assessed by examining smears of cells taken from the patient's blood and/or BM. Malignant hematopoietic cells often morphologically resemble the blast stages of hematopoietic cells and so are termed "blast-like" (Plate 20-1). In a healthy individual, such blast-like cells account for up to 5% of cells developing in BM but are virtually absent from the blood, so that the presence of cells of this morphology in the blood is often a sign of cancer. After an HC patient has been treated, if more than 5% of cells in a BM smear are still blast-like and blasts are still present in a peripheral blood smear, the treatment has failed and there has been no clinical response. A *partial response* to treatment means that between 5% and 20% of cells in the BM smear may be blast-like but the number of blast-like cells in the blood smear has been reduced to almost zero. In addition, neutrophil and platelet counts are close to normal and the patient does not require a blood transfusion. A *complete response* means that <5% blast-like cells remain in the patient's post-treatment BM smear, and that the blood smear is essentially clear of blast-like cells. If these levels hold over a period of at least 4 weeks following treatment, the patient's disease is said to be in *remission*. (Rarely, remission occurs due to a natural resolution of the HC in the absence of treatment.) A *relapse* is the reappearance of disease in a patient who was previously in remission, and is due to the survival and proliferation of cancer cells. In general, a patient who has not suffered a relapse of the same tumor for a period of 5 years is said to be a *long term survivor*.

iii) Immunotherapy

Immunotherapy has greatly contributed to recent progress in treating HCs. Monoclonal antibodies directed against molecules expressed predominantly on the surfaces of tumor cells are now used as therapeutic agents on a regular basis. For example, the tumor cells of many B cell lymphomas and leukemias show elevated expression of the marker proteins CD20 and/or CD22, so that mAbs directed against these molecules can be used to induce tumor cell death by either complement activation or ADCC. Conjugation of these antibodies to toxins or radioisotopes to form immunotoxins or immunoradioisotopes (refer to Ch. 16) can also be used to kill cancerous cells bearing the appropriate markers.

iv) Hematopoietic Cell Transplants

Hematopoietic cell transplants can provide a cure for some HCs. In a case where a patient's own hematopoietic cells are used for the transplant (an *autologous* HCT), the patient first undergoes a round of non-myeloablative conditioning (refer to Ch. 17) that kills the rapidly dividing HC cells and so leads to remission. The patient's normal HSCs, which are quiescent, are spared. The patient is then treated with an agent such as *granulocyte colony stimulating factor* (G-CSF) to mobilize the HSCs into the peripheral blood. The blood is collected and the HSCs are purified by leukapheresis and other techniques that depend on surface marker expression. The purified HSCs are stored at −150°C while the patient undergoes a round of myeloablative chemotherapy that empties the patient's BM and periphery of virtually all hematopoietic cells. The patient is then infused with his/her stored HSCs to reconstitute a hopefully cancer-free immune system. A relatively new source of cells for an autologous HCT is umbilical cord blood. Some families are now having the umbilical cord blood of their newborn babies stored as a source of HSCs in case the child needs an autologous HCT later in life. These cells can also be of use to other family members or even to unrelated individuals in need of an allogeneic HCT.

Allogeneic HCTs intended to reconstitute an HC patient's immune system most often involve leukocytes prepared from the peripheral blood or BM of a donor who is matched with the patient at the HLA loci but differs at various minor histocompatibility loci. The donated leukocyte preparation that is infused into the BM-ablated patient contains not only HSCs but also immunocompetent donor lymphocytes. These lymphocytes recognize allogeneic differences and can therefore destroy residual HC cells in the patient, leading to a form of the beneficial GvL effect described in Chapter 17. As a result, relapse rates are lower in patients who receive an allogeneic HCT compared to those who receive an autologous HCT. However, because the donor lymphocytes also attack non-cancerous cells in the recipient, the trade-off is a higher chance of potentially fatal GvHD (refer to Ch. 17). Studies are currently under way to find ways of eliminating GvHD while preserving GvL.

Another approach that is in early clinical trials is the vaccination of leukemia patients with proteins or peptides representing antigens specific to their malignancies (i.e., TSAs). For example, a group of CML patients has been vaccinated with a peptide derived from the aberrant Bcr–Abl fusion protein (see later) that drives their disease. Similarly, a group of AML patients has been vaccinated with a modified peptide called PR1 that was derived from a protein aberrantly overexpressed by their AML cells. *In vitro* testing has shown that T cells recognizing such TSAs can be generated in the vaccinated patients, and clinical responses have been observed in some cases.

The rest of this chapter outlines the clinical and genetic features of the major subtypes of leukemias, myelomas and lymphomas as well as their various treatment options.

B. Leukemias

The majority of leukemias are broadly classified by whether the transformed cell is derived from the myeloid or lymphoid lineage and by whether disease onset and course are acute or chronic. The four major classes of leukemias are *acute myeloid leukemia* (AML), *chronic myelogenous leukemia* (CML), *acute lymphoblastic leukemia* (ALL), and *chronic lymphocytic leukemia* (CLL). The relative frequencies of these disorders in North American adults and children are given in Figure 20-3. Other leukemias include those involving transformed cells of erythroid, megakaryocytic or NK cell lineages. Acute leukemias can strike both children and adults, whereas chronic leukemias tend to arise in individuals 50 years of age and older. The overall incidence of leukemia is relatively low compared to NHCs and accounts for only 3% of all malignancies. Nevertheless, in persons under 20 years of age, acute leukemia is the leading fatal cancer.

I. ACUTE MYELOID LEUKEMIA (AML)

AMLs are acute cancers of the myeloid lineage. The target cells are often early stage myeloblasts, as shown in Plate 20-2.

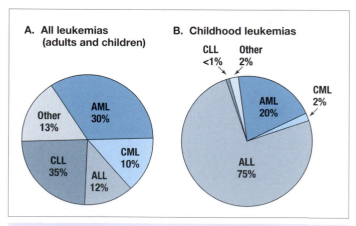

Fig. 20-3 **Relative Frequencies of Leukemias in North America**

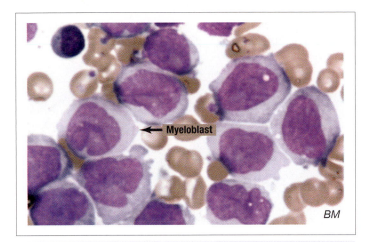

Plate 20-2 Acute Myeloid Leukemia
[Reproduced by permission of Doug Tkachuk, Department of Pathobiology and Laboratory Medicine, University of Toronto.]

Although AML mainly affects patients over 60 years of age, it also accounts for 20% of childhood leukemias.

i) Clinical Features

Like most leukemia sufferers, AML patients present with symptoms of malaise, fatigue, fever, increased susceptibility to infection and weight loss. Bone pain is present in rare cases. The high numbers of leukemic cells disrupt hematopoiesis in the BM such that the BM is said to have "failed". Anemia and thrombocytopenia result and are manifested as fatigue, dizziness, shortness of breath, bleeding and easy bruising. Innate immune responses are compromised such that respiratory infections are common. *Hepatosplenomegaly* (enlargement of the liver and spleen) arises when leukemic cells accumulate in these organs. The leukemic cells may also migrate to *extramedullary tissues*, a term used by hematologists to refer to tissues and organs in the body that are neither lymphoid tissues nor BM.

ii) Genetic Aberrations

About half of AML cases show gross chromosomal abnormalities that are easily detected. In some of these patients, a whole chromosome has been gained or lost, but why these aberrations result in AML is not known. A common recurring translocation called (t8;21)(q22;q22) leads to AML because a part of the gene encoding the CBF2A subunit of the transcription factor CBF is fused to part of the ETO gene encoding a DNA binding protein (refer to Box 20-2). The failure to produce normal CBF2A disrupts the normal function of CBF, which is to regulate many genes involved in myeloid cell development, proliferation and function. As a result of the translocation, myeloid precursors developing in the BM get "stuck" at an early stage of development and never receive the appropriate signals to stop proliferating and start differentiating. The excess cells spill into the blood and are detected as AML. Another subset of AML cases involves disruption of the TEL gene, which encodes a transcriptional repressor. The recurring translocation t(12;22)(p13;q11) fuses part of the TEL gene to part of the MN1 gene. The TEL gene is required for hemato-

poiesis in the BM and MN1 encodes a transcriptional coactivator.

In about 50% of AML cases, the leukemic cells do not display a gross chromosomal anomaly. An interesting subgroup of these AML cases is associated with two primary immunodeficiencies called *Bloom syndrome* and *ataxia–telangiectasia* (refer to Box 2-1). Both involve hereditary defects in DNA repair genes. A second subgroup comprises patients with *Li-Fraumeni syndrome*, most of whom have germline mutations of the p53 TSG. Other patients with essentially normal karyotypes have point mutations or small deletions of the p15 gene essential for controlling cell cycle progression; the nucleophosmin gene involved in ribosome synthesis and intracellular transport of these organelles; or the FLT3 gene, which encodes a receptor with tyrosine kinase activity that is involved in myeloid differentiation and stem cell survival.

iii) Treatment

Despite over 45 years of steady advances in cancer therapies, the prognosis for AML patients remains relatively poor. Better clinical management of anemia, bleeding and infections, as well as careful monitoring of cardiac and other complications, has helped to improve the survival of AML patients. Unfortunately, AML has a high rate of relapse despite apparently successful initial treatment.

AML patients diagnosed with very high leukemic cell counts need to be treated immediately with large doses of standard chemotherapeutics (refer to Ch. 16). Several new types of drugs are in clinical trials for treatment of patients whose AML has recurred. Topotecan inhibits the topoisomerase enzyme required for DNA replication, whereas flavopiridol inhibits cyclin-dependent kinases required for cell division. Decitabine decreases the methylation of DNA and disrupts DNA replication. Cloretazine cross-links DNA and thus also inhibits DNA replication. Cloretazine appears to be quite effective in elderly patients with newly diagnosed AML.

Most cases of AML have been refractory to the mAb-based therapies helpful in other types of HCs. An exception may be cases of AML in which the leukemic blasts express the surface marker CD33. These blasts are killed if they internalize an immunotoxin composed of anti-CD33 mAb conjugated to a toxic drug. The success of treatment with this anti-CD33 immunotoxin has been enhanced by cotreatment with an agent designed to block the expression of Bcl-2, a critical anti-apoptotic protein that promotes cell survival.

AML patients for whom all the above treatments have failed can try HCT. About 50% of AML patients receiving an allogeneic HCT are alive and disease-free 5 years after the transplant. Clinicians are also exploring ways of enhancing the GvL effect to reduce the chance of cancer recurrence after transplant. Some AML patients have been treated with an autologous HCT but the rate of relapse has been relatively high.

II. CHRONIC MYELOGENOUS LEUKEMIA (CML)

Chronic myelogenous leukemia (CML) is characterized by increased numbers of mature and immature granulocytes in

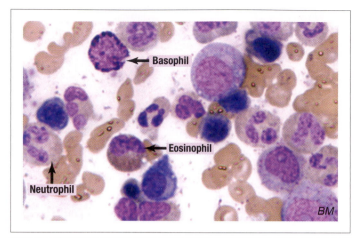

Plate 20-3 **Chronic Myelogenous Leukemia: Chronic Phase**
[Reproduced by permission of Doug Tkachuk, Department of Pathobiology and Laboratory Medicine, University of Toronto.]

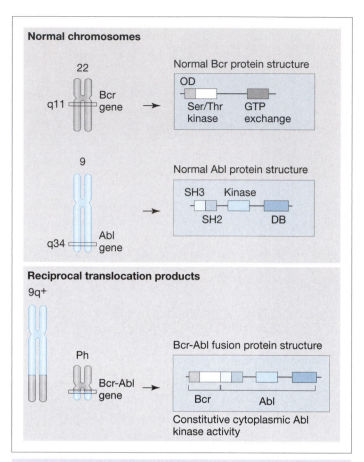

Fig. 20-4 **The Philadelphia (Ph) Chromosome**

the blood. Although CML accounts for a mere 0.3% of all cancers, it represents about 10% of adult leukemias. This malignancy usually strikes adults with a median age of 50 years, and men are affected almost twice as often as women.

i) Clinical Features

The onset of CML is relatively benign and the disease can remain "silent" in many patients for 3–5 years. This stage of the disease is known as the *chronic phase*, in which blood smears show increased numbers of precursor and mature granulocytes (Plate 20-3). However, the disease eventually progresses to an *accelerated phase* and finally to a *blast crisis*. In the accelerated phase, there is a sharp increase in the numbers of immature leukemic cells in the blood or BM. Splenomegaly is often present. In the blast crisis, CML blast cells make up more than 30% of cells in the blood or BM. Once the blast crisis has started, the leukemic cells start to infiltrate extramedullary tissues and an untreated patient survives only a few months. Symptoms of blast crisis CML include weight loss, abdominal discomfort, bleeding, weakness, lethargy and night sweats. Histological examination of the blood and BM reveals mild anemia with increased numbers of mostly granulocyte precursors but also B lymphoid (rarely T lymphoid), erythroid and megakaryocytic lineage cells, plus connective tissue-forming cells. This spectrum of leukemic cell types indicates that the original transformation event must have taken place in a very early progenitor in the BM that retained its ability to divide and differentiate into multiple hematopoietic lineages. As a result, despite the name of this disorder, cells of either the myeloid or lymphoid lineage may predominate in the blast crisis phase. Indeed, phenotypically, blast crisis CML can bear a striking resemblance to either AML or ALL.

ii) Genetic Aberrations

The tumor cells in over 90% of CML patients possess the *Philadelphia* (Ph) *chromosome*. The Ph chromosome arises from a reciprocal translocation that occurs between chromosomes 9 and 22 (Fig. 20-4). Normally, the Bcr gene on chromosome 22 encodes a protein with serine-threonine kinase and GTP exchange activities as well as an oligomerization domain (OD). The normal Abl gene on chromosome 9 encodes a protein that possesses tyrosine kinase activity, SH2 and SH3 protein interaction domains, and a DNA binding domain (DB). During leukemogenesis, a t(9;22)(q34;q11) reciprocal translocation occurs in which the chromosomal breakpoints are located in the Bcr and Abl genes. One of the products of this translocation is a shortened version of chromosome 22 (the Ph chromosome) containing a new fusion gene made up of parts of the Bcr and Abl genes. The chimeric protein encoded by the Bcr–Abl fusion gene has constitutive Abl tyrosine kinase activity and acts in the cytoplasm due to regulatory effects exerted by the Bcr moiety. The Bcr–Abl protein directly and indirectly activates other genes involved in cell division, cell survival and myeloid differentiation. A less frequent recurring chromosomal translocation found in *atypical CML* is t(9;12)(q34;p13). This fusion event joins the Abl kinase gene to the TEL gene described earlier for AML.

iii) Treatment

The prognosis of CML patients varies according to the phase of the disease at diagnosis, with patients in blast crisis having the shortest survival times. In the past, CML patients were treated with chemotherapy or with the anti-proliferative cytokine IFNα. IFNα administration is effective in over 50% of chronic phase CML patients but it is not clear exactly how it works. In the early 2000s, the drug imatinib mesylate (Gleevec[R]) was introduced to treat CML patients whose tumor cells bear Bcr–Abl or TEL–Abl translocations. Imatinib mesylate potently inhibits the activity of the chimeric Abl kinase that drives the growth and survival of the leukemic cells; normal cells are not affected at the doses used. Not only is imatinib mesylate highly efficacious for early chronic phase CML, it also has the advantages of oral administration and few systemic side effects. If a patient's CML develops resistance to imatinib mesylate, treatment with other kinase inhibitors can be effective.

Allogeneic HCT can be used to treat chronic phase CML patients who do not respond to imatinib mesylate, particularly if the patient is under 30 years of age. More than half of chronic phase CML patients who receive such transplants achieve over 5 years disease-free survival. However, if the disease has progressed to the accelerated or blast crisis phase, the chance of HCT success is reduced to about 15%.

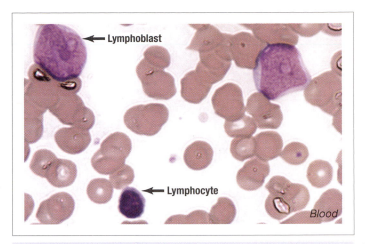

Plate 20-4 **Acute Lymphoblastic Leukemia**
[Reproduced by permission of Doug Tkachuk, Department of Pathobiology and Laboratory Medicine, University of Toronto.]

III. ACUTE LYMPHOBLASTIC LEUKEMIA (ALL)

ALL is characterized by increased numbers of lymphoblasts in the blood and BM. Cases of ALL account for 1.5% of all cancers. ALL can affect adults but occurs most often in children of age 3–5 years. Indeed, in industrialized countries, the disease represents close to 75% of childhood leukemias and is the most frequent form of cancer in children. The only firm cause identified for ALL is exposure to ionizing radiation, as occurred in Japan in 1945 and in Chernobyl in 1986. Other links sometimes cited as causes of ALL (exposure to electromagnetic fields or industrial pollutants) have yet to be definitively proven. Interestingly, like allergy and autoimmune disease, ALL is more common in developed countries than in less developed ones. Some researchers therefore speculate that skewed T cell responses to common infections (as proposed in the hygiene hypothesis) may be to blame for at least some cases of childhood ALL. However, to date, no infectious agent has been directly linked to ALL.

i) Clinical Features

In general, ALL patients first present with fever, fatigue, weight loss, infections, hemorrhages, dizziness, easy bruising and joint pain. High numbers of lymphoblast-like leukemic cells are found in the blood, BM and extramedullary tissues. Most cases of ALL involve cells of the B lineage whereas less common forms involve cells of the T lineage. Patients with B cell ALL (B-ALL) often have masses in the abdomen that can compromise kidney function. Patients with T cell ALL (T-ALL) may have a mass in the chest that can lead to wheezing and cardiac complications if it is sufficiently large. CNS involvement, characterized by leukemic cell invasion of the spinal cord and brain, is rare at ALL presentation but can be a source of relapsed disease if not addressed promptly. Symptoms of CNS involvement include headaches, nausea, lethargy and irritability.

A diagnosis of ALL is reached when analysis of the patient's BM smear reveals that at least 20% of cells present are leukemic lymphoblasts with large nuclei and minimal cytoplasm (Plate 20-4). Under the microscope, cases of B-ALL and T-ALL are morphologically alike so that immunophenotyping is used to distinguish them. B-ALL cells often exhibit surface expression of CD19 and CD20, whereas T-ALL cells tend to express CD2, CD3, CD5 and CD7.

ii) Genetic Aberrations

Some of the genetic changes leading to ALL development are similar to those found in AML and CML patients. The tumor cells in about 25% of adult and 5% of childhood ALL cases exhibit a variant of the chimeric Bcr–Abl protein first described for CML. About 30% of childhood ALL cases show the t(12;21)(p13;q22) translocation in which the gene encoding the CBF transcription factor subunit CBF2A is translocated to the TEL transcriptional repressor locus. Other recurring chromosomal translocations leading to ALL involve the introduction of gene fragments into either the Ig loci or the TCR loci. In cases of mature cell B-ALL, the c-myc gene is inserted into the Igh locus such that c-myc becomes constitutively activated and deregulates the expression of genes involved in cell proliferation, apoptosis and differentiation. Similarly, the insertion of the c-myc gene into the TCRA locus results in cases of T-ALL.

Other genetic aberrations associated with ALL involve too many or too few chromosomes in the leukemic cells. In many cases of precursor cell B-ALL in children, the leukemic cells show more than 50 chromosomes per cell (normal cells have 46 chromosomes). These cases have a relatively favorable prognosis. Trisomy (three copies of a chromosome, rather than the usual two) for chromosomes 4, 10 or 17 is particularly common in childhood ALL. Other ALL leukemic cells show fewer than 45 chromosomes; these cases have a relatively poor prognosis. Intriguingly, with or without a chromosomal abnormality, more than half of T-ALL cases show inactivating mutations in the gene encoding the important cell fate-determining protein Notch1.

iii) Treatment

Chemotherapy is the first line of treatment for ALL. Modern drug regimens are quite successful and the majority of treated patients become long term survivors. Although mAb-based immunotherapy has been explored as a treatment for ALL, it has not been as effective as for other types of leukemias. An exception is the success of imatinib mesylate treatment for those ALL patients whose leukemic cells bear the Ph chromosome. Allogeneic HCT is a viable option for relapsing patients but the success rate is low for adult cases of ALL.

IV. CHRONIC LYMPHOCYTIC LEUKEMIA (CLL)

CLL arises from the transformation of a peripheral (mature) blood lymphocyte. As its name suggests, CLL is a chronic form of lymphocytic leukemia and is characterized by a prolonged disease course. CLL mostly affects individuals over 60 years of age and occurs about twice as often in men as in women. Curiously, although CLL represents about 35% of adult leukemias in North America, it accounts for fewer than 5% of adult cases in Asia.

i) Clinical Features

Some CLL patients have no clinical symptoms of this disease at diagnosis and are only identified when a blood analysis is conducted to investigate other health concerns. These patients are considered at "low risk" and generally survive at least another 10 years even without treatment. Other CLL patients show splenomegaly or hepatomegaly and are considered to be at "intermediate risk". These patients live an average of 7 years following diagnosis. Patients considered at "high risk" have more advanced disease and display *lymphadenopathy* (swelling of the lymph nodes) and hepatosplenomegaly. Bacterial and viral infections are frequent in high risk CLL patients. Furthermore, the chance of a CLL patient developing an NHC is twice that of the general population. As a result of all these difficulties, CLL patients with advanced disease can expect to live only 2–4 years after diagnosis.

CLL is mainly a disorder of B lineage differentiation and proliferation. Normal B cell production in the BM and B cell functions in the periphery are disrupted. Two subgroups of these patients have been identified at the molecular level, each accounting for about half of all cases. In one subgroup, the malignant B cells have Ig genes showing evidence of somatic hypermutation; the prognosis for these patients is relatively favorable. In patients in the other subgroup, the Ig genes of the malignant cells have not undergone somatic hypermutation; these patients have a poor prognosis. Alterations to the normal levels of many cytokines have also been noted in both subgroups. Perhaps as a result of these cytokine abnormalities, helper T cell responses to antigens are also impaired. All of these factors culminate in reduced antibody responses in CLL patients. To add insult to injury, about 20% of all CLL patients develop some type of autoimmune disease such as autoimmune hemolytic anemia or ITP (refer to Ch. 19).

About 5% of CLL cases involve transformed mature T lymphocytes. Patients with this HC respond only weakly to chemotherapy and thus have a very poor prognosis.

ii) Genetic Aberrations

Leukemic cells in about 80% of CLL cases show overt cytogenetic alterations. The most common mutation, found in over 50% of CLL patients, is a deletion at 13q14. The deletion affects two non-coding "micro-RNA" genes called miR15 and miR16 that are involved in tissue-specific gene regulation. The Rb TSG found in this chromosomal region is normal in these patients. Another chromosomal deletion found in 20% of CLL cases occurs at 11q23 and removes the ATM TSG. The disease course is particularly aggressive in 11q23 CLL patients. About 15% of CLL patients have trisomy 12. In half of trisomy 12 CLL cells, a translocation at t(14;19)(q32;q13) joins the Igh locus to that encoding the transcriptional coactivator Bcl-3. Overexpression of the Bcl-3 gene stimulated by elements in the Igh locus may in turn drive excessive expression of survival genes, causing these lymphocytes to accumulate.

Atypical CLL occurs in about 15% of patients. In these cases, the CLL cells acquire new mutations that drive the conversion of the disease into a lymphoma. The disease course is aggressive and the prognosis poor. This group of patients exhibits abnormalities of chromosome 17, some of which affect the p53 TSG located at 17p13.

iii) Treatment

Untreated CLL patients survive for an overall average of 6 years following diagnosis. Because CLL patients are generally over 60 years of age, it is important to weigh the benefits of treatment against the significant side effects of chemotherapy. Infections due to chemotherapy-induced immunosuppression pose a real threat to the older patient. Consequently, one of the most important management issues of CLL is to decide when to simply watch disease progression and when to medically intervene.

When CLL requires treatment, chemotherapy using a combination of fludarabine and cyclophosphamide is the preferred first line of attack. Clinicians are also turning to combinations of chemotherapy and immunotherapy. An immunotherapeutic agent that has been useful for treatment of CLL is the CAMPATH-1H mAb that binds to CD52 expressed on most B and T lymphocytes (refer to Ch. 19). About 30% of CLL patients who do not benefit from chemotherapy will respond to CAMPATH-1H therapy. Another subset will respond to high doses of anti-CD20 mAb. IFNα, which works well for CML, appears to be of minimal benefit to CLL patients. HCT is rarely attempted as a treatment for CLL because most patients are at an age when they are less likely to survive the HCT procedure. However, some younger CLL patients have enjoyed prolonged remission following HCT.

C. Myelomas

Normal plasma cells cannot divide and so die soon after doing their job of secreting antigen-specific antibody for a few days.

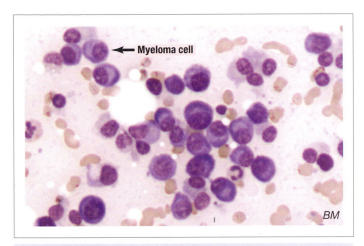

Plate 20-5 Myeloma
[Reproduced by permission of Doug Tkachuk, Department of Pathobiology and Laboratory Medicine, University of Toronto.]

In contrast, cancerous plasma cells divide uncontrollably and express huge quantities of antibodies or single Ig chains of unknown antigenic specificity. The Ig protein produced by the malignant plasma cell is called a *paraprotein* (as in "IgM paraprotein" or "para-IgM"). As a group, HCs involving transformed plasma cells are called *plasma cell dyscrasias*. Plasma cell dyscrasias are not considered leukemias because, although their paraprotein products may enter the blood, the cancerous plasma cells themselves are (at least initially) confined to the BM and do not enter the circulation. There is a broad spectrum of plasma cell dyscrasias whose severity ranges from the almost asymptomatic to life-threatening. At the mildest end of this spectrum is the colorfully named "monoclonal gammopathy of undetermined significance" (MGUS). About 70% of MGUS patients show no clinical symptoms. If symptoms do occur, they take the form of mild anemia and usually disappear on their own with time. The most serious plasma cell dyscrasias are the myelomas (Plate 20-5). Myelomas can arise from pre-existing MGUS disease when a minimally transformed plasma cell acquires additional mutations that cause it to divide aggressively and synthesize high levels of a particular paraprotein. In its most advanced stages, large numbers of para-Ig-secreting tumor cells leave the BM and take up residence in multiple body sites, such that this disease is often referred to as *multiple myeloma*.

I. CLINICAL FEATURES

Myelomas represent about 1% of all malignancies in the United States, with over 80% of cases occurring in persons over 60 years of age. Myeloma incidence is slightly higher in men than women and occurs in twice as many blacks as whites. Exposure to ionizing radiation has been implicated in myelomagenesis. For example, epidemiological studies of individuals exposed to atomic radiation in Hiroshima and Nagasaki demonstrated a five-fold increase in myelomagenesis beginning about two decades later. Exposure to environmental carcino-

gens or certain chemicals also increases the rate of myeloma development.

Myeloma cells cause disease because their high numbers clog the BM and disrupt hematopoiesis such that the production of erythrocytes is decreased, resulting in anemia. Normal B cell production may also be impaired, leaving myeloma patients very vulnerable to infections. As well, the physical presence of high concentrations of para-Ig in the blood can cause coagulation and circulatory difficulties. Pulmonary and neurological effects may also be observed. In many myeloma cases, the transformed cells produce an abundance of Ig light chains in the absence of Ig heavy chains. The free Ig light chains can form aggregates with a glycoprotein in the urine, and deposition of these aggregates in the kidney can result in the failure of this organ.

Myeloma disease is classified according to the staging system shown in Table 20-2. These stages are defined by the number of malignant cells present, their pattern of distribution in the body, and blood levels of hemoglobin, calcium and paraproteins. In stage I myeloma disease, the myeloma clone remains partially reliant on BM stromal cell factors for its survival and so is confined to the BM. Serum hemoglobin and calcium levels are normal, the number of malignant cells is low and the production of Ig paraproteins is relatively low. Clinical symptoms are often mild and may go unnoticed. Stage II myeloma patients show subnormal hemoglobin, elevated serum calcium, moderate levels of malignant cells in the BM and moderate levels of Ig paraprotein the blood. In patients with stage III myeloma disease, the number of malignant cells is high and large masses may form in the BM. In addition, the myeloma cells accumulate additional oncogenic mutations such that they lose their dependence on the BM stroma. Large numbers of malignant cells migrate out of the BM and into the blood, infiltrating a variety of extramedullary tissues and disrupting their functions. Serum calcium is highly elevated and hemoglobin is severely decreased, frequently leading to renal complications and anemia. Levels of Ig paraproteins may be three times those in stage I patients.

Table 20-2 Stages of Myeloma

	Stage I	Stage II	Stage III
Number of malignant cells	Low	Moderate	High
Location	Bone marrow	Bone marrow	Bone marrow, blood, extramedullary tissues
Serum Hb and Ca^{2+}*	Both normal	$\downarrow$ Hb $\uparrow$ Ca^{2+}	$\downarrow\downarrow$ Hb $\uparrow\uparrow$ Ca^{2+}
Ig paraprotein	Low	Moderate	High
Bone pain	No	No	Yes
Mean survival (untreated)	5 year	3 year	1 year

*Hb, hemoglobin; Ca^{2+}, calcium.

One of the biochemical consequences of myelomagenesis is the deregulation of several cytokines and chemokines, the most prominent of which is IL-6. When overexpressed either by the myeloma cells themselves or by BM stromal cells, IL-6 appears to drive the development and maintenance of the cancer. When this cytokine is combined with IL-1, TNF and a chemokine called MIP-1α (all secreted by myeloma cells), bone metabolism is disrupted such that lesions develop in the bone structure. These lesions cause significant pain to many stage III patients. A protein called DKK1 also contributes to bone lesion development. DKK1 inhibits the intracellular signaling that supports bone formation. Myeloma cells overexpress DKK1 such that bone formation is suppressed and bone resorption is stimulated. Additional cytokines implicated in myeloma disease progression include IL-2, IL-7, IL-11, LT and GM-CSF. The upregulation of these cytokines and their receptors appears tends to suppress Ig production by normal B cells and to inhibit helper T and NK cell functions, thereby increasing the susceptibility of these patients to infection.

II. GENETIC ABERRATIONS

The genetic aberrations leading to myelomagenesis are complex and have been much harder to identify than those associated with leukemias. There are some recurring chromosomal translocations associated with myelomas but no one genetic change appears to be dominant or essential. Trisomy 3, 5, 6, 7, 9, 11, 15 or 19 has been reported in various myeloma clones. For unknown reasons, aberrations of chromosomes 6 and 9 are associated with the most favorable disease outcomes. In contrast, a deletion in the chromosomal region 13q14 that affects the Rb TSG is associated with a very poor prognosis. Cases of stage III myeloma often show reciprocal translocations that affect the Igh locus (refer to Box 20-2). The genes involved in these translocations include the fibroblast growth factor receptor FGFR3 and the cell cycle regulator cyclin D1. More subtle genetic abnormalities include mutations of the c-myc or p53 genes or activation of the Ras oncogene. These alterations are frequently reported in stage II and III myeloma patients. Other myeloma patients have mutations in a group of cyclin-dependent kinases associated with cell cycle control. Interestingly, regardless of the underlying genetic defect, almost all myeloma clones show upregulation of the anti-apoptotic genes Bcl-2 and Bcl-xL. Overexpression of these cell survival proteins, which is also seen in certain lymphomas, is thought to drive myeloma expansion.

III. TREATMENT

Myelomas are among the hardest HCs to treat. Chemotherapy is the first line of treatment for myeloma patients with progressing disease but the 5-year survival rate is only about 30%. At later stages or in the most aggressive cases, magnetic resonance imaging (MRI) may be helpful in evaluating the number and location of myeloma masses. Localized irradiation can be effective for patients with isolated myeloma masses. As was true for CLL treatment, the timing and choice of myeloma therapy are especially important for older patients, for whom the side effects of medical intervention have to be balanced against quality of life.

The drug *thalidomide*, which was originally prescribed for the relief of nausea in pregnant women, became infamous in the 1960s for its devastating effects on fetuses. In the twenty-first century, this drug is finding a new use as a chemotherapeutic agent for myeloma. Thalidomide promotes the apoptosis of myeloma cells and blocks the secretion of IL-6 and angiogenic factors that support myeloma growth. However, the side effects of thalidomide use are significant. Researchers are now investigating the efficacy of several less toxic thalidomide analogues, including lenalidomide. Lenalidomide is a more potent immunoregulator than thalidomide and has shown considerable efficacy in treating myeloma. Another approach to myeloma treatment has been the use of a proteasome inhibitor called bortezomib (Velcade^R). Bortezomib inhibits the degradation of proteins by the proteasome, including the degradation of an inhibitor of the transcription factor NF-κB that is required for IL-6 expression. If the inhibitor is not degraded, NF-κB cannot be activated, and the production of IL-6 that drives myeloma cell proliferation is blocked. Bortezomib also directly induces the apoptosis of myeloma cells and decreases angiogenesis.

Autologous HCT is a realistic option for many myeloma patients, especially for those younger than 70 years of age. The mean survival of myeloma patients after autologous HCT is about 5 years. Allogeneic HCTs have also been attempted, in order to take advantage of a postulated GvL-like effect against myeloma cells, but these efforts have been associated with high morbidity and mortality.

D. Lymphomas

Lymphomas are solid cancers that initiate from the malignant transformation of a single lymphocyte. The affected lymphocyte is usually located in a lymph node but may be resident in another organized lymphoid tissue outside the BM such as the spleen or thymus. When the transformed lymphocyte is located in a diffuse lymphoid tissue such as the GALT, the lymphoma that develops is said to be *extranodal*. Lymphomas are generally dependent on surrounding stromal cells for survival and growth factors as well as vital intercellular contacts, and so are generally restricted to sites within tissues. Thus, from its initiation site, a lymphoma tends to spread to additional secondary lymphoid tissues and eventually to non-lymphoid organs. Occasionally, a lymphoma cell undergoes additional mutations that allow it to survive and circulate in the blood; i.e., it becomes a leukemic cell. The disease may then be called a "leukemia/lymphoma".

The progression of any lymphoma can be described in four stages, as set out in Table 20-3. The identification of tumor sites and the staging of lymphomas have been made more accurate in recent years by the application of a diagnostic imaging technique called ^{18}F-FDG PET/CT. PET/CT stands for *positron emission tomography/computerized tomography* and ^{18}F-FDG is a radioactive form of glucose that emits positrons. Tumor cells use glucose at a much higher rate than normal

Table 20-3 Stages of Lymphoma

Stage	Diagnostic Features
I	One or more diseased lymph nodes are present in a single group of lymph nodes in one particular lymphoid tissue of the body.
II	Diseased lymph nodes are present in more than one group of lymph nodes but all diseased nodes are contained either above or below the diaphragm. Tumor cells may also be present in a single organ near an affected node.
III	Diseased lymph nodes are present in two or more groups on both sides of the diaphragm. Tumor cells may also be present in the spleen and/or another organ near an affected node.
IV	Wide dissemination of tumor cells into multiple lymph nodes, bone marrow, liver and multiple organs.

cells, and so stand out in images of the body when the patient has been administered minute amounts of ^{18}F-FDG as a tracer and PET/CT imaging has been applied.

Lymphomas are broadly classified into *Hodgkin's lymphomas* (HLs) and *non-Hodgkin's lymphomas* (NHLs). Of all HCs, 12.5% are HLs and 44.5% are NHLs. HLs and NHLs are distinguished by the architecture of the malignant mass, the morphology of its component cells, and the surface marker phenotype of these cells. In HL, only a tiny percentage of cells in the tumor mass are actually cancerous, and these cells are usually a peculiar B lineage-like cell type called *Reed-Sternberg* cells. The remainder of the tumor mass is made up of a so-called *reactive infiltrate* composed of <u>non-transformed</u> lymphocytes, fibroblasts and other cell types. In NHL, the solid mass consists almost entirely of transformed lymphocytes. The malignant cells are most often derived from a peripheral B cell but sometimes from a peripheral T cell. Occasionally, the cellular origin of the lymphoma cannot be clearly defined.

I. HODGKIN'S LYMPHOMA (HL)

In about 1900, researchers studying the enlarged lymph nodes of "Hodgkin's disease" patients found that each affected node contained a few large, multinucleated cells within an infiltrate of normal-looking cells. These unusual multinucleated cells became known as Reed-Sternberg (RS) cells in honor of their discoverers. RS cells were subsequently shown to be clonal in their growth, suggesting that they were transformed. Indeed, Hodgkin's disease was eventually demonstrated to be a true HC in which the RS cells were the tumor cells of the malignant mass. Hodgkin's disease was later renamed "Hodgkin's lymphoma" (HL) and tissues showing the infiltration of an HL mass are said to be "HL-involved".

i) Clinical Features

HL is a rare disorder, representing 0.7% of all malignancies in the United States. Unlike most solid NHCs, HL usually affects relatively young patients between the ages of 15 and 35

years. Males are affected slightly more often than females. Most patients initially present with a lump in the neck region due to one or more enlarged lymph nodes, although chest masses may appear upon radiography. Patients also experience a constellation of systemic complaints known as *B symptoms*, which are defined as unexplained rapid weight loss, fatigue, cyclic bouts of fever, and night sweats frequently accompanied by chest pain. *Pruritis* (intense itching) is present in about 20% of patients. HL patients are highly vulnerable to fungal and viral infections.

At least in the early stages of HL, RS cells do not establish just anywhere in the body. Instead, these cells appear to travel via the lymphatics in a sequential fashion from one lymph node to the next lymph node in the anatomical chain. It is not understood exactly how this unique pattern of HL spreading occurs but it appears to depend on some type of close range interaction either between cells or between cells and cytokines. Some scientists have speculated that cytokines accumulating in an affected lymph node spill down the efferent lymphatic, exiting that node to enter the next node in the chain. These cytokines may then alter conditions among the cells in the second node such that a reactive infiltrate develops. When an RS cell does break away from the original node and reaches the second node, conditions may then be ripe for its continued survival and proliferation. Histological studies have confirmed that lymph nodes near an HL-involved lymph node frequently contain reactive infiltrate but no RS cells. In the later stages of HL, however, the RS cells acquire additional mutations that allow them to become independent of the support of cytokines and the reactive infiltrate. The disease then spreads to additional lymph nodes in a less systematic fashion and invades non-lymphoid organs and tissues.

ii) Genetic Aberrations

It has been difficult to ascertain what types of mutations are associated with HL because of the challenges in obtaining adequate amounts of relatively pure populations of fresh RS cells for biochemical and molecular analyses. Although recurring chromosomal translocations are not common, the tumor cells of 25–75% of HL patients have cytogenetic anomalies such as trisomy 1, 2, 5, 12 or 21. Mutations of known TSGs and oncogenes have been found in only a tiny percentage of HL cases. Most researchers believe that the unknown genetic changes underlying HL somehow lead to the constitutive activation of a key transcription factor in RS cells. This transcription factor would then drive abnormal expression of genes promoting proliferation and cell survival.

iii) Role of Cytokines in HL

Whatever the defect in the genome of an RS cell, the abnormality has a profound effect on the regulation of cytokines and their receptors that appears to be crucial for the development of HL disease. HL patients succumb readily to fungal and viral infections, indicating that Th1 and CTL responses are impaired. However, antibody responses to pathogens are functional. This reduction in cell-mediated immunity coupled with normal humoral immunity suggests that immune deviation to a Th2 response may be occurring during HL development.

A multitude of cytokines has been implicated in HL in various studies but it is IL-13 that plays the most important role in HL pathogenesis. Only IL-13 is consistently and abundantly expressed by RS cells from different HL patients. The IL-13 receptor is also upregulated on RS cells, establishing a feedback loop in which the IL-13 produced by RS cells binds to IL-13R on the same (and other) RS cells. The persistent IL-13 signaling results in the constitutive activation of transcription factors that in turn activate the transcription of genes driving RS cell proliferation and survival. Some immunologists theorize that a minimally transformed B cell that starts to overproduce IL-13 may respond to that IL-13 by becoming a full-fledged RS cell. IL-13 and other cytokines frequently elevated in HL may also contribute to the recruitment of the cells composing the reactive infiltrate, and this infiltrate may secrete still more cytokines and/or supply intercellular contacts that promote RS survival and proliferation.

iv) Viral Involvement?

The exact etiology of HL is unknown but some scientists theorize that infection by EBV, a virus capable of transforming lymphocytes, may be a causative factor in some cases. Two EBV proteins can be found on the surfaces of RS cells in 40% of HL tumors, suggesting that these viral proteins might have a role in HL development. In addition, clinical studies have found that HL is two- to five-fold more frequent in patients who have had infectious mononucleosis, a disease known to be caused by EBV. However, 60% of HL tumors show no evidence of EBV proteins. Moreover, at least in North America, over 80% of the population has experienced an EBV infection at some point in their lives, but only a fraction of this population ever develops HL. Whether EBV is a contributing factor to HL therefore remains controversial.

v) Subtypes of HL

Two major subtypes of HL exist: *classical HL* (95% of cases) and *nodular lymphocyte predominant HL* (NLPHL; 5%). Classical HL is characterized by the presence of RS cells at a frequency of 0.1–1% (Plate 20-6). As well as by their multiple nuclei, RS cells are identified by their distinctive shape and their expression of the markers CD30 and CD15. RS cells are thought to be derived from B cell precursors at an early stage of differentiation. Accordingly, the Ig genes are often rearranged in RS cells but there is no evidence of somatic hypermutation. Rarely, there is some expression of Ig light chains. In about 40% of classical HL cases, the RS cells express CD20.

NLPHL is characterized by the presence of malignant *lymphocytic* and/or *histiocytic* (L&H) cells rather than RS cells. L&H cells are also called "popcorn cells" due to their unique appearance (Plate 20-7). L&H cells do not usually express CD30 or IL-13 but almost always express CD19 and CD20. Unlike the Ig genes in RS cells, the Ig genes in L&H cells show evidence of considerable somatic hypermutation. Ig light chains are more frequently expressed than in classical HL cases.

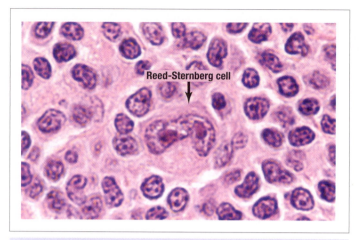

Plate 20-6 Classical Hodgkin's Lymphoma
[Reproduced by permission of Doug Tkachuk, Department of Pathobiology and Laboratory Medicine, University of Toronto.]

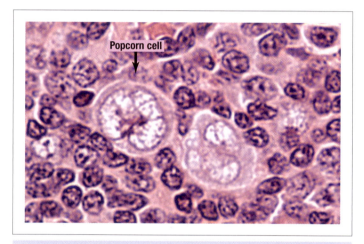

Plate 20-7 "Popcorn cells" in Nodular Lymphocyte Predominant Hodgkin's Lymphoma
[Reproduced by permission of Doug Tkachuk, Department of Pathobiology and Laboratory Medicine, University of Toronto.]

vi) Treatment of HL

The prognosis of HL patients was improved considerably in the early 1980s by the discovery that HL disease spread sequentially through the lymphatics to adjacent lymph nodes. Prophylactic irradiation of lymphatic channels in the immediate area of the affected node, a procedure called *extended field radiation*, became routine and increased the survival of many HL patients. Today, although some early stage HL patients may require only radiation treatment, chemotherapy is advised for patients with more aggressive subtypes of the disease. Patients with advanced HL undergo multiple rounds of chemotherapy combined with aggressive local irradiation. These regimens, coupled with good clinical management to control infections, have enabled about 80% of both early and later stage HL patients to enjoy long term disease-free survival.

Unfortunately, the increased doses of chemotherapy and radiation required to achieve these excellent results for advanced HL are associated with several serious side effects. The impact of these problems is made more devastating by the fact that HL patients are generally quite young (20-40 years of age). Sterility and hypothyroidism are not uncommon after extensive chemotherapy, and pulmonary and cardiac complications are frequent. However, the most sobering unintended consequence of aggressive HL treatment is a high rate of secondary tumorigenesis. AML, NHL and even some NHCs such as lung and breast cancers have appeared in HL patients within 5–15 years after initial treatment. Indeed, about 15% of conventionally treated HL patients develop a secondary cancer within 20 years.

The existence of HLs that resist aggressive chemotherapy and irradiation, coupled with the desire to eliminate secondary cancers and severe complications, drives the development of alternative therapies for HL. Patients now have several other options, some of which are still experimental in nature. Anti-CD20 mAb treatment has been helpful for HL cases featuring CD20+ RS cells. In addition, clinical trials are ongoing to evaluate the effectiveness of immunotoxins directed against the CD30 protein found on the RS cells of a majority of HL cases. The administration of neutralizing antibodies against IL-13, the growth factor postulated to drive RS cell proliferation and survival, is also being assessed as a novel therapy for HL. Autologous HCT has been used to successfully treat some HL patients, and allogeneic HCT is under investigation.

II. NON-HODGKIN'S LYMPHOMA (NHL)

The NHLs are a family of heterogeneous cancers. Unlike HL, the cancerous mass of an NHL is composed almost entirely of malignant lymphoid cells. In addition, NHLs lack unique tumor cell types such as RS cells or L&H cells. NHLs account for about 3–5% of all human cancers, occurring twice as often in men as in women. The median age of NHL patients is 65 years, considerably older than that for HL patients. Nevertheless, significant numbers of younger people between the ages of 30 and 40 develop NHL disease. Most NHLs are generally rare in children with the exception of *Burkitt's lymphoma* (see later) in African children. Interestingly, the incidence of NHL in adults has doubled over the last three decades. Although a small percentage of this "epidemic" is due to a 10-fold increase in lymphomas in AIDS patients, an across-the-board elevation in incidence has been found for the general population. At least part of this increase has been attributed to a combination of factors, including: (1) an expanding population of older individuals; (2) new imaging technologies that can detect previously unnoticed lymphomas; and (3) the reclassification of some HL malignancies under the NHL umbrella. Although specific genetic determinants promoting the development of most NHLs have yet to be conclusively identified, there is accumulating evidence that infection with certain pathogens can cause at least some NHLs. Prolonged exposure to various noxious substances has been cited as a possible cause of NHLs but there is as yet no definitive proof of this link.

i) Clinical Features

NHL patients usually present with B symptoms (unexplained weight loss, fever, night sweats) accompanied by fatigue and pain in the bones, chest and abdomen. Physical examination may reveal lymphadenopathy and/or hepatosplenomegaly. The lymphadenopathy may "wax and wane" over time, meaning that the lymph nodes swell and resolve repeatedly. Sometimes the NHL presentation is extranodal, in that the lymph nodes are not obviously affected and the suspicious lumps are present in other tissues such as the skin and GI tract. Most NHLs eventually go on to involve the BM.

Unlike HL, NHL does not usually spread sequentially from lymph node to lymph node. Rather, NHL cells may migrate from the initial affected lymph node and travel via the blood and lymphatics to scattered nodes in various regions of the body. It is postulated that the NHL tumor cells that have left the original lymph node are no longer reliant on a particular microenvironment or cytokine gradient, making local and distant lymph nodes equally suitable for invasion.

ii) Pathogen Involvement

Although the causes of most cases of NHL are unknown, infection by a pathogen has been linked to several types of NHLs. In some cases, an oncogenic virus transforms lymphocytes directly, leading to malignancy and lymphomagenesis. In other cases, chronic inflammation induced in response to a persistent pathogen is thought to interfere with normal lymphocyte proliferation and apoptosis. The affected cells may then accumulate mutations to DNA repair genes, oncogenes and/or TSGs that promote lymphomagenesis. Thus, different NHLs are prevalent in different geographic areas, depending on the global distribution of the relevant associated pathogen. For example, the greatest numbers of T cell lymphoma cases occur in Japan, in the Caribbean islands and in the countries surrounding the Mediterranean Sea, where there is a high rate of infection by human T cell leukemia virus-1 (HTLV-1). HTLV-1 is an oncogenic retrovirus capable of inducing T cell malignancies (both leukemias and lymphomas) several decades after infection. Similarly, the incidence of gastric lymphoma is high in regions of the world (such as Italy) where *Helicobacter pylori* bacteria abound. Burkitt's lymphoma, which is linked to EBV infection, is most often found in equatorial Africa where this virus is highly prevalent. Consistent with the involvement of a pathogen in NHL initiation, SCID and AIDS patients infected with EBV have a dramatically enhanced risk of developing NHL compared to patients that escape EBV infection.

iii) Subtypes of NHL

The NHLs occur in a wide range of subtypes that are classified based on a complex portfolio of morphological and immunophenotypic criteria, clinical features, and genetic aberrations. However, within a given subtype, there may be variations of disease with features that may substantially overlap with those of another subtype or even with a leukemia. In the majority of adult NHLs, the target cell is a peripheral B cell but some are derived from a peripheral T cell. In children, B cell and T cell lymphomas occur at almost equal frequency. The clinical features and relative frequencies of the major B and T cell NHL subtypes are summarized in Tables 20-4 and 20-5, respectively. The distinguishing histological features of three NHLs are shown in Plate 20-8. In panel A of Plate 20-8, a lymph node has been completely filled by lymphoma cells in a patient

Table 20-4 **NHL Subtypes: B Cell Lymphomas**

Abbr.	Name	Frequency (% of all NHLs)	Distinguishing Features
B-LBL	Precursor B cell lymphoblastic leukemia/lymphoma	<1% (A*) 2.5% (C)	Early[†] onset, very aggressive Tissue invasion by immature B cells Can have leukemia component Ig genes may be rearranged Good prognosis with vigorous therapy
MCL	Mantle cell lymphoma	7% (A)	Late[†] onset, often aggressive Invading B cells are from follicular mantle Often shows t(11;14)(q13;q32) [cyclin D1; Igh] Colon often shows invasion Improved prognosis with mAb therapy
B-CLL	B cell chronic lymphocytic lymphoma	7% (A)	Late onset, indolent[†] Has leukemia component Good prognosis with therapy
FL	Follicular lymphoma	22% (A)	Late onset, indolent Invading B cells are from follicular center Often shows t(14;18)(q32;q21) [Igh; Bcl-2] Poor prognosis even with therapy

MALT-L	Mucosa-associated lymphoid tissue lymphoma	8% (A)	Late onset; usually indolent but can be aggressive
			Extranodal, usually in GI mucosa
			Some are associated with t(11;18)(q21;q21) [IAP-2[§]; MALT1]
			    or t(1;14)(p22;q32) [Bcl-10; Igh]
			    or t(14;18) (q32;q21) [Igh; MALT1]
			Others are associated with *H. pylori* infection
			Sometimes progresses to DLCL
			Good prognosis with therapy

DLCL	Diffuse large cell lymphoma	33% (A) 12% (C)	Early or late onset, aggressive Tumor cells are large and exhibit variable morphology Extranodal growth is common Often associated with t(3;14)(q26;q32) [Bcl-6; Igh] or t(3;22)(q26;q11) [Bcl-6; MN1] Fair prognosis with therapy
BL	Burkitt's lymphoma	2% (A) 36% (C)	Usually early onset, very aggressive Often associated with t(8;14)(q24;q32) [c-myc; Igh] Sometimes associated with t(2;8)(p11;q24) [Igk; c-myc] or t(8:22)(q24;q32) [c-myc; Igl] Good prognosis with vigorous therapy

*A, in adults; C, in children.

[†]Early onset, in children; late onset, in older adults.

[‡]Indolent, slow-growing low grade lymphomas. Percentages do not add up to 100% due to the occurrence of unclassified NHLs.

[§]Description of genes not in Box 20-2: IAP-2, apoptosis inhibitor; MALT1, scaffold protein required for intracellular signaling; Bcl-10, scaffold protein required for intracellular signaling; Igk, Ig kappa light chain; Igl, Ig lambda light chain; ACK, tyrosine kinase; nucleophosmin, nucleolar export protein.

Table 20-5 NHL Subtypes: T Cell Lymphomas

Abbr.	Name	Frequency (% of all NHLs)	Distinguishing Features
T-LBL	Precursor T cell lymphoblastic leukemia/lymphoma	1.7% (A*) 30% (C)	Usually early[†] onset, very aggressive Tissue invasion by immature T cells Can have leukemia component TCRB and TCRG may be rearranged Good prognosis with vigorous therapy
ATL	Adult T cell leukemia/lymphoma	6% (A)	Late onset Indolent at first, then aggressive[‡] Tumor cells are mature T cells of heterogeneous morphology Often has leukemia component Often associated with HTLV-1 Fair prognosis with vigorous therapy
ALCL	Anaplastic large cell lymphoma	2% (A) 13% (C)	Usually early onset, aggressive Tumor cells are large mature T cells of distinctive morphology TCR genes are rearranged in most cases Often associated with t(2;5)(p23;q35) [ALK[§]; nucleophosmin] Good prognosis with therapy
AITL	Angioimmunoblastic T cell lymphoma	4% (A)	Late onset, very aggressive Tumor cells are mature T cells Some association with EBV infection Poor prognosis even with therapy
MF	Mycosis fungoides	5% (A)	Late onset, indolent Appearance and symptoms resemble fungal skin infection Tumor cells are mature T cells Sometimes associated with bacterial or HTLV-1 infections Good prognosis with early therapy

*A, in adults; C, in children.
[†]Early onset, in children; late onset, in older adults.
[‡]Indolent, slow-growing low grade lymphomas. Percentages do not add up to 100% due to the occurrence of unclassified NHLs.
[§]Description of genes not in Box 20-2: IAP-2, apoptosis inhibitor; MALT1, scaffold protein required for intracellular signaling; Bcl-10, scaffold protein required for intracellular signaling; Igk, Ig kappa light chain; Igl, Ig lambda light chain; ACK, tyrosine kinase; nucleophosmin, nucleolar export protein.

with mantle cell lymphoma (MCL). Panel B shows a BM biopsy in a Burkitt's lymphoma (BL) patient in which rapidly growing tumor cells have displaced all normal hematopoietic cells. The "starry sky" appearance of the sample is caused by the presence of vacuole-bearing macrophages that have phagocytosed apoptotic tumor cells and tumor debris. Panel C shows a skin biopsy of a mycosis fungoides (MF) patient. Aggressive infiltration of the epidermis by malignant T cells can be seen.

iv) Treatment of NHLs

The heterogeneity of NHL subtypes means that there is wide variability in their response to treatment. However, there are some common approaches that are used at least initially for therapy of these malignancies.

Chemotherapy involving combinations of several anti-cancer drugs is the weapon of choice used to fight most NHLs, although irradiation of a localized mass can also be utilized. Several rounds of chemotherapy and irradiation may be required to achieve remission, but repeated relapses may still occur if the lymphoma is particularly aggressive, like precursor B cell lymphoblastic leukemia/lymphoma (B-LBL) or follicular lymphoma (FL). If a lymphoma threatens to invade the CNS and brain, a procedure called *CNS prophylaxis* may be carried out. CNS prophylaxis relies on *intrathecal chemotherapy* in which a drug that is slightly less toxic to the CNS than to proliferating lymphocytes is introduced just under the sheath covering the spinal cord. Localized irradiation of the CNS and brain may also be applied. CNS prophylaxis has also proven effective in preventing the spread of BL and some forms of diffuse large cell lymphoma (DLCL).

If a patient is younger than 60 years, lacks CNS involvement and is in a second remission, an autologous or allogeneic HCT may be beneficial. For example, autologous and allogeneic HCTs have been successful in prolonging the lives of some younger FL, DLCL and adult T cell leukemia/lymphoma (ATL) patients. HCTs can be a particularly attractive option (even considering the risk of GvHD) for young patients with very aggressive cancers, such as MCL. If left untreated, this type of lymphoma is associated with such a short life expectancy that the risks associated with the disease usually outweigh the dangers of GvHD.

Immunotherapy (often in combination with chemotherapy) is now the standard of care for treatment of several NHLs. For

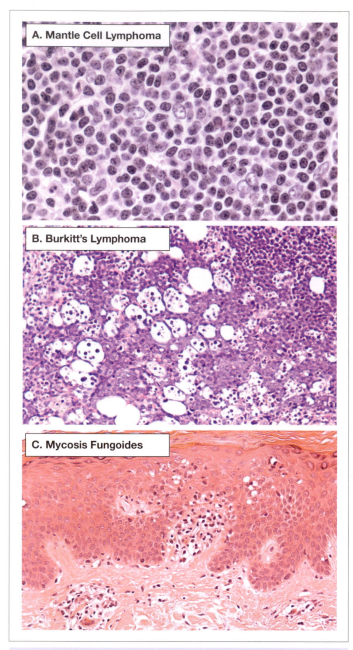

A. Mantle Cell Lymphoma

B. Burkitt's Lymphoma

C. Mycosis Fungoides

Plate 20-8 Examples of Subtypes of Non-Hodgkin's Lymphoma
[Reproduced by permission of Doug Tkachuk, Department of Pathobiology and Laboratory Medicine, University of Toronto.]

example, some older MCL or FL patients suffering a relapse have achieved a partial remission after administration of anti-CD20 mAb or IFNα. Anti-CD20 mAb and CAMPATH-1H have also been employed with some success for cases of B cell chronic lymphocytic lymphoma (B-CLL). Some DLCL patients for whom chemotherapy has not worked show long-lasting responses to IFNα treatment, and about 40% of patients (usually those with small tumors) benefit to some degree from treatment with anti-CD20 mAb. Patients with MF often find that IFNα or CAMPATH-1H is helpful. As well, about half of the malignant cells in MF patients express IL-2R so that an immunotoxin created by fusing the receptor-binding domain of IL-2 to diphtheria toxin has been investigated as a potential therapy. This agent, called denileukin difitox, binds to IL-2R-expressing cells (including the cancerous T cells of MF) and kills them by shutting down their protein synthesis. Some success with this approach has been achieved in early clinical trials. Studies of mAbs targeting CD4 and CD30 as treatments for MF are also ongoing.

If a pathogen is associated with the onset of a lymphoma, treatment with an anti-pathogen agent may be effective. For example, about 80% of mucosa-associated lymphoid tissue lymphoma (MALT-L) cases are associated with *H. pylori* infection, and about 70% of patients in this group (those that lack translocations) will respond to antibiotics. As the bacteria are killed, the pathogen antigen putatively driving the lymphomagenesis disappears and the tumor slowly resolves. Similarly, because ATL is often associated with HTLV-1 infection, survival rates of ATL patients have been greatly improved by the use of a combination of IFNα and the anti-retroviral drug zidovudine (AZT).

We have come to the end, not only of this chapter, but also of this book. Hopefully we have given our readers a clear introduction to the principles of basic and clinical immunology, and have perhaps inspired them to pursue further study in this field. Many intriguing problems in immunology remain to be resolved. More importantly, the growing links between the immune system and the nervous and endocrine systems, and between immune responses and cancer, mean that a solid understanding of immunology is more useful than ever before. Multidisciplinary approaches in both the laboratory and the clinic are the way of the future, and immunologists can expect to be valuable contributors to these endeavors.

- **Hematopoietic cancers (HCs) are malignancies of immune system cells.**

- **Unlike non-hematopoietic cancers, HCs are more commonly associated with gross chromosomal abnormalities such as recurring translocations.**

- **The three major types of HCs are leukemias, myelomas and lymphomas.**

- **Leukemias are "liquid tumors" in the blood and are derived from the transformation of either a hematopoietic precursor in the bone marrow or a mature hematopoietic cell in the blood. Leukemias can be lymphoid or myeloid, and acute or chronic.**

- **In myelomas, the transformed cell is a fully differentiated plasma cell. The proliferating malignant clone may be present either in dispersed form or as a solid mass in the bone marrow.**

- **In lymphomas, the transformed cell is a lymphocyte resident in a lymphoid tissue outside the bone marrow. The proliferating malignant clone creates a solid mass in this tissue.**

- **A lymphoma is classified as either a Hodgkin's lymphoma (HL) or a non-Hodgkin's lymphoma (NHL).**

- **In an HL, a reactive infiltrate of non-transformed cells is drawn to form a mass around a malignant clone of Reed-Sternberg cells. In an NHL, the entire cancerous mass develops from a transformed B or T lineage cell.**

- **Subtypes of HL and NHL are defined based on architecture of the tumor, cell morphology, state of differentiation, surface marker expression and genetic aberrations.**

- **Identification of the genetic aberration in an HC guides clinicians in optimizing treatment. Improvements in diagnosis and therapy have led to increased patient survival over the last few decades.**

Section A.I–II

1) Can you define these terms? target cell, leukemogenesis, myelomagenesis, lymphomagenesis

2) What is the most common type of HC in North America?

3) Briefly distinguish between the three main types of HCs.

4) What is a recurring chromosomal translocation?

5) Interpret the following: t(9;12)(q34;p13).

Section A.III

1) Can you define these terms? blast-like, relapse, remission, FISH

2) Compare the differentiation stage of the cells involved in high grade versus low grade HCs.

3) What is immunophenotyping and how is it helpful?

4) How is a response to chemotherapy or radiation treatment determined?

5) Give an example of how immunotherapy can treat HCs.

6) Compare/contrast autologous and allogeneic HCTs as treatments for HCs.

Section B Introduction–I

1) Can you define these terms? hepatosplenomegaly, extramedullary

2) Name the four major classes of leukemias and distinguish between them.

3) AML is dominated by what type of transformed cell?

4) Describe a recurring chromosomal translocation in AML.

5) Name a primary immunodeficiency associated with AML.

Section B.II

1) Describe the three phases of CML.

2) What is the Philadelphia chromosome and how does it arise?

3) What is Gleevec[R] and how does it work?

Section B.III

1) ALL is the most frequent form of cancer in what population?

2) ALL is dominated by what type of transformed cell?

3) Describe a recurring chromosomal translocation in ALL.

Section B.IV

1) CLL is dominated by what type of transformed cell?

2) Describe two chromosomal anomalies associated with CLL.

3) What is atypical CLL?

4) Why is quality of life especially important to consider for CLL patients?

5) Describe two mAbs that have been useful for CLL therapy.

Section C

1) Can you define these terms? paraprotein, plasma cell dyscrasia, MGUS, multiple myeloma

2) Why are plasma cell dyscrasias not generally considered leukemias?

3) How do normal plasma cells differ from myeloma cells?

4) Give three ways in which myeloma cells cause disease.

5) Outline the staging system used to classify myelomas.

6) What is the role of IL-6 in myeloma disease?

7) Describe three chromosomal anomalies associated with myelomagenesis.

8) How are thalidomide and Velcade[R] thought to be useful for myeloma treatment?

Section D Introduction–I

1) Can you define these terms? extranodal, Reed-Sternberg cell, L&H cell, reactive infiltrate, HL-involved tissue, B symptoms, [18]FDG-PET/CT

2) Why do lymphoma cells not usually migrate into the blood?

3) Distinguish between Hodgkin's and non-Hodgkin's lymphomas.

4) Outline the staging system used to classify lymphomas.

5) Describe the unique pattern of HL spreading in a patient.

6) Which cytokine is thought to be most important for HL development and why?

7) Distinguish between the two major subtypes of HL.

8) What is extended field radiation and why does it improve the prognosis of HL patients?

9) What risk is associated with aggressive treatment for HL?

Section D.II

1) What is the incidence of NHL and what age group is most often affected?

2) Give two reasons why the incidence of NHLs has increased over the last three decades.

3) How does the migration pattern of NHL cells differ from that of HL cells?

4) Give two examples of pathogens associated with NHL development.

5) What are the two most prevalent NHL subtypes in adults? In children?

6) What is CNS prophylaxis?

7) Give two examples of immunotherapeutics that have been helpful in NHL treatment.

8) Why are antibiotics effective in the treatment of some cases of MALT-L?

Appendix A: Selected Landmark Discoveries in Immunology

1798	Vaccination to prevent smallpox
1880s	Attenuated vaccines Phagocytic theory of immune defense Complement
1890s	Antibodies (antitoxins)
1900s	ABO blood groups Anaphylaxis Opsonization
1920s	Tuberculosis vaccine based on bacillus Calmette-Guérin (BCG) Bacterial toxin vaccine for diphtheria Delayed hypersensitivity
1930s	Histocompatibility antigens in mice Characterization of antibodies as the gamma globulin subset (immunoglobulins)
1940s	Adjuvant for vaccination (Freund's) Transplantation immunology Plasma cell production of antibodies
1950s	Lymphocytes Tolerance Protein structure and function of antibodies Clonal selection theory of antibody formation
1960s	Role of the thymus in immunity Hematopoietic stem cells Lymphokines (cytokines) B cell/T cell cooperation Helper T cell subset Primary sequence of an immunoglobulin molecule
1970s	Hypervariable regions of immunoglobulins Monoclonal antibody production by hybridomas Generation of antibody diversity by immunoglobulin gene rearrangement Link between immune responsiveness and histocompatibility genes Major histocompatibility complex (MHC) restriction of immune responses Processing and presentation of exogenous antigen Membrane attack complex in the complement cascade NK cells Drug-mediated immunosuppression (cyclosporine A)
1980s	Declaration of the worldwide eradication of smallpox Isolation of HIV from an AIDS patient Cloning of the T cell receptor genes Lymphocyte migration Role of coreceptors (CD4 and CD8) in T cell activation Processing and presentation of endogenous antigen Recognition of stress proteins by the immune system Th1/Th2 subsets of T helper cells Crystal structure of an MHC molecule
1990s	Thymic selection of T cells in establishing self tolerance Inhibition of NK cell inhibitory receptors by MHC class I Pattern recognition in innate immunity
2000s	Role of dendritic cells in T cell activation NKT cells Th17 cells Role of regulatory T cells in controlling immune responses

Appendix B: Nobel Prizes Awarded for Work in Immunology

1901	Emil von Behring, for his discovery of serum antitoxins (antibodies) and serum therapy, and its application to the treatment of diphtheria.
1905	Robert Koch, for his investigations and discoveries in regard to tuberculosis. Koch developed tuberculin reactivity tests that later were important in the development of our current understanding of cellular immunity.
1908	Elie Metchnikoff, for his discovery of phagocytosis, and Paul Ehrlich, for his work on fundamental immunology.
1913	Charles Robert Richet, for his discovery of anaphylaxis.
1919	Jules Bordet, for his studies in regard to immunology, particularly complement-mediated lysis.
1930	Karl Landsteiner, for his discovery of the human blood groups.
1951	Max Theiler, for his development of vaccines against yellow fever.
1957	Daniel Bovet, for his development of antihistamines in the treatment of allergy.
1960	Frank MacFarlane Burnet and Peter Brian Medawar, for the discovery of acquired immunological tolerance.
1972	Gerald Maurice Edelman and Rodney Robert Porter, for their discoveries concerning the chemical structure of antibodies.
1977	Rosalyn Yalow, for the development of radioimmunoassays for peptide hormones.
1980	Jean Dausset, George Davis Snell, and Baruj Benacerraf, for the discoveries of the histocompatibility antigens on human and animal cells and their role in tissue and blood transplantation rejection (Dausset and Snell), and for work on the genetic control of immune responses (Benacerraf).
1984	Georges J. F. Köhler and César Milstein, for their development of cell hybridization as a technique to produce monoclonal antibodies, and Niels K. Jerne, for his many fundamental contributions to theoretical immunology.
1987	Susumu Tonegawa, for his work on the immunoglobulin genes and the mechanism by which antibody diversity is generated.
1990	Joseph E. Murray and E. Donnall Thomas, for their work on organ and bone marrow transplantation.
1996	Peter C. Doherty and Rolf M. Zinkernagel, for their discovery of the MHC restriction of T cell responses.

Appendix C: Comparative Immunology

All multicellular forms of life have a mechanism to distinguish self from non-self, and the means to preserve that self in the face of pathogen attack or competition for nutrients. These mechanisms and means were shaped by the various evolutionary and developmental pressures experienced by each organism. The elements of innate and specific immunity possessed by various phyla of the Animal Kingdom from the protozoans through the birds and mammals are summarized in Figure C1, Parts 1 and 2. For reference, an evolutionary tree illustrating the hierarchical relationship among groups of phyla is shown in Figure C-2.

The most basic of multicellular organisms, such as the sponges of the phylum Porifera, grow in colonies of relatively undifferentiated cells. The functions of food gathering, waste product disposal and host defense are all carried out by the same type of cell. Nevertheless, these organisms can detect the encroachment of cells from another colony and kill the invaders. With the evolution in the lower invertebrates of circulatory systems and multiple body layers, the functions of host nutrition and host defense are carried out by separate cell types. Cells in the circulation of lower invertebrates and pre-vertebrates are specialized for detecting infected cells via the expression of a small number of PRMs of relatively broad specificity. The primitive phagocytes bearing these PRMs are tasked with the disposal of non-self entities. Lectins, antimicrobial molecules, and at least some complement or complement-like components are also present in lower invertebrates and pre-vertebrates.

The range of PRMs expands as one proceeds higher up the pre-vertebrate and higher invertebrate evolutionary branches. More sophisticated phagocytes patrol the body and engulf invaders, and large arrays of antimicrobial peptides and proteins are produced. Many early protostomes and deuterostomes have molecules resembling complement components of the lectin pathway, but these proteins function only as opsonins. Some higher invertebrates possess a pathway called the *prophenoloxidase-activating (ProPO) system* which mediates cytotoxicity by coating and paralyzing a pathogen in the pigment melanin. Vertebrates do not have the ProPO system, making this defense mechanism unique to higher invertebrates.

No true lymphocytes, antibodies or MHC molecules are present in either the lower or higher invertebrates or in pre-vertebrates. Thus, it seems that adaptive immunity is not needed for the survival of these species. These organisms have a limited habitat range, relatively short life spans (short reproductive cycle) and large reproductive capacities. The immune repertoire is limited, as must be true when the genes encoding defense molecules do not undergo somatic recombination and are "hardwired" in the germline. However, this repertoire is sufficient for survival because a limited habitat means that the range of pathogens encountered is generally narrow. A short life span means the total number of pathogens encountered is relatively low, and a large reproductive capacity means that the huge numbers of offspring produced offset the loss of substantial numbers of them to pathogens.

Vertebrate species show enhanced anatomical complexity accompanied by increased mobility, such that these animals very often wander over great distances. In addition, vertebrates have longer life spans that correlate with an increased time to reach reproductive maturity. As a result, vertebrates frequently encounter a large number and wide variety of pathogens prior to successful reproduction. Vertebrates also produce many fewer offspring than either invertebrates or pre-vertebrates, such that severe losses to pathogens could threaten the species as a whole. These evolutionary pressures are thought to have promoted the development of adaptive immunity in vertebrates.

Although no conventional antibody can be detected in the lowest vertebrates (such as the jawless fish, Agnatha), primitive GALT is present. Cartilaginous fish like the sharks have hinged jaws, making them better predators able to take advantage of a broader range of nutritional opportunities. However, with such a diet comes an increased chance of internal injury and/or infection. Even with innate immunity in place, a strictly germline-encoded repertoire of non-self recognition molecules would not be sufficiently diverse to counter all the pathogens such vertebrates meet. Cartilaginous fish are thus the first organisms in which there exists a mechanism to somatically diversify immune system genes and expand the immune repertoire. A distinct thymus and spleen and true lymphocytes expressing forms of Ig and TCR molecules are present, with IgM and non-mammalian antibody isotypes being produced. The terminal complement components and MAC-mediated lysis are also first seen in the cartilaginous fish. In the bony fish, an IgD-like antibody joins IgM as an antibody isotype, and complement components unique to the alternative pathway appear. The anterior kidney serves as a bone marrow equivalent.

Vertebrates like amphibians that move from the sea to the land require additional host defense mechanisms to cope with the new environment. These animals have the limbs necessary to move on land and sophisticated vascular systems containing multiple types of circulating cells. Skins designed to shield the exposed animal from the sun's harmful rays are present, providing a physical barrier against pathogens. Bone marrow-like tissue serves as a source of distinct T and B cells, and lymphoid tissues of increased complexity and wider distribution are present. In the reptiles, there are advanced lymphoid tissues plus eggs in which the young develop in a self-contained aqueous system enclosed by multiple membranes. This innovation frees these cold-blooded animals from having to return to the water to reproduce, and increases barrier protection for the offspring. However, the cold-bloodedness of the amphibians and reptiles affects their immune responses, causing measurable seasonal variations in the proliferative capacity of lymphocytes and the production of cytokines. The warm-bloodedness of birds and mammals removes this variation and allows these animals to forage and hunt at night when cold-blooded vertebrates are less active. However, with a permanently warm body comes an environment favoring pathogen growth. To counter the onslaught, birds and mammals have highly differentiated and structured lymphoid tissues, complete with distinct germinal centers and lymph nodes. Cell-mediated and humoral responses are optimally coordinated and controlled to deliver the most efficient response possible with the least amount of collateral damage to the host. It is these responses that have been the subject of this book.

	Mammals	Birds	Reptiles	Amphibians	Bony fish	Cartilaginous fish	Jawless fish	Protochordates	Echinoderms	Higher invertebrates	Lower invertebrates	Protozoa
PRM Recognition												
PRMs	+	+	+	+	+	+	+	+	+	+	+	+
Anti-microbials	+	+	+	+	+	+	+	+	+	+	+	+
Cytokines	+	+	+	+	+	+	+	+	+	+	+*	+*
Chemokines	+	+	+	+	+	+	+	+	?	+	−	−
ProPO System												
ProPO	−	−	−	−	−	−	−	−	−	+	−	−
Complement System												
CR	+	+	+	+*	+*	+*	+*	+*	+*	−	−	−
C3	+	+	+	+	+	+	+*	+*	+*	−	−	−
MBL/MASP	+	+	+?	+	+*	+*	+*	+*	+*	−	−	−
Terminal C'	+	+	+	+	+	+	−	−	−	−	−	−
Factor I	+	+	+	+	+	+	−	−	−	−	−	−
Factor H	+	+	+	+	+	−	−	−	−	−	−	−
Lectin pathway	+	+	+	+	+	+	+	+	−	−	−	−
Classical pathway	+	+	+?	+	+	+	−	−?	−	−	−	−
Alternative pathway	+	+	+	+	+	+	−	−	−	−	−	−
Lymphoid Tissue												
Lymphocytes	T, B	T, B	T, B	T, B	T, B	T*, B*	T/B	+*	+*	+*	+*	−
GALT	+	+	+	+	+	+	+	−	−	−	−	−
Spleen and thymus	+	+	+	+	+	+	−	−	−	−	−	−
Bone marrow	+	+	+	+*	−	−	−	−	−	−	−	−
Lymph nodes (+ GC)	+	+	−	−	−	−	−	−	−	−	−	−
MALT/SALT	+	+	−	−	−	−	−	−	−	−	−	−

Fig. C-1, Part 1 **Elements of Innate and Specific Immunity through Evolution.**
Key: *, like; ?, not definitively proven; +/−, limited; C′, complement components; GC, germinal center; IgN, non-mammalian Ab isotype; •, non-MHC histocompatibility molecules exist; F, fast; S, slow; VS, very slow.

	Mammals	Birds	Reptiles	Amphibians	Bony fish	Cartilaginous fish	Jawless fish	Protochordates	Echinoderms	Higher invertebrates	Lower invertebrates	Protozoa
Ig Responses												
RAG	+	+	+?	+	+	+	−	−	−	−	−	−
Tdt	+	+	+	+	+	+	−	−	−	−	−	−
Isotype switching	+	+	+	+	−	−	−	−	−	−	−	−
Somatic hypermutation	+	+	+	+	+	+	−	−	−	−	−	−
Affinity maturation	+	+	+?	+/−	+/−	−	−	−	−	−	−	−
Memory	+	+	+/−	+/−	+/−	−	−	−	−	−	−	−
Ig Isotypes												
IgM	+	+	+	+	+	+	−	−	−	−	−	−
IgD	+	+	−	−	+	−	−	−	−	−	−	−
IgG	+	−	−	−	−	−	−	−	−	−	−	−
IgA	+	+	−	−	−	−	−	−	−	−	−	−
IgE	+	−	−	−	−	−	−	−	−	−	−	−
IgN	−	+	+	+	+	+	−	−	−	−	−	−
MHC												
MHC class I	+	+	+	+	+	+	−•	−•	−	−	−	−
MHC class II	+	+	+	+	+	+	−	−	−	−	−	−
MHC class III	+	+*	+?	+	−	−	−	−	−	−	−	−
MHC class Ib	+	+	−	+	+	+	−	−	−	−	−	−
β2-microglobulin	+	+	+	+	+	+	−	−	−	+*	+*	−
Allograft rejection	F	F	F	F	F	S	VS	+*	+*	+*	+*	+*
TCR												
TCR genes	+	+	+?	+	+	+	−	−	−	−	−	−
Td responses	+	+	+	+	+	−	−	−	−	−	−	−
Ti responses	+	+	?	+	+	+	−	−	−	−	−	−

Fig. C-1, Part 2 **Elements of Innate and Specific Immunity through Evolution**
Key: *, like; ?, not definitively proven; +/−, limited; C′, complement components; GC, germinal center; IgN, non-mammalian Ab isotype; •, non-MHC histocompatibility molecules exist; F, fast; S, slow; VS, very slow.

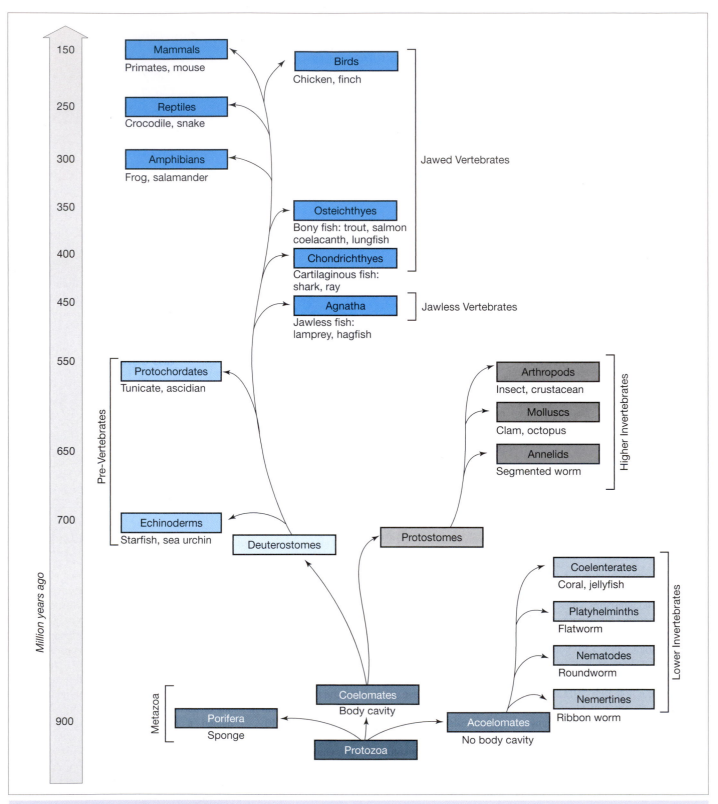

Fig. C-2 Evolutionary Tree of Kindom Animalia.
The major phyla and the approximate times when animals classified in these phyla first appeared are shown, starting with the protozoa at 900 million years ago. Phyla are grouped into the metazoa, lower invertebrates, higher invertebrates, pre-vertebrates and vertebrates as indicated. Time scale is approximate.

Appendix D: Selected CD Markers

CD Number	Common Names	Cellular/Tissue Expression	Function
CD1a	R4	APCs*, cortical thymocytes	Structurally similar to MHC class I Presents lipid/glycolipid antigens to T cells
CD1b	R1	APCs, cortical thymocytes	Structurally similar to MHC class I Presents lipid/glycolipid antigens to T cells
CD1c	R7	APCs, cortical thymocytes	Structurally similar to MHC class I Presents lipid/glycolipid antigens to T cells, especially $\gamma\delta$ T cells
CD1d	R3	B cells, DCs, iIELs, cortical thymocytes	Structurally similar to MHC class I Presents glycolipid antigens to NKT cells
CD1e	R2	APCs, cortical thymocytes	Structurally similar to MHC class I Presents lipid/glycolipid antigens to T cells
CD2	LFA-2	Most T cells, thymocytes, NK cells	Ig superfamily glycoprotein Adhesion molecule binding to LFA-3 (CD58) Minor T cell costimulatory function
CD3d	CD3δ	T lineage cells	Ig superfamily glycoprotein Part of CD3 complex Required for cell surface expression and signal transduction of TCR
CD3e	CD3ε	T lineage cells	Non-glycosylated Ig superfamily protein Part of CD3 complex Required for cell surface expression and signal transduction of TCR
CD3g	CD3γ	T lineage cells	Ig superfamily glycoprotein Part of CD3 complex Required for cell surface expression and signal transduction of TCR
CD4	OKT4, T4	Subsets of thymocytes and Th cells; some DCs, monocytes and macrophages	Ig superfamily protein Coreceptor binding to MHC class II Required for thymocyte development and Th effector differentiation Binds to HIV gp120 protein
CD5	Leu-1	T lineage cells, neonatal B cells	Transmembrane glycoprotein CD5$^+$ B cells are implicated in autoimmunity CD5$^+$ T cells are seen in T-ALL
CD7	Leu-9	T lineage cells	Ig superfamily glycoprotein CD7$^+$ T cells are seen in T-ALL
CD8a	OKT8, T8	Subsets of thymocytes and Tc cells, CTLs, some $\gamma\delta$ T cells and NK cells, iIELs	Ig superfamily protein Coreceptor subunit binding to MHC class I Required for thymocyte development and CTL differentiation
CD8b	Lyt3, Ly3	Thymocyte and Tc cell subsets, CTLs	Ig superfamily protein Coreceptor subunit binding to MHC class I Required for thymocyte development and CTL differentiation
CD10		Pre-B cells, GC B cells, malignant B cells	Membrane zinc metalloproteinase family Function is unclear
CD11a	LFA-1 α chain	Neutrophils, monocytes, macrophages, lymphocytes	CAM glycoprotein Complexes with CD18 to form LFA-1 Binds to ICAM1, ICAM2, ICAM3, ICAM4 on endothelial cells Adhesion and signal transduction during inflammation

continued overleaf

CD Number	Common Names	Cellular/Tissue Expression	Function
CD11b	CR3 α chain	Neutrophils, monocytes, macrophages, NK cells	CAM glycoprotein Complexes with CD18 to form CR3 Binds to iC3b, ICAM1, CD23 Adhesion and signal transduction during inflammation
CD11c	CR4 α chain	Neutrophils, monocytes, macrophages, NK cells, DCs	CAM glycoprotein Complexes with CD18 to form CR4 Binds to iC3b, LPS, ICAM-1, fibrinogen Adhesion and signal transduction during inflammation
CD14	LPS receptor	Monocytes, macrophages, granulocytes	Membrane protein Binds to TLR4 to transduce LPS signaling Mediates inflammation and endotoxic shock
CD15	Le-X	Neutrophils, eosinophils, monocytes, RS cells	Pentasaccharide linked to protein or lipid Binds to CD62 (selectins)
CD16a	FcγRIIIa	NK cells, macrophages, activated monocytes, mast cells	Ig superfamily glycoprotein Associates with CD3ζ chain or FcϵRIγ chain to form low affinity IgG Fc receptor Mediates phagocytosis and ADCC
CD16b	FcγRIIIb	Neutrophils, activated eosinophils	Ig superfamily protein Does not mediate phagocytosis or ADCC Traps immune complexes
CD18	Integrin β2 β-chain	Broadly expressed on leukocytes	Transmembrane glycoprotein Associates with CD11a, b or c to form LFA-1, CR3 or CR4, respectively Binds to ICAMs, complement components Mediates cell adhesion to matrix proteins
CD19	B4	All B lineage cells except plasma cells, FDCs, malignant B cells	Ig superfamily protein Promotes BCR signal transduction
CD20		All B lineage cells, including plasma cells	Non-glycosylated protein Therapeutic target for lymphomas
CD21	CR2	Mature B cells, FDCs, DCs, epithelial cells	Glycosylated RCA protein Complement receptor for C3d Receptor for Epstein-Barr virus
CD22	BL-CAM	Mature B cells, not plasma cells	Ig superfamily glycoprotein Adhesion and signal transduction Therapeutic target for lymphoma treatment
CD23	FcϵRII	Mature B cells, monocytes, eosinophils, FDCs, TECs	Transmembrane glycoprotein Low affinity IgE Fc receptor Triggers NO and cytokine production by monocytes
CD25	IL-2Rα	Activated T and B cells, T_{reg} cells, thymocytes	Transmembrane glycoprotein Associates with IL-2Rβ (CD122) and IL-2Rγ (CD132) to form the high affinity IL-2 receptor
CD28	T44	Activated T and B cells	Transmembrane protein Binds to B7-1 (CD80) and B7-2 (CD86) on APCs Mediates T cell costimulation and stimulates T cell survival, proliferation, production of IL-2 and other cytokines TCR engagement in the absence of CD28 signaling can lead to anergy
CD29	VLA β chain	Most leukocytes	Transmembrane protein Associates with CD49$_{a-f}$ to form VLA molecules Mediates cellular adhesion to VCAM-1, MAdCAM-1 and matrix proteins Critical for leukocyte migration

CD Number	Common Names	Cellular/Tissue Expression	Function
CD30		Activated B, T, and NK cells, monocytes, RS cells	TNFR superfamily glycoprotein May regulate development of thymocytes and proliferation of activated lymphocytes
CD33	My9	Myeloid progenitors and mature cells	Sialoadhesin family protein Binds to sialic acids May inhibit signal transduction
CD34	My10	Hematopoietic precursors, BM stromal cells	Transmembrane glycoprotein Marker for human HSCs May mediate adhesion between hematopoietic precursors and BM stromal cells or matrix
CD35	CR1	Most blood cells (not platelets)	RCA glycoprotein Complement receptor for C3b and C4b bound to immune complexes Mediates opsonized phagocytosis, immune complex clearance
CD36		Platelets, monocytes, macrophages	Scavenger receptor family (PRR) Binds to collagen, long-chain fatty acids and other ligands Mediates phagocytosis, platelet adhesion and aggregation
CD38	T10	Developing T and B cells, GC B cells, some hematopoietic progenitors	Adhesion glycoprotein Some hydrolase activity
CD40	TNFR5	Mature B cells, some pre-B cells, monocytes, DCs, T cell subsets	TNFR superfamily glycoprotein Binds to CD40L (CD154) on Th cells to co-stimulate B cell activation, GC formation and isotype switching Important for DC licensing
CD43	Leukosialin	Most leukocytes	Membrane glycoprotein May bind to ICAM-1 and selectins
CD44	pgp1	Leukocytes, erythrocytes, endothelial and epithelial cells	Heterogeneous glycoprotein Binds hyaluronic acid, collagen, fibronectin Mediates cell–cell and cell–matrix adhesion and signaling Important for lymphocyte homing, recirculation
CD45	B220	All leukocytes, hematopoietic progenitors	Heterogeneous glycoprotein Tyrosine phosphatase functioning in signal transduction
CD46	MCP	Most leukocytes, endothelial and epithelial cells	RCA glycoprotein Controls complement activation by promoting cleavage of C3b and C4b
CD49a	VLA-1 α chain	Activated T cells, monocytes	Integrin glycoprotein Binds to CD29 to form VLA-1 Mediates adhesion to collagen, laminin Upregulated during inflammation
CD49b	VLA-2 α chain	Platelets, megakaryocytes, monocytes, epithelial and endothelial cells	Integrin glycoprotein Binds to CD29 to form VLA-2 Mediates cell and platelet adhesion to collagen, laminin, fibronectin, E-cadherin Promotes wound healing
CD49c	VLA-3 α chain	B cells, monocytes	Integrin glycoprotein Binds to CD29 to form VLA-3 Mediates adhesion to collagen, laminin, fibronectin, thrombospondin
CD49d	VLA-4 α chain	Most leukocytes (not platelets or neutrophils)	Integrin glycoprotein Binds to CD29 to form VLA-4 Mediates adhesion to VCAM-1, MAdCAM-1, fibronectin, thrombospondin Important for inflammation, HSC migration

continued overleaf

CD Number	Common Names	Cellular/Tissue Expression	Function
CD49e	VLA-5 α chain	Monocytes, DCs, NK cells	Integrin glycoprotein Binds to CD29 to form VLA-5 Mediates adhesion to fibronectin, fibrinogen
CD49f	VLA-6 α chain	Platelets, megakaryocytes, monocytes, T cells, thymocytes	Integrin glycoprotein Binds to CD29 to form VLA-6 Mediates adhesion to laminin, other ligands Mediates interaction between epithelial cells and basement membrane during wound healing Involved in tumor metastasis
CD50	ICAM-3	Most leukocytes	Ig superfamily glycoprotein Binds to LFA-1 (CD11a/CD18) Contributes to APC adhesion, minor T cell costimulation
CD52		Lymphocytes, monocytes, eosinophils, sperm	Small membrane protein Physiological function unknown Anti-CD52 mAbs (eg. CAMPATH-1H) are used therapeutically to treat certain leukemias and lymphomas and to reduce GvHD
CD54	ICAM-1	Activated T cells, B cells, monocytes, endothelial cells	Ig superfamily glycoprotein Binds to LFA-1, CR3, fibrinogen Signaling and adhesion in inflammatory and adaptive immune responses
CD55	DAF	Most hematopoietic cells	RCA protein Binds C3b and C4b to inhibit C3 convertase formation, blocking complement activation Binds C3bBb and C4b2a to accelerate decay of C3 convertases
CD58	LFA-3	Leukocytes, erythrocytes, endothelial and epithelial cells, fibroblasts	Ig superfamily protein Binds to CD2 Mediates adhesion between Th cells and APCs, Tc cells and target cells, thymocytes and thymic epithelial cells Minor T cell costimulation costimulation
CD59	MIRL	Most hematopoietic cells	RCA glycoprotein Binds to C8 and/or C9 and inhibits final steps of MAC formation
CD62E	E-selectin	Activated endothelial cells, platelets, megakaryocytes	Adhesion glycoprotein Required for leukocyte extravasation and activation during inflammation Binds to glycoproteins and glycolipids, including CD162
CD62L	L-selectin	Neutrophils, monocytes, NK cells, memory T cells	Adhesion glycoprotein Required for leukocyte extravasation and activation during inflammation Mediates lymphocyte adherence to HEVs in lymph nodes Binds to glycoproteins and glycolipids, including CD162
CD62P	P-selectin	Megakaryocytes, activated platelets and endothelial cells	Adhesion glycoprotien Required for leukocyte extravasation and activation during inflammation Binds to glycoproteins and glycolipids, including CD162
CD64	FcγRI	Monocytes, macrophages, neutrophils, DC subsets	Ig superfamily protein High affinity IgG Fc receptor Mediates phagocytosis, ADCC Mediates transfer of IgG across the placenta
CD74	Invariant chain, Ii	Cells expressing MHC class II	Membrane glycoprotein Critical for MHC class II stabilization and antigen presentation Surface function not known

CD Number	Common Names	Cellular/Tissue Expression	Function
CD79a	Igα	All B cells and plasma cells	Ig superfamily glycoprotein with ITAMs Associates with Igβ (CD79b) and participates in the BCR complex to carry out signal transduction Essential for BCR function, B cell development
CD79b	Igβ	All B cells but not plasma cells	Ig superfamily glycoprotein with ITAMs Associates with Igα (CD79a) and participates in the BCR complex to carry out signal transduction Essential for BCR function, B cell development
CD80	B7-1	Professional APCs, activated T cells	Ig superfamily glycoprotein Binds to CD28 to costimulate naïve T cells Binds to CTLA-4 (CD152) to inhibit T cell activation Highly induced on stimulated APCs
CD86	B7-2	Professional APCs	Ig superfamily protein Binds to CD28 to costimulate naïve T cells Binds to CTLA-4 (CD152) to inhibit T cell activation
CD88	C5aR	Most leukocytes, epithelial and endothelial cells, hepatocytes	Transmembrane protein Binds to anaphylatoxin C5a Stimulates granule release, chemotaxis, ROI and RNI production during inflammation
CD89	FcαR	Neutrophils, monocytes, macrophages, activated eosinophils	Ig superfamily glycoprotein IgA Fc receptor Stimulates phagocytosis, respiratory burst, degranulation, ADCC, release of inflammatory mediators and cytokines
CD91		Monocytes, macrophages, hepatocytes	Scavenger receptor (PRR) Binds to HSPs Mediates clearance of necrotic cells
CD94		NK cells, some γδ T cells, some αβ CD8$^+$ T cells	Glycoprotein with C-type lectin domain Associates with NKG2A (CD159a) and NKG2C (CD159c) to form NK inhibitory receptors Associates with NKG2D (CD314) to form an NK activatory receptor
CD95	Fas	Activated T and B cells	TNF superfamily glycoprotein Binds to FasL (CD178) to initiate apoptosis Important for peripheral tolerance and lymphocyte homeostasis
CD102	ICAM-2	Most leukocytes, endothelial cells	Ig superfamily glycoprotein Binds to LFA-1 (CD11a/CD18) to mediate adhesion and signal transduction important for T cell interactions, lymphocyte recirculation, NK cell migration
CD106	VCAM-1	Endothelial cells, DCs, FDCs	Ig superfamily glycoprotein Binds to VLA-4 (CD49d/CD29) to mediate leukocyte extravasation during inflammatory and adaptive immune responses
CD114	G-CSFR	Granulocytes, monocytes, platelets, endothelial cells	Growth factor receptor Binds to G-CSF and regulates granulocyte differentiation and proliferation Stimulates mobilization of HSCs from BM into blood
CD115	M-CSFR	Myeloid cells	Growth factor receptor Binds to M-CSF and regulates monocyte and macrophage differentiation and proliferation Promotes adhesion to BM stroma
CD116	GM-CSFR	Macrophages, neutrophils, eosinophils, DCs, myeloid and erythroid progenitors	Growth factor receptor subunit Associates with βc (CD131) to bind GM-CSF with high affinity Transduces signals promoting differentiation, proliferation and activation of myeloid and erythroid cells

continued overleaf

CD Number	Common Names	Cellular/Tissue Expression	Function
CD117	c-kit, SCFR	HSCs and other progenitors, mast cells	Growth factor receptor Binds to SCF to mediate signal transduction promoting differentiation
CD119	IFNγRα	Ubiquitous except RBCs	Cytokine receptor subunit Associates with IFNγR β subunit to form complex transducing IFNγ signaling Triggers numerous antiviral, antiproliferative and immunomodulatory effects during innate and adaptive responses
CD120a	TNFRI	T and B cells; upregulated on most other cell types	TNFR superfamily protein Binds to TNF and LT Mediates signaling promoting inflammation, fever, shock, tumor necrosis, cell proliferation, differentiation and apoptosis
CD120b	TNFRII	Most hematopoietic cells, especially myeloid cells	TNFR superfamily protein Binds to TNF and LT Promotes mainly cell survival but sometimes necrosis
CD121a	IL-1RI	Almost ubiquitous	Ig superfamily glycoprotein Binds to IL-1 and mediates its pro-inflammatory activities
CD122	IL-2Rβ	Constitutive low levels on T, B and NK cells; upregulated by activation	Cytokine receptor subunit Associates with CD132 (γc; IL-2Rγ) to form low affinity IL-2 receptor Associates with CD25 (IL-2Rα) and CD132 to form high affinity IL-2 receptor Binding to IL-2 promotes lymphocyte proliferation, differentiation and regulation contributing to peripheral tolerance
CD123	IL-3Rα	HSCs and other progenitors, NK cells, myeloid and B cell subsets	Cytokine receptor subunit Binds to βc (CD131) to form IL-3 receptor Mediates signaling influencing growth and differentiation
CD124	IL-4R	T and B cells, hematopoietic precursors, fibroblasts, endothelial cells	Cytokine receptor subunit Binds to γc (CD132) to form IL-4 receptor IL-4 binding promotes growth of B and T cells Required for Th2 differentiation, IgE production and allergic inflammation Also associates with IL-13Rα (CD213) to form IL-13 receptor
CD125	IL-5Rα	Eosinophils, basophils	Cytokine receptor subunit Binds with βc (CD131) to form IL-5 receptor Promotes eosinophil generation and activation
CD126	IL-6Rα	Mature T cells, B cell subsets, epithelial cells	Cytokine receptor subunit Binds to gp130 (CD130) to form IL-6 receptor Stimulates acute phase response in liver, regulates hematopoiesis Mediates growth signals for myelomas
CD127	IL-7Rα	Pro- and pre-B cells, thymocytes	Cytokine receptor subunit Binds to γc (CD132) to form IL-7 receptor Promotes T and B lymphopoiesis, γδ T cell development and survival (especially in mice)
CD129	IL-9Rα	Mast cells, macrophages, erythroid and myeloid precursors	Cytokine receptor Binds to γc (CD132) to form IL-9 receptor Promotes growth of mast cells and erythroid and myeloid progenitors
CD130	gp130	Almost ubiquitous	Cytokine receptor subunit Common signaling chain for receptors binding IL-6, IL-11 or IL-27 (among others) Does not bind to cytokines itself

CD Number	Common Names	Cellular/Tissue Expression	Function
CD131	βc Common β chain	HSCs and other progenitors, monocytic lineage cells	Cytokine receptor subunit Common signaling chain for receptors binding IL-3, IL-5 or GM-CSF Does not bind to cytokines itself Upregulated in response to inflammatory cytokines
CD132	γc Common γ chain	T, B and NK cells, monocyte lineage cells	Cytokine receptor subunit Common signaling chain for receptors binding IL-2, 4, 7, 9, 15 or 21 Does not bind to cytokines itself
CD134	OX40	Activated T cells, hematopoietic precursors	TNFR superfamily protein Binds to OX40L (CD252) to promote T cell interaction with APCs Mediates adhesion of T cells to endothelium
CD152	CTLA-4	Activated T cells, T$_{reg}$ cells	Ig superfamily protein Structurally similar to CD28 High avidity receptor for B7–1 (CD80) and B7–2 (CD86) Downregulates T cell activation, promotes T$_{reg}$ generation, contributes to maintenance of peripheral tolerance and lymphocyte homeostasis
CD153	CD30L	Activated T cells and macrophages, neutrophils, eosinophils, B cells	TNF superfamily glycoprotein Ligand for CD30 Precise function is unclear but may stimulate T-B cell interaction, proliferation and apoptosis
CD154	CD40L	Activated T cells (CD4$^+$), mast and NK cells, granulocytes, monocytes, activated platelets	TNF superfamily glycoprotein Ligand for CD40 Provides major costimulatory signal for B cell activation and survival signal to GC B cells Required for isotype switching, DC licensing
CD158	KIRs	NK cells	Ig superfamily glycoproteins Family of NK inhibitory receptors Bind to classical MHC class I molecules and inhibit natural cytotoxicity
CD159a	NKG2A	NK cells, subset of CD8$^+$ T cells	Glycoprotein with C-type lectin domain Associates with CD94 to form NK inhibitory receptor binding to HLA-E Blocks natural cytotoxicity
CD159c	NKG2C	NK, NKT and γδ T cells	Glycoprotein with C-type lectin domain Associates with CD94 to form NK activatory receptor binding to HLA-E May activate NK, NKT and γδ T cells
CD178	FasL	T cells, NK cells	TNF superfamily glycoprotein Ligand for Fas (CD95); induces apoptosis Involved in maintenance of peripheral tolerance
CD179a	V pre-β	Pro-B, pre-B cells	Protein homologous to Ig V region Associates with lambda 5 chain (CD179b) to form surrogate light chain required for pre-BCR expression and early B cell development
CD179b	Lambda 5	Pro-B, pre-B cells	Protein homologous to Ig lambda C region Associates with V pre-β chain (CD179a) to form surrogate light chain required for pre-BCR expression and early B cell development
CD181	CXCR1 IL-8RA IL-8R1	Neutrophils, some T cells	Chemokine receptor Binds to IL-8 (only) and mediates signaling for neutrophil activation and chemotaxis

continued overleaf

CD Number	Common Names	Cellular/Tissue Expression	Function
CD182	CXCR2 IL-8RB IL-8R2	Neutrophils, granulocytes, monocytes, some T cells	Chemokine receptor Binds to IL-8 plus two related growth factors Mediates IL-8 signaling promoting neutrophil activation and chemotaxis Transduces signals for chemokine-mediated angiogenesis and possibly metastasis
CD183	CXCR3	Activated T cells, NK cells, Th1 effectors	Chemokine receptor Binds to chemokines IP-10, Mig, I-TAC Facilitates T cell homing to the lung
CD184	CXCR4	Hematopoietic cells, endothelial cells, breast cancer cells	Chemokine receptor Binds primarily to chemokine SDF-1 Directs T cells to T cell-rich zones of secondary lymphoid tissues Coreceptor for HIV infection
CD185	CXCR5	B cells	Chemokine receptor Binds primarily to chemokine BCA-1 Directs B cells to B cell-rich zones of secondary lymphoid tissues Required for GC formation
CD191	CCR1	Activated T cells, macrophages, neutrophils, basophils, myeloid precursors	Chemokine receptor Binds to chemokines RANTES, MIP-1α, MCP-3, MCP-4 Facilitates early events in inflammation
CD192	CCR2	Macrophages, monocytes, activated T cells, basophils, endothelial cells	Chemokine receptor Binds to chemokines MCP-2, 3, 4 Required for normal trafficking of APCs
CD193	CCR3 Eotaxin receptor	Eosinophils, DCs, activated T cells, Th2 effectors, basophils, airway epithelial cells	Chemokine receptor Binds to chemokines eotaxin, MCP-2, 3, 4 Required for eosinophil trafficking Facilitates allergic reactions
CD195	CCR5	Lymphocytes, monocytes, macrophages, DCs	Chemokine receptor Binds to chemokines MIP-1, MIP-1β and RANTES Coreceptor for HIV infection
CD197	CCR7	Naïve T and B cells, mature DCs, Th1 effectors, some memory T cells	Chemokine receptor Binds to chemokines SLC, ELC and MIP-3β Directs homing to splenic PALS, lymph nodes Directs activated B cells to move to T cell zones of secondary lymphoid tissues
CD212	IL-12Rβ	Activated T and NK cells, some B cells and DCs	Cytokine receptor subunit Binds to IL-12Rα chain to form IL-12 receptor Promotes Th1 differentiation and IFNγ production
CD213	IL-13Rα	Vascular endothelial cells, monocytes, macrophages, B cells, RS cells	Cytokine receptor subunit Binds to IL-13 with low affinity Binds to the IL-4R chain (CD124) to form the high affinity IL-13 receptor Inhibits Th1 cytokine production but does not promote Th2 differentiation Promotes isotype switching to IgE
CD230	PrP^c	Brain cells, DCs, T and B cells, monocytes	Membrane glycoprotein Function of normal PrP^c is unknown PrP^{res} is abnormal form that forms cytotoxic aggregates PrP^{res} is infective agent in spongiform encephalopathies
CD247	CD3ζ	All T lineage cells	Ig superfamily protein Part of CD3 complex Required for cell surface expression and signal transduction of TCR
CD252	OX40L	Activated B cells, macrophages, DCs, endothelial cells	TNF superfamily membrane protein Ligand for OX40 (CD134) Mediates adhesion of T cells to APCs; minor costimulation

CD Number	Common Names	Cellular/Tissue Expression	Function
CD257	BAFF	Myeloid cells	Membrane TNF superfamily protein Can be cleaved to give secreted cytokine form Essential for the T1 to T2 transition during B cell development
CD275	ICOSL	B cells, monocytes, macrophages, some stimulated DCs	Ig superfamily protein related to B7 Ligand for ICOS (CD278) on T cells Upregulated by inflammatory cytokines Contributes to T cell costimulation for proliferation, cytokine secretion, Th2 responses, Td antibody responses Required for T–B cooperation, GC formation, isotype switching, IgE and IgG1 production
CD278	ICOS	Activated T cells, Th2 cells, thymocyte subsets	Ig superfamily glycoprotein related to CD28 Binds to ICOSL (CD275) on APCs Induced on activated T cells Contributes to T cell costimulation for proliferation, cytokine secretion, Th2 responses, Td antibody responses Required for T–B cooperation, GC formation, isotype switching, IgE and IgG1 production
CD282	TLR2	Monocytes, neutrophils, granulocytes, macrophages, DCs	TLR glycoprotein (PRR) Cooperates with TLR1 or TLR6 to bind to lipoprotein and lipoglycan in bacterial cell walls Triggers innate antibacterial response Stimulates respiratory burst and production of NO and IL-12 by macrophages
CD283	TLR3	DCs, fibroblasts	TLR glycoprotein (PRR) Binds to double-stranded viral RNA Triggers innate antiviral response (IFN production)
CD284	TLR4	Myeloid cells, endothelial cells, B cells	TLR glycoprotein (PRR) Binds to LPS of bacterial cell walls Triggers innate antibacterial response Induces phagocytosis, inflammatory cytokines
CD314	NKG2D	NK cells, $\gamma\delta$ T cells, some CD8$^+$ $\alpha\beta$ T cells	Glycoprotein with C-type lectin domain Associates with DAP10 to form NK activatory receptor Binds to stress antigens MICA, MICB Activates natural cytotoxicity and costimulates some T cells
CD335	NKp46 NCR1	Resting and activated NK cells	Ig superfamily glycoprotein Associates with CD3ζ or FcϵRIγ to form NK activatory receptor Binds to a non-MHC ligand Activates natural cytotoxicity, cytokine production
CD336	NKp44 NCR2	Activated NK cells, some activated $\gamma\delta$ T cells	Ig superfamily glycoprotein Associates with DAP12 to form NK activatory receptor Binds to a non-MHC ligand Increases efficiency of natural cytotoxicity, cytokine production
CD337	NKp30	Resting and activated NK cells	Ig superfamily glycoprotein Associates with CD3ζ to form NK activatory receptor Binds to a non-MHC ligand Increases efficiency of tumor cell killing by activated NK cells

*Major abbreviations: ADCC, antibody-dependent cell-mediated cytotoxicity; ALL, acute lymphoblastic leukemia; APC, antigen-presenting cells; BCR, B cell receptor; BM, bone marrow; CAM, cellular adhesion molecule; CR, complement receptor; CTLA-4, cytotoxic T lymphocyte antigen-4; DAF, decay accelerating factor; DC, dendritic cells; FDC, follicular dendritic cells; GC, germinal center; G-CSF, granulocyte colony stimulating factor; GM-CSF, granulocyte–macrophage colony stimulating factor; GvHD, graft-versus-host disease; HEV, high endothelial venules; HSC, hematopoietic stem cell; HSP, heat shock protein; ICAM, intercellular adhesion molecule; ICOS, inducible costimulator; ICOSL, inducible costimulator ligand; IFN, interferon; Ig, immunoglobulin; iIEL, intestinal intraepithelial lymphocyte; IL, interleukin; LFA, leukocyte function associated-1; ITAM, immunoreceptor tyrosine-based activation motif; LPS, lipopolysaccharide; LT, lymphotoxin; mAb, monoclonal antibody; MAC, membrane attack complex; MAdCAM, mucosal addressin cellular adhesion molecule; M-CSF, macrophage colony stimulating factor; MCP, monocyte chemotactic protein; MIRL, membrane inhibitor of reactive lysis; NKG, natural killer gene; NO, nitric oxide; PALS, periarteriolar lymphoid sheath; PrP^c, prion protein, cellular; PrP^{pres}, prion protein resistant; PRR, pattern recognition receptor; RBC, red blood cells; RCA, regulator of complement activation; RNI, reactive nitrogen intermediate; ROI, reactive oxygen intermediate; RS, Reed-Sternberg cells of Hodgkin's lymphoma; SCF, stem cell factor; SDF, stromal cell-derived factor; TEC, thymic epithelial cell; TLR, Toll-like receptor; TNF, tumor necrosis factor; TNFR, tumor necrosis factor receptor; VCAM, vascular cellular adhesion molecule; VLA, very late antigen.

Appendix E: Cytokines, Chemokines and Receptors

Table 1 Major Human Cytokines and Cytokine Receptors

Abbreviation	Name of Cytokine	Function of Cytokine	Producers of Cytokine	Cytokine Receptor: Component Chains	Cells/Tissues Expressing Cytokine Receptor
Interferons					
IFNα*	Interferon alpha	Induces antiviral state ↓ Cell proliferation ↑ NK cell and CTL functions Influences isotype switching	Activated macrophages, monocytes; some activated T cells	IFNα/βR (type 1 IFN receptor)	Virtually all cells
IFNβ	Interferon beta	Induces antiviral state ↓ Cell proliferation ↑ NK cell and CTL functions Influences isotype switching	Fibroblasts	IFNα/βR	Virtually all cells
IFNγ	Interferon gamma	Induces antiviral state ↓ Cell proliferation ↑ NK cell and CTL functions Influences isotype switching and apoptosis ↑ APC production of IL-12 (promotes Th1 differentiation) ↓ IL-4 production and Th2 differentiation	Activated Th1 cells, CTLs, NK cells	IFNγR (type 2 IFN receptor)	Virtually all cells; not erythrocytes
Interleukins					
IL-1	Interleukin-1	Pro-inflammatory ↑ Acute phase response Induces fever and wasting Mediates endotoxic shock	Macrophages, neutrophils, keratinocytes, epithelial cells, endothelial cells	IL-1R	Most cell types
IL-2	Interleukin-2	Th1 cytokine ↑ T and B cell activation, proliferation, differentiation ↑ NK cell proliferation and production of TNF, IFNγ Required for T cell peripheral tolerance and homeostasis	Activated T cells	IL-2R	Activated T, B and NK cells
IL-3	Interleukin-3	Primarily a mast cell and basophil growth factor ↑ T cell production of IL-10, IL-13 Promotes anti-parasite response	Activated T cells and mast cells	IL-3R	Early hematopoietic cells, most myeloid lineages, some B cells
IL-4	Interleukin-4	Th2 cytokine Required for Th2 differentiation ↓ Macrophage and IFNγ functions ↑ B cell proliferation, differentiation, isotype switching ↑ Mast cell proliferation	Activated T cells, basophils, mast cells and NKT cells	IL-4R	Hematopoietic cells
IL-5	Interleukin-5	Th2 cytokine ↑ Eosinophil chemotaxis and activation ↑ Mast cell histamine release	Activated Th2 cells, mast cells, NK cells, B cells and eosinophils	IL-5R	Eosinophils, mast cells, basophils

(handwritten annotation near IFNα row: "Immune regatory")

continued overleaf

Abbreviation	Name of Cytokine	Function of Cytokine	Producers of Cytokine	Cytokine Receptor: Component Chains	Cells/Tissues Expressing Cytokine Receptor
IL-6	Interleukin-6	Pro-inflammatory ↑ Acute phase response Induces fever ↑ Neutrophil microbicidal functions ↑ B cell terminal differentiation ↑ Th17 cell differentiation	Activated phagocytes, fibroblasts and endothelial cells; some activated T cells	IL-6R	Hepatocytes, monocytes, neutrophils, activated B cells, mature T cells
IL-7	Interleukin-7	Promotes lymphopoiesis ↑ Development of $\alpha\beta$ T cells, $\gamma\delta$ T cells, B cells ↑ Generation and maintenance of memory T cells	Primarily BM and thymic stromal cells	IL-7R	T, B, NK and NKT precursors
IL-8	Interleukin-8	CXC chemokine ↑ Neutrophil chemotaxis ↑ Neutrophil degranulation and microbicidal functions	All cell types encountering TNF, IL-1 or bacterial endotoxin	CXCR1 CXCR2	Neutrophils, NK cells, T cells, basophils
IL-9	Interleukin-9	Promotes erythroid, myeloid and neuronal precursor differentiation ↑ Mast cell proliferation and differentiation ↑ Anti-helminth worm defence (with IL-4)	Activated Th2 cells, memory $CD4^+$ T cells	IL-9R	Many hematopoietic cell types
IL-10	Interleukin-10	Anti-inflammatory, immunosuppressive ↓ Activation of macrophages, neutrophils, mast cells, eosinophils ↓ Th1 cytokine production ↓ APC function	Activated macrophages, monocytes, Th2 cells, B cells, eosinophils, mast cells	IL-10R	Most hematopoietic cell types
IL-11	Interleukin-11	Promotes erythroid, myeloid and megakaryocyte precursor proliferation ↑ T and B cell and neutrophil proliferation ↓ Macrophage functions ↑ Fibroblast growth and collagen deposition	BM stromal cells, osteoblasts, cells in brain, joints, testes	IL-11R	Many hematopoietic and non-hematopoietic cell types
IL-12	Interleukin-12	Required for Th1 differentiation ↑ Production of IFNγ by macrophages, activated Th1 cells, NK cells ↑ DC and macrophage cytokine secretion ↑ CTL and NK cytotoxicity ↑ Memory T cell differentiation into Th1 cells Influences isotype switching	Activated macrophages, DCs; neutrophils, monocytes, B cells	IL-12R	Activated T and NK cells, B cells, DCs
IL-13	Interleukin-13	Th2 cytokine ↑ Th2 cell production of IL-4, IL-5, IL-10 Does not induce Th2 differentiation ↓ Macrophage cytokine secretion ↑ B cell proliferation and switching to IgE ↑ Anti-nematode defence	Activated T cells, mast cells, basophils	IL-13R	Monocytes, macrophages, B cells, endothelial cells

Abbreviation	Name of Cytokine	Function of Cytokine	Producers of Cytokine	Cytokine Receptor: Component Chains	Cells/Tissues Expressing Cytokine Receptor
IL-15	Interleukin-15	Required for NK cell development, proliferation and production of TNF, IFNγ ↑ $\gamma\delta$T cell development ↑ T cell activation, proliferation, differentiation, homing and adhesion ↑ Memory CD8$^+$ T cell survival ↑ Mast cell proliferation	Activated APCs	IL-15R (T, B and NK cells) IL-15RX (mast cells)	Lymphoid precursors; mature T, B, NK and mast cells
IL-17	Interleukin-17	Family of 6 closely related cytokines: IL-17A–F Structurally unique interleukin May induce endothelial cells and monocytes to secrete pro-inflammatory cytokines May mobilize neutrophils in allergic and autoimmune responses	Th17 cells	IL-17R (binds IL-17A and IL-17F) IL-17RB (IL-17B, E)	Widely expressed, including peripheral T and B cells Some non-hematopoietic tissues
IL-18	Interleukin-18	Synergizes with IL-12 functions during later stages of Th1 response ↑ Th1 cell proliferation, production of IFNγ, IL-2R ↑ NK cytotoxicity and production of IFNγ, TNF	Widely expressed	IL-18R	Virtually all cells
IL-23	Interleukin-23	Promotes expansion of Th17 cells Promotes memory CD4$^+$ T cell response ↑ Memory CD4$^+$ T cell proliferation and differentiation into Th1 cells ↑ IFNγ production by DCs and memory Th1 effectors	Activated APCs	IL-23R	Memory CD4$^+$ T cells, Th17 cells, DCs, NK cells
IL-27	Interleukin-27	Required for early stages of Th1 response ↑ Th1 and NK cell activation and production of IFNγ ↑ Pro-inflammatory cytokine production by mast cells, monocytes ↑ Activity of anti-tumor CTLs and NK cells ↑ Th17 cell differentiation	Activated APCs	IL-27R	Naïve CD4$^+$ T cells, NK cells

TNF-Related Cytokines

Abbreviation	Name of Cytokine	Function of Cytokine	Producers of Cytokine	Cytokine Receptor: Component Chains	Cells/Tissues Expressing Cytokine Receptor
TNF	Tumor necrosis factor	Potent inflammatory, immunoregulatory, cytotoxic, antiviral, pro-coagulatory, and growth stimulatory effects ↑ Proliferation, activation, adhesion, extravasation, cytokine production of hematopoietic cells ↑ Acute phase response ↑ Macrophage and neutrophil microbicidal functions ↑ APC functions ↑ B cell proliferation, Ab production, GC formation ↑ Tumor cell apoptosis and hemorrhagic necrosis High concentrations induce wasting, endotoxic shock, fibrosis, bone destruction	Many types of activated hematopoietic and non-hematopoietic cells	TNFRI TNFRII	Widely expressed (but not resting T and B cells)

continued overleaf

Abbreviation	Name of Cytokine	Function of Cytokine	Producers of Cytokine	Cytokine Receptor: Component Chains	Cells/Tissues Expressing Cytokine Receptor
LT	Lymphotoxin	Secreted molecule with TNF-like activities	Activated Th1, B and NK cells	TNFRI TNFRII	Widely expressed (but not resting T and B cells)
BAFF	B cell activating factor	Secreted molecule essential for survival of transitional B cells during B cell development	Myeloid lineage cells	BAFF-R TACI BCMA	B lineage cells B lineage cells, some T cells B lineage cells, plasma cells

Transforming Growth Factor

TGFβ	Transforming growth factor-β	Anti-inflammatory, immunosuppressive Chemoattractant for T cells, monocytes, neutrophils ↓ Activation, homing and effector functions of macrophages, DCs, T and B cells, CTLs, NK cells ↑ Angiogenesis and extracellular matrix protein production ↑ Th17 cell differentiation	Most activated hematopoietic cells; some non-hematopoietic cells	TGFβR	Widely expressed

Hematopoietic Growth Factors

SCF	Stem cell factor	Promotes HSC survival, self-renewal and differentiation into hematopoietic progenitors ↑ Proliferation of lymphoid and myeloid precursors	Stromal cells in fetal liver, bone marrow, thymus	c-kit	HSCs in BM, cells in CNS and gut
GM-CSF	Granulo-monocyte colony stimulating factor	Promotes generation and differentiation of monocyte and granulocyte precursors	BM stromal cells, activated T cells, endothelial cells, macrophages	GM-CSFR	Myeloid lineage precursors
G-CSF	Granulocyte colony stimulating factor	Acts on monocyte/granulocyte precursors to generate granulocytes ↑ Steady state and emergency production of neutrophils	BM stromal cells, activated T cells, endothelial cells, fibroblasts, macrophages	G-CSFR	Monocyte-granulocyte precursors
M-CSF	Monocyte colony stimulating factor	Acts on monocyte/granulocyte precursors to generate monocytes and macrophages ↑ Generation of bone-resorbing cells	BM stromal cells, endothelial cells, fibroblasts, macrophages	M-CSFR	Monocyte-granulocyte precursors

*Major abbreviations: BM, bone marrow; HSC, hematopoietic stem cell; R, receptor

Table 2 Major Human Chemokines and Chemokine Receptors

Chemokine Systematic Name	Chemokine Common Name	Chemokine Receptor	Cells/Tissues Expressing Chemokine Receptor
CCL Chemokines			
CCL1	I-309*	CCR8	Act T, NK, Th2
CCL2	MCP-1	CCR4	Act T, Bas, Th2
CCL3	MIP-1α	CCR4	Act T, Bas, Th2
CCL4	MIP-1β	CCR5	Mac, Act T, Th1
CCL5	RANTES	CCR1	Mac, Act T, Neu, Bas
CCL11	Eotaxin	CCR3	Eo, Act T, Bas, Th2
CCL21	SLC	CCR7	Mature DCs; naïve T and B
CCL26	MIP-4α	CCR3	Eo, Bas
CCL27	CTACK	CCR10	Memory T
CCL28	MEC	CCR3	Mucosal T and B
CXCL Chemokines			
CXCL6	GCP-2	CXCR2	Neu, NK
CXCL7	NAP-2	CXCR2	Neu, NK
CXCL8	IL-8	CXCR1	Neu, NK
CXCL10	IP-10	CXCR3	Act T, NK, Th1
CXCL12	SDF-1	CXCR4	T
CXCL13	BCA-1	CXCR5	B
CXCL14	BRAK	?	Mon, DCs, Act NK
CXCL16	SCYB16	CXCR6	T, NKT

*Major abbreviations: Act T, activated T cells; B, B cells; Bas, basophils; BCA-1, B cell attracting chemokine 1; BRAK, breast and kidney-expressed chemokine; CTACK, cutaneous T cell-attracting chemokine; DCs, dendritic cells; Eo, eosinophils; GCP-2, granulocyte chemotactic protein 2; IL-8, interleukin 8; IP-10, interferon-inducible protein 10; Mac, macrophages; MCP, monocyte chemotactic protein; MEC, mucosae-associated epithelial chemokine; MIP, macrophage inflammatory protein; NAP-2, neutrophil activating peptide-2; Neu, neutrophils; NK, natural killer cells; RANTES, regulated on activation normal T cell expressed and secreted; SCY, small cytokine; SDF-1, stromal cell-derived factor 1; SLC, secondary lymphoid tissue chemokine; T, T cells; Th1, T helper cells type 1; Th2, T helper cells type 2.

Appendix F: Laboratory Uses of Antibodies

Scientists interested in dissecting a biological system often take advantage of the properties of the antigen–antibody bond because this interaction is highly specific. Where standard techniques may not be able to distinguish between very closely related molecules, specific antibodies for distinct epitopes on those molecules can do so with ease. Identification and purification of a single component from a complex mixture becomes a ready possibility, because an antibody can detect one antigenic molecule among 10^8 other molecules. In addition, the antigen–antibody interaction is reversible and does not alter the antigen. For these reasons, techniques employing antibodies are used to purify, characterize and quantitate antigens, and to pinpoint their expression in cells or tissues. Below we discuss the sources of antibodies used in experimental work and provide brief descriptions of various experimental techniques that use antibodies. Several figures illustrating these techniques are included at the end of this appendix.

A. Sources of Antibodies

I. ANTISERA

Serology is the study of antibodies present within a given *antiserum*—the clear liquid serum fraction of clotted blood obtained from an individual who has been immunized or exposed to a foreign substance or infectious agent. An antiserum is first tested for its *titer* (relative concentration of antigen-specific antibodies) by serially diluting samples of the antiserum until binding to specific antigen can no longer be detected. An antiserum that can be diluted extensively and still shows binding activity is said to have a "high titer" of antigen-specific antibodies. Often an antiserum can be used without further purification. If necessary, non-Ig proteins can be removed from the antiserum by biochemical methods.

Antisera are *polyclonal*, meaning that, when an animal is exposed to an antigen, many B cell clones respond to the antigen's entire collection of epitopes. A plethora of different antibodies specific for different epitopes of the antigen is produced, with each specificity present in a relatively small quantity. This mixture is an advantage to an organism *in vivo* because it offers multiple ways to attack a pathogen. Similarly, a researcher will use an antiserum to identify an antigen as a whole (as opposed to one particular epitope of that antigen) because, even if some epitopes on the antigen have been denatured during handling or altered due to mutation, the mixed population of antibodies in the antiserum will still likely contain at least some antibodies capable of binding to the antigen.

However, the heterogeneity of an antiserum is a problem when one wants to limit antibody binding to a specific epitope. Removal of undesired antibodies to other epitopes is time-consuming, expensive, less than 100% effective (leaves cross-reacting antibodies behind), and can result in a significant decrease in the concentration of the desired antibody. In addition, antisera vary in composition and titer even among inbred animals, and even when the same protocol is followed for the preparation of different batches.

II. MONOCLONAL ANTIBODIES FROM HYBRIDOMAS

An enormous technical breakthrough occurred in the 1970s when immunologists discovered a way to derive antibodies of a single defined specificity from a clone of antibody-producing cells that could live indefinitely in culture. These cells were called *hybridomas* because they resulted from the hybridization of two cell types: an antibody-secreting B cell and a *myeloma*. Myelomas are B cell cancers that arise from the malignant transformation of a single plasma cell. A myeloma clone secretes antibodies of a single specificity like any B cell clone. However, unlike a normal plasma cell clone that dies after a few days, a myeloma clone has an unlimited life span: it is said to be "immortal". This means that unlimited numbers of antibody-producing cells can be grown in culture and manipulated in the laboratory. However, the antigenic specificities of myeloma antibodies are for the most part unknown because the B cells involved are not selected with any defined experimental antigen. A *hybridoma* is created by artificially fusing the plasma membrane of a myeloma cell with the plasma membrane of an isolated B cell of <u>known</u> antigenic specificity, such that the cells are combined. As a result, a hybridoma has the immortality and production capacity of the myeloma but the known antibody specificity of the B cell. The hybridoma grows to form a colony secreting large quantities of *monoclonal antibody* (mAb); that is, antibody derived from a single, defined B cell clone.

Monoclonal antibodies are typically employed to identify a specific protein marker on a cell surface or in a tissue or serum sample, or to map individual epitopes on an antigen. Large quantities of mAbs can be used to purify proteins to be used in research studies, or in industrial or clinical applications. Because hybridomas can be clonally expanded and maintained indefinitely, they provide a permanent and uniform source of antibody. However, because mAbs recognize only a single determinant, a virus that manages to mutate that precise epitope can escape detection by a mAb where it would not escape detection by a polyclonal antiserum. In addition, mAbs are not ideal for tests based on the detection of large immune complex networks formed between multiple antigen and antibody molecules. A mAb recognizes only a single epitope, which may be represented only once on an antigen molecule that occurs naturally as a monomer. Thus, networking between multiple antigen and antibody molecules may not occur. It should also be noted that a mAb will be of a single defined isotype that will give it isotype-specific characteristics. These characteristics may make the mAb suitable for use in some applications but not others.

The mAbs used for the immunotherapies described in Chapters 16 to 20 have usually been "humanized" prior to use. Technically, it is far easier to produce a mAb of the desired specificity in a mouse than in a human. However, a human's immune system will normally mount a response to mouse-specific epitopes on a mouse mAb, decreasing its effectiveness. To avoid this anti-mouse response, a "humanized" mAb is produced by isolating the gene encoding a mouse mAb of the desired specificity and using genetic engineering techniques to combine the sequences encoding the specificity-defining V region of the mouse mAb with the sequences encoding the structural elements of a human Ig. The framework regions of the V region of the Ig may also be "humanized". Such manipulations create a chimeric gene that is translated *in vitro* to yield a mouse/human antibody protein in which most of the mouse Ig sequence has been replaced with human Ig sequence. The patient's immune system therefore does not recognize the humanized mAb as foreign and does not generate large quantities of anti-mAb antibodies that would clear the therapeutic mAb before it could do its job.

B. Experimental Techniques using Antibodies

There are two broad categories of techniques that use antibodies to detect antigen. One category encompasses assays based on detection of large, visible immune complexes containing antibody and antigen trapped in a network. Examples of these techniques are the *precipitin reaction*, *agglutination*, and *complement fixation*. The other category includes techniques based on the formation of individual antigen–antibody pairs, in which detection relies on a "tag" chemically introduced onto either the antigen or the antibody molecule. Such tags are usually radioactive, enzymatic or fluorescent and give rise to easily detectable assay signals. These assays tend to be more sensitive than those based on immune complex formation. Techniques of this type include *radioimmunoassay (RIA)*, *enzyme-linked immunosorbent assay (ELISA)*, *immunofluorescence* and *flow cytometry*. A third category of antibody-based techniques involves the use of antibodies to isolate and characterize antigens, and includes *immunoprecipitation*, *affinity chromatography* and *Western blotting*.

I. TECHNIQUES BASED ON IMMUNE COMPLEX FORMATION

Like all chemical reactions, the kinetics of antigen–antibody binding are driven by relative concentration. When complementary antibody and antigen (i.e., a binding pair) are mixed in a fluid in approximately equal amounts, non-covalent bonds are rapidly formed between individual molecules, resulting in a small soluble complex. However, because of bivalency, a single antibody molecule may use one antigen-combining site to bind to its epitope on one molecule of antigen molecule, and the other antigen-combining site to bind the identical epitope on a second antigen molecule. Each of these two antigen molecules may possess additional epitopes for additional antibody binding, so that different antibody molecules mutually binding to this antigen are said to be *cross-linked*. Further cross-connections between additional antigen and antibody molecules result in the formation of an *immune complex* or *lattice*. As more and more antigen and antibody molecules become cross-linked, they form lattices large enough to precipitate out of solution and to become visible. These are the properties that were first exploited to examine antigen–antibody interactions, in the form of the *precipitin* and *immunodiffusion* assays. The collection of precipitated immune complexes by centrifugation also provided the first means of isolating and purifying antigens.

As illustrated in the *precipitin curve* shown in Figure F-1, when increasing amounts of soluble antigen are added to a fixed amount of soluble complementary antibody, the amount of lattice or immune complex precipitated increases up to a broad peak called the *zone of equivalence* and then slowly declines again. In the zone of equivalence, the number of antigenic epitopes and antibody-combining sites is approximately equal and the binding is optimal. Enough antigenic epitopes are present such that each individual antibody Fab site binds to a different antigen molecule, but the antigen concentration is still low enough that each antigen molecule is shared between (on average) two antibodies. Under these conditions, cross-linking readily occurs and immune complex precipitation is maximal. Where the amount of antigen added is very low and an excess of antibody over antigen is present, on average only one Fab site of an antibody is in use, meaning that cross-linking and immune complex formation are minimal. Free antibody remains in the supernatant after antigen–antibody complexes are removed by centrifugation. This area of the precipitin curve is known as the *zone of antibody excess*. When the antigen concentration increases past the zone of equivalence, a point is reached at which every available Fab site is bound to a separate antigen molecule: no antigen molecule is shared between two antibodies and a large immune complex cannot form. Free antigen remains in the supernatant after antigen–antibody complexes are removed by centrifugation, giving rise to the area of the precipitin curve known as the *zone of antigen excess*. Figures F-2, F-3, F-4 and F-5 illustrate several experimental techniques based on immune complex formation.

II. TECHNIQUES BASED ON "TAGGING" ANTIGEN–ANTIBODY PAIRS

Techniques based on immune complex formation exploit the fact that antigen–antibody immune complexes are relatively large in size and thus readily make themselves "visible" in assays. In contrast, unitary antibody–antigen pairs by definition are not found in extensive complexes. Thus, they cannot be made visible for quantitation unless the pair is somehow labeled with an easily detected *tag*. Techniques that make use of detectable tags to track antigens or antibodies of interest have greatly increased the scope of antibody-based assays. Tags are generally radioisotopes, enzymes, or fluorochromes that are covalently bound to either the antigen or the antibody. In addition, one partner of the antigen–antibody pair must be immobilized in some way so that any tag that is not part of an antigen–antibody pair can be removed from the assay system by washing. As well as for detection, tag assays are frequently used for the quantitation of an antigen or antibody because the amount of tag detected is proportional to the number of antigen–antibody pairs in the sample. Tag assays can either be *direct* or *indirect*.

i) Direct Tag Assays

Direct tag assays refer to single step procedures in which a tagged antigen (or antibody) is used to detect the presence of its untagged antibody (or antigen) binding partner. The tagged antigen or antibody is incubated with the test sample to allow antigen–antibody pairs to form. Unbound tagged molecules are removed by washing and the remaining tagged molecules are quantitated by measuring the amount of tag present relative to a standard curve. In many cases, the absolute number of antigen–antibody pairs present is not as important as the relative amount of tag (pairs) present compared to controls.

ii) Indirect Tag Assays

When the antibody or antigen of interest is not available in pure form or is chemically difficult to tag, antigen–antibody binding can be detected by tagging a third component that binds to the unlabeled antigen–antibody pair of interest. Three reagents commonly used as the third component in indirect assays are secondary antibodies, *Staphylococcus aureus* Protein A or Protein G, and the biotin–avidin system (Fig. F-6). Secondary antibodies are usually prepared by immunizing another species with the primary antibody protein (used as the antigen).

iii) Common Tag Assays for Antigen Detection

Several types of assays use tagged antigens or antibodies either directly or indirectly to detect antigens of interest in soluble, cellular or tissue samples. *Binder–ligand* assays include the radioimmunoassay (RIA) and the enzyme-linked immunosorbent assay (ELISA), in which the tag is a radioisotope in the former and a chromogenic substrate in the latter. How a binder–ligand assay can be used to detect antigen is illustrated for ELISA in Figure F-7. *Immunofluorescence* assays allow scientists to examine the presence of an antigen *in situ* in a blood or tissue sample. In this case, the tag is a fluorochrome that emits light of a specific color after excitation by UV radiation (Fig. F-8 and Plate F-1). Fluorochrome tag assays are also used for *flow cytometric analysis*, in which living cells bind antibodies tagged with particular fluorochromes. The cells can then be tracked and separated according to the tagged antibodies they have bound (Plate F-2).

III. TECHNIQUES FOR THE ISOLATION AND CHARACTERIZATION OF ANTIGENS

Antigen–antibody interaction is frequently used to purify antigens and characterize their physical properties. The specificity of antibodies often makes extensive preliminary purification of the antigen unnecessary. The technique of *immunoprecipitation* by specific antibody can be used to isolate antigens present at low concentrations in complex mixtures of proteins. Specific antibody is added to the protein mixture (such as a cell extract), and the antigen–antibody complex is caused to precipitate out of solution by adding an insoluble agent to which the complexes will bind (Fig. F-9). Antibodies can also be used to isolate antigens using *affinity chromatography*. The protein mixture is passed over a column of agarose beads to which specific antibody has been covalently fixed. The antigen is retained on the column, whereas other proteins are washed through (Fig. F-10). The *Western blot* uses the specificity of antigen–antibody binding in conjunction with gel electrophoresis to detect very small quantities of a protein of interest in a complex mixture. The proteins are first separated by size via electrophoresis and transferred to a nitrocellulose membrane. The membrane is then incubated with a solution containing tagged specific antibody to identify the position of the antigen of interest (Fig. F-11).

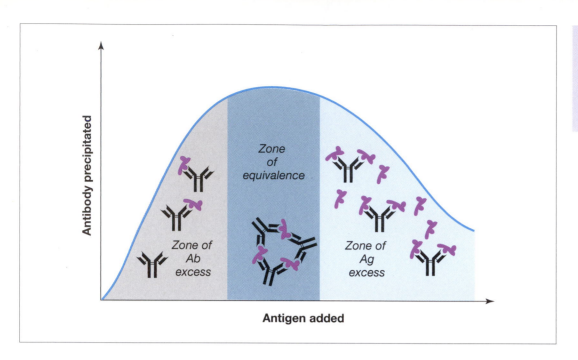

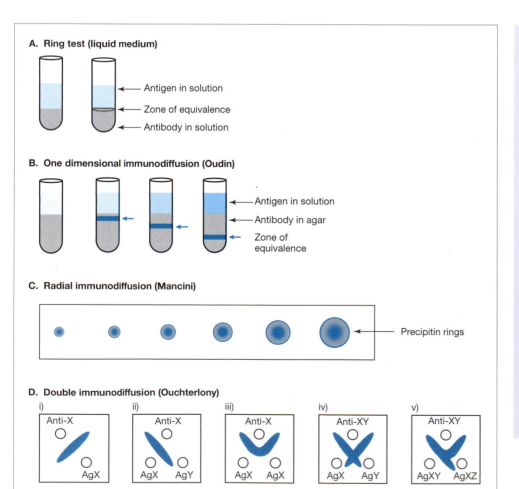

Fig. F-2 **Assays Based on the Precipitin Reaction**
(A) Ring test. Left-hand tube, no complementary antigen and antibody are present. Right-hand tube, complementary antigen and antibody cause a precipitin line to form at the solution interface (zone of equivalence). (B) One-dimensional immunodiffusion (Oudin) assay. With antibody-impregnated agar as the medium, increasing concentrations of antigen (left to right) form a series of descending precipitin lines after diffusion into the agar. (C) Radial immunodiffusion (Mancini) assay. Various amounts of antigen are placed in wells in antibody-impregnated agar. The greater the antigen concentration, the greater the diameter of the precipitin ring formed. (D) Double immunodiffusion (Ouchterlony) assay. Antibody and antigen samples diffuse toward each other from wells made in agar. A precipitin line will form wherever a zone of equivalence exists between a complementary antibody and antigen (i). In panels ii–v, a single antiserum is used to give information about the relationship between test antigens. The resulting patterns show the following relationships between test antigens: (ii) non-identity (no epitopes shared between antigens), (iii) identity (all epitopes shared), (iv) non-identity, and (v) partial identity (some epitopes shared).

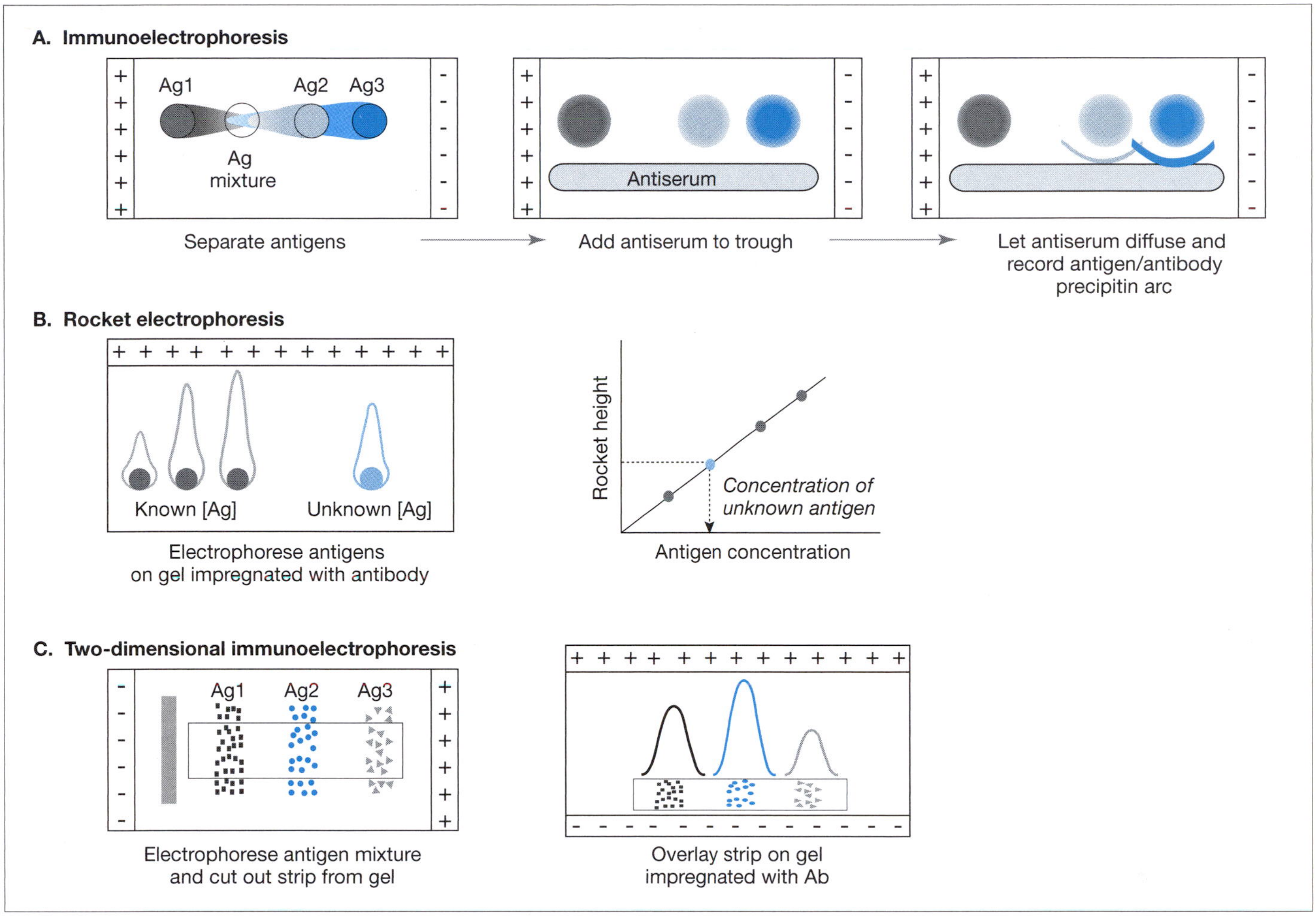

A. Immunoelectrophoresis

Separate antigens → Add antiserum to trough → Let antiserum diffuse and record antigen/antibody precipitin arc

B. Rocket electrophoresis

Known [Ag]   Unknown [Ag]

Electrophorese antigens on gel impregnated with antibody

Rocket height

Concentration of unknown antigen

Antigen concentration

C. Two-dimensional immunoelectrophoresis

Ag1   Ag2   Ag3

Electrophorese antigen mixture and cut out strip from gel

Overlay strip on gel impregnated with Ab

Fig. F-3 Types of Immunoelectrophoresis
(A) Immunoelectrophoresis. Separation through a gel on the basis of charge is followed by characterization using antigen–antiserum interaction. For example, known antigens can be separated and used to determine the identity of antibodies in a sample of antiserum. The first panel shows some residual antigen mixture in the well of the electrophoresis plate; in reality, all antigen would be gone from the well by the time electrophoresis was complete. (B) Rocket electrophoresis. Electrophoresis of an antigen sample in an antibody-impregnated gel results in a precipitin line whose height is proportional to the antigen concentration (left panel). Using known antigen concentrations, a standard curve can be constructed to determine the sample concentration (right panel). (C) Two-dimensional immunoelectrophoresis. If a sample contains a mixture of antigens, they can first be separated in a gel by immunoelectrophoresis and then subjected to analysis by rocket electrophoresis.

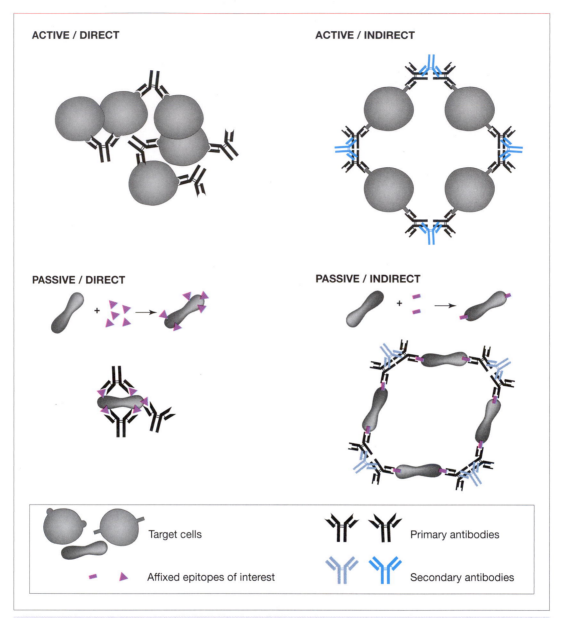

ACTIVE / DIRECT

ACTIVE / INDIRECT

PASSIVE / DIRECT

PASSIVE / INDIRECT

Target cells

Primary antibodies

Affixed epitopes of interest

Secondary antibodies

Fig. F-4 **Types of Agglutination**
Agglutination assays are used to test for the presence of an antibody of interest in a serum sample. In agglutination, the binding of specific antibody mediates the visible clumping of target cells (shown) or test particles (not shown). The epitope recognized by the antibody may occur naturally on the target cells (active) or may need to be affixed to target cells artificially (passive). The specific antibody itself may be able to agglutinate the target cells (direct). If not, a secondary antibody that recognizes and binds to the Fc region of the primary antibody must be used to mediate agglutination (indirect).

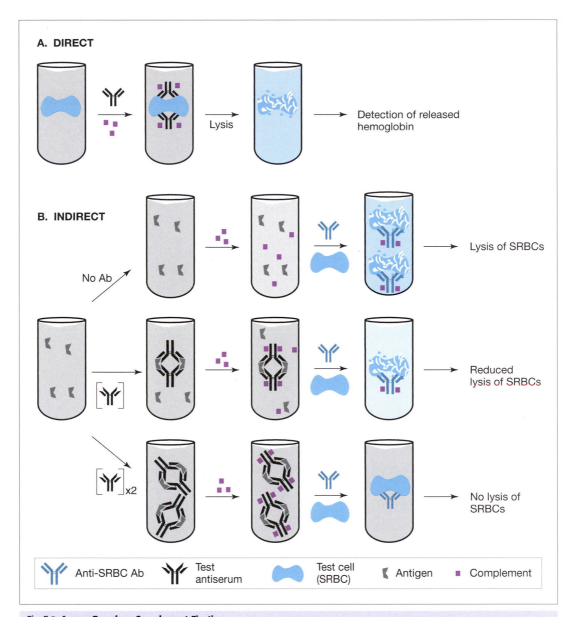

A. DIRECT

Lysis → Detection of released hemoglobin

B. INDIRECT

No Ab → Lysis of SRBCs

Reduced lysis of SRBCs

[Y]x2 → No lysis of SRBCs

Anti-SRBC Ab | Test antiserum | Test cell (SRBC) | Antigen | Complement

Fig. F-5 **Assays Based on Complement Fixation**
Classical complement-mediated lysis of cells can be used to evaluate the presence of specific antibody. In these examples, the complement-mediated lysis of sheep red blood cells (SRBCs) used as test cells results in the release of hemoglobin that can be measured and directly related to the concentration of specific antibody present. Test cells other than SRBCs can also be used, in which case evaluation of lysis can be measured by the uptake of dyes or by the release of an internalized radioisotope. (A) Direct assay; the antigenic epitope of interest occurs on the test cell. (B) Indirect assay; the antigenic epitope of interest does not occur on a test cell. After allowing the antigen and antibody to interact and consume ("fix") complement, the amount of complement remaining unfixed is measured by adding anti-SRBC antibodies and SRBCs. The amount of hemoglobin released is inversely proportional to the amount of antibody–antigen binding occurring in the first step of the assay. To simplify the illustration, the antigen–antibody complexes present since the first step of the indirect assay are not shown in the final tube, as they play no direct role at this point.

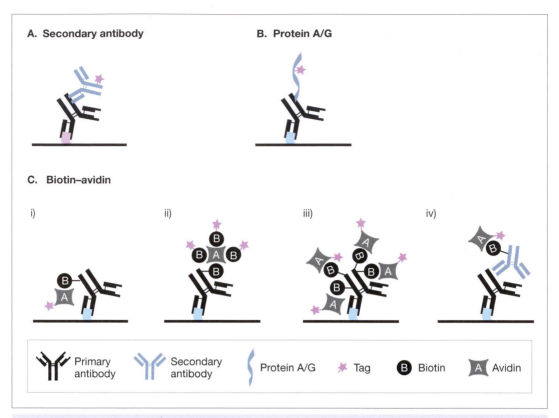

A. Secondary antibody

B. Protein A/G

C. Biotin–avidin

i)   ii)   iii)   iv)

Primary antibody   Secondary antibody   Protein A/G   Tag   B Biotin   A Avidin

Fig. F-6 Third Components in Antigen–Antibody Assays
(A) A tagged secondary antibody is specific for the Fc region of the primary antibody. (B) Tagged Protein A or Protein G purified from the bacterium *Staphylococcus aureus* binds to the Fc region of primary IgG antibodies. (C) *Biotin* is a small molecule that binds with extremely high affinity to an egg white glycoprotein called *avidin*. Avidin contains four biotin-binding sites. A primary antibody can be conjugated with biotin before use in an assay so that the presence of antigen–antibody pairs can be later detected using tagged avidin (i). The assay sensitivity can be increased by using the avidin molecule as a multivalent bridge between the biotinylated antibody and tagged biotin (ii), or by ensuring that the biotinylated antibody has undergone biotin conjugation at multiple sites (iii). If desired, biotin/avidin-based approaches can also be applied using a secondary antibody that is biotinylated (iv).

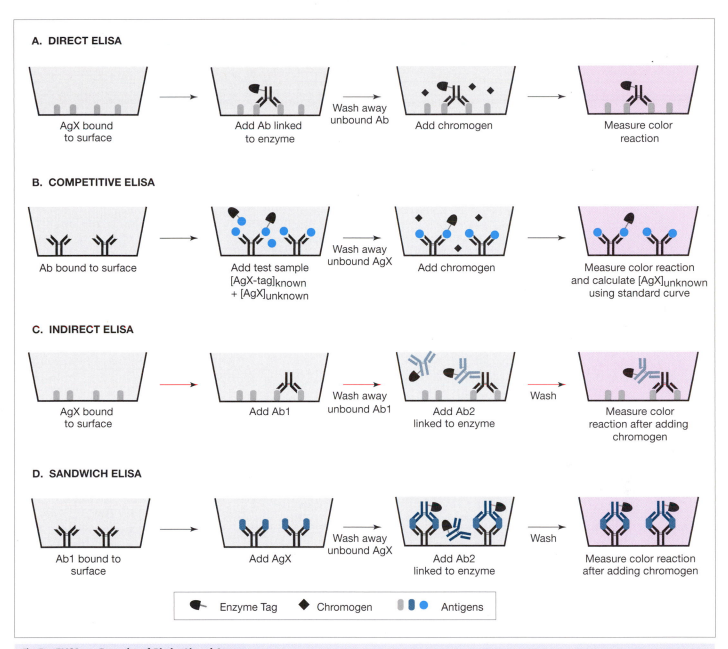

A. DIRECT ELISA

AgX bound to surface → Add Ab linked to enzyme → *Wash away unbound Ab* → Add chromogen → Measure color reaction

B. COMPETITIVE ELISA

Ab bound to surface → Add test sample $[AgX\text{-}tag]_{known}$ + $[AgX]_{unknown}$ → *Wash away unbound AgX* → Add chromogen → Measure color reaction and calculate $[AgX]_{unknown}$ using standard curve

C. INDIRECT ELISA

AgX bound to surface → Add Ab1 → *Wash away unbound Ab1* → Add Ab2 linked to enzyme → *Wash* → Measure color reaction after adding chromogen

D. SANDWICH ELISA

Ab1 bound to surface → Add AgX → *Wash away unbound AgX* → Add Ab2 linked to enzyme → *Wash* → Measure color reaction after adding chromogen

Enzyme Tag   ◆ Chromogen   Antigens

Fig. F-7 ELISAs as Examples of Binder-Ligand Assays
In all binder–ligand assays, one component of interest, either the antigen or antibody, is immobilized so that antigen–antibody pairs formed during the course of the assay are not lost during subsequent washing steps. If a radioisotope is used as the tag, the assay is an RIA (not shown). In an ELISA, the tag is an enzyme that acts on a chromogenic substrate to yield a detectable colored product. In the direct ELISA shown in (A), antigen X of unknown concentration is coated onto the surface of a microtiter well and tagged antibody is added to detect it. The concentration of antigen X is then calculated using a standard curve. In a competitive ELISA (B), antibody specific for antigen X is coated onto the surface of a microtiter well. The presence of AgX in the test sample leads to a proportional displacement of tagged AgX from the immobilized antibody, decreasing the intensity of the final color reaction proportionally. For the indirect ELISA shown in (C), a secondary antibody has been used as the third component; however, Protein A, Protein G or the biotin–avidin system could also be used for this step. The sandwich ELISA (D) is a variation of the indirect ELISA in which the secondary antibody is also specific for the antigen of interest.

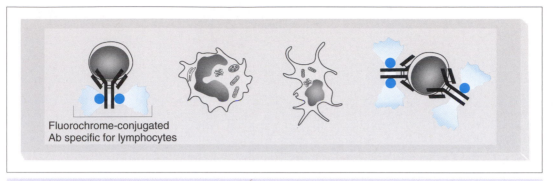

Fig. F-8 Immunofluorescence

In this schematic example, a blood cell sample (containing a spectrum of leukocytes) is tested for the presence of lymphocytes using direct immunofluorescence. The sample is fixed on a microscope slide and incubated with fluorochrome-conjugated antibody that binds to an antigen present only on lymphocytes. After washing away excess unbound antibody, the slide is viewed under a fluorescence microscope. Fluorescing sites indicate the location of lymphocytes on the slide.

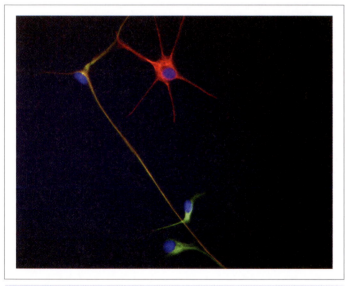

Plate F-1 Immunofluorescent Examination of Human Neural Cells

In this example, a neural filament protein (GFAP) is bound by an antibody tagged with a fluorochrome that emits red light. A different filament protein (nestin) is bound by an antibody tagged with a fluorochrome that emits green light. The yellow color associated with one of the neurons indicates that it expresses both GFAP and nestin. The nuclei of all the neurons are stained blued with Hoechst dye, a non-proteinaceous molecule that binds specifically to DNA [Reproduced by permission of Radha Chaddah, Neurobiology Research Group, University of Toronto.]

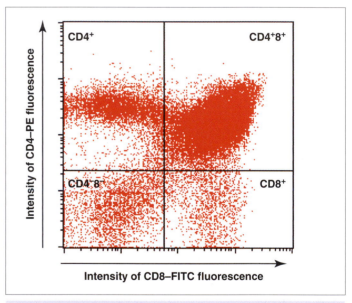

Plate F-2 Flow Cytometry

In this example, a mixed population of T lineage cells has been tested for levels of expression of the T cell coreceptors CD4 and CD8. The T cells have been incubated with two antibodies: anti-CD4 antibody tagged with the fluorochrome PE, and anti-CD8 antibody tagged with the fluorochrome FITC. The cells are then fed into a machine called a *flow cytometer* that separates the cells on the basis of the wavelength and intensity of the fluorescence emanating from the antibodies that have bound to each cell. These data indicate which antibodies (if any) each cell has bound and quantitates the amount of antibody bound by each cell (which is a measure of the level of expression of CD4 or CD8 by that cell). The results are presented as a quadrant graph. Mature T cells express either CD4 (upper left quadrant) or CD8 (lower right). Very early T cell precursors express neither CD4 nor CD8 (lower left), while slightly more advanced precursors express both CD4 and CD8 (upper right). [Reproduced by permission of Juan-Carlos Zúñiga-Pflücker, Department of Immunology, University of Toronto and Sunnybrook Research Institute, Toronto.]

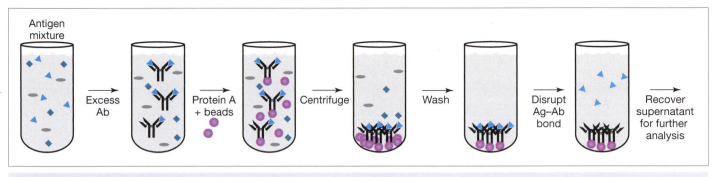

Fig. F-9 **Immunoprecipitation.**
A protein mixture is incubated with specific antibody. Any antigen–antibody complexes that form are precipitated from solution by the addition of Protein A-coated beads that bind to the antibodies and collect at the bottom of the tube under the force of centrifugation. After washing, the desired antigen is released from the antibody-bound beads using altered pH and/or high salt concentration.

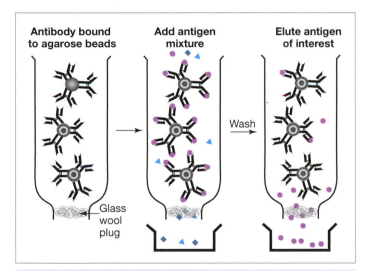

Fig. F-10 **Affinity Chromatography**
Agarose beads bearing immobilized specific antibody are placed into a column with a semi-permeable plug at the bottom. A solution containing antigen is passed slowly through the column, allowing the binding of specific antigen to the immobilized antibody. Unbound entities pass through the plug and any molecule that binds to the beads non-specifically is removed by extensive washing. A solution with the appropriate pH and salt concentration to disrupt antigen–antibody binding is then passed through the column to elute (wash off) the antigen of interest.

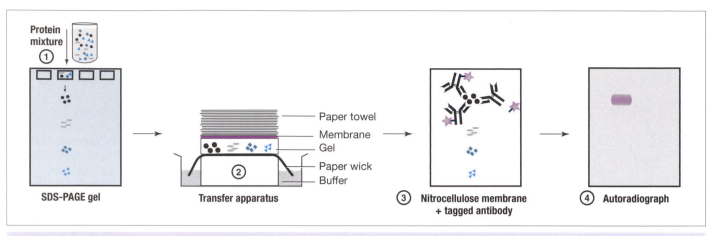

Fig. F-11 **Western Blotting**
Proteins are separated by charge using gel electrophoresis (1) and transferred onto a nitrocellulose membrane by the capillary action of
buffer that is drawn up through a filter paper wick toward absorbent paper towel (2). (Electric current can be used to speed the transfer.) As the proteins are drawn
upward by the buffer, they adhere to the nitrocellulose membrane. The antigenic protein of interest is then detected on the membrane by the binding of tagged antibody
(3). If the tag is a radioisotope, its presence is detected by subjecting the membrane to autoradiography (4). If the tag is an enzyme (not shown), the membrane is exposed
to the appropriate chromogenic substrate. A colored band develops where the protein is bound to the membrane.

Glossary

Term	Definition
αβ TCR checkpoint	Second major checkpoint in T cell development. *Positive selection* of *DP thymocytes* expressing a fully functional αβTCR promotes the survival of thymocyte clones recognizing self-MHC alleles with moderate *avidity*, while *negative selection* deletes *autoreactive* thymocyte clones.
ABO antigens	Family of glycoprotein antigens expressed on the surfaces of RBCs that define the ABO blood group system. Differences in ABO blood types can cause *hyperacute graft rejection* and severe blood *transfusion reactions*.
activation-induced cell death (AICD)	Apoptotic death of lymphocytes subjected to prolonged antigenic stimulation. A means of eliminating effector cells after they are no longer needed.
acute cellular rejection (ACR)	A form of *acute graft rejection* in which allogeneic graft cells are destroyed by effector functions of recipient leukocytes.
acute graft rejection	Graft rejection occurring within days or weeks of a transplant. Can be cell-mediated (*acute cellular rejection*) or antibody-mediated (*acute humoral rejection*).
acute humoral rejection (AHR)	A form of *acute graft rejection* due to production of antibodies directed against allo-MHC in the graft.
acute infection	An infection causing disease symptoms that appear rapidly but remain for only a short time.
acute phase proteins	Early inflammatory proteins made by hepatocytes. Acute phase proteins bind to cell wall components of microbes and activate complement. See also *inflammation*.
adaptive immunity	Uniquely specific recognition of a non-self entity by lymphocytes whose activation leads to elimination of the entity and the production of specific memory lymphocytes. Because these memory lymphocytes forestall disease in subsequent attacks by the same pathogen, the host immune system has "adapted" to cope with the entity.
ADCC	=Antibody-dependent cell-mediated cytotoxicity. The lysis of a cell that occurs when the Fc region of surface-bound antibody interacts with *FcR* expressed on lytic cells, such as eosinophils and NK cells, triggering their degranulation. Macrophages, neutrophils and monocytes may also mediate ADCC.
adenoids	See *tonsil*.
ADEPT	=Antibody-directed enzyme/pro-drug therapy. Therapy employing an *immunoconjugate* in which the mAb is linked to an enzyme capable of converting an inert pro-drug into an active cytotoxic drug. Used for anti-cancer therapy.
adjuvant	A substance that, when mixed with an isolated antigen, increases its immunogenicity. Adjuvants provoke local inflammation, drawing immune system cells to the site and triggering maturation of DCs.
adoptive transfer	The transfer of donor lymphocytes into a recipient.
afferent lymphatic vessel	A vessel that conveys *lymph* into a *lymph node*.
affinity	A measure of the strength of the association established at a single point of binding between a receptor and its ligand. For *antibody*, the strength of the non-covalent association between a single antigen-binding site on the antibody and a single epitope on an *antigen*.
affinity chromatography	Isolation of a protein from a mixture in solution based on its ability to bind to a particular ligand immobilized in a column.

Term	Definition
affinity maturation	Positive selection of developing B cells with BCRs that have undergone *somatic hypermutation* resulting in increased *affinity* for specific antigen. Memory cells with this increased affinity for antigen are generated.
agglutination	Aggregation of cells to form a visible particle.
AIDS	=Acquired immunodeficiency syndrome. Failure of adaptive immunity due to T cell destruction caused by HIV infection.
alleles	Two slightly different sequences of the same gene. Proteins produced from alleles have the same function.
allergen	An antigen that is innocuous in most individuals but provokes *type I hypersensitivity* in some.
allergy	Clinical manifestation of *type I hypersensitivity*. Mediated by mast cells armed with allergen-specific IgE.
alloantibodies	Antibodies distinguishing between the different forms of a protein encoded by different alleles of a given gene. Often refers to antibodies specific for MHC molecules.
allogeneic	Having different alleles at one or more loci in the genome compared with another individual of the same species.
allograft	Tissue transplanted between allogeneic members of the same species.
allorecognition	Recognition of allelic differences expressed by cells of one individual by the lymphocytes of another individual. Most often refers to recognition of MHC-encoded differences.
alternative complement activation	Activation of the complement cascade initiated by direct binding of complement component C3b to a stabilizing ligand on a microbe. Involves cleavage and activation of factors B and D and properdin.
anaphylatoxins	Complement component cleavage products that have pro-inflammatory and chemoattractant effects; includes C3a, C4a and C5a.
anaphylaxis	Systemic *type I hypersensitivity* response that may be fatal due to a catastrophic drop in blood pressure induced by the release of large quantities of inflammatory mediators.
anergy	State of lymphocyte non-responsiveness to specific antigen. Induced by an encounter of the lymphocyte with cognate antigen under less than optimal conditions, such as in the absence of *costimulation*.
angiogenesis	Process by which new blood vessels are formed.
antibody	Secreted *immunoglobulin* that is produced by B lineage *plasma cells* and binds to specific antigens. Able to recognize antigens that are either soluble or fixed in a tissue or on a cell surface.
antigen	Entity (such as an element of an infectious pathogen, cancer or inert injurious material, or of a self tissue) that can bind to the antigen receptor of a T or B cell. This binding does not necessarily lead to lymphocyte activation.
antigenic determinant	See *epitope*.
antigenic drift	Subtle modification of pathogen antigens through point mutations. Usually involves surface proteins that would normally be the target of neutralizing antibodies.
antigenic shift	Dramatic modification of viral antigens due to reassortment of genomic segments of two different strains of a virus (that simultaneously infect the same individual) to generate progeny virions with new combinations of genome segments and thus new proteins.
antigen presentation	Cell surface display of peptides in association with either MHC class I or II molecules allows T cell recognition of antigenic peptides on the surface of target cells or APCs, respectively.

Term	Definition
antigen-presenting cell (APC)	Cell expressing MHC class II and thus capable of presenting peptides to CD4$^+$ Th cells. APCs are usually DCs, B cells or macrophages.
antigen processing	Degradation of a protein into peptides suitable for binding in the peptide binding grooves of either MHC class I or II molecules.
antiserum	The clear liquid (serum) fraction of clotted blood containing antibodies produced when an individual or animal is exposed to a foreign substance or infectious agent.
antitoxins	Antibodies made against bacterial *exotoxins* and *endotoxins*.
antiviral state	A metabolic state of viral resistance induced in a cell by exposure to IFNs released by neighboring virus-infected cells.
APC licensing	Concept that full activation of an antigen-stimulated Tc cell is made possible through costimulation by a DC that has upregulated its costimulatory molecules in response to prior CD40–CD40L-mediated contact with an antigen-activated Th cell.
apoptosis	The controlled death of a cell mediated by intracellular proteases that cause the orderly breakdown of the cell nucleus and its DNA. Death occurs without the release of internal contents and without triggering inflammation.
Arthus reaction	Localized *type III hypersensitivity* characterized by redness and swelling in the site of immune complex deposition.
armed CTL	A mature CTL that has yet to encounter antigen but that has synthesized the chemical mediators that will be used to carry out target cell destruction.
armed mast cell	A mast cell that has bound *allergen*-specific IgE to its FcεR but has yet to encounter the allergen.
atopic	Involving symptoms or reactions caused by *type I hypersensitivity*.
autoantigen	Self antigen inducing an *autoreactive* response.
autoimmune disease	Pathophysiological state in which the host's own tissues are damaged as a result of *autoimmunity*.
autoimmunity	The response of an individual's immune system against self tissues. May cause *autoimmune disease* if unrestrained by mechanisms of *peripheral tolerance*.
autologous graft	Transplantation of tissue from one part of an individual's body to another part of his or her body.
autoreactive	Lymphocytes that recognize and are activated by self-antigens are autoreactive.
avidity	A measure of the total strength of all the associations established between a multivalent receptor and its multivalent ligand.
β2-microglobulin (β2m)	Invariant chain of MHC class I proteins; encoded by a gene outside the *MHC*.
β-selection	First major checkpoint in T cell development. Process in which the TCRβ chain expressed by a DN3 thymocyte becomes part of a *pre-TCR* that allows the thymocyte to receive a survival/proliferation signal and become committed to the αβ T lineage.
bacillus Calmette-Guérin	Strain of *Mycobacterium bovis* used as the BCG vaccine against tuberculosis.
basement membrane	A layer composed of collagen and other proteins and sugars that supports an overlying layer of epithelial or endothelial cells.
basophil	Circulating granulocyte with an irregularly shaped nucleus. Contains granules that stain a dark blue color with basic dyes. Basophil granules contain heparin and vasoactive amines, as well as many enzymes capable of promoting inflammation.
B cell receptor (BCR) complex	Antigen–receptor complex of B lineage cells. Composed of a membrane-bound Ig (mIg) monomer plus the *Igα–Igβ* complex required for *intracellular signaling*.

Term	Definition
B lymphocyte (B cell)	A *leukocyte* that matures in a *primary lymphoid tissue* (*bone marrow*), becomes activated in *secondary lymphoid tissues*, and mediates *adaptive immunity* by differentiating into antibody-producing *plasma cells*.
benign tumor	A mass formed by abnormally dividing cells that are well differentiated and well organized. Resembles the normal tissue from which it originated and is securely encapsulated and relatively slow growing. Causes death only by indirect means.
bone marrow	The *primary lymphoid tissue* (in the adult) in which all lymphocytes and other hematopoietic cells arise.
bone marrow transplant (BMT)	Replacement of a patient's dysfunctional immune system by *transplantation* of healthy whole *bone marrow* from a donor.
booster	A second or subsequent immunization (*vaccination*) to stimulate *memory cell* production.
brachytherapy	Implantation of a metal "seed" containing a radioisotope to mediate radiation therapy internally.
bronchi-associated lymphoid tissue (BALT)	Lymphoid patches and cells in the mucosae of the trachea and lungs. See *mucosa-associated lymphoid tissue* (MALT).
brush border	A layer of dense microvilli that cover the apical surfaces of gut epithelial cells, greatly increasing the surface area.
cachexia	Wasting of the body due to uncontrolled cellular catabolism. Cachexia in cancer patients is induced by their high levels of TNF.
cadaveric organ	Transplantable organ recovered from a recently deceased human (a cadaver).
cancer	See *malignant tumor*.
cancer-testis antigens	Proteins that are normally expressed solely in spermatogonia and spermatocytes but become *tumor-associated antigens* (TAAs) when transformation causes them to be expressed on other cell types.
capsid	Protein coat of a virus.
capsule	Polysaccharide coating of certain bacteria.
carcinogen	Any substance or agent that significantly increases the incidence of *malignant tumors* by mutating or deregulating *oncogenes*, *tumor suppressor genes* or *DNA repair genes*.
carcinogenesis	Multistep process by which malignant transformation of a cell culminates in cell growth deregulation and the formation of a *malignant tumor*. Steps are initiation, promotion, progression and malignant conversion.
CD marker	=Cluster of differentiation marker. A cell surface protein identified by the binding to that protein of a cluster of different antibodies. Each CD marker is given a unique numeric designation.
CD1 molecules	Unconventional *antigen presentation* molecules that present non-peptide antigens (lipid or glycolipid antigens) to subsets of αβ and γδ T cells and NKT cells.
CD3 complex	Family of five ITAM-containing accessory chains necessary for TCR signaling and insertion of the TCR complex in the membrane. See *TCR complex*.
cell-mediated immunity	Originally, adaptive immune responses mediated by effector actions of cytotoxic T lymphocytes. Now includes effector actions of NK and NKT cells
cellular adhesion molecules (CAMs)	Molecules facilitating cell–cell and cell–matrix adhesion as well as extravasation. See *homing receptors* and *vascular addressins*.

Term	Definition
central tolerance	Mechanism that eliminates most *autoreactive* T and B cells during lymphocyte development. Established by *negative selection* processes in the thymus and bone marrow.
centroblasts	Rapidly proliferating, antigen-activated B cells that fill the *dark zone* of a *GC* and undergo *somatic hypermutation*.
centrocytes	Smaller, non-dividing cells arising from centroblasts that migrate to the *light zone* of the *GC* and undergo *affinity maturation* and *isotype switching*. Give rise to *plasma cells* and memory cells.
chemokine receptors	Receptors expressed on leukocytes that allow them to follow appropriate chemokine gradients for migration.
chemokines	Cytokines that induce leukocyte migration. See also *chemotaxis*.
chemotactic factors	Factors derived from either host or bacterial cells that can induce *chemotaxis*. Includes chemokines and the *anaphylatoxin* C5a.
chemotaxis	The directed movement of cells along the concentration gradient of a *chemotactic factor*. Serves to draw neutrophils, lymphocytes, and other leukocytes from the circulation into an injured or infected tissue.
chemotherapy	Use of chemical drugs to kill tumor cells. Chemotherapeutic agents generally affect only those cells that are growing faster than most normal cells, or those that have a metabolic imbalance.
chromosomal translocation	The abnormal exchange of genetic material between two different chromosomes or two different regions on the same chromosome.
chronic disease	A disease in which symptoms are experienced on an ongoing or recurring basis. See *persistent infection*.
chronic graft rejection (CGR)	Immune response against an *allograft* that occurs several months after transplantation and leads to loss of allograft function.
clades	Original subtypes of HIV based on sequence diversity.
classical complement activation	Initiated by the Fc region of an antigen-bound antibody binding to C1, followed by recruitment and cleavage of C4, C2, C3 and assembly of the terminal complement components. See *complement*.
clathrin	A protein component of the microtubule network involved in endocytosis mediated by cell surface receptors.
clathrin-mediated endocytosis	Process in which binding of a soluble macromolecule to its complementary cell surface receptor induces cellular internalization via *clathrin* polymerization.
clinical trials	A series of controlled tests of a drug, vaccine or treatment in human volunteers that is used to determine if its safety and efficacy warrant licensing and use in the general population. Generally involves four phases from I to IV.
CLIP	=Class II invariant chain peptide. A small fragment of *invariant chain* that sits in the MHC class II groove and prevents the premature binding of peptides.
clonal deletion	The induction of *apoptosis* in B or T lymphocytes that have bound their *cognate* antigen.
clonal exhaustion	A process by which an activated lymphocyte divides so quickly in the face of persistent antigen that its progeny burn out before memory cells have been generated.
clonal selection	The activation of only those clones of lymphocytes bearing receptors specific for a given antigen.
codominance	Expression of a given gene from both the maternal and paternal chromosomes.
cognate antigen	An antigen known to be recognized by a given lymphocyte antigen receptor because it was used for the original activation of that lymphocyte.

Term	Definition
cold chain (vaccination)	Constant refrigeration of a labile vaccine from the point of manufacture and shipping to the point of use in the field.
collectins	Soluble *pattern recognition molecules* that mediate pathogen clearance by *opsonization*, *agglutination* or the *lectin pathway of complement activation*. Includes *MBL*.
commensal organisms	Beneficial microbes that normally inhabit the surfaces of skin and body tracts.
common lymphoid progenitor (CLP)	Early descendant of *MPPs* that gives rise to B lineage cells.
common mucosal immune system (CMIS)	A system of extended mucosal defense established when lymphocytes activated in one mucosal *inductive site* migrate to a large number of *effector sites* in various mucosal tissues.
common myeloid progenitor (CMP)	Early descendant of *MPPs* that gives rise to *myeloid cells*.
complement	System of over 30 soluble and membrane-bound proteins that act through a tightly regulated cascade of pro-protein cleavage and activation to mediate cell lysis through assembly of the *MAC* in a target cell membrane. Intermediates in the complement cascade play a variety of roles in antigen clearance.
complementarity-determining region (CDR)	See *hypervariable region*.
complement fixing antibodies	Antibodies that efficiently trigger the classical complement activation pathway. In humans: IgM, IgG1, IgG2 and IgG3.
conformational determinant	A protein *epitope* in which the contributing amino acids are located far apart in the linear sequence but which become juxtaposed when the protein is folded in its native shape.
conjugate	A structure resulting from the physical joining of two other structures. Two cells may form a conjugate, such as Th and B cells during the provision of B–T cooperation, or CTLs and target cells during cytolysis. Two molecules may also form a conjugate, such as the covalent joining of a small inorganic molecule to a protein carrier to induce an immune response.
conjugate vaccine	*Vaccine* based on the covalent linkage of a pathogen carbohydrate epitope to a protein carrier such that it induces a B cell response to *Td antigens* rather than *Ti antigens*.
constant domain	A domain of an Ig or TCR chain that is encoded by the corresponding *constant exon*. The constant domains have very little amino acid variability.
constant exon	Exon encoding a *constant domain* of either an Ig or a TCR protein. A C exon is spliced at the mRNA level to a rearranged variable (V) exon to produce a transcript of a complete Ig or TCR gene.
constant region	The relatively invariant C-terminal portion of an Ig or TCR molecule. Comprises the constant domains of all the polypeptides involved.
contact hypersensitivity	*Type IV hypersensitivity* caused by an immune response to a chemically reactive molecule that has bound covalently to self proteins in the skin to form a *neo-antigen*.
coreceptor	Protein that enhances the binding of a primary receptor to a ligand. The T cell coreceptors CD4 and CD8 bind to non-polymorphic sites on MHC class II and I, respectively, that are outside the peptide-binding groove. This binding stabilizes the contact between the pMHC and the TCR, and also recruits intracellular signaling enzymes.
cortex	The outer layer of an organ such as the *thymus*.
cortical thymic epithelial cells (cTECs)	See *thymic epithelial cells*.
corticosteroids	Molecules that mediate powerful anti-*inflammatory* and *immunosuppressive* effects when they bind to glucocorticoid receptors and inactivate transcription factors.

Term	Definition
costimulation	The second signal required for completion of lymphocyte activation and prevention of anergy. Supplied by engagement of CD28 by B7 (T cells), and of CD40 by CD40L (B cells).
costimulatory blockade	Artificial interference with *costimulation* that anergizes antigen-specific T cells.
cross-presentation	Presentation of peptides from exogenous antigens on the MHC class I molecules of DCs and macrophages; required for naïve Tc cell activation.
cross-reactivity	Recognition by a lymphocyte or antibody of an antigen other than the *cognate antigen*. Cross-reactivity results either when the same epitope is found on two different antigens, or when two epitopes on separate antigens are similar.
cytokines	Low molecular weight, soluble proteins that bind specific cell surface receptors whose engagement leads to intracellular signaling triggering the activation, proliferation, differentiation, effector action, or death of the cell. Synthesized by leukocytes and some non-hematopoietic cells under tight regulatory controls.
cytotoxic cytokines	Cytokines that directly induce the death of cells. For example, TNF and LT.
cytotoxic T lymphocytes (CTLs)	*Effector* progeny of an activated Tc cell. CTLs recognize and destroy target cells displaying foreign peptide complexed to MHC class I. Target cell killing occurs via *cytotoxic cytokine* secretion and *perforin/granzyme-mediated cytotoxicity*.
danger signals	Molecules released by stressed or dying cells of host or pathogen origin that bind to *PRMs* of innate response cells and induce inflammation.
dark zone	Region of a *germinal center* where *somatic hypermutation* occurs.
delayed-type hypersensitivity (DTH)	Immunopathological damage occurring 24–72 hours after exposure of a sensitized individual to an antigen. Cell-mediated rather than antibody-mediated. See also *type IV hypersensitivity*.
delivery vehicle	Inert, non-toxic structure designed to protect *vaccine* antigens from nuclease- or protease-mediated degradation. May also act as an *adjuvant* or increase antigen display.
dendritic cells (DCs)	Irregularly shaped phagocytic leukocytes with finger-like processes resembling neuronal dendrites (except plasmacytoid subset). DC subsets arise from both the myeloid and lymphoid lineages and include conventional and plasmacytoid DCs.
dermis	Lower layer of skin beneath the epidermis and basement membrane. Contains lymphatics and blood vessels.
desmosomes	Junctions between keratinocytes that hold a layer together.
diapedesis	Second step of *extravasation* in which flattened leukocytes squeeze between venule endothelial cells and enter a tissue.
differential RNA processing	Polyadenylation-mediated mechanism by which two structural versions of a protein can be derived from a single gene.
differentiation	Functional specialization of a developing precursor cell or an activated cell.
direct allorecognition	A transplant recipient's T cells recognize peptide/allo-MHC epitopes presented on the surfaces of *allogeneic* donor cells in the graft. See *allorecognition*.
disease	Detectable clinical damage caused by the replication of the pathogen; not synonymous with *infection*.
DN thymocytes	=Double negative thymocytes. Thymocytes that express neither CD4 nor CD8 molecules. Includes DN1–4 subsets.
DNA repair	Intracellular pathways that correct mutations in DNA to maintain genomic stability.
dome	Region of mucosal lymphoid follicle overlying a *germinal center*.
donor cell infusion (graft tolerance)	Donor BM or a purified donor HSC preparation is administered to a *solid organ transplant* recipient to promote acceptance of donor cells.

Term	Definition
DP thymocytes	=Double positive thymocytes. Thymocytes that express both CD4 and CD8 molecules.
draining lymph node	Nearest lymph node that receives the lymph, antigen, lymphocytes and APCs emanating from a particular tissue.
early phase reaction	In the *effector* phase of *type I hypersensitivity*, the rapid onset of clinical symptoms induced by preformed inflammatory mediators immediately released via *mast cell* degranulation.
effector functions	The actions taken by effector cells and antibodies to eliminate foreign entities. Includes cytokine secretion, cytotoxicity, and antibody-mediated clearance.
effector cells	The differentiated progeny of an activated leukocyte that act to eliminate a non-self entity. Includes *plasma cells*, *Th effector cells* and *CTLs*.
effector site	Remote mucosal location where lymphocytes activated in a mucosal *inductive site* differentiate and exert effector actions.
effector stage (hypersensitivity)	The excessive, abnormal secondary response to a *sensitizing* agent that results in inflammatory tissue damage. See *type I–IV hypersensitivities*.
efferent lymphatic vessel	A vessel that conveys *lymph* away from a *lymph node*.
efficacy (vaccine)	The ability of a *vaccine* to effectively protect individuals from disease. Expressed as the percentage of individuals vaccinated that develop immune responses to the pathogen. Also called "coverage."
ELISA	=Enzyme-linked immunosorbent assay. *Binder–ligand* assay in which the antibody or antigen used is linked to an enzyme. The presence of binder–ligand pairs can be determined by adding a chromogenic substrate whose enzymatic conversion causes a detectable color change.
embryonic antigens	A protein or carbohydrate antigen normally expressed solely in the embryo. Inappropriate expression in a tumor cell can make it a *tumor-associated antigen*.
endocytic pathway	Intracellular system of membrane-bound endosomes and endolysosomes. These vesicles contain hydrolytic enzymes and other substances that digest internalized materials. Responsible for *exogenous antigen processing and presentation*.
endogenous antigen	An antigenic protein that originates within a cell in the host, as in a protein synthesized in a cell infected by a virus or intracellular bacterium.
endogenous antigen processing and presentation	A mechanism by which endogenous antigens in the cytosol are degraded into peptides via *proteasomes* and complexed to MHC class I in the rER. The peptide–MHC class I complex is then displayed on the cell surface. This pathway operates in almost all nucleated cell types.
endotoxic shock	A sometimes fatal collapse of circulatory and metabolic systems induced by overwhelming amounts of cytokines (particularly IL-1 and TNF) released in response to bacterial *endotoxins*.
endotoxin	Toxins released from the cell walls of damaged *gram-negative bacteria*; lipid portion of *lipopolysaccharide (LPS)*.
eosinophils	Connective tissue granulocytes with granules that stain reddish with acidic dyes. The granules contain highly basic proteins and enzymes effective in the killing of larger parasites. Eosinophils also play a role in *allergy*.
epidermis	Top layers of skin that contain mainly *keratinocytes* plus elements of *SALT* but no blood vessels.
epitope	The small region of a macromolecule that specifically binds to the antigen receptor of a B or T lymphocyte. B cell epitopes can be composed of almost any structure. T cell epitopes are a complex of antigenic peptide bound to either MHC class I or II.
epitope spreading	An immune response against one *epitope* causes tissue destruction that exposes previously hidden epitopes, activating additional lymphocyte clones.

Term	Definition
exocytosis	Process by which the membrane of an exocytic vesicle fuses with the plasma membrane, everting the contents into the extracellular fluid.
exogenous antigen	An antigenic protein that originates outside the cells of the host, as in a bacterial toxin.
exogenous antigen processing and presentation	*Exogenous antigens* are internalized into an APC, degraded within the *endocytic pathway*, and complexed to MHC class II in an endolysosomal vesicle. The peptide–MHC class II complex is then displayed on the cell surface. This pathway operates almost exclusively in *APCs*.
exotoxin	Toxic protein actively secreted by a *gram-positive* or *gram-negative bacterium*.
experimental peripheral tolerance	Lack of an immune response to a foreign antigen induced by treatment of a mature animal with either a non-immunogenic form of the antigen and using the usual dose and route of immunization, or an immunogenic form of the antigen and using a non-immunogenic dose or route of administration.
extracellular pathogen	A pathogen that does not enter host cells but reproduces in the interstitial fluid, blood or lumens of the respiratory, urogenital and gastrointestinal tracts.
extramedullary tissues	Tissues and organs that are neither *lymphoid tissues* nor *bone marrow*.
extrathymic T cell development	Development and maturation of T cells in tissues outside the *thymus*.
extravasation	Exit of leukocytes from the blood circulation into the tissues in response to inflammatory signals. The two steps of this process are *margination* and *diapedesis*.
Fab fragment/region	=Fragment, antigen binding. Originally, the N-terminal portion of Ig molecule left after digestion with papain. The Fab region contains the two antigen-binding sites of the antibody.
familial cancer	Cancer that develops with increased frequency in genetically related individuals.
Fc fragment/region	=Fragment, crystallizable. Originally, the C-terminal portion of Ig molecule left after digestion with papain; crystallizes at low temperature. The Fc region contains the *constant region* of the antibody.
Fc receptor (FcR)	Family of structurally diverse leukocyte receptors that bind to the *Fc regions* of a specific antibody isotype. Engagement can trigger *clathrin-mediated endocytosis*, *phagocytosis*, ADCC, degranulation and cytokine release.
fetal–maternal tolerance	Tolerance of a mammalian mother for her fetuses despite their expression of *allogeneic* paternal MHC.
follicle-associated epithelium (FAE)	Epithelium lying directly over single or aggregated lymphoid follicles that is specialized for *transcytosis* due to the presence of *M cells*.
follicular dendritic cells (FDCs)	Distinct lineage of DCs found in B cell-rich areas of lymphoid organs. FDC do not internalize antigen and do not function as APCs but rather trap antigen–antibody complexes on their cell surfaces and display them for extended periods.
framework regions	Relatively invariant parts of the V domain of an Ig or TCR chain that are outside the *hypervariable regions*.
gamma–delta (γδ) T cells	T lymphocytes bearing γδ TCRs. Considered cells of innate immunity.
gamma–delta (γδ) TCRs	Heterodimer of TCRγ and δ chains plus the *CD3 complex*. Rather than pMHC, γδ TCRs recognize antigens such as stress proteins and *heat shock proteins*. See *TCR complex*.
gene segment	A short, germline sequence of DNA from either the variable (V), diversity (D), or joining (J) families that randomly joins via *V(D)J recombination* with one or two other gene segments to complete a *V exon* in either the Ig or TCR loci.
germinal centers (GC)	Aggregations of rapidly proliferating B cells and differentiating memory B and plasma cells that develop in the *secondary lymphoid follicles*. Site of *isotype switching*, *somatic hypermutation* and *affinity maturation*.

Term	Definition
glycocalyx	A thick glue-like layer of mucin-related molecules anchored in the *brush border* of gut epithelial cells. Contains hydrolytic enzymes and has a negative charge that repels pathogens.
goblet cells	*Mucus*-producing cells in the gut epithelium.
graft rejection	Attack by recipient immune system on transplanted donor tissue (a graft). See *hyperacute, acute* and *chronic graft rejection*.
graft-versus-host disease (GvHD)	Attack by immunocompetent cells in transplanted donated tissue on recipient tissues due to *MHC* or *MiHA* mismatches.
graft versus leukemia (GvL) effect	In BMT/HCT for cancer treatment, destruction of residual recipient leukemia cells by T cells from an *allogeneic* donor.
gram-negative bacteria	Bacteria that have thin cell walls containing LPS.
gram-positive bacteria	Bacteria that have thick cell walls containing peptidoglycan and lipotechoic and techoic acids.
granulocytes	*Myeloid* leukocytes that harbor large intracellular granules containing microbe-destroying hydrolytic enzymes. Include *neutrophils, basophils* and *eosinophils*.
granuloma	Structure formed by a group of *hyperactivated macrophages* that fuse together to wall off a persistent pathogen from the rest of the body. Also contains CD4$^+$ and CD8$^+$ T cells. Formation depends on TNF produced by activated Th1 effectors.
gut-associated lymphoid tissue (GALT)	The Peyer's patches, appendix and diffuse collections of immune system cells in the linings of the small and large intestine. See *mucosa-associated lymphoid tissue* (MALT).
H-2 complex	Murine *major histocompatibility complex (MHC)*.
haplotype (MHC)	The set of MHC alleles contained on a single chromosome of an individual.
heat shock proteins (HSPs)	Proteins whose expression is upregulated in cells subjected to environmental stresses such as heat or inflammation.
hematopoiesis	The generation of hematopoietic cells from *HSCs* in the bone marrow or fetal liver.
hematopoietic cells	Red and white blood cells (erythrocytes and leukocytes).
hematopoietic cell transplant (HCT)	Replacement of a damaged immune system by transplantation of isolated healthy HSCs from a donor.
hematopoietic stem cell (HSC)	Pluripotent hematopoietic precursor that either self-renews or differentiates into lymphoid, myeloid or mast cell precursors.
herd immunity	Protection of non-immune individuals in a population from a given pathogen due to effective vaccination of the majority of the population.
high endothelial venules (HEV)	Specialized post-capillary venules in most secondary lymphoid tissues that allow lymphocyte extravasation from the blood into these sites.
high zone tolerance	*Experimental peripheral tolerance* induced by a very large dose of an immunogen.
hinge region	Site in the Ig monomer where the Fab region joins the Fc region.
histamine	Mast cell granule component that causes vasodilation and contraction of intestinal and bronchial smooth muscles.
histocompatibility	Ability of a recipient to accept a tissue graft from another individual.
HLA complex	=Human leukocyte antigen complex. Human *major histocompatibility complex (MHC)*.
Hodgkin's lymphoma	*Lymphoma* in which the tumor mass is made up non-transformed lymphocytes, macrophages and fibroblasts plus scattered, malignant *Reed-Sternberg cells*.
homing receptors	Receptors expressed by lymphocytes that bind to specific *vascular addressins* and direct lymphocyte trafficking.

Term	Definition
humanized antibodies	Engineered antibodies in which the V domains of non-human antibodies that recognize a human antigen are combined genetically with the C domains of human antibodies.
humoral immunity	Adaptive immune responses mediated by B cells that differentiate into plasma cells producing antibodies.
hybridoma	An immortalized cell secreting large amounts of pure *monoclonal antibody* of a single known specificity; created by fusion of a *myeloma* cell with an activated B cell producing antibody of known specificity.
hygiene hypothesis	The theory that excessive zeal in preventing exposure to pathogens in infancy leads to a lack of activation of the immature immune system. The resulting bias toward Th2 responses may predispose an individual to hypersensitivity and/or autoimmunity.
hyperactivated macrophage	See *macrophage*.
hyperacute graft rejection (HAR)	Extremely rapid destruction of a graft almost immediately after transplantation. Mediated by classical *complement* activation initiated by preformed antibodies recognizing *allogeneic* epitopes on the graft vasculature. A form of *type II hypersensitivity*.
hypersensitivity	Excessive immune reactivity to a generally innocuous antigen that results in inflammation and/or tissue damage. Includes *type I–IV hypersensitivities*.
hypervariable region	Region of extreme amino acid variability in the V domain of an Ig or TCR chain. The hypervariable regions largely form the antigen-binding site. Also known as complementarity-determining regions (CDRs).
hyposensitization	Allergy therapy in which the patient receives subcutaneous injections of purified *allergen* regularly for 3–5 years. Production of allergen-specific IgE is converted to production of allergen-specific IgG4, which does not trigger mast cell degranulation.
Igα/Igβ complex	Accessory heterodimer within the *BCR complex*. Required to transduce intracellular signaling initiated by mIg engagement by antigen.
immune complex	Latticelike structure composed of interlinked antigen–antibody complexes. Large immune complexes are insoluble and can become trapped in vessel walls or narrow body channels, provoking inflammation and *type III hypersensitivity*.
immune deviation	Conversion of an adaptive immune response that is harmful to a less harmful response via a switch between Th1 and Th2 responses. A form of peripheral T cell control.
immune privileged sites	Anatomical sites in which immune responses are actively or passively suppressed.
immune response	A coordinated action by numerous cellular and soluble components in a network of tissues and circulating systems that combats pathogens, injury by inert materials, and cancers.
immunity	The ability to rid the body successfully of a foreign entity.
immunoconjugate	Chimeric protein in which a whole *mAb* (or a structural derivative of that mAb) is linked to a cytokine, radioisotope or toxin either chemically or at the DNA level.
immunodiffusion	*Precipitin*-based assay in which antigen and antibody diffuse toward each other within an agar gel.
immunodominant epitope	The *epitope* against which the majority of antibodies is raised, or to which the majority of T cells responds.
immunoelectrophoresis	Separation of proteins by nondenaturing electrophoresis plus determination of protein identity by *immunodiffusion*.
immunofluorescence	Specific antibody tagged with a fluorochrome binds to an antigen and is detected by fluorescence microscopy.
immunogen	An antigen that can induce lymphocyte activation.

Term	Definition
immunoglobulin (Ig)	Antigen-binding protein expressed by B lineage cells. An Ig monomer is composed of two identical light and two identical heavy chains (H_2L_2). In its plasma membrane-bound form, an Ig is the antigen-binding component of the *BCR complex*. In its secreted or secretory form, an Ig is an antibody.
immunoglobulin (Ig) domain	A protein domain of 70–100 amino acids found in all Ig molecules. Forms a characteristic structure known as the Ig fold. Mediates inter- and intramolecular interactions.
immunological memory	During the *primary adaptive response* to a pathogen, each antigen-specific lymphocyte clone activated generates many memory lymphocytes of identical specificity and greater affinity. When the same pathogen attacks the body a second time, it is eliminated more rapidly and efficiently by the *secondary response* mediated by the activation of these memory lymphocytes.
immunological synapse	The contact zone between a T cell and an APC.
immunopathic damage	Collateral damage to tissues caused by an immune response.
immunoprecipitation	Use of specific antibody to isolate antigens in solution. The antibody is complexed to insoluble particles that can be collected by centrifugation.
immunosuppression	The reduction or elimination of immune responses. Mediated by cytokines (IL-10, TGFβ) or drugs.
immunosurveillance	Concept that the immune system monitors the body for pathogens and tumor cells and destroys them.
immunotherapy	The manipulation of the immune system to fight diseases.
indirect allorecognition	Peptides derived from *allogeneic* proteins of the graft are presented to recipient T cells by recipient APCs.
induced fit	Antigen influences the conformation of the antigen binding site of an antibody such that it better accommodates the antigen.
inducible nitric oxide synthetase (iNOS)	Enzyme induced mainly in phagocytes by the presence of microbial products or pro-inflammatory cytokines. Generates *nitric oxide*, which is toxic to endocytosed pathogens.
inductive site	A local area of the *mucosae* where antigen is encountered and a primary adaptive response initiates.
infection	The attachment and entry of a pathogen into a host's body such that the organism successfully reproduces.
inflammation	A local response at a site of infection or injury initiated by an influx of innate leukocytes that fight infections using broadly specific recognition mechanisms. Clinically, inflammation is characterized by heat and pain as well as swelling and redness.
innate immunity	Non-specific and broadly specific mechanisms that deter entry or result in elimination of foreign entities. Innate immunity is mediated by (1) physical, chemical and molecular barriers that exclude antigens in a totally non-specific way, and (2) receptors (*PRMs*) that recognize a limited number of molecular patterns (*PAMPs*) that are common to a wide variety of pathogens.
interfollicular region	Tissue between *lymphoid follicles* positioned in a group. Contains mature T cells surrounding *HEVs*.
intestinal cryptopatches	Small regions of lymphoid tissue in the *lamina propria* of the crypts of the small intestine.
intestinal follicles	Lymphoid follicles located in the intestinal *lamina propria* either grouped in *Peyer's patches* or in the appendix, or scattered singly. Composed of a *germinal center* containing B cells and *FDCs* topped by a dome containing *APCs* and T cells.
intracellular pathogen	Pathogen that spends a significant portion of its life cycle within a host cell; reproduction is usually within the host cell.

Term	Definition
intracellular signaling	The binding of a ligand to its receptor initiates a series of interactions between various proteins which culminate in the activation of transcription factors that enter the nucleus and alter the transcription patterns of genes controlling cellular proliferation, differentiation and effector functions.
intraepithelial lymphocytes (IEL)	αβ and γδ T cells dispersed among the epithelial cells lining a body tract.
intraepithelial pocket	Pocket within the *FAE* created by invagination of the basolateral surface of an *M cell*.
invariant chain (Ii)	Transmembrane protein that binds to peptide-binding groove of newly synthesized *MHC class II molecules* and chaperones them out of the rER into the endocytic system. Upon cleavage, Ii gives rise to *CLIP*.
ISCOM	=Immunostimulating complex. A vaccine *delivery vehicle* in which an antigen is mixed with cholesterol, phospholipid and detergent to form a ball-shaped structure.
isograft	Graft between two genetically identical individuals.
isotypes	Classes of Igs defined on the basis of the amino acid sequences of their *constant regions*. Include IgM, IgD, IgA, IgG and IgE heavy chain isotypes and Igκ and Igλ light chain isotypes.
isotype switching	Mechanism by which a B cell producing an Ig heavy chain of one isotype can then switch to producing Ig of the same variable region but a different constant region. See *switch recombination*.
ITAM	=Immunoreceptor tyrosine-based activation motif. Activatory sequence in receptor tails that facilitates recruitment of kinases promoting intracellular signaling.
ITIM	=Immunoreceptor tyrosine-based inhibition motif. Inhibitory sequence in receptor tails that facilitates recruitment of phosphatases downregulating intracellular signaling.
J chain	=Joining chain. Small polypeptide that bonds to the tail pieces of α and μ Ig heavy chains; stabilizes polymeric IgA or IgM.
junctional diversity	Variation in the amino acid sequences of Igs and TCRs that arises during *V(D)J recombination*. Due to imprecise joining of gene segments as well as deletion and/or addition of nucleotides to the joint.
keratinocytes	Epidermal squamous epithelial cells that produce keratin. Stratified layers are held together as units by *desmosomes*.
lamina propria	Layer of loose connective tissue between the basolateral surface of the mucosal epithelium and the underlying muscle layer.
Langerhans cells (LCs)	Epidermal DCs.
latent infection	A pathogen is present for an extended time but is non-infectious and does not cause clinical symptoms.
late phase reaction	In the effector phase of *type I hypersensitivity*, leukocytes (especially eosinophils) that entered the site of allergen penetration in the *early phase reaction* release additional cytokines, enzymes and inflammatory mediators that do further damage.
lectin	A protein that binds to particular carbohydrate moieties on membrane glycoproteins or glycolipids on cell surfaces.
lectin-mediated complement activation	Complement activation initiated by the binding of *mannose-binding lectin* (MBL) to pathogen monosaccharides. Involves the MASP complex and C4, C2 and C3. See *complement*.
leukapheresis	Removal of leukocytes from a donor's blood, followed by return of the remaining blood products to the donor.
leukemia	"Liquid" malignancy of hematopoietic cells. Manifested as greatly increased cell numbers in the blood and bone marrow.

Term	Definition
leukocytes	White blood cells, including lymphocytes, granulocytes, monocytes, macrophages, NK and NKT cells.
light zone	Region of the *germinal center* where *isotype switching* and *affinity maturation* occur.
linear determinant	A protein epitope defined by a series of consecutive amino acids.
linked recognition	The requirement that B and T cell *epitopes* be physically linked to induce an efficient humoral response.
lipopolysaccharide (LPS)	Component of *gram-negative* bacterial cell walls that generates *endotoxin* and induces *endotoxic shock*.
live attenuated vaccine	*Vaccine* based on a weakened version of the pathogen that has reduced virulence but still acts as an *immunogen*.
low zone tolerance	*Experimental peripheral tolerance* induced by administration of very small doses of immunogen over an extended period.
lymph	Nutrient-rich interstitial fluid that bathes all cells in the body. Lymph filters slowly through the tissues, collects antigen, and eventually enters the *lymphatic system*.
lymphatic system	A network of lymphatic capillaries, vessels and trunks through which lymph and lymphocytes travel. The lymphatic trunks connect with certain veins of the blood circulation.
lymph nodes	Bean-shaped, encapsulated *secondary lymphoid tissues* located in clusters along the length of the lymphatic system. Contain the concentrations of the T and B lymphocytes and DCs required for primary adaptive immune responses.
lymphocyte	Small round leukocytes with a large nucleus and little cytoplasm. Concentrated in the *secondary lymphoid tissues* while in the resting state. Two major types exist, T cells and B cells, which are responsible for adaptive immune responses.
lymphocyte recirculation	Continual migration of lymphocytes from the tissues back into the blood via the *lymphatic system*, followed by return to the tissues via *extravasation* from the blood circulation.
lymphocyte specificity	The restricted and unique range of *epitopes* potentially bound by a lymphocyte due to its expression of a single type of randomly generated antigen receptor gene.
lymphocyte tolerization	The induction of *anergy* in a lymphocyte.
lymphoid cells	T and B *lymphocytes*, *NK cells* and *NKT cells*.
lymphoid follicles	Organized spherical aggregates of lymphocytes.
lymphoid organs	Encapsulated groups of lymphoid follicles.
lymphoid patches	Groups of lymphoid follicles that are not encapsulated.
lymphoma	Solid malignancy of *lymphoid cells* arising in a structured or diffuse secondary lymphoid tissue rather than in the blood or bone marrow. Classified as *Hodgkin's* or *non-Hodgkin's*.
lymphopoiesis	The process of *HSC* differentiation into *lymphoid cells*.
lysozyme	Protease that digests particular bacterial cell wall components. Found in lysosomes, neutrophil granules, tears and body secretions.
macrophage	Powerful *phagocyte* that also secretes a large array of proteases, cytokines and growth factors and can act as an *APC*. In the presence of high levels of IFNγ, an activated macrophage becomes "hyperactivated" and acquires enhanced anti-pathogen activities and the capacity to kill tumor cells.
macropinocytosis	Internalization of extracellular fluid containing soluble macromolecules. Droplets are engulfed and form *macropinosomes* that enter the *endocytic pathway*.
macropinosome	Membrane-bound vesicle formed by plasma cell invagination around a droplet of extracellular fluid. See *macropinocytosis*.

Term	Definition
major histocompatibility complex (MHC)	Region of the genome containing genes encoding the chains of the *MHC class I, class II* and *class III proteins*. MHC class I and class II proteins combine with antigenic peptides and display them on the surface of host cells for recognition by T cells. The MHC class III genes encode various proteins important in complement activation, inflammation and stress responses.
malignant tumor	=Cancer. A mass formed by abnormally dividing cells. Appears disorganized, is rarely encapsulated, and is subject to *metastasis*. Directly lethal to the host unless removed or killed.
mannose-binding lectin (MBL)	A serum *collectin* that specifically binds to distinctive mannose structures on microbial pathogens in the blood. Engagement of MBL triggers *lectin-mediated complement activation*.
margination	First step of *extravasation*. Receptor-mediated adherence of leukocytes to endothelial cells in the postcapillary venules.
mast cells	Leukocytes with granules containing preformed mediators such as *histamine*. Mast cell degranulation is important for *inflammation* and *allergy*.
mast cell committed progenitor (MCP)	Early descendant of *MPPs* that gives rise to mast cells.
M cells	=Membranous or microfold cells. Large epithelial cells with an *intraepithelial pocket*. *Transcytose* antigens from a body tract lumen across the epithelial layer. Most M cells reside in the dome overlying *intestinal follicles*.
medulla	Inner region of an organ such as the *thymus*.
medullary thymic epithelial cells (mTECs)	See *thymic epithelial cells*.
megakaryocytes	Multinucleate *myeloid* leukocytes giving rise to platelets.
membrane attack complex (MAC)	Pore-shaped structure assembled in the membrane of a pathogen or target cell as a consequence of *complement* activation. Facilitates osmotic imbalance and lysis.
membrane-bound Ig (mIg)	Cell surface form of Ig molecule. *BCR* antigen-binding moiety.
memory cells	Lymphocytes generated during a *primary response* that remain in a quiescent state until fully activated by a subsequent exposure to specific antigen (a *secondary response*).
metastases	Secondary tumors established by *metastasis* of a *primary tumor* to sites in a different organ or tissue.
metastasis	Process by which malignant cells break away from a *primary tumor* and spread via the blood to secondary sites.
MHC class I proteins	Cell surface proteins composed of a polymorphic MHC class I α chain non-covalently associated with the invariant β2-microglobulin (β2m) chain. Expressed by most nucleated cells. Present peptide antigens to CD8+ T cells.
MHC class Ib and IIb proteins	Non-classical MHC proteins with restricted polymorphism. Most are not involved in antigen presentation.
MHC class II proteins	Cell surface proteins composed of a polymorphic MHC class II α chain and a polymorphic MHC class II β chain. Expressed on APCs. Present peptide antigens to CD4+ T cells.
MHC class III proteins	Proteins encoded by the MHC class III loci. Include certain complement proteins, heat shock proteins, TNF and LT.
MHC-like proteins	Non-polymorphic proteins encoded outside the MHC (e.g., CD1). Structurally and functionally similar to MHC proteins.
MHC restriction	The principle that a T cell recognizing a given antigenic peptide presented by a particular MHC molecule will not recognize the same peptide if presented by a different MHC molecule.

Term	Definition
MICA, MICB	=MHC class I chain related A and B. Human stress ligands upregulated in response to heat, infection or transformation.
MIICs	=MHC class II compartments. Late endosomal compartments that are part of the *exogenous antigen processing* pathway.
minor histocompatibility antigens (MiHA)	Proteins that exist in a small number of different allelic forms in a population. In a transplant situation, peptides derived from *allogeneic* MiHA of the donor are recognized as non-self by the recipient's T cells. Generally invoke slower, weaker *graft rejection* than MHC incompatibilities.
minor H peptides	Peptides derived from donor *minor histocompatibility antigens*.
mixed chimerism	An recipient receives (non)-myeloablative conditioning followed by a *hematopoietic cell transplant* (HCT). Hematopoiesis is then carried out by both recipient and donor HSCs.
modulated DC	A DC that, due to exposure to immunosuppressive cytokines, *anergizes* T cells and/or induces generation of T_{reg} cells.
molecular mimicry	Concept that a pathogen *epitope* may resemble a self epitope closely enough to activate an *autoreactive* lymphocyte, if the appropriate *cytokine* microenvironment is present.
monoclonal antibody (mAb)	Antibodies of a single, known specificity produced by a single B cell clone. See *hybridoma*.
monocyte	Myeloid cell in the blood that enters the tissues and matures into a macrophage.
mucosae	Mucosal epithelial layers that cover the luminal surfaces of the gastrointestinal, respiratory and urogenital tracts.
mucosa-associated lymphoid tissue (MALT)	Collections of APCs and lymphocytes in the mucosae. Includes the *gut-associated lymphoid tissue* (GALT), the *bronchi-associated lymphoid tissue* (BALT) and the *nasopharynx-associated lymphoid tissue* (NALT).
mucosal immunity	Immunity mediated by physical barriers, substances (such as *mucus*), and cells in the *MALT* that can respond to antigen attacking the *mucosae* without involving a *lymph node*.
mucus	Viscous fluid that coats the surface of cells of the mucosae. Contains secretory antibodies and antimicrobial molecules.
multipotent progenitor (MPP)	Descendant of *HSCs* that gives rise to *NK/T, CLP* and *MCP* precursors.
myeloablative conditioning	Complete elimination of a patient's hematopoietic cells in the bone marrow using chemotherapy and irradiation, leading to eventual depletion of immune system cells from the peripheral blood and all secondary lymphoid tissues.
myeloid cells	Cells that develop from common myeloid progenitors (CMPs). Include erythrocytes, *neutrophils, monocyte/macrophages, eosinophils, basophils* and *megakaryocytes*.
myeloma	*Plasma cell* tumor that secretes large quantities of an Ig protein of (usually) unknown specificity.
myelopoiesis	The process of *HSC* differentiation into myeloid cells.
naïve lymphocytes	Resting B and T cells that have never interacted with specific antigen in the periphery.
naked DNA vaccine	*Vaccine* based on an isolated DNA plasmid (no vector) encoding the vaccine antigen.
nasopharynx-associated lymphoid tissue (NALT)	Mucosal lymphoid elements in the tonsils and upper respiratory epithelium. See *MALT*.
natural cytotoxicity	*Perforin/granzyme-mediated cytotoxicity* of target cells carried out by *NK cells*.

Term	Definition
natural killer (NK) cells	*Lymphoid* lineage cells that recognize non-self entities with broad specificity. Activated when a target cell expresses ligands that bind to *NK activatory receptors* but lacks sufficient MHC class I to adequately engage *NK inhibitory receptors*. Sentinels of innate immunity.
natural killer T (NKT) cells	*Lymphoid* lineage cells that express a *semi-invariant TCR* recognizing glycolipid or lipid antigens presented on CD1d. Activated NKT cells quickly secrete cytokines that affect DCs, and NK, T and B cells. Sentinels of innate immunity.
necrosis	Sudden, uncontrolled cell death due to *infection* or trauma. The cell spills its contents into the surrounding milieu, releasing *danger signal* mediators that trigger *inflammation*.
negative selection	A *central tolerance* process that removes *autoreactive* cells from the developing lymphocyte pool destined for the *periphery*. Based on high *affinity/avidity* of antigen receptors for self antigens.
neo-antigen	An antigen formed when a small, chemically reactive molecule binds to a self protein. See *contact hypersensitivity*.
neonatal immunity	Immunity in the newborn due to maternal circulating antibodies passed on to the fetus via the placenta, or maternal secretory antibodies consumed by the newborn in breast milk. A form of *passive immunization*.
neonatal tolerance	Phenomenon that *tolerance* to an antigen is established more easily in neonatal than mature animals. Due to functional immaturity and low numbers of neonatal T and B cells, DCs, macrophages and FDCs, and altered lymphocyte recirculation.
neoplasm	An abnormal tissue mass (tumor). May be *benign* or *malignant*.
neutralization	Ability of an antibody to bind to an antigen and physically prevent it from binding to and harming a host cell.
neutrophils	Most common *leukocytes*. Function as both *granulocytes* and *phagocytes*. Enter tissues from the circulation immediately in great numbers in response to injury or pathogen attack. Neutrally staining cytoplasmic granules.
NK activatory receptors	Receptors whose engagement induces *natural cytotoxicity* and *cytotoxic cytokine* secretion by NK cells if not counteracted sufficiently by inhibitory receptor engagement.
NK inhibitory receptors	Receptors whose engagement by self MHC class I molecules on a potential target counteracts the effects of *NK activatory receptor* engagement, preventing target cell destruction.
NK/T precursor	*MPP*-derived precursor that can develop into T, NKT or NK cells but not B cells.
N nucleotides	=Non-templated nucleotides. Nucleotides that are added randomly by *TdT* onto the ends of two antigen receptor gene segments undergoing *V(D)J recombination*.
NOD proteins	=Nucleotide-binding oligomerization domain proteins. Cytoplasmic *PRMs* that detect products of intracellular pathogens. Engagement induces pro-inflammatory cytokines.
non-Hodgkin's lymphoma	Heterogeneous group of *lymphomas* in which the solid tumor mass consists almost entirely of malignant lymphocytes.
non-myeloablative conditioning	Partial elimination of a patient's hematopoietic cells in the bone marrow using chemotherapy and irradiation, leading to eventual reductions of immune system cells in the peripheral blood and secondary lymphoid tissues. Less toxic than myeloablative conditioning.
non-responder	Individual that fails to mount an immune response to a foreign protein that provokes a strong response in other individuals.
non-selection	In *central tolerance*, the apoptotic death of CD4$^+$CD8$^+$ thymocytes expressing TCRs with little or no affinity for self MHC. Also called "neglect".
oncogene	A gene whose deregulation is directly associated with *carcinogenesis*. Oncogenes often encode positive regulators of cell growth.

Term	Definition
opportunistic pathogen	A pathogen that does not cause disease unless offered an unexpected opportunity by a failure in host defense.
opsonin	A host protein that coats a foreign entity such that it binds more easily to phagocyte receptors, enhancing *phagocytosis*.
oral tolerance	Experimental *tolerance* induced by feeding of an *immunogen*.
organ-specific autoimmune disease	An *autoimmune* disease targeting a specific anatomical site.
Paneth cells	Cells located at the bottom of intestinal crypts. Produce anti-microbial proteins.
parasite	A pathogen that depends on a host organism for both habitat and nutrition at some point in its life cycle.
parenteral	Administration of a substance by a non-oral route (injection).
passive immunization	Transfer of *antibodies* to a non-immune recipient to provide immediate protection against a particular pathogen.
pathogen	Organism that causes *disease* in its host as it attempts to reproduce. Includes extracellular bacteria, intracellular bacteria, viruses, *parasites*, fungi and *prions*.
pathogen-associated molecular patterns (PAMPs)	Structural patterns present in components or products common to a wide variety of microbes (but not host cells). Ligands for *pattern recognition molecules* (PRMs).
pattern recognition molecules (PRMs)	Proteins recognizing *PAMPs*. Soluble PRMs include the *collectins* (MBL), *acute phase proteins* and *NOD proteins*. Membrane-bound PRMs are *pattern recognition receptors* (PRRs).
pattern recognition receptors (PRRs)	Widely distributed membrane-bound *PRMs* fixed in either the plasma membrane of a cell or in the membranes of its endocytic vesicles. Includes *Toll-like receptors* (TLRs) and scavenger receptors. Engagement of PRRs induces pro-inflammatory cytokines.
peptide vaccine	A *vaccine* that uses a small antigenic peptide for immunization.
perforin/granzyme-mediated cytotoxicity	Mechanism of apoptotic target cell destruction triggered when CTLs or NK degranulate to release granzymes (proteases) and perforin (pore-forming protein).
periarteriolar lymphoid sheath (PALS)	A cylindrical lymphoid tissue surrounding each splenic arteriole. Populated by mature T cells, some B cells, plasma cells, macrophages and DCs.
peripheral tolerance	Functional silencing or deletion of *autoreactive* peripheral lymphocytes that escaped elimination via *central tolerance*.
periphery	Tissues and organs other than the bone marrow and thymus.
persistent infection	An infection in which the pathogen remains in the body for a prolonged period. May be *latent* or cause *chronic disease*.
Peyer's patches	*Secondary lymphoid tissue* in the intestine. See *intestinal follicle*.
phagocyte	Cell capable of carrying out *phagocytosis*.
phagocytosis	Process by which a *phagocyte* captures particulate entities by membrane-mediated engulfment.
phagolysosome	Vesicle formed by fusion of a phagosome with a lysosome during *phagocytosis*.
phagosome	Intracellular vesicle in which a captured entity is first sequestered during *phagocytosis*.
plasmablasts	Proliferating progeny of an activated B cell. Become plasma cells.
plasma cell dyscrasias	Hematopoietic cancers of plasma cells.
plasma cells	Terminally differentiated B cells that secrete *antibody*.

Term	Definition
plasmapheresis	An individual's blood is withdrawn and passed through a machine designed to remove antibody proteins. The machine then returns the treated blood to the patient.
platelet-activating factor (PAF)	A lipid inflammatory mediator that activates *platelets*.
platelets	Small non-nucleated cells in blood derived from *megakaryocytes*. Promote blood clotting.
pluripotency	Capacity to differentiate into a variety of cell types.
P nucleotides	If two gene segments undergoing *V(D)J recombination* are nicked elsewhere than at their precise ends, a recessed strand end and an overhang are generated. The nucleotides added to fill the gaps on both strands are considered P nucleotides.
polyclonal antiserum	Antiserum that contains antibodies produced by many different B cell clones responding to different epitopes of an antigen.
poly-Ig receptor (pIgR)	=Polymeric immunoglobulin receptor. Receptor positioned on the basolateral surface of mucosal epithelial cells. Binds to the *J chain* in secreted polymeric Ig and facilitates *transcytosis* of the Ig into external secretions. Cleavage of pIgR leaves *secretory component* attached to the Ig molecule.
polymorphism	Existence of different alleles of a gene in a population.
positive cross-match	Occurs when an individual's blood contains preformed antibodies to one or more HLA molecules expressed on an *allogeneic* donor organ.
positive selection	A *central tolerance* process that promotes the survival and maturation of developing thymocytes that bind to self-antigen with only low *affinity/avidity*.
preactivation	The state of a provirus that remains untranscribed in the host cell genome due to the absence of host cell stimulation.
pre-BCR	Complex composed of *surrogate light chain* (SLC), a candidate μ chain, and the *Igα/Igβ* complex. The pre-BCR is inserted transiently in the membrane of a developing B cell to test the functionality of a particular heavy chain VDJ combination.
precipitin curve	A graph showing the amount of antibody precipitated out of solution by varying amounts of antigen.
pre-T alpha chain (pTα)	Invariant TCRα-like chain expressed only in DN thymocytes. Used to test the functionality of candidate TCRβ chains. Not required for γδ T cell development. See *pre-TCR*.
pre-TCR	Transient complex composed of a candidate TCRβ chain plus the *pTα* chain plus the CD3 chains. Used to test functionality of a particular V(D)J rearrangement in the TCRβ gene.
primary follicles	Spherical aggregates of resting mature B cells, macrophages, and FDCs within B cell-rich regions of *secondary lymphoid tissues* such as spleen, lymph nodes and Peyer's patches.
primary immunodeficiency (PI)	Failure of a component of the immune system due to an inborn genetic mutation.
primary lymphoid tissue	Lymphoid tissues (*bone marrow* and *thymus*) where lymphocytes are generated and mature.
primary response	The adaptive immune response mounted upon a first exposure to a non-self entity. The primary response is slower and weaker than secondary (or subsequent) responses.
primary tumor	Original tumor mass established by the first transformed cell.
priming	First encounter of a naïve lymphocyte with specific antigen. Leads to a *primary response*
prions	Infectious proteins of abnormal conformation. Prions spread by altering the conformation of their normal protein counterparts in infected brain, causing *spongiform encephalopathies*.
proteasome	Cytoplasmic organelle containing multiple proteases that digest proteins into peptides. Integral component of the *endogenous antigen processing and presentation pathway*.

Term	Definition
protective epitopes	In *vaccination*, *epitopes* of a pathogen that induce an immune response preventing subsequent infection by that pathogen.
provirus	Viral DNA that has been integrated in the host cell genome.
pulsing	Loading of APCs with antigen by use of electric current.
purpura	Areas of purplish or brownish red discoloration on the skin caused by leakage of blood (hemorrhage) into the skin layers.
pus	Cream-colored substance at a site of injury or infection. Accumulation of leukocytes that have died fighting infection.
pyrogenic infections	Infections that induce an acute, high fever.
R5 viruses	HIV strains that bind to CCR5 and infect macrophages as well as CD4$^+$ T cells.
radiation therapy	Use of ionizing radiation to kill tumor cells.
RAG recombinases	=Recombination activation gene recombinases. RAG-1 and RAG-2 mediate *V(D)J recombination*.
reactivation	Occurs when a *latent* virus resumes replication and causes disease.
reactive nitrogen intermediates (RNIs)	Nitrogen-derived free radicals that kill microbes.
reactive oxygen intermediates (ROIs)	Oxygen-derived free radicals that kill microbes.
receptor blockade	Anergization of a B cell due to very large amounts of antigen that persistently occupy the BCRs without cross-linking them.
recombinant vector vaccine	*Vaccine* in which the DNA encoding the vaccine antigen is incorporated into a vector that enters host cells and promotes translation of the vaccine antigen directly within them.
recurring chromosomal translocation	A chromosomal translocation that appears in many different patients. Most often observed in hematopoietic cancers.
Reed-Sternberg cells	The tiny percentage of cancerous cells present in a classical *Hodgkin's lymphoma* tumor mass.
regulators of complement activation (RCA)	Soluble and membrane proteins that bind to C4 and C3 products and interfere with *complement* activation.
regulatory T cells	T cells that inhibit the responses of other immune system cells by intercellular contact and/or immunosuppressive cytokine secretion. Includes T_{reg}, Tr1 and Th3 cells.
relapse	Reappearance of clinical disease in a patient who was previously in *remission*.
remission	A patient's disease is in remission if it can no longer be detected clinically.
repertoire	Pool of antigenic specificities represented in the total population of T or B lymphocytes.
reservoir	Non-human species or environmental niche in which a *pathogen* that normally infects humans can survive.
respiratory burst	Significant increase in oxygen utilization by the NADPH oxidases that generate ROIs. Observed during degranulation.
retrovirus	Virus that uses reverse transcriptase to synthesize a DNA copy of its RNA genome and integrates it into the host cell DNA.
Rh disease	Destruction of the erythrocytes of a fetus during the pregnancy of an Rh$^-$ mother carrying her second (or subsequent) Rh$^+$ fetus.
RSS	=Recombination signal sequence. The 12-RSS and the 23-RSS flank germline V, D and J gene segments and align them for *V(D)J recombination*.

Term	Definition
SCID	=Severe combined immunodeficiency disease. Family of *PIs* characterized by a lack of T and B cell functions.
secondary follicles	Once the *primary follicles* are infiltrated by activated T and B cells, they become secondary follicles that form *germinal centers* and foster terminal differentiation of activated B cells into *memory* B and *plasma cells*.
secondary lymphoid tissues	Peripheral lymphoid tissues inhabited by mature lymphocytes. Includes the *spleen*, *lymph nodes*, *mucosa-associated lymphoid tissues* and *skin-associated lymphoid tissues*. Adaptive immune responses are initiated in these sites.
secondary response	A secondary response is mounted by antigen-specific *memory* lymphocytes activated by a subsequent exposure to a given non-self entity. Faster and stronger than the primary response.
secreted antibody (sIg)	Soluble form of Ig serving as circulating antibody in the blood.
secretory antibody (SIg)	Secreted antibodies containing secretory component. Present in body secretions such as tears and mucus.
secretory component (SC)	Protein fragment of the *poly-Ig receptor* that remains associated with the Ig after cleavage of the receptor during Ig *transcytosis.*
self tolerance	Lack of an immune response to self antigens. See *peripheral tolerance*.
sensitization (hypersensitivity)	An abnormal primary response to a *sensitizing agent* such that, in a subsequent exposure, the individual mounts an excessive or abnormal secondary response that causes disease rather than immunity. See *effector stage* and *hypersensitivity*.
sensitizing agent	A generally innocuous antigen inducing *hypersensitivity*. Includes *allergens*.
serology	The study of the antibodies present within a given *antiserum*.
skin-associated lymphoid tissue (SALT)	The diffuse collections of αβ and γδ T cells and DCs (*Langerhans cells*) in the epidermis, and αβ T cells, fibroblasts, DCs, macrophages and lymphatic vessels in the dermis.
somatic hypermutation	Introduction of random point mutations at an unusually high frequency into the V exons of Ig genes. Increases V region variability in Ig proteins.
spleen	An organ in the abdomen containing *secondary lymphoid tissue*. Traps blood-borne antigens.
spongiform encephalopathies	Fatal diseases caused by *prions*. Characterized by CNS lesions that render the brain "sponge-like".
sporadic cancer	A *malignant tumor* in humans that is caused by transformation of a somatic cell of a tissue and is not inherited.
SP thymocytes	=Single positive thymocytes. Thymocytes that express either CD4 or CD8 but not both.
subunit vaccine	*Vaccine* based on an isolated pathogen component such as a viral protein or bacterial polysaccharide.
surrogate light chain (SLC)	Two polypeptides called V_{preB} and λ5 assemble non-covalently to form the SLC. The SLC participates in the *pre-BCR*.
switch recombination	Mechanism of Ig *isotype switching*. The switch region of the C_H exon originally in a VDJ-C gene pairs with the switch region of a downstream C_H exon such that the intervening C_H exons are excised. The VDJ exon is then joined to the new C_H exon. Cytokines influence which C_H exon is selected.
syngeneic	Individuals are syngeneic at a given genetic locus if they have the same alleles at that locus.
systemic autoimmune disease	An *autoimmune* disease in which the destruction is not restricted to a specific organ or tissue.
tag	A labeling molecule used to make unitary antigen–antibody pairs detectable. Radioisotopes, enzymes or fluorochromes that are covalently bound to either the antigen or the antibody.

Term	Definition
tail piece	Short C-terminal domain in the heavy chains of *secreted antibody*.
TAP	=Transporter of antigen processing. Complex positioned in the rER membrane that transfers peptides from the cytosol into the rER lumen for loading onto MHC class I.
tapasin	rER protein that links TAP and the MHC class I α chain.
target cell (cancer)	The first cell that undergoes a tumorigenic mutation during the initiation step of *carcinogenesis*.
target cell (cytolysis)	An altered self cell, such as an infected cell, a cancer cell or a graft cell, that is destroyed by cytotoxicity or cytokine secretion mediated by CTLs or NK cells.
Tc cells	Cytotoxic T cells that generally express the CD8 coreceptor and recognize non-self peptide presented on MHC class I. Upon activation, Tc cells differentiate into CTL effectors that kill *target cells* by *perforin/granzyme-mediated cytotoxicity* or by secretion of *cytotoxic cytokines*.
T cell help	Provision of cytokines and/or costimulatory contacts by *Th cells* to B and *Tc cells* to support their activation.
T cell receptor (TCR)	Antigen receptor expressed by T cells. Composed of an α and β chain, or a γ and δ chain.
TCR complex	Complete antigen receptor of T lineage cells. Contains αβ or γδ TCR plus five CD3 chains.
T-dependent (Td) antigens	Antigens that bind to BCRs and initiate B cell activation but cannot induce B cell differentiation or Ig production without direct contact between the B cell and an activated helper T cell. Td antigens are proteins containing both B and T cell epitopes.
terminal complement components	Complement components C5–9. Required for *MAC* formation.
Th cells	Helper T cells that generally express the CD4 coreceptor and recognize non-self peptide presented on MHC class II molecules expressed by an APC. Effector Th cells are generated that secrete cytokines and are primarily *Th1* or *Th2* in phenotype.
Th1/Th2 cells	Th1 cells secrete IFNγ and IL-2, generally combat intracellular bacteria and viruses, and induce isotype switching in humans to IgG1 and IgG3. Th2 cells secrete IL-4, IL-5 and IL-10, act against extracellular bacteria and parasites, and induce isotype switching to IgA, IgE and IgG4.
Th3 cells	CD4+ *regulatory T cells* that secrete large amounts of TGFβ.
Th17 cells	Th17 cells secrete IL-17 and IL-6 and may mediate immune defense against certain bacteria and fungi. May be involved in autoimmune diseases.
therapeutic vaccination	A *vaccine* given to ameliorate a pre-existing condition.
thymic epithelial cells (TECs)	Stromal epithelial cells in the cortex (cTECs) or medulla (mTECs) of the thymus. Involved in positive and negative thymocyte selection.
thymic involution	After puberty, the replacement of the lymphoid components of the thymus with fatty connective tissue.
thymocytes	T cell precursors developing in the thymus.
thymus	A small bilobed organ located above the heart, consisting of the medulla, the cortex, and the subcapsule. *Primary lymphoid tissue* for T cell maturation.
T-independent (Ti) antigens	Antigens that can stimulate B cells in the absence of T cell help but induce only very limited isotype switching, somatic hypermutation and memory B cell generation. Usually large polymeric proteins or carbohydrates with repetitive elements.
tissue typing	Identification of HLA alleles expressed on an individual's cells. Used to determine HLA mismatching between a donor and recipient in a transplant situation.
titer	Concentration of antigen-specific antibodies in an *antiserum*.

Term	Definition
T lymphocyte (T cell)	A *leukocyte* that matures in a *primary lymphoid tissue* (*thymus*), becomes activated in *secondary lymphoid tissues*, and mediates *adaptive immunity* by differentiating into either *Th* or *Tc* effector cells.
tolerance	The absence of an immune response to a given antigen. See also *self tolerance*, *experimental tolerance*, *central tolerance*, and *peripheral tolerance*.
tolerogen	An experimental foreign antigen that is recognized by a T or B lymphocyte but *anergizes* rather than activates these cells.
tolerogenic DCs	A DC that *anergizes* rather than activates naïve T cells. Tolerogenic DCs may express costimulatory molecules but cannot deliver T cell activation signal 2.
Toll-like receptors (TLR)	Membrane-bound *pattern recognition receptors* (PRRs) that bind to pathogen products such as *lipopolysaccharide* (LPS).
tonsil	A network of cells that supports *lymphoid follicles* and *interfollicular regions* that are part of the *NALT*. The nasopharyngeal tonsil is called "the adenoids".
toxin	Pathogen-derived molecule that damages or kills host cells. See *exotoxin* and *endotoxin*.
toxoid	Chemically inactivated pathogen *toxin*.
Tr1 cells	CD4$^+$ *regulatory T cells* that secrete IL-10 plus low amounts of TGFβ.
transcytosis	Transport of a molecule from one surface of a cell to its opposite surface, via an intracellular transport vesicle.
transformation	Malignant conversion. See *carcinogenesis*.
transfusion reaction	Destruction of transfused blood cells by preformed recipient antibodies specific for donor blood group antigens.
T$_{reg}$ cells	CD4$^+$CD25$^+$ *regulatory T cells* that *anergize* other T cells non-specifically via intercellular contacts.
tumor-associated antigen (TAA)	Structurally normal protein or carbohydrate expressed in a tumor at a concentration, location or time that is abnormal relative to its status in the healthy, fully differentiated cells. TAAs are encoded by normal cellular genes that are dysregulated.
tumor-infiltrating lymphocytes (TILs)	CTLs found within a tumor.
tumor regression	Disappearance of a *malignant tumor*. May be induced by anti-cancer treatment, or occur spontaneously.
tumor-specific antigen (TSA)	A macromolecule that is unique to a tumor and not produced by any type of normal cell. TSAs are encoded by mutated cellular genes or by viral oncogenes.
tumor suppressor gene	Gene encoding a protein whose absence promotes *carcinogenesis*. Often encode negative regulators of cell growth or survival.
type I hypersensitivity	*Hypersensitivity* arising from the synthesis of IgE antibodies directed against an antigen (*sensitizing agent*). These antibodies arm mast cells via FcεR binding. See *allergy* and *anaphylaxis*.
type II hypersensitivity	*Hypersensitivity* arising from direct antibody-mediated cytotoxicity. Target cells may be mobile (blood cells) or fixed as part of a solid tissue. Antibodies may be IgM or IgG. If autoantibodies are involved, the reaction may be a component of an *autoimmune disease*.
type III hypersensitivity	Immune complex-mediated *hypersensitivity*. A soluble antigen forms large insoluble immune complexes with IgM or IgG in the circulation. Deposition of these complexes in narrow body channels triggers damaging inflammation. If autoantibodies are involved, the reaction may contribute to *autoimmune disease*.

Term	Definition
type IV hypersensitivity	Cell-mediated *hypersensitivity* in which Th cells activate Tc cells and macrophages that then damage or destroy host cells.
vaccination (prophylactic)	Administration of a non-pathogenic form of a pathogen or its components (the *vaccine*) to prime an adaptive response and generate pathogen-specific memory T and B cells. A natural exposure to the pathogen then triggers a secondary, rather than primary, response and protects against overt disease.
vaccine	A modified, non-pathogenic form of a natural *immunogen*. May be a killed, inactivated or *attenuated* form of the pathogen, or composed of pathogen proteins, DNA or other molecules.
variable (V) domain	The domain of an Ig or TCR chain that is encoded by the corresponding *variable (V) exon*. The variable domains have a high degree of amino acid variability.
variable (V) exon	Exon encoding the *variable domain* of an Ig or TCR protein. V exons in the Igk, Igl, TCRA and TCRD loci are randomly assembled from V and J *gene segments*, while the V exons in the Igh, TCRB and TCRG loci contain V, D and J segments.
variable (V) region	The highly variable N-terminal portion of an Ig or TCR molecule. Composed of the variable domains of all the polypeptides involved. Responsible for antigen recognition.
vascular addressins	Cellular adhesion molecules expressed on *high endothelial venules* (HEVs) that mediate lymphocyte extravasation at particular sites in the body.
V(D)J recombination	Site-specific recombination of pre-existing V, D and J *gene segments* in the Ig and TCR loci to generate unique *variable (V) exons*.
virulence	Ability of a *pathogen* to invade host tissues and cause *disease*.
X4 viruses	HIV strains that bind to CXCR4 and infect $CD4^+$ T cells but not macrophages.
xenograft	Tissue transplanted between members of two different species.
zone of equivalence	Area of the *precipitin curve* where the presence of approximately equal numbers of antigenic *epitopes* and antibody-combining sites allow the formation of extended, insoluble immune complexes.

Index

List of Abbreviations

Ab	antibody	HC	hematopoietic cancer	NKG	natural killer gene
ACR	acute cellular graft rejection	HCT	hematopoietic cell transplant	NKT	natural killer T cell
ADCC	antibody-dependent cell-mediated cytotoxicity	HEV	high endothelial venule	NO	nitric oxide
Ag	antigen	HIGM	hyper IgM	PAF	platelet-activating factor
AHR	acute humoral graft rejection	HIV	human immunodeficiency virus	PALS	periarteriolar lymphoid sheath
AICD	activation-induced cell death	HL	Hodgkin's lymphoma	PAMP	pathogen-associated molecular pattern
AIDS	acquired immunodeficiency syndrome	HLA	human leukocyte antigen	PC	plasma cell
AL	activatory ligand	HPV	human papilloma virus	PI	primary immunodeficiency
ALL	acute lymphocytic leukemia	HS	hypersensitivity	pIgR	poly-Ig receptor
AML	acute myeloid leukemia	HSC	hematopoietic stem cell	pMHC	peptide–MHC complex
APC	antigen-presenting cell	HSP	heat shock protein	PP	Peyer's patches
AR	activatory receptor	HTLV	human T cell leukemia virus	PRM	pattern recognition molecule
AT	ataxia–telangiectasia	IC	immune complex	PRR	pattern recognition receptor
β2m	beta2-microglobulin	ICAM	intercellular adhesion molecule	PTK	protein tyrosine kinase
BALT	bronchi-associated lymphoid tissue	ICOS	inducible costimulatory molecule	RA	rheumatoid arthritis
BCR	B cell receptor	IDC	immature dendritic cell	RAG	recombination activation gene
BMT	bone marrow transplant	IEL	intraepithelial lymphocyte	RBC	red blood cell
C	constant	IFN	interferon	RCA	regulator of complement activation
CCR	CC chemokine receptor	Ig	immunoglobulin	rER	rough endoplasmic reticulum
CD	cluster of differentiation	Ii	invariant chain	RNI	reactive nitrogen intermediate
CDR	complementarity-determining region	IL	interleukin	ROI	reactive oxygen intermediate
CGR	chronic graft rejection	IL-2R	interleukin-2 receptor	RSS	recombination signal sequence
CLL	chronic lymphocytic leukemia	iNOS	inducible nitric oxide synthase	S	switch region
CLIP	class II associated invariant chain peptide	IR	inhibitory receptor	SALT	skin-associated lymphoid tissue
CLP	common lymphoid progenitor	ITAM	immunoreceptor tyrosine-based activation motif	SCF	stem cell factor
CML	chronic myelogenous leukemia			SCID	severe combined immunodeficiency
CMP	common myeloid progenitor	ITIM	immunoreceptor tyrosine-based inhibition motif	sIg	secreted Ig
CNS	central nervous system			SIg	secretory Ig
CR	complement receptor	IV-IG	intravenous immunoglobulin	SLC	surrogate light chain
CRP	C-reactive protein	J	joining	SLE	systemic lupus erythematosus
cTEC	cortical thymic epithelial cell	KSHV	Kaposi's sarcoma herpesvirus	SP	single positive (CD4$^+$ or CD8$^+$)
CTL	cytotoxic T lymphocyte (effector)	KIR	killer Ig-like receptor	SR	scavenger receptor
CTLA-4	cytotoxic T lymphocyte antigen-4	L	ligand	ssRNA	single stranded RNA
CXCR	CXC chemokine receptor	LC	Langerhans cell	T1DM	type 1 diabetes mellitus
CYT	cytoplasmic domain	LPS	lipopolysaccharide	TAA	tumor-associated antigen
D	diversity	LT	lymphotoxin	TAP	transporter associated with antigen processing
DAF	decay accelerating factor	mAb	monoclonal antibody	TB	tuberculosis
DC	dendritic cell	MAC	membrane attack complex	Tc	cytotoxic T cell
DETC	dendritic epidermal T cell	MAdCAM	mucosal addressin cellular adhesion molecule	TCR	T cell receptor
DN	double negative (CD4$^-$CD8$^-$)	MALT	mucosa-associated lymphoid tissue	Td	T-dependent
DP	double positive (CD4$^+$CD8$^+$)	MASP	MBL-associated serine protease	TdT	terminal dideoxy transferase
dsDNA	double-stranded DNA	MBL	mannose-binding lectin	TEC	thymic epithelial cell
dsRNA	double-stranded RNA	MCP	mast cell committed progenitor	TGFβ	transforming growth factor beta
DTH	delayed type hypersensitivity	MDC	mature dendritic cell	Th	helper T cell
EBV	Epstein-Barr virus	MHC	major histocompatibility complex	Ti	T-independent
ECM	extracellular matrix	mIg	membrane-bound Ig	TIL	tumor-infiltrating lymphocyte
ELISA	enzyme-linked immunosorbent assay	MiHA	minor histocompatibility antigen	TLR	Toll-like receptor
FAE	follicle-associated epithelium	MIIC	MHC class II compartment	TM	transmembrane
FcR	Fc receptor	MPP	multipotent progenitor	TNF	tumor necrosis factor
FDC	follicular dendritic cell	MR	mannose receptor	TNFR	tumor necrosis factor receptor
GALT	gut-associated lymphoid tissue	MS	multiple sclerosis	TSA	tumor-specific antigen
GC	germinal center	mTEC	medullary thymic epithelial cell	TSG	tumor suppressor gene
GM-CSF	granulocyte–monocyte colony stimulating factor	NALT	nasopharynx-associated lymphoid tissue	V	variable
		NCR	natural cytotoxicity receptor	VCAM	vascular cellular adhesion molecule
GvHD	graft-versus-host disease	NHC	non-hematopoietic cancer	WHO	World Health Organization
GvL	graft-versus-leukemia effect	NHL	non-Hodgkin's lymphoma	XP	xeroderma pigmentosum
HAR	hyperacute rejection	NK	natural killer cell	XSCID	X-linked severe combined immunodeficiency